Charles Seale-Hayne Library
University of Plymouth
(01752) 588 588
LibraryandITenquiries@plymouth.ac.uk

Archaeological Chemistry
Second Edition

Archaeological Chemistry
Second Edition

A. Mark Pollard
Research Laboratory for Archaeology and the History of Art, University of Oxford, Oxford, UK

Carl Heron
Department of Archaeological Sciences, University of Bradford, Bradford, UK

RSCPublishing

ISBN: 978-0-85404-262-3

A catalogue record for this book is available from the British Library

Published by The Royal Society of Chemistry,
Thomas Graham House, Science Park, Milton Road,
Cambridge CB4 0WF, UK

Registered Charity Number 207890

For further information see our web site at www.rsc.org

Foreword to the First Edition

Archaeological science is a discipline which is growing rapidly in its scope and maturity. The application of techniques deriving from the natural sciences to archaeological research is, of course, not new and the present volume rightly emphasizes some of the earlier initiatives in the field of archaeological chemistry. But amongst the indicators of this increasing maturity are the quality of the science, on the one hand, and a growing awareness of the problems of interpretation on the other.

This work by Pollard and Heron performs a valuable service in defining more carefully than hitherto and then richly exemplifying the field of archaeological chemistry, and it is fitting that it should be published by The Royal Society of Chemistry. Hitherto it is perhaps the discipline of physics which has come in for a good deal of the limelight since it holds what is almost a monopoly in two important fields of archaeological science: dating and prospecting. Some of the techniques of dating may indeed be included within the province of chemistry: amino acid racemization is dealt with very fully here, and obsidian hydration is mentioned. But many of the most useful dating methods are dependent upon radioactivity, a subject area often relinquished to physics. And again many of the techniques of field prospection – resistivity surveys, use of the proton magnetometer, *etc.* – fall within the same province.

As defined here it is, very reasonably, chemical *analysis* which lies at the heart of archaeological chemistry. Therefore, following the historical introduction, a wide-ranging survey of analytical techniques is undertaken which certainly offers the best overview currently available to the archaeologist. This is followed by what at first sight seems a comprehensive survey of examples and applications. On closer examination, however, it emerges that nearly every chapter is based in good measure upon the original research of one of the authors, supplemented by a further overview. Thus Heron's work on organic residues, notably resins, is put to very good use. Pollard's researches in the characterization of ceramics, in glass studies, and in the use of brass and other metals likewise form the basis for much wider reviews of important fields.

One of the great strengths of the book in my view is its clear perception that archaeology and archaeological science, although interrelated and interdependent, are not at all the same thing. For the interpretation of the data of archaeological science is a complicated and difficult matter, and it involves two separate operations. The first, the elucidation of rather concrete, factual questions – 'how old is this sample?', 'where did it come from?', 'how has it been modified through burial in the ground?' – fall within archaeological science proper. But this does not make them easy questions: the difficulties in the interpretation of the data from lead isotope analysis obtained in order to determine the places of origin of metals used in the Aegean is a case in point. Another such question would be the age and authenticity of the materials from the French site of Glozel – perhaps wisely not dealt with here, but still an unresolved problem in the field of archaeological science. The second operation is to achieve a clearer realization of the implications of such concrete conclusions for the understanding of human behaviour in the past. There the archaeologists and the archaeological scientists have yet to find more effective ways of working together: the thoughtful caution in the book makes an excellent start.

Archaeological chemistry is clearly not the same field as materials science, although the two overlap, nor as metallurgy: Chapter 6 draws a distinction between 'the chemical study of metals' and (I presume) the application of metallurgical techniques to elucidate the history of the working of individual artefacts. And this suggests to me that there lies here another sub-field which has not yet emerged, fully fledged, in its own right: the archaeology of technology, or archaeo-technology. For the time will come when we shall wish to focus more clearly upon processes and procedures of manufacture. The concept of the '*chaine operatoire*' now commonly applied to the study of lithic artefacts and to the sequence of operations undertaken in their manufacture has been applied already to ceramics and can certainly be applied also to other products of pyrotechnology. Texts are available from Mesopotamia which give recipes for various preparation procedures, for instance those of metals, and very odd reading they make for modern eyes. This suggests to me that the study of ancient technology, giving more attention to what the ancient smiths, potters, glassblowers, and dyers *thought* they were doing – this is to say, laying greater emphasis upon cognitive aspects – will soon be a possibility. But in order to do this a prerequisite is to get the modern science right, and that is what Pollard and Heron are so systematically doing. It is a great pleasure to introduce a work which, more clearly than ever before, delineates an important field within archaeological science and thus makes an important contribution to the discipline as a whole.

Colin Renfrew
Cambridge
January 1996

Preface to the First Edition

Archaeological chemistry is the application of chemical knowledge in order to help solve problems in archaeology. It is, however, much more than a straight-forward application of existing techniques to new and interesting questions. Many of the chemical questions posed are unique to archaeology, although, somewhat surprisingly, many other disciplines share, from time to time, similar concerns. Our premise is that archaeological chemistry requires a thorough understanding of the background of both halves of the story, and often mastery of information from related disciplines such as biochemistry and geochemistry. This book is therefore aimed at two groups – chemists who are interested in new applications, and also archaeologists, particularly those on undergraduate and postgraduate courses encompassing aspects of scientific archaeology. It will also be of interest to geochemists, materials scientists and forensic scientists. Perhaps the most important message it contains is the need to tackle fundamental issues of chemical change in archaeological materials if scientific analysis is to make major contributions to the study of the past.

The continued expansion of scientific applications in the study of the past is one reason for writing this book. If the title of the book is not entirely new, then we believe our approach is – the adoption of a thematic structure. There are several reasons for this. Firstly, the majority of previous texts in archaeological science have tended to emphasize techniques at the expense of applications. Nowadays, there are so many techniques that such an approach would be unduly laborious, although some of the more important techniques are summarized in Chapter 2. Secondly, through an examination of particular themes, it is possible to document the successes and failures of past applications and assess the impact of scientific analysis on specific archaeological problems. It is also possible to see contemporary debates in terms of competing scientific views and to suggest how these might be resolved from a knowledge of the underlying principles. In Chapter 1, we provide a short historical context and return to general issues and future challenges in the final chapter.

Archaeological Chemistry, Second Edition
By A. Mark Pollard and Carl Heron
© The Royal Society of Chemistry 2008

We are aware of a Eurocentric bias in most of the chapters, although the issues raised will be applicable to other contexts. The majority of chapters feature the results of primary research carried out separately by the authors in collaboration with other colleagues. It is not possible to do justice to every application of chemistry to archaeology in a single volume; some of the criteria for inclusion are discussed above and in Chapter 1. As such, this book does not pretend to be a truly balanced review of archaeological chemistry and some readers may feel that some heinous crime of omission has been committed. We hope, however, that there is enough of interest to justify the attempt to adopt a unified approach to archaeological chemistry as a worthwhile area of endeavour in its own right.

Mark Pollard and Carl Heron
Bradford, December 1995

Preface to Second Edition

Much has happened in the world of archaeological chemistry since we sat down to write the Preface to the First Edition of this volume in December 1995. New centres of research have emerged, whilst some established centres, sadly, have closed down as senior figures retire or equipment is phased out. Analytical and interpretative techniques have developed immeasurably – compare the description of optical emission spectroscopy with that of laser-ablation inductively coupled plasma mass spectrometry. Although we might miss (in a masochistic sort of way!) the care and attention necessary to ensure that the graphite electrode was just the right distance above the cup, and that the photographic plate was developed for just the right length of time, there is no doubting the improvements made in analytical sensitivity, reproducibility and sample throughput.

We referred in the first edition to the 'golden age of archaeometry', being that period in the 1960s and 1970s, when, for some archaeologists at least, the chemical analysis of inorganic materials was an essential component of the investigation of prehistoric contact and exchange. Curiously, even though such analysis has become generally easier, better, cheaper and more available, there has been a perceptible decline of interest in such studies. This is perhaps a consequence of the debate about the use of lead isotope data in the Aegean during the 1990s. Archaeologists have been left unsure about how to use these data, given the various competing interpretations presented from, apparently, the same data. They may also be slightly bemused by the whole spectacle, and certainly not prepared to invest any more time in the venture until greater consensus is achieved. Given the fact that, with new instrumentation, such measurements are easier and faster to make than ever before, this is clearly regrettable, and we certainly encourage a new generation of archaeological chemists to return to this and other related issues. We hope this volume will provide the stimulus for some to do so.

Archaeological Chemistry, Second Edition
By A. Mark Pollard and Carl Heron
© The Royal Society of Chemistry 2008

It is, however, possible to perceive the growth of a new 'golden age' of archaeometry – this time, however, it is focussed more on the biochemistry of archaeological materials. The growth of biomolecular archaeology on all fronts has been dramatic in the 12 years since the First Edition was published. Foremost has been the development of the isotopic study of human bone, for the reconstruction of diet, status and mobility in the past. More recently, building on developments in mass spectrometry, a whole new range of organic molecules, including small proteins, has been detected in archaeological contexts. These organic molecules have been used to answer questions of major archaeological significance, such as the nature of animal husbandry in the early history of domestication and agriculture.

We have, therefore, added two new chapters on these subjects to this Second Edition. We have also updated all the other chapters to reflect developments over the past ten years, and added many new references. The volume, inevitably, remains a partial reflection of the still-expanding field of archaeological chemistry, and also, equally inevitably, a reflection of personal preferences and interests. We therefore apologise in advance for any omissions, oversimplifications or errors which may occur in this Second Edition.

Mark Pollard and Carl Heron
Oxford and Bradford
July 2007

Acknowledgements

The authors are grateful to many people over the past 20 years, all of whom have contributed in various ways to the production of this book. These include Dr G.A. Cox, Professor O.S. Heavens and Professor J.A.D. Mathew (Department of Physics, University of York); Professor Teddy Hall, Professor Robert Hedges, Dr Richard Gillespie, Professor John Gowlett, and Helen Hatcher (Research Laboratory for Archaeology and the History of Art, Oxford University); Dr Robin Symonds (Museum of London); Dr Cath Mortimer (Ancient Monuments Laboratory, English Heritage); Professor Bob Gillard, Professor Pete Edwards and the troops (School of Chemistry, University of Wales, Cardiff); Professor Pete Williams and Dr Richard Thomas (School of Chemistry, University of Western Sydney); Professor Richard Evershed (Department of Chemistry, University of Bristol); Dr John Goad (Department of Biochemistry, University of Liverpool); Dr Gerry McDonnell, Dr Paul Budd, Professor Arnold Aspinall, and Geoff Gaunt (Department of Archaeological Sciences, University of Bradford). We are also indebted to the following individuals: Miranda Schofield for drawing the figures; Rebecca Stacey for compiling the chemical structures; Professor Curt Beck (Vassar College, New York), Dr Vasily Kilikoglou (Demokritos, Athens), Professor Dr Klaus Ruthenberg (Coburg, Germany) and Dr Robert Tykot (Harvard University) for providing papers in advance of publication; Paula Mills, Dr Cathy Batt, Dr Randolph Haggerty, Dr Vikki Carolan, Dr Suzanne Young (Harvard University) and Dr Olwen Williams-Thorpe (Open University) for reading Chapters or parts thereof; and the many authors and publishers who allowed reproduction of figures used in this book. Finally, our thanks go to our respective families for considerable forbearance during the production of this book.

In addition to the above, many people have assisted in producing the second edition, including Dr Ron Hancock in Toronto, Professor Ian Freestone in Cardiff and Professor Matthew Collins and Dr Oliver Craig in York for making available unpublished material and offering helpful comments.

Contents

Archaeological Chemistry, Second Edition
By A. Mark Pollard and Carl Heron
© The Royal Society of Chemistry 2008

Chapter 7   The Chemistry and Use of Resinous Substances

Chapter 8   Amino Acid Stereochemistry and the First Americans

The Development of Archaeological Chemistry

1.1 INTRODUCTION

In its endeavour to understand human behaviour primarily through the material remains of past societies, archaeology has interacted more and more with the sciences of physics, chemistry, biology and of the Earth. In truth, it is a test to conjure the name of any scientific discipline which has not at one time or another provided information of direct use for the archaeologist (Pollard, 1995). Indeed, many would consider archaeology itself, a discipline which involves the systematic collection, evaluation and analysis of data and aims to model, test and theorize the nature of past human activity, to be a science. Furthermore, they might argue that it is possible to arrive at an objective understanding of past human behaviour, and in that sense archaeology is no different from other scientific disciplines, given the obvious differences in methodology. As Trigger (1988; 1) has reminded us, from a different perspective, archaeologists have a unique challenge:

'*Because archaeologists study the past, they are unable to observe human behaviour directly. Unlike historians, they also lack access to verbally encoded records of the past. Instead they must attempt to infer human behaviour and beliefs from the surviving remains of what people made and used before they can begin, like other social scientists, to explain phenomena.*'

The claim that archaeology is a science is clearly not universally held. Many archaeologists suggest that the study of human behaviour in the past is restricted by science, with its apparent rigidity of scientific method and dubious claims of certainty, and must continue to reside with the humanities. Undoubtedly, archaeology is one of the few disciplines which straddles the gulf between the humanities and the sciences.

Archaeological Chemistry, Second Edition
By A. Mark Pollard and Carl Heron
© The Royal Society of Chemistry 2008

In our view, one of the fundamental enquiries in archaeology is the relationship between residues, artefacts, buildings and monuments, and human behaviour. From the period of production, use or modification of materials (whether natural or synthetic) to the time when traces are recovered by archaeologists, the material output of humans is altered by a plethora of physical, chemical and biological processes including those operating after deposition into the archaeological record. A significant part of the evidence is lost, displaced or altered significantly. Inferring the activities, motivations, ideas and beliefs of our ancestors from such a fragmentary record is no small task. In fact, it is a considerable challenge. Although there are notable exceptions, archaeology in the past 150 years has been transformed from a pastime pre-occupied with the embellishment of the contemporary world (or at least a minuscule portion of it) with treasure recovered from 'lost civilizations' (still a view which predominates in some media representations of archaeology, such as the cinema) to a discipline which relies on painstaking and systematic recovery of data followed by synthesis and interpretation. However, the development of archaeology has not been one uniform trajectory. There have been, and still are, numerous agendas which encompass the broad range of archaeological thought, and many uncertainties and disagreements concerning the direction of the discipline remain. Collectively, the sciences provide archaeology with numerous techniques and approaches to facilitate data analysis and interpretation, enhancing the opportunity to extract more information from the material record of past human activity. Specifically, chemistry has as much to offer as any other scientific discipline, if not more.

The sheer diversity of scientific analysis in archaeology renders a coherent and comprehensive summary intractable. Tite (1991) has packaged archaeological science rather neatly into the following areas:

- Physical and chemical dating methods which provide archaeology with absolute and relative chronologies.
- Artefact studies incorporating (i) provenance, (ii) technology, and (iii) use.
- Environmental approaches which provide information on past landscapes, climates, flora and fauna as well as diet, nutrition, health and pathology of people.
- Mathematical methods as tools for data treatment also encompassing the role of computers in handling, analysing and modelling the vast sources of data.
- Remote sensing applications comprising a battery of non-destructive techniques for the location and characterization of buried features at the regional, microregional and intrasite levels.
- Conservation science involving the study of decay processes and the development of new methods of conservation.

Although in this volume we focus on the interaction between chemistry and archaeology or *archaeological chemistry*, chemistry is relevant to most if not all of the areas proposed by Tite. For example, although many subsurface prospecting techniques rely on (geo)physical principles of measurement (such as localized variations in electrical resistance and small variations in Earth magnetism), geochemical prospection methods involving the determination of inorganic and biological markers of anthropogenic origin (*i.e.*, chemical species arising as a direct consequence of human action) also have a role to play. Throughout this book, archaeological chemistry is viewed not as a straightforward application of routine methods to archaeological material but as a challenging field of enquiry, which requires a deep knowledge of the underlying principles in order to make a significant contribution.

1.2 EARLY INVESTIGATIONS

It would not be possible to write a history of chemistry without acknowledging the contribution of individuals such as Martin Heinrich Klaproth (1743–1817), Humphry Davy (1778–1829), Jöns Jakob Berzelius (1779–1848), Michael Faraday (1791–1867), Marcelin Berthelot (1827–1907) and Friedrich August von Kekulé (1829–1896). Yet these eminent scientists also figure in the early history of the scientific analysis of antiquities. Perhaps the primary motivation for their work was curiosity, which resulted from their dedication to the study and identification of matter and the way in which it is altered by chemical reaction. In addition to his significant contributions to analytical and mineralogical chemistry, Martin Heinrich Klaproth determined the approximate composition of some Greek and Roman coins, a number of other metal objects and a few pieces of Roman glass. Klaproth was a pioneer in *gravimetry* – the determination of elemental composition through the weighing of an insoluble product of a definite chemical reaction involving that element. His first paper entitled '*Mémoire de numismatique docimastique*' was presented at the Royal Academy of Sciences and Belles-Lettres of Berlin on July 9th, 1795. The coins were either copper or copper alloy. In producing compositional data on ancient materials, Klaproth had first to devise workable quantitative schemes for the analysis of copper alloys and glass. His scheme for coins has been studied by Caley (1949; 242–43) and is summarized briefly below:

'*After the corrosion products had been removed from the surface of the metal to be analysed, a weighed sample was treated with 'moderately concentrated' nitric acid and the reaction mixture was allowed to stand overnight ... the supernatant liquid was poured off and saved, and any undissolved metal or insoluble residue again treated with nitric acid ... If tin was present as shown by the continued presence of a residue insoluble in nitric acid, this was collected on filter paper ... (this) was simply dried in an oven and weighed ... a parallel control experiment was made with a known weight of pure tin. It was found from this that 100 parts of dried residue contained 71 parts metallic tin, in other words the gravimetric factor was 0.71.*

> The filtrate from the separation of the tin was tested for silver by the addition of a saturated solution of sodium chloride to one portion and the introduction of a weighed copper plate into another.
>
> Lead was separated from the solutions...by evaporation to a small volume. The separated lead sulfate was collected and either weighed as such or reduced to metallic lead in a crucible for direct weighing as metal.
>
> (Copper) was determined as metal from the filtrate from the lead separation by placing in it a clean iron plate. The precipitated copper was then collected, dried and weighed.'

In addition to Klaproth's pioneering work in quantitative analysis, he made a major contribution to mineralogical chemistry and discovered many elements in the process. His efforts did not go unrewarded, since he became Berlin's first Professor of Chemistry.

In 1815, Humphry Davy published a paper on the examination of ancient pigments collected at Rome and Pompeii. In addition to reviewing evidence for natural pigments, he was also able to identify a synthetic pigment, later to be called *Egyptian Blue*, formed by fusing copper, silica and naturally occurring natron (sodium carbonate). A report by H. Diamond, published in the journal *Archaeologia* in 1867, includes a section on a Roman pottery glaze studied by Michael Faraday in which the presence of lead in the sample provided the first indications on chemical grounds of the use of lead glaze in antiquity. In addition to his significant contributions to modern chemistry during the first half of the 19th Century, Berzelius became interested in the composition of ancient bronzes. Similarly, Kekulé carried out analysis of an ancient sample of wood tar that may have comprised, in part, compounds with aromatic or benzene rings, the structure of which he subsequently proposed in 1865.

In addition to the diverse activities of these well-known scientists, efforts made by a number of other investigators during the 19th Century are worthy of note. Frequently they sought to examine ancient metal objects (Caley, 1949, 1951, 1967) with a view initially to understanding their composition and the technology needed to produce the artefacts, although other questions began to emerge. As these investigations continued, mostly in isolation from one another, prehistoric archaeology was making its first steps towards a systematic enquiry into the study and chronology of early materials. In 1819, Christian Thomsen assigned the artefacts in the Danish national collection into successive ages of stone, stone and copper, bronze, early iron and later iron. This relative chronology was based on comparisons of material type, decoration and the context of recovery, and it marked a major development in the study of ancient materials which prevails in archaeology today [see Trigger (2006; 121–129) for a more detailed consideration].

As early as the mid-19th Century, the Austrian scholar J.E. Wocel suggested that correlations in chemical composition could be used to 'provenance' or identify the source of archaeological materials and even to provide relative dates of manufacture and use. During the 1840s, C.C.T.C. Göbel, a chemist at the University of Dorpat (Tartu) in Estonia, began a study of large numbers of copper alloy artefacts from the Baltic region, comparing those recovered from excavations with known artefacts of prehistoric, Greek and Roman date. He concluded that the artefacts were probably Roman in origin. With the work of Göbel, scientific analysis progressed beyond the generation of analytical data on single specimens to, as Harbottle (1982; 14) has emphasized, '*establishing a group chemical property.*' The French mineralogist Damour proposed that the geographical source of stone axes could be located by considering the density and chemical composition of a number of rock types, including jade and obsidian found '*dans les monuments celtiques et chez les tribus sauvages*', as his papers of 1864 and 1866 were entitled (Caley, 1951; 66). Damour also exhorted archaeologists to work with specialists from other disciplines such as geology, zoology and palaeontology (Harbottle, 1982; 14). Perhaps he was aware of the interdisciplinary research programmes comprising zoologists, geologists and archaeologists then being carried out in Scandinavia on ancient shell mounds along the coast of Denmark (Klindt-Jensen, 1975; 71–73). Damour's primary interest was jade. He proposed that prehistoric jade axes were fashioned from outcrops in the Mont Viso massif in northern Italy. It is only recently that his ideas have been confirmed by archaeological fieldwork (Pétrequin *et al.*, 2006).

The appearance of the first appendices of chemical analysis and references to them in the text of a major excavation report represents the earliest significant collaboration between archaeologists and chemists. Examples include the analysis of four Assyrian bronzes and a sample of glass in Austen Henry Layard's '*Discoveries in the Ruins of Nineveh and Babylon*' published in 1853 and Heinrich Schliemann's '*Mycenae*' first published in 1878 (so distinguished was the publication of the Arno Press edition of 1880 that William Gladstone, the then British Prime Minister, wrote the preface!). The reports in the appendices of both these works were overseen by the metallurgist, John Percy, at the Royal School of Mines in London. Between 1861 and 1875, Percy wrote four major works on metallurgy which included significant sections on the early production and use of metals (Percy, 1861, 1864, 1870, 1875). These books remain important sources even today, because they contain first-hand descriptions of now lost metallurgical processes. Analysis of metal objects from Mycenae showed the extensive use of native gold and both copper and bronze, the latter used predominantly for weapons. Percy wrote in a letter to Schliemann dated August 10, 1877 that '*Some of the results are, I think, both novel and important, in a metallurgical as well as archaeological point of view.*'

The effort made by Otto Helm, an apothecary from Gdansk, Poland, to source amber towards the end of the 19th Century constitutes one of the earliest systematic applications of the natural sciences to archaeology. It truly can be said that this enquiry was advanced with a specific archaeological problem in mind – determining the geographical source of over 2000 amber beads excavated by Schliemann at Mycenae. In the excavation monograph, Schliemann noted that '*It will, of course, for ever remain a secret to us whether this amber is derived from the coast of the Baltic or from Italy, where it is found in several places, but particularly on the east coast of Sicily.*' Helm based his approach on the succinic (butanedioic) acid content of Baltic amber [known since the mid-16th Century from the studies by Georg Bauer (1494–1555), who is better known to metallurgists as *Agricola*], but did not undertake a systematic study of fossil resins from other sources in Europe. His motivation lay, at least partly, in disproving the hypothesis of an Italian mineralogist, Capellini, who suggested that some of the earliest finds of amber in the south could have been fashioned from local fossil resins. A full account of the investigations made and the success claimed by Helm along with the eventual shortcomings has been compiled by Curt Beck (Beck, 1986) who in the 1960s published, with his co-workers, the results of some 500 analyses using infrared (IR) spectroscopy which demonstrated for the first time successful discrimination between Baltic and non-Baltic European fossil resins (Beck *et al.*, 1964; 1965). Unless severely weathered, it is usually possible to demonstrate that the vast majority of amber finds throughout prehistoric Europe derive from the Baltic coastal region.

The French chemist Marcelin Berthelot was active in chemical analysis in the late 19th Century, investigating some 150 artefacts from Egypt and the Near East. According to Caley (1967; 122), Berthelot may have been '*less interested in the exact composition of ancient materials than in obtaining results of immediate practical value to archaeologists.*' This was coupled with an interest in the corrosion of metals and the degradation of organic materials, which prompted a series of experimental studies based on prolonged contact of metal objects with air and water. Although Berthelot published some 42 papers in this field, many of them remained unaltered, in title or content, from journal to journal. For this, at least, he perhaps deserves credit from contemporary academics for enterprise!

Towards the end of the 19th Century, as archaeological excavation became a more systematic undertaking, the results of chemical analysis became more common in reports and new suggestions began to appear. As early as 1892, A. Carnot suggested that fluorine uptake in long-buried bone might be used to provide an indication of the age of the bone (Caley, 1967; 122), although the feasibility of the method was not tested until the 1940s. The increasing numbers of antiquities brought about more emphasis on their restoration and conserva-tion. The pioneer in this field was Friedrich Rathgen, who established a laboratory at the State Museum in Berlin and later published the first book ('*Die Konservierung von Alterthumsfunden*') dealing with practical procedures for the conservation of antiquities, including electrolytic removal of corrosion from ancient artefacts and the use of natural consolidants (such as pine resin and gelatin) in the conservation process. Developments in the examination of archaeological materials in Europe began to be applied to New World artefacts. In Sweden, Gustav Nordenskiöld submitted pottery sherds collected at Mesa Verde, Colorado, for petrological examination (thin section analysis). The results appeared in his volume '*Cliff Dwellers of the Mesa Verde*' published in 1893. One of the first wet chemical investigations of ancient ceramics (Athenian pottery from the Boston Museum of Fine Arts) was carried out at Harvard and published in the *American Chemical Journal* in 1895 by T.W. Richards (Harbottle, 1982; 17).

1.3 THE GROWTH OF SCIENTIFIC ARCHAEOLOGY IN THE 20TH CENTURY

The 1920s and 1930s saw the addition of instrumental measurement techniques, such as optical emission spectroscopy (OES; see Chapter 2), to the repertoire of the analyst. The principal archaeological interest at the time was understanding the level of technology represented by finds of ancient metalwork, especially in terms of alloying, and systematic programmes of analysis were initiated in Britain and Germany leading to substantial analytical reports (*e.g.*, Otto and Witter, 1952). As a result of the rapid scientific and technological advances precipitated by the Second World War, the post-war years witnessed a wider range of scientific techniques being deployed in the study of the past. Eventual reconstruction as a result of war damage was preceded by a major expansion of archaeological excavation which produced very large quantities of artefacts. The development of radiocarbon dating by Willard Libby in 1949 paved the way for establishing absolute chronologies throughout the world. Although the impact was not immediate, radiocarbon dating eventually allowed sites to be dated in relation to one another and enabled cultural sequences to be established independent of cross-cultural comparisons (based on artefact typologies) with areas dated by historical methods (Renfrew, 1973).

Other materials such as faience beads and ceramics were incorporated into analytical programmes. Faience comprises a core of finely powdered quartz grains cemented by fusion with a small amount of alkali and lime. The core is coated with a glaze of soda lime and coloured in the range blue to green with copper compounds. Faience was first produced in the Near East although Egyptian faience became very important between the 4th and 2nd Millennia BC. During the 2nd Millennium BC, faience was distributed across prehistoric Europe and occurred in England and Scotland. In 1956 Stone and Thomas reported on the use of OES to '*find some trace element, existent only in minute quantities, which might serve to distinguish between the quartz or sand and the alkalis used in the manufacture of faience and glassy faience in Egypt and in specimens found elsewhere in Europe*' (Stone and Thomas, 1956; 68). This study represented a clear example of the use of chemical criteria to determine whether faience beads recovered from sites in Britain were imported from Egypt or the eastern Mediterranean. For many years, it had generally been assumed that faience manufacture and other technological innovations originated in the east and diffused westwards. Although the initial results suggested that OES could not be used unequivocally, the data were subsequently re-evaluated statistically by Newton and Renfrew (1970) who suggested a local origin on the basis of the quantities of tin, aluminium and magnesium present in the beads. This was augmented by re-analysis of most of the beads using neutron activation analysis (NAA) by Aspinall *et al.* (1972). They confirmed that the tin content of British beads is significantly higher than that found in groups of beads from elsewhere and that a number of other trace elements showed promise. However, the belief in a local origin for the British beads is by no means universally held and only investigations of larger sample groups can hope to resolve the issue. This use of the highly sensitive technique of NAA was by no means the first. Sayre and Dodson (1957) applied NAA in their study of ancient Mediterranean ceramics and, in conjunction with gamma ray spectroscopy, NAA was used by Emeleus and Simpson (1960) to test the applicability of trace element analysis to locate the region of origin of Roman Samian sherds which could not be placed stylistically (see Chapter 4).

In Britain, the term '*archaeometry*' was coined in the early 1950s by Christopher Hawkes in Oxford to describe the increased emphasis on dating, quantification and physicochemical analysis of archaeological materials. A journal with the same name was launched in 1958 and textbooks by Martin Aitken (1961) and Mike Tite (1972) illustrated the full potential of emerging applications. In 1974, the first volume of another periodical dedicated to scientific work in archaeology (*Journal of Archaeological Science*) was published.

During the late 1950s and early 1960s, a number of individuals advocated strongly a new and refreshing approach to archaeology, although the roots of this impetus are evident in earlier writings. Progressive thinking in anthropology and the social sciences as well as the explicit use of models by geographers had largely left archaeology lagging behind. This transformation, which became known as the '*New Archaeology*', represented an explicit effort on the part of a number of archaeologists who emphasized optimistically the potential for explaining past human action rather than simply describing it. Such was the optimism of the early 1960s that it was felt that all human behaviour could be embodied within laws of cultural phenomena. Patterning in '*material culture*' and the '*archaeological record*' could be used to explore behavioural correlates regardless of time or place. Not surprisingly, the philosophy of science played a significant role in providing the terminology for statistical and quantitative approaches in archaeology (see Trigger, 2006). The New Archaeology rejuvenated research into prehistoric trade and exchange. Invasion or diffusion of peoples was no longer viewed as the principal instigator of cultural change; instead, internal processes within society were emphasized. Evidence for 'contact' arising from exchange of artefacts and natural materials (as well as the transmission of ideas) was still seen as an important factor. Scientific analysis might therefore be used to evaluate change in economic and social systems. This increased interest in the distribution of materials initiated a 'golden era' in archaeometry as a wide range of scientific techniques were deployed in the hope of chemically characterizing certain rock types, such as obsidian (see Chapter 3) and marble (Rybach and Nissen, 1964) as well as ceramics (*e.g.*, Catling *et al.*, 1963), metals (Junghans *et al.*, 1960), glass (Sayre and Smith, 1961) and natural materials, such as amber (Beck *et al.*, 1964). These characterization studies were aimed at '*the documentation of culture contact on the basis of hard evidence, rather than on supposed similarities of form*' (Renfrew, 1979; 17). The substantial data sets generated by these techniques could now also be subjected to statistical treatment using computers. Characterization studies remain an important research area in archaeological science, utilizing a range of chemical properties incorporating trace element composition, biomarker composition, mineralogy and scientific dating, including isotopic measurements. In a review of chemical characterization, Harbottle (1982; 15) reminded practitioners that:

> '... *with a very few exceptions, you cannot unequivocally source anything. What you can do is characterize the object, or better, groups of similar objects found in a site or archaeological zone by mineralogical, thermoluminescent, density, hardness, chemical, and other tests, and also characterize the equivalent source materials, if they are available, and look for similarities to generate attributions. A careful job of chemical characterization, plus a little numerical taxonomy and some auxiliary archaeological and/or stylistic information, will often do something almost as useful: it will produce groupings of artefacts that make archaeological sense. This, rather than absolute proof of origin, will often necessarily be the goal.*'

The geographical source of the materials under investigation included quarries, mines or clay deposits and sites of production where materials are modified or fabricated. If the material remains unaltered during preparation or modification, for example, when flakes of obsidian are removed from a large core of the rock, then the bulk composition of the artefact is unaltered from the source material, although subtle changes may occur (such as in the case of a *hydration layer* on obsidian – see Chapter 3). However, in the case of synthetic materials such as ceramics, metals and glass, production may bring about significant changes in the composition of the finished artefact with respect to the composition of the raw materials. The whole question of provenance becomes a complex issue (*e.g.*, Tite, 1991; 143–144; Cherry and Knapp, 1991; Wilson and Pollard, 2001).

Until two or three decades ago, archaeology generally paid more attention to the analysis of inorganic artefacts – natural stone, metal, glass, ceramic material and so on – reflecting an interest in the most obviously durable artefacts in the archaeological record. In subsequent years, increasing attention has been directed at biological materials; natural products such as waxes and resins, accidental survivals, such as food residues, and, above all, human remains, including bone, protein, lipids and, most recently of all, DNA. Some of the methodology for this work has been imported not only from chemistry, biochemistry and molecular biology, but also from organic geochemistry, which has grown from a discipline with a principal interest in elucidating the chemical origins of oil and coal into one which studies the short-term alteration and long-term survival of a very wide range of biomolecules (*e.g.*, Engel and Macko, 1993). Another related discipline in this quest for ancient biomolecular information is molecular palaeontology. These disciplines are widely recognized as having much to offer each other, particularly in the recovery of genetic information from animals and plants. In terms of specific archaeological interest, the ability to extract DNA has considerable significance. Hitherto, extraction of nucleic acids from bone some 25 000 years old has been claimed. Preserved soft tissue and seed remains also yield extractable DNA. Specific DNA sequences can be targeted, amplified using the *polymerase chain reaction*

(PCR) and compared with sequences in other individuals and modern specimens. However, ancient DNA is severely damaged and fragmented. Contamination of aged samples and extracts with modern DNA is a serious problem and, whilst the study of DNA in archaeological samples will constitute a major area of future activity in the discipline, current research will continue to focus on the authentication of samples of ancient DNA advances have been so rapid that perusal of the appropriate scientific journals is essential. For somewhat more recent views of the state of ancient DNA research, see Willerslev and Cooper (2005).

Preservation of a wider range of biomolecules has been demonstrated in number of archaeological contexts. In particular, proteins preserved in human bone have been subject to immunological investigation (*e.g.*, Smith and Wilson, 1990; Cattaneo *et al.*, 1992, Gernaey *et al.*, 2001). The survival of protein residues on stone tool surfaces (*e.g.*, Loy, 1983) hints at the possibility of characterizing artefact use and identifying utilization of specific resources and dietary items, although the specificity and replicability of the approaches used remains contentious (*e.g.*, Eisele *et al.*, 1995; Tuross and Dillehay, 1995; Smith and Wilson, 2001). Certain biomolecules, such as lipids, are more durable than nucleic acids and, although some chemical alteration is to be expected, specific identifications can be made on aged samples (*e.g.*, Evershed, 1993; Heron *et al.*, 1994; Evershed *et al.*, 2001: see Chapters 7 and 11 herein).

The voluminous literature on bone chemical investigations generated during the past three decades represents one of the significant growth areas of archaeological chemistry (*e.g.*, Price, 1989; Grupe and Lambert, 1993; Sandford, 1993; Sealy, 2001: see Chapter 10 herein). Quantitative analysis of certain trace elements (such as strontium, barium, zinc and lead) incorporated into bone mineral and, less frequently, teeth and hair has been used to assess diet, nutrition, health status and pathology. Similarly, dietary inferences have been made through measurement of light stable isotope ratios of carbon and nitrogen in bone collagen and the carbon isotope composition of bone apatite. The realization that maize and other grasses have a high ^{13}C content has been used to characterize long-term consumption of maize in the Americas, specifically its introduction and expansion in eastern North America (van der Merwe and Vogel, 1978). However, the recognition of significant compositional and mineralogical alteration during long-term burial (often labelled *diagenesis* – a term which, unfortunately, has different meanings in different branches of the historical sciences) has brought about a re-evaluation of bone chemical investigations; the onus of proof is now on the analyst to demonstrate that the data are not geochemical artefacts that reflect more on the complex interaction between bone and the burial environment than on any dietary or other signal which may have accumulated during life. Undoubtedly, trace element compositions are highly susceptible to a wide range of post-depositional alterations including exchange between ions in the soil solution and bone mineral (Sandford, 1993).

The limitations of the types of sample analysed in archaeological chemistry can be considerable. Typically samples are far from ideal from the analytical point of view – small, fragmentary and, particularly in the case of biological samples, considerably degraded. Contamination during deposition is also a major problem, as it is once the sample is recovered (post-excavation), due to storage media, handling and airborne particles. These ubiquitous problems of degradation and contamination make archaeological chemistry a challenging field, and not one which can be regarded as just another routine application of analytical chemistry. Strong parallels have been drawn with forensic science (Hunter *et al.*, 1996), in the sense that both disciplines use a wide range of scientific approaches to extract information from the (often non-ideal) material record, with a view to reconstructing activities and intentions in the past.

1.4 CURRENT STATUS AND SCOPE OF ARCHAEOLOGICAL CHEMISTRY

Although archaeological science is recognized as a fundamental component of the inquiry into past human behaviour and development, the demand for more relevant data has been echoed on many occasions. Scientific analysis should be much more than a descriptive exercise which simply documents the date, morphology or composition of ancient materials. As DeAtley and Bishop (1991; 371) have pointed out (see also Trigger, 1988) no analytical technique has '*built-in interpretative value for archaeological investigations; the links between physical properties of objects and human behaviour producing the variations in physical states of artefacts must always be evaluated.*' This demand for meaningful scientific data also needs to be viewed against the changing approaches to the study of the past as the discipline of archaeology evolves. The historical relationship between scientific approaches and techniques and pre-vailing theoretical views regarding past human behaviour has been reviewed by Trigger (1988; 1) who states that '*archaeologists have asked different questions at different periods. Some of these questions have encouraged close relations with the biological and physical sciences, while other equally important ones have dis-couraged them.*' During the 1960s, archaeology embraced the sciences, not only the techniques but also the terminology of scientific method for explaining human behaviour. During the past two decades, general attitudes in society towards science have been shifting towards a more critical stance. Against this backdrop, the contribution of scientific analysis to the study of the past has come under increasing scrutiny (Pollard, 2004). Certain approaches to archae-ological thinking from the mid-1980s onwards, commonly labelled *post-processual*, have stressed relativism and subjectivism in interpretation, as well as the ideological and symbolic roles of material culture, to a much greater extent than before. This has often been accompanied by an exposition on the limited contribution that scientific data have to make to this inquiry (see, for example, Hodder, 1984; Thomas, 1991).

Although the majority of archaeologists acknowledge the contribution of scientific dating and analytical techniques to increasing the information poten-tial of the past, the central concern prevails that scientific studies often proceed in a context devoid of a specific archaeological problem (Yoffee and Sherratt,

1993; 4–5). However, it would be misleading to suggest that chronological, compositional or locational data generated by scientific techniques have no role to play in providing foundations for interpretations of past human behaviour. As Cherry and Knapp (1991; 92) have remarked, scientific analysis could '*help arbitrate amongst competing cultural hypotheses*' although they could find little evidence for the adoption of such an approach. Perhaps success in archaeological science is difficult to measure. For some, it may be the implementation of a robust or elegant scientific methodology; for others it may be the degree of integration within archaeological problems. Ideally, it should display characteristics of both. The promotion of genuine interdisciplinary studies rather than multidisciplinary investigation lies at the heart of modern archaeological endeavour (DeAtley and Bishop, 1991; Pollard, 2004).

1.5   THE STRUCTURE OF THIS VOLUME

It is hoped that the foregoing provides a short historical context to this volume. Indeed we return to this discussion briefly in the final chapter. In the intervening chapters, we have selected a number of themes which exemplify past and current research in archaeological chemistry. The themes presented in this volume (representing only a small component of what is called archaeological science) span many diverse areas of chemistry. There are several reasons for adopting a thematic approach. Firstly, the majority of previous texts in archaeological science have tended to emphasize techniques at the expense of applications which have produced relevant archaeological information. Nowadays, there are so many techniques that such an approach would be unduly laborious. Some of the more relevant techniques of analysis are summarized as briefly as possible in Chapter 2. The remaining chapters range in scope, but each includes a discussion of some of the underlying science. Chapter 3 reviews and updates the classic characterization studies undertaken with the aim of locating the source of the volcanic glass, obsidian. Chapter 4 reviews the structural chemistry of clays, and illustrates the power of chemical studies of ceramics with an example from Roman Britain and Gaul. Chapter 5 discusses the structure and chemistry of archaeological glass, together with a review of some work on the atmospheric corrosion of Medieval window glass, and an introduction to isotopic methods of provenancing glass. Two chapters (6 and 9) focus on the chemical study of metals. Chapter 6 considers the chemical analysis of European Medieval and later brass objects, including a discussion of the use of this knowledge for the study of brass scientific instruments, and the use of copper alloy analyses to help document the early contact period in North America. Chapter 7 focuses on the chemistry of resins and aims to consider the future role of analytical organic chemistry applied to amorphous deposits surviving on artefact surfaces. Chapter 8 continues the theme of organic chemistry in archaeology, with a consideration of the racemization of amino acids in bones and teeth, and an example drawn from the intractable question of dating the arrival of the earliest humans in the New World. Chapter 9 returns to metals, and tackles the controversial field of lead isotope geochemistry, in particular with a critical review of its role in locating the source of metals in the Mediterranean Bronze Age.

The second edition adds two further chapters. Chapter 10 reviews the applications of 'isotope archaeology' to the study of human bone for purposes of dietary reconstruction, which leads on to questions of health, status and mobility. This is an area of research which has exploded since the 1990s, and is now regarded as an essential – and virtually routine – part of the study of human remains. The second area to be added (Chapter 11) is the study of food lipids from archaeological contexts. This has also developed rapidly in the past ten years, partly because of advances in the field of biomolecular mass spectrometry, which affords the potential to identify small biomolecules relatively easily, but also because some of the pioneering work discussed in Chapter 11 has demonstrated that, if one is prepared to look, these molecules do survive in the archaeological record. The most striking application to date has been the elucidation of the role of dairying in the 'secondary products revolution' – in early Old World agricultural societies, it is not clear from the faunal record whether domesticated animals were reared primarily for meat or milk and cheese, or, in the case of cattle, for ploughing. The identification of milk proteins in Early Neolithic pottery is beginning to unravel this question.

Every reader will probably feel that some heinous crime of omission has been committed. This book does not pretend to be a completely balanced review of archaeological chemistry – the size of a single volume precludes any serious attempt to do that. What we have tried to do is present a range of studies which have been important archaeologically, are interesting from a chemical standpoint and have interested the authors at one time or another.

1.6 FURTHER READING

A comprehensive history of scientific analysis applied to the study of past people and materials is lacking. The papers by Caley (1949; 1951; 1967) remain useful for summaries of the early applications of chemistry to archaeology and the paper by Trigger (1988) is essential reading. The contributions of Berzelius, Davy, Faraday and others to the development of chemistry are summarized by Hudson (1992). The text by Renfrew and Bahn (2004) serves as a very useful general introduction, covering the scope and aims of modern archaeology, including many scientific applications. For a more detailed consideration of the development of archaeology, see Trigger (2006). Debates on the theory of archaeology can be found in a series of essays in Yoffee and Sherratt (1993), Hodder *et al.* (1995) and Bintliff (2004).

A collection of recent scientific studies, largely relating to museum objects, including dating, authenticity, metalwork, ceramics and glass, can be found in the edited volume by Bowman (1991). Henderson (2000) provides an overview of the information derived from scientific studies of a similar range of inorganic archaeological materials. Many conference proceedings [especially those entitled *Archaeological Chemistry*, latterly produced by the American Chemical Society (Levey, 1967; Brill, 1971; Beck, 1974; Carter; 1978; Lambert, 1984;

Allen, 1989; Orna, 1996; Jakes, 2002), and also the published proceedings of the *International Archaeometry Symposia* and *Materials Issues in Art and Archaeology* meetings] contain a very wide range of chemical studies in archaeology. Goffer (1980) gives a broad introduction to archaeological chemistry, covering basic analytical chemistry, the materials used in antiquity, and the decay and restoration of archaeological materials. More recent publications include Lambert (1997) which has eight chapters, each one based on the study of a particular archaeological material, and Pollard *et al.* (2007), which, in addition to case studies, contains more information on the science underlying analytical chemistry as applied to archaeology. The 'standard works' on science in archaeology in general include Brothwell and Higgs (1963), Ciliberto and Spoto (2000), and Brothwell and Pollard (2001).

Scientific dating methods, such as radiocarbon dating, electron spin resonance, luminescence dating and so on, are not covered in this volume, even though chemistry is intimately involved in many of the methods. There is a long history of relevant texts from Zeuner's *Dating the Past* (four editions between 1945 and 1957) to Aitken's *Science-based Dating in Archaeology* (1990) and also Taylor and Aitken (1997).

REFERENCES

Aitken, M.J. (1961). *Physics and Archaeology*. Interscience Publishers, New York (2nd edn 1974, published by Oxford University Press, Oxford).

Aitken, M.J. (1990). *Scientific Dating Techniques in Archaeology*. Longman, London.

Allen, R.O. (ed.) (1989). *Archaeological Chemistry IV*. Advances in Chemistry Series 220, American Chemical Society, Washington, D.C.

Aspinall, A., Warren, S.E., Crummett, J.G. and Newton, R.G. (1972). Neutron activation analysis of faience beads. *Archaeometry* **14** 41–53.

Beck, C.W. (ed.) (1974). *Archaeological Chemistry*. Advances in Chemistry Series 138, American Chemical Society, Washington, D.C.

Beck, C.W. (1986). Spectroscopic studies of amber. *Applied Spectroscopy Reviews* **22** 57–110.

Beck, C.W., Wilbur, E. and Meret, S. (1964). Infrared spectra and the origins of amber. *Nature* **201** 256–257.

Beck, C.W., Wilbur, E., Meret, S., Kossove, D. and Kermani, K. (1965). The infrared spectra of amber and the identification of Baltic amber. *Archaeometry* **8** 96–109.

Bintliff, J. (ed.) (2004). *A Companion to Archaeology*. Blackwell, Oxford.

Bowman, S. (ed.) (1991). *Science and the Past*. British Museum Press, London.

Brill, R.H. (ed.) (1971). *Science and Archaeology: Symposium on Archaeological Chemistry 1968, Atlantic City*. MIT Press, Cambridge, Mass.

Brothwell, D. and Higgs, E. (ed.) (1963). *Science in Archaeology: A Survey of Progress and Research*. Thames and Hudson, London (2nd expanded edn 1969).

Brothwell, D.R. and Pollard, A.M. (ed.) (2001). *Handbook of Archaeological Sciences*. John Wiley and Sons, Chichester.

Caley, E.R. (1949). Klaproth as a pioneer in the chemical investigation of antiquities. *Journal of Chemical Education* **26** 242–247; 268.

Caley, E.R. (1951). Early history and literature of archaeological chemistry. *Journal of Chemical Education* **28** 64–66.

Caley, E.R. (1967). The early history of chemistry in the service of archaeology. *Journal of Chemical Education* **44** 120–123.

Carter, G.F. (ed.) (1978). *Archaeological Chemistry II*. Advances in Chemistry Series 171, American Chemical Society, Washington, D.C.

Catling, H.W., Blin-Stoyle, A.E. and Richards, E.E. (1963). Correlations between composition and provenance of Mycenaean and Minoan pottery. *Annual of the British School at Athens* **58** 94–115.

Cattaneo, C., Gelsthorpe, K., Phillips, P and Sokol, R.J. (1992). Reliable identification of human albumin in ancient bone using ELISA and monoclonal antibodies. *American Journal of Physical Anthropology* **87** 365–372.

Cherry, J.F. and Knapp, A.B. (1991). Quantitative provenance studies and Bronze Age trade in the Mediterranean: some preliminary reflections. In *Bronze Age Trade in the Mediterranean*, ed. Gale, N.H., Studies in Mediterranean Archaeology XC, Paul Åström's Förlag, Jønsered, pp. 92–119.

Ciliberto, E. and Spoto, G. (ed.) (2000). *Modern Analytical Methods in Art and Archaeology*. Wiley, New York.

DeAtley, S.P. and Bishop, R.L. (1991). Toward an integrated interface for archaeology and archeometry. In *The Ceramic Legacy of Anna O. Shepard*, ed. Bishop, R.L. and Lange, F.W., University Press of Colorado, Boulder, Colorado, pp. 358–380.

Eisele, J.A., Fowler, D.D., Haynes, G. and Lewis, R.A. (1995). Survival and detection of blood residues on stone tools. *Antiquity* **69** 36–46.

Emeleus, V.M. and Simpson, G. (1960). Neutron activation analysis of ancient Roman potsherds. *Nature* **185** 196.

Engel, M.H. and Macko, S.A. (ed.) (1993). *Organic Geochemistry: Principles and Applications*. Plenum Press, New York.

Evershed, R.P. (1993). Biomolecular archaeology and lipids. *World Archaeology* **25** 74–93.

Evershed, R.P., Dudd, S.N., Lockheart, M.J. and Jim, S. (2001). Lipids in archaeology. In *Handbook of Archaeological Sciences,* ed. Brothwell, D.R. and Pollard, A.M., John Wiley and Sons, Chichester, pp. 331–349.

Gernaey, A.M., Waite, E.R., Collins, M.J., Craig, O.E. and Sokal, R.J. (2001). Survival and interpretation of archaeological proteins. In *Handbook of Archaeological Sciences,* ed. Brothwell, D.R. and Pollard, A.M., John Wiley and Sons, Chichester, pp. 323–329.

Goffer, Z. (1980). *Archaeological Chemistry*. Wiley-Interscience, New York.

Grupe, G. and Lambert, J.B. (ed.) (1993). *Prehistoric Human Bone: Archaeology at the Molecular Level*. Springer-Verlag, Berlin.

Harbottle, G. (1982). Chemical characterization in archaeology. In *Contexts for Prehistoric Exchange,* ed. Ericson, J.E. and Earle, T.K., Academic Press, New York, pp. 13–51.

Henderson, J. (2000). *The Science and Archaeology of Materials: An Investigation of Inorganic Materials.* Routledge, London.

Heron, C., Nemcek, N., Bonfield, K.M., Dixon, J. and Ottaway, B.S. (1994). The chemistry of Neolithic beeswax. *Naturwissenschaften* **81** 266–269.

Hodder, I. (1984). Archaeology in 1984. *Antiquity* **58** 25–32.

Hodder, I., Shanks, M., Alexandri, A., Buchli, V., Carman, J., Last, J. and Lucas, G. (ed.) (1995). *Interpreting Archaeology: Finding Meaning in the Past.* Routledge, London.

Hudson, J. (1992). *The History of Chemistry.* Macmillan, London.

Hunter, J.R., Roberts, C.A. and Martin, A. (ed.) (1996). *Studies in Crime: An Introduction to Forensic Archaeology.* Seaby/Batsford, London.

Jakes, K.A. (ed.) (2002). *Archaeological Chemistry: Materials, Methods, and Meaning.* ACS symposium series no. 831, American Chemical Society, Washington, D.C.

Junghans, S., Sangmeister, E. and Schroder, M. (1960). *Kupfer und Bronze in der fruhen Metallzeit Europas.* Mann Verlag, Berlin.

Klindt-Jensen, O. (1975). *A History of Scandinavian Archaeology.* Thames and Hudson, London.

Lambert, J.B. (ed.) (1984). *Archaeological Chemistry III.* Advances in Chemistry Series 205, American Chemical Society, Washington, D.C.

Lambert, J.B. (1997). *Traces of the Past: Unraveling the Secrets of Archaeology through Chemistry.* Addison-Wesley, Reading, Mass.

Levey, M. (ed.) (1967). *Archaeological Chemistry: A Symposium.* University of Pennsylvania Press, Philadelphia.

Loy, T.H. (1983). Prehistoric blood residues: detection on stone tool surfaces and identification of species of interest. *Science* **220** 1269–1271.

Newton, R.G. and Renfrew, C. (1970). British faience beads reconsidered. *Antiquity* **44** 199–206.

Orna, M.V. (ed.) (1996). *Archaeological Chemistry: Organic, Inorganic, and Biochemical Analysis.* ACS symposium series no. 625, American Chemical Society, Washington, D.C.

Otto, H. and Witter, W. (1952). *Handbuch der altesten vorgeschichtlichen metallurgie in Mitteleuropa.* Johann Ambrosius Barth, Leipzig.

Percy, J. (1861). *Metallurgy. Volume I: Fuel; Fire-Clays; Copper, Zinc; Brass.* Murray, London.

Percy, J. (1864). *Metallurgy. Volume II: Iron; Steel.* Murray, London.

Percy, J. (1870). *Metallurgy. Volume III: Lead.* Murray, London.

Percy, J. (1875). *Metallurgy. Volume IV: Silver; Gold.* Murray, London.

Pétrequin, P., Errera, M., Pétrequin, A.-M. and Allard, P. (2006). The Neolithic quarries on Mont Viso, Piedmont, Italy: Initial radiocarbon dates. *European Journal of Archaeology* **9** 7–30.

Pollard, A.M. (1995). Why teach Heisenberg to archaeologists? *Antiquity* **69** 242–247.

Pollard, A.M. (2004). Putting infinity up on trial: a consideration of the role of scientific thinking in future archaeologies. In *A Companion to Archaeology*, ed. Bintliff, J., Blackwell, Oxford, pp. 380–396.

Pollard, A.M, Batt, C.M., Stern, B. and Young, S.M.M. (2007). *Analytical Chemistry in Archaeology*. Cambridge University Press, Cambridge.

Price, T.D. (ed.) (1989). *The Chemistry of Prehistoric Bone*. Cambridge University Press, Cambridge.

Renfrew, C. (1973). *Before Civilization: The Radiocarbon Revolution and Prehistoric Europe*. Jonathan Cape, London.

Renfrew, C. (1979). *Problems in European Prehistory*. Edinburgh University Press, Edinburgh.

Renfrew, C. and Bahn, P. (2004). *Archaeology: Theories, Methods and Practice*. Thames and Hudson, London, 4th edn.

Rybach, L. and Nissen, H.U. (1964). Neutron activation of Mn and Na traces in marbles worked by the Ancient Greeks. In *Proceedings of Radiochemical Methods of Analysis*. International Atomic Energy Agency, Vienna, pp. 105–117.

Sandford, M.K. (ed.) (1993). *Investigations of Ancient Human Tissue*: *Chemical Analyses in Anthropology*. Food and Nutrition in History and Anthropology, Volume 10, Gordon and Breach, Langhorne, Pennsylvania.

Sayre, E.V. and Dodson, R.W. (1957). Neutron activation study of Mediterranean potsherds. *American Journal of Archaeology* **61** 35–41.

Sayre, E.V. and Smith, R.V. (1961). Compositional categories of ancient glass. *Science* **133** 1824–1826.

Sealy, J. (2001). Body tissue and palaeodiet. In *Handbook of Archaeological Sciences,* ed. Brothwell, D.R. and Pollard, A.M., John Wiley and Sons, Chichester, pp. 269–279.

Smith, P.R. and Wilson, M.T. (1990). Detection of haemoglobin in human skeletal remains by ELISA. *Journal of Archaeological Science* **17** 255–268.

Smith, P.R. and Wilson, M.T. (2001). Blood residues in archaeology. In *Handbook of Archaeological Sciences,* ed. Brothwell, D.R. and Pollard, A.M., John Wiley and Sons, Chichester, pp. 313–322.

Stone, J.F.S. and Thomas, L.C. (1956). The use and distribution of faience in the Ancient East and Prehistoric Europe. *Proceedings of the Prehistoric Society* **22** 37–84.

Taylor, R.E. and Aitken, M.J. (ed.) (1997). *Chronometric Dating in Archaeology*. Plenum Press, New York.

Thomas, J. (1991). Science or anti-science? *Archaeological Review from Cambridge* **10** 27–36.

Tite, M.S. (1972). *Methods of Physical Examination in Archaeology*. Seminar Press, London.

Tite, M.S. (1991). Archaeological science – past achievements and future prospects. *Archaeometry* **31** 139–151.

Trigger, B.G. (1988). Archaeology's relations with the physical and biological sciences: a historical review. In *Proceedings of the 26th International*

Archaeometry Symposium, eds Farquhar, R.M., Hancock, R.G.V. and Pavlish, L.A., University of Toronto, Toronto, pp. 1–9.

Trigger, B.G. (2006). *A History of Archaeological Thought.* 2nd Edition Cambridge University Press, Cambridge.

Tuross, N. and Dillehay, T.D. (1995). The mechanism of organic preservation at Monte Verde, Chile, and one use of biomolecules in archaeological interpretation. *Journal of Field Archaeology* **22** 97–110.

van der Merwe, N.J. and Vogel, J.C. (1978). ^{13}C content of human collagen as a measure of prehistoric diet in Woodland North America. *Nature* **276** 815–816.

Willerslev, E. and Cooper, A. (2005). Ancient DNA. *Proceedings of the Royal Society of London, Series B, Biological Sciences* **272** 3–16.

Wilson, L. and Pollard, A.M. (2001). The provenance hypothesis. In *Handbook of Archaeological Sciences,* eds Brothwell, D.R. and Pollard, A.M., John Wiley and Sons, Chichester, pp. 507–517.

Yoffee, N. and Sherratt, A. (ed.) (1993). *Archaeological Theory: Who Sets the Agenda?* Cambridge University Press, Cambridge.

Zeuner, F.E. (1957). *Dating the Past.* Methuen, London, 4th edn.

CHAPTER 2

Analytical Techniques Applied to Archaeology

2.1 INTRODUCTION

The purpose of this chapter is to give a brief but largely non-mathematical introduction to some of the many analytical techniques used in modern archaeological chemistry. The vast majority of applications use standard instrumentation, which is well-described elsewhere in the analytical chemistry literature (*e.g.*, Ewing, 1985; Sibilia, 1988; Skoog and Leary, 1992; Christian, 1994; Cazes, 2005), or, as applied to archaeology, in Pollard *et al.* (2007). Often, however, some accommodation has to be made for the unique nature of many archaeological samples, as discussed in Chapter 1. The analytical techniques are described in groups, which are based on factors such as the region of the electromagnetic spectrum they employ, or the type of information obtained. Inevitably, most attention has been given to describing those techniques which have appeared most often in the archaeological literature. Although this might be slightly 'behind the times' analytically, attempts have been made where appropriate to refer to recent developments and to compare newer methods with established techniques.

In order to give some background, Section 2.2 gives a description of the relationship between the electronic structure of the atom, electronic transitions and the electromagnetic spectrum, since this is the basis of analytical spectroscopy. For those unfamiliar with the simple Bohr theory of the structure of the atom, a summary is given in Appendix 1. The Section 2.3 then describes the techniques which use the visible (or near visible) region of the spectrum – optical emission spectroscopy (OES), atomic absorption spectroscopy (AAS) and inductively coupled plasma emission spectroscopy (ICP). The Section 2.4 focuses on techniques which use the more energetic X-ray region of the spectrum [X-ray fluorescence (XRF), scanning electron microscopy (SEM) and proton-induced X-ray emission (PIXE)]. Before leaving the electromagnetic

Archaeological Chemistry, Second Edition
By A. Mark Pollard and Carl Heron
© The Royal Society of Chemistry 2008

spectrum, we consider neutron activation analysis (NAA), which uses extremely energetic photons, called gamma rays, to identify the elements present. We then look at those techniques which use mass spectrometry, including a discussion of 'hyphenated techniques', which employ a mass spectrometer (MS) as a more sensitive detector for an established technique, such as ICP-MS. This is followed by a consideration of chromatographic methods, which have become important with the growing interest in the analysis of organic and biological remains in archaeology. Continuing this theme, there is a discussion of the related techniques of infrared (IR) and Raman spectroscopy, which have grown rapidly in importance over the past ten years. Finally, Section 2.5 gives a very brief review of some other analytical techniques which, although mainstream and important elsewhere, have received less attention in archaeological chemistry. These include electron spin resonance (ESR) and nuclear magnetic resonance (NMR), and some of the thermal methods of analysis, including differential thermal analysis (DTA) and differential scanning calorimetry (DSC).

2.2   THE STRUCTURE OF THE ATOM, THE ELECTROMAGNETIC SPECTRUM AND ANALYTICAL SPECTROSCOPY

Appendix 1 reviews the traditional Bohr model of the atom, and the associated electronic structure. In this model, atoms are made up of a positively charged nucleus surrounded by a 'cloud' of orbital electrons carrying an equal negative charge. The nucleus contains both positively charged particles (*protons*) and electrically neutral particles (*neutrons*). The number of protons in the nucleus is called the *atomic number*, and is given the symbol Z. It is this number which gives all the elements their different chemical characteristics, and distinguishes one element from another. It also dictates the number of electrons orbiting the nucleus – in a neutral atom, the number of electrons is identical to the number of protons in the nucleus, since the charge on the proton and electron is identical but opposite. All elements have a unique proton number, but most do not have a unique number of neutrons – the vast majority of natural stable elements exist with two or more different neutron numbers in their nucleus, termed *isotopes*. The number of neutrons is given the symbol N, and the combined number of protons and neutrons in the nucleus is A, which is referred to as the *atomic mass number*. Isotopes of the same element have the same number of protons in their nucleus (and hence orbital electrons, and hence chemical properties), but different numbers of neutrons, and hence different atomic mass numbers. A table of the elements is shown in Appendix 4, listing the chemical symbol, atomic number (Z) and approximate atomic weight for each element.

Appendix 1 also shows how the periodic table of the elements (Appendix 5) can be built up from the known rules for filling up the various electron energy levels. The Bohr model shows that electrons can only occupy orbitals whose energy is 'fixed' (quantized), and that each atom is characterized by a particular set of energy levels. These energy levels differ in detail between atoms of different elements, because each element, by definition, has a different nuclear charge and orbital electron configuration. Interaction between the nuclear charge and the orbital electrons ensures that each element has a unique pattern of energy levels, despite the fact that the same notation is used to label the energy levels within all atoms (1s, 2s, 2p, *etc.*). Electron transitions between allowed energy levels, as described in Appendix 1, either absorb or emit a fixed amount of energy, corresponding to the energy difference between the energy levels of that atom. This fixed amount of energy, symbolized as ΔE, where $\Delta E = E_2 - E_1$, is the energy difference between the two levels E_1 and E_2. This quantum of energy manifests itself as electromagnetic radiation, whose energy and wavelength are related by the equation:

$$E = h\nu = hc/\lambda$$

where ν is the *frequency*, λ is the *wavelength*, c is the speed of light and h is Planck's constant (see Appendix 3).

A key prediction from the Bohr model is therefore that atoms can only emit or absorb electromagnetic radiation in fixed units or quanta, corresponding to the energy differences between electron orbitals. If the energy and the number of electrons occupying each orbital is known, then it should be possible to predict exactly the allowed transition energies between orbitals, and therefore, via the above equation, the wavelengths corresponding to these electronic transitions for any particular atom. Furthermore, at least some of these transitions give rise to radiation which falls into the visible region of the spectrum. Early spectroscopic experiments on the simplest atom (hydrogen) showed that the emission spectrum obtained by passing an electrical discharge through hydrogen gas and dispersing the light emitted with a prism did indeed yield a relatively small number of discrete lines in the visible region, whose wavelengths were calculable using a formula (the *Rydberg equation*), derived from the models outlined above. Figure 2.1 shows a schematic diagram of the electronic energy levels in the hydrogen atom and the allowed transitions (Figure 2.1a) and the resulting spectral lines (Figure 2.1b). More complicated (multi-electron) atoms give more emission lines, and corrections need to be applied to the Rydberg formula, but the theory is generally found to hold for such atoms. Thus it was realized that each element in the periodic table has a unique line emission or absorption spectrum in the visible region of the spectrum, directly reflecting the unique orbital electronic structure of the atom. Spectral analysis of elements became an established 'fingerprinting' technique in a range of applications, although the heavier elements can display a bewildering total number of emission lines – up to 4500 for iron, for example, but not all in the visible region.

An important observation was that the emission lines are not confined to the narrow visible region of the electromagnetic spectrum. Instrumental detection showed that discrete lines are also present in the infrared and ultraviolet wavelengths, and eventually it showed this in the X-ray region also. It became clear that the wavelength of the line simply corresponded to the energy

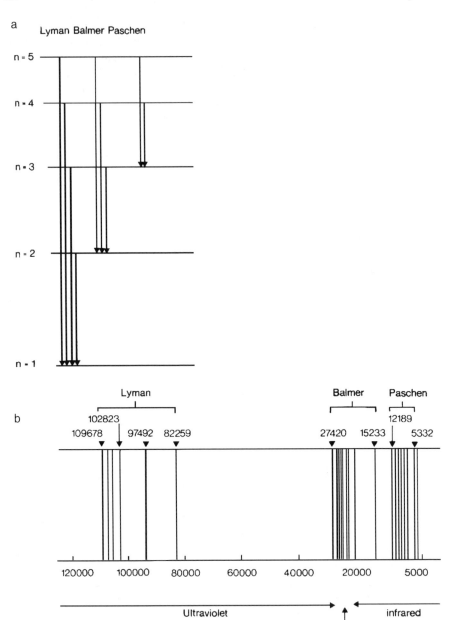

Figure 2.1 Electronic orbitals and the resulting emission spectrum in the hydrogen atom. (a) Bohr orbitals of the hydrogen atom and the resulting spectral series, (b) emission spectrum of atomic hydrogen. The spectrum in (b) is calibrated in terms of wavenumber ($\bar{\nu}$), which is reciprocal wavelength. The Balmer series, which consists of those transitions terminating on the second orbital, give rise to emission lines in the visible region of the spectrum. (© 1990 John Wiley & Sons, Inc. Reprinted from Brady, 1990, by permission of the publisher.)

difference between the particular pair of electron energy levels involved, and therefore that different regions of the electromagnetic spectrum give information from different energy 'depths' within the orbital structure of the atom. The correspondence between these regions and the wavelength is shown in Figure A1.2 in Appendix 1. It ranges from microwave frequencies which relate to the relatively low energy rotations of molecules around their bonds, down to X-rays, which arise from high energy transitions between the deepest electronic orbitals in large atoms. Spectroscopy therefore provides a wide range of tools capable of giving a great deal of information about the electronic structure of the atom, as well as quantitative analytical data.

Most pertinent to the current discussion is that each atom has a unique pattern of emission lines around the visible region of the spectrum, thereby allowing chemical identity to be established by simply comparing these patterns of emission or absorption. Moreover, via a relationship known as *Beer's Law* (or the *Beer–Lambert Law*), a quantitative link is established between the number of atoms involved and the intensity of the emission lines. Thus, in the hydrogen experiment described above, if the number of hydrogen atoms in the electrical discharge chamber was doubled, the intensity (brightness) of the observed emission lines theoretically would also double. Under normal circumstances, therefore, there is a simple linear relationship between the intensity of an emission line and the number of atoms which give rise to that emission. This is Beer's Law, and is the basis of quantitative analytical spectroscopy.

A number of further considerations are necessary. Atoms which lose or gain one or more electrons in the course of chemical bonding become electrically charged and are called *ions*: an ion will not have exactly the same emission spectrum as its parent atom, since the electronic energy levels will adjust to allow for the change in the number of electrons orbiting the nucleus. It is therefore important in spectroscopy to distinguish between atomic and ionic spectra. By extension, when atoms or ions combine to form *molecules*, the molecular energy levels are different to both the ionic and atomic levels, and molecular emission spectra are consequently different again. In general, molecular spectra do not show the sharp lines characteristic of atomic or ionic spectra, tending more towards broad bands. In all of this, however, there is a very simple relationship between the wavelengths which atoms, ions or molecules will emit when excited by some external stimulus, such as an electrical discharge or an increase in temperature, and the wavelengths which will be absorbed if the same atom, ion or molecule is exposed to electromagnetic radiation. They all absorb at exactly the same wavelengths as they emit. In the hydrogen example, therefore, if instead of passing an electric discharge to cause emission, the gas was illuminated by light of all wavelengths, the transmitted light would be exactly the same as that illuminating the gas, but without the discrete wavelengths seen in the emission spectrum. Thus the normal 'rainbow' of dispersed white light would appear to have a number of dark lines in it, at wavelengths identical to those which occur in the emission spectrum. Beer's law also applies to the absorption case – *i.e.*, the strength of the absorption is directly proportional to the number of atoms, ions or molecules involved in the absorption process. This simple concept underpins much of quantitative analytical chemistry, and often allows the identity and relative amounts of a wide range of elements present in a sample to be determined from a single sample of a few milligrams.

2.3 TECHNIQUES BASED ON OPTICAL WAVELENGTHS

2.3.1 Optical Emission Spectroscopy

Although now completely redundant as an analytical technique in its original form, it is useful to start our consideration of the three intimately related techniques of optical emission, atomic absorption and inductively coupled plasma emission with a brief description of OES (Britton and Richards, 1969). This is partly because the principles involved in OES transplant simply to more modern techniques, but also because there is a considerable body of OES data in the older archaeological literature, which it is important to be able to evaluate. An outline diagram of the components of a large quartz spectrograph is shown in Figure 2.2. The sample in powder form is placed in a hollow graphite cup and a graphite electrode is brought close to the cup, allowing an

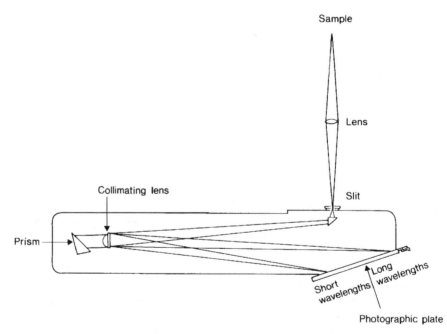

Figure 2.2 Schematic drawing of an optical emission spectrograph. Light from the sample is focused onto the input slit of the spectrograph and is then dispersed via a prism (or diffraction grating) and recorded on a photographic plate. (Adapted from Britton and Richards, 1969; Fig. 108, by permission of Thames and Hudson Ltd.)

electric spark to be struck between the two, thus volatilizing the sample and causing it to emit light. This light is focused through a series of lenses onto a large quartz prism (or, in later models, a diffraction grating) which disperses the light into its constituent wavelengths. This light is then focused onto the recording medium, which, in the original spectrometers, was simply a photographic plate which was subsequently developed in the dark room. Careful measurement from the plate enables the wavelength of each recorded line to be determined, allowing the elements present to be identified, and optical densitometry techniques allow the intensity of particular emission lines to be measured, enabling quantification of the amount of each element present. In practice, a single line from each element of interest is selected for quantification, taking into account the relative theoretical intensities of all the lines from the element, and the number of spectral overlaps known to occur in the relevant sample. By calibrating the spectrometer against known concentrations of each element, it is therefore possible to produce quantitative data. One particular advantage of this technique is that elements which may be present but might not have been expected can be detected by careful observation of unusual lines on the photographic plate. This is not the case with absorption techniques. In the more recent terminology of instrumental analytical techniques, OES has a *simultaneous* detection system, since all elements are recorded at the same time. The plates can also be stored, thus enabling reference at a later date should it be necessary. The disadvantages, however, are many. The exact positioning of the graphite electrodes affects the reproducibility of the measurements, and the wet photographic procedure is notoriously difficult to standardize. Both of these problems can be allowed for to some extent by 'spiking' the sample with a known concentration of an element not present in the sample (often lithium), and referencing all intensity measurements to that element. Using this approach, it is possible to measure virtually any element present in a sample of 10 mg, at concentrations between 0.001% and 10%. In practice, because of the complexity of the emission spectra of the elements, the maximum number of elements measurable from a single exposure is about 20, with a coefficient of variation (the standard deviation of a set of measurements divided by the mean, multiplied by 100 – a measure of the reproducibility of the analysis) of between 5 and 25% for major and minor elements, respectively (Jones, 1986; 24 and 883–888).

2.3.2 Atomic Absorption Spectrometry

OES was a standard method of analysis in archaeological chemistry for pottery, obsidian, faience and metals from the 1950s through to about 1980, and there is therefore a great deal of OES data in the literature. It was gradually replaced in most laboratories by AAS. This differs principally in that it requires the sample to be in a liquid (normally aqueous, *i.e.*, dissolved in water) form, thus making the sample preparation stage somewhat more complicated, and that it can only measure one element at a time (a so-called *sequential* operation). A schematic diagram of a simple atomic absorption spectrometer is shown in Figure 2.3.

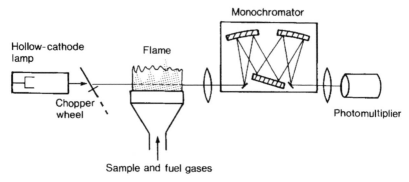

Figure 2.3   Schematic diagram of an atomic absorption spectrometer. Light of
wavelengths characteristic of the sample of interest is generated in the
lamp and passes through the flame containing the atomized sample. The
light is quantitatively absorbed in the flame, and the wavelengths are
separated in the monochromator. The intensity of the transmitted light is
measured in a photomultiplier. (Adapted from Ewing,1985; Fig. 5–9, with
permission of The McGraw-Hill Companies.)

Since in its basic mode of operation it is an absorption (as opposed to emission)
technique, light which is characteristic of the element to be determined has to be
passed through the sample. This is produced in a special lamp called a *hollow-
cathode lamp*, which consists of a glass (or quartz) envelope filled with a noble
gas at low pressure. Within the envelope are two electrodes, one a wire and the
other a cup made from (or lined with) the element of interest. On application of
a few hundred volts between the electrodes, the atoms of the cup are excited
and emit their characteristic radiation. In the spectrometer, the light from this
lamp is then guided towards the photomultiplier detector through the long axis
of a long thin gas flame produced at a specially shaped burner. The liquid
sample is aspirated into the flame along with the combustible gases. The fuel
and oxidant gases are pre-mixed in a chamber designed to ensure complete
homogenization of the gases, and the flow of these gases causes the liquid
sample to be sucked into the chamber via a capillary tube – no additional
pumping is necessary. The design of the mixing chamber also ensures that the
sample liquid is effectively atomized into a fine aerosol before it enters the flame
(see Ewing, 1985; Chapter 5).

    On entering the flame, the sample is almost immediately converted into an
atomic vapour. This is then in the ideal form to absorb characteristic radiation
from the light source shining through the flame, and the amount of radiation
absorbed at a particular wavelength is directly proportional to the concentration of that particular element in the flame. It is sometimes necessary, for certain
elements, to increase the temperature of the flame to ensure complete atomization. This is done by changing the mixture of gases used, from the normal
compressed air–acetylene mixture, which burns at about 2200 °C, to nitrous
oxide–acetylene, giving a temperature of up to 3000 °C. The latter gases require
a slightly different design of burner, so the changeover cannot be made at will.

In the simplest version of AAS, the light from the hollow-cathode lamp shines directly through the flame onto a monochromator device (prism or diffraction grating), which disperses the light into its constituent wavelengths. This is followed by a slit which selects a particular wavelength for transmission onto the detector, which can be a photomultiplier tube capable of quantitatively converting the light intensity into an easily measured electric current. In operation, the light intensity passing through the flame is firstly measured without any sample being aspirated, and then with the sample introduced. The difference between the two is the absorption due to the atoms in the sample, and this absorption is calibrated by measuring the absorption due to the aspiration of a solution containing a known concentration of that element. The concentration of the element in the sample solution can then be calculated, which in turn allows the calculation of the concentration in the original solid sample, since the weight of sample dissolved to make the solution is carefully recorded. This is a rather laborious procedure, and most modern AAS spectrometers are *double-beam instruments*, which allow the simultaneous comparison of absorbed and unabsorbed intensities. The light from the hollow-cathode lamp is split by a Maltese cross rotating at an angle to the beam, which splits it into a sample and a reference beam, one of which goes through the flame containing the sample and one of which does not. Signal noise reduction techniques ('signal chopping', using the Maltese cross as a 'chopper' and also the mirror) are now also used, so that 'flicker' in the light produced by the flame (a major source of noise) can be eliminated. Further background–noise reduction techniques are also routinely employed, such as the use of polarized light. Details of these can be found in Ewing (1985; 109–123). It has even proved possible to remove altogether the biggest source of signal noise and irreproducibility – the flame itself – by using what is known as an *electrothermal* or *graphite furnace*. In this the sample solution is injected directly into a small electrically heated chamber which replaces the burner, where the temperature is rapidly increased in a programmed manner so as to produce fast and reproducible atomization. Light from the hollow-cathode lamp passes directly through the furnace chamber, and the absorption is measured as before. This improves the sensitivity and detection levels of many elements (see below).

Once a flame atomic absorption spectrometer has been set up for a particular element and all of the instrumental variables (*e.g.*, gas flow, burner position relative to the light beam, *etc.*) have been optimized to produce the best conditions, it provides a cheap, rapid and effective means of analysis, capable of analysing many tens of samples per hour (especially if the instrument is equipped with an auto-sampler). Analytical precision is generally good, with a coefficient of variation of between 1 and 5% for a wide range of elements. Detection limits in solution are typically between 1 and 100 parts per million (ppm: 1 ppm is equivalent to 1 μg of element per litre of solution), depending on element, analytical conditions and the particular absorption line selected (Hughes *et al.*, 1976). One advantage of solution techniques of analysis in general is that the standards used for calibration can be accurately made up from commercially available standard solutions, allowing precise matching between the composition of the standards and the unknowns. This allows the analyst to undertake good *quality assurance* (QA) procedures by having separate calibration standards and quality control samples (usually samples of known composition) to validate the analysis (Pollard *et al.*, 2007; 319).

Disadvantages include problems of reproducibility between 'runs' on the same element, due to inconsistencies in the setting of instrumental parameters such as gas flow, although this can usually be minimized (or at least quantified) by extensive use of quality control standards. Problems of calibration drift can also be encountered during a 'run' as a result of small changes in operating conditions. These are monitored by regular referral to one or more of the standards being used. In computer-controlled instruments this drift can be automatically compensated for by adjusting the calibration curve being used, but this is an unsatisfactory procedure if the drift is significant. Other limitations of AAS derive from the sequential nature of the operation. Although multi-element lamps exist (*e.g.*, a combined lamp for calcium and magnesium), they are not always as spectrally clean as single element lamps, and the basic procedure remains the sequential use of a single lamp for each element to be determined in the sample. This results in a lengthy analytical procedure if several elements are required from the same samples. One further consequence is that, unlike OES, it is very unlikely that any unexpected elements would be noticed in a sample, since the procedure is designed only to give information about the elements sought. As with most analytical techniques, a great deal of skill and experience is needed to produce reliable data, taking into account problems such as chemical and spectral interferences in the flame, *etc.* One further major consideration is that the sample must be in liquid form, and some archaeological materials (*e.g.*, glass) require aggressive dissolution conditions involving hydrofluoric acid. In the analysis of metals, some elements such as tin are difficult to keep in solution once the sample has been dissolved. Standard texts, such as van Loon (1980; 1985), are available covering a wide range of dissolution techniques for various samples.

Despite the name, atomic absorption spectrometers can also be used as emission spectrometers. This is simply achieved by switching off the lamp, and letting the light emitted by the excited atoms in the flame pass through the monochromator onto the detector. Using Beer's Law, the intensity of the light emitted is proportional to the concentration of that element in the flame. It might be asked why the machine is not permanently used in this so-called *emission mode*, since it is simpler than using it in the absorption mode, but the answer lies in the behaviour of different elements in flames of different temperatures. For atomic absorption, the optimum state of the elements in the sample is completely atomized, but with all the electrons in their lowest possible energy levels – the *ground state*. In this condition, the maximum possible number of atoms is available to absorb the characteristic frequency from the hollow-cathode lamp, thus giving maximum analytical sensitivity. If conditions in the flame are slightly more energetic, the flame itself will excite a proportion of the atoms into a higher energy state, preventing them from absorbing the incoming light, and actually causing them to emit light of the correct atomic frequency as they return to their ground state. This will lead to reduced analytical sensitivity. If there are a significant number of excited atoms in the flame, then it is better to switch off the lamp and do the analysis in the emission mode. If conditions in the flame are even more energetic, then some of the atoms will lose (or gain) electrons and become ionized, thus changing the emission spectrum completely, and making the ions 'invisible' to the light used in the analysis, which is characteristic of atoms. In this situation, the sensitivity of the method deteriorates badly. Some elements, such as the alkali metals sodium and potassium, are relatively easily excited and ionized, and these are often best measured in the emission mode. Others, such as aluminium and titanium, require a great deal of energy to get them even into the atomic state, and can only be measured using the absorption mode with the hottest flame available. There is a comprehensive literature available listing optimal measuring conditions for each element and for each type of sample (*e.g.*, van Loon, 1980).

2.3.3 Inductively Coupled Plasma Emission Spectrometry

As noted above, there is sometimes a need to raise the temperature of the flame well above that easily obtained in a conventional gas burner, in order to ensure that some of the more refractory compounds are fully dissociated. In atomic absorption, the upper limit is about 3000–4000 °C, but temperatures in excess of 8000–10 000 °C are achievable using an instrument called an *inductively coupled plasma atomic emission spectrometer* (ICP-AES – sometimes referred to as ICP-OES, meaning 'optical emission spectrometer'). This is a development of the OES instrument described above, and is essentially an atomic absorption spectrometer operated in the emission mode, but with the gas burner replaced by a *plasma torch*, capable of supporting the combustion of argon at these very high temperatures. Clearly, at such high temperatures, any normal material would rapidly melt and fail, and the ingenuity of the plasma torch is that even though it is made out of ordinary materials (silica tubing, with a melting point around 1700 °C), it is designed in such a way as to support the hot plasma. A typical plasma torch consists of three concentric silica tubes, with copper coils wound around the outside at the top (Figure 2.4). The argon gas which forms the plasma is injected vertically through the central tube, but a larger volume enters between the two outer envelopes at a tangential angle, and spirals up between the outer casing, acting as a coolant. When ignited, the plasma passing through the centre of the torch glows white-hot, but it is lifted away from the silica tubing by the toroidal flow of cooling gas. The heating is maintained by a high-power radio frequency (RF) alternating current which is passed through the copper coils surrounding the torch, which causes the charged particles in the plasma to flow through the gas in a circular path by induction. The friction caused by this rapid motion through the gas holds the temperature at several thousand degrees, and ensures that the plasma is sufficiently ionized to respond to the RF heating. One complication is that the ignition of argon is not easy – an external spark is passed through the argon

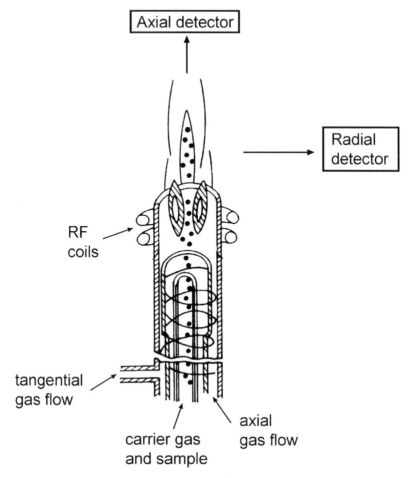

Figure 2.4   Schematic diagram of an ICP torch. The sample is carried into the torch by
the carrier argon gas, and is ignited by radio-frequency heating from the
RF coils. The tangential argon flow lifts the flame from the burner,
preventing melting. The position of the detector in axial or radial mode
is shown. (From Pollard *et al.*, 2007; Fig. 3–3, by permission of Cambridge
University Press.)

which causes some of the gas to ionize, and enables it to respond to the RF
heating. Frictional heating soon raises the temperature sufficiently for the
ionization to become self-sustaining.

As with atomic absorption, a liquid sample is injected into the hot plasma as
a solution carried by the fuel argon, although the design usually involves a
pump to suck up the sample and inject it into the argon stream. At the high
temperatures of the torch, all compounds are usually completely dissociated
and in an excited state, so that they strongly emit characteristic lines, which are
subsequently dispersed using a diffraction grating and slit system similar to that
used in atomic absorption. Detection used to be by photomultiplier tube, as

with AAS, but is now often a solid state *charge-coupled device* (CCD: Pollard *et al.*, 2007; 75). Other innovations are also incorporated into modern instrumentation. Some machines used for quality control in industry, or for the routine analysis of a particular material, are set up with a bank of detectors, each one positioned (at a fixed angle relative to the diffraction grating) for a particular wavelength corresponding to a particular element. In this manner simultaneous measurements of up to 20 elements in a single sample can be achieved, making the instrument analogous to the old optical emission spectrometers with their simultaneous detection capability. More usually for research purposes the ICP is equipped with a single computer-controlled detector, which automatically performs sequential analysis of several tens of elements whilst the sample is being continuously aspirated into the machine. The detector is moved to the correct position for a particular element, measures the emission intensity for a pre-set time, and then moves on to the next element according to a pre-set programme in the computer. Although strictly sequential in operation, the combination of software-controlled analysis and automated sampling makes the machine quasi-simultaneous. It would be fair to say that over the past 20 years or so ICP has almost completely replaced AAS as the industrial standard for the multi-element analysis of solution samples (Thompson and Walsh, 1989). Obviously, for archaeology, there is still sometimes a problem of getting the sample into solution, and keeping it there.

An early improvement in ICP-AES instrumentation occurred in the 1990s with the introduction of *axially viewed* spectrometers. Conventional spectrometers (now referred to as *radial ICP-AES*) employ a detector which views the plasma radially or side-on. Axial detectors are aligned directly along the major axis of the plasma torch. There are a number of different approaches to achieving this configuration (Brenner and Zander, 2000). Temperature gradients exist within the plasma so the outer layers are cooler than the centre. Axial ICP-AES eliminates emission from cooler parts of the flame, resulting in improved limits of detection by a factor of 2 to 20 depending on element, configuration, matrix, *etc.*

A major benefit of the ICP torch which has been exploited over the past 20 years is that it can be connected up to a MS detector (see Section 2.6) to give the very powerful technique of *inductively coupled plasma mass spectrometry* (ICP-MS: see Jarvis *et al.*, 1992). The temperature of the torch (up to 10 000 °C) is sufficient to ionize approximately 50% of the atoms in the sample, making the plasma an ideal source of ions which, given an appropriate interface to deal with the high temperatures and vacuum considerations involved, can be extracted from the plasma and injected directly into a MS. When used in this mode, individual charged ions are separated according to their mass and charge, and effectively can be counted individually. Not only is this more sensitive than the measurement of emission intensities (thus lowering even more the detection limits for certain elements), but also it allows the measurement of the abundance ratios of individual isotopes of a particular element, such as lead. Some of the advantages and uses of this approach are discussed in Chapter 9. Further developments have enhanced its performance even more,

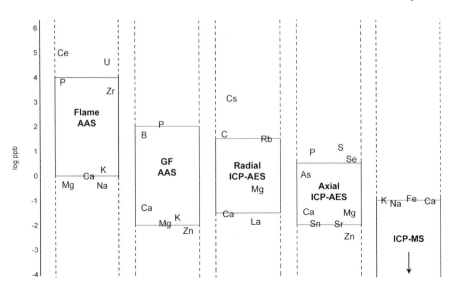

Figure 2.5   Schematic comparison of limits of detection in solution (log ppb) for
various absorption and emission spectrometries. For each technique, the
solid 'box' encompasses the majority of elements reported. A few relevant
elements have been marked specifically at the upper and lower end of the
range for each technique. (From Pollard *et al.*, 2007; Fig. 3–4, by
permission of Cambridge University Press.)

such as the development of a laser ablation system (LA-ICP-MS), which allows
a solid sample to be precisely vaporized into the MS by a high-power laser.
These are described in Section 2.6.

Figure 2.5 shows a comparison of the analytical sensitivity measured as the
lowest amount detectable in log parts per billion (ppb) in the sample solution
(1 in 10^{-9}, or $\mu g\,l^{-1}$) for five techniques – flame AAS, furnace AAS (AAS using
electrothermal atomization), radial ICP-AES, axial ICP-AES and ICP-MS.
In general, flame AAS is the least sensitive of these, with graphite furnace AA,
radial and axial ICP-AES all approximately equal, and ICP-MS substantially
more sensitive and comparable to NAA (see below). Element-by-element
comparison across the methods, however, can often reveal significant exceptions. In general, the refractory elements (Al, Ti, *etc.*) are better detected by
ICP-AES, whereas the heavy (non-refractory) elements are better detected by
flame AAS. For example, barium by ICP-AES has a detection limit of around
0.5 ppb, compared with about 8 ppb by flame AAS, whereas Pb is better
detected by AAS (10 ppb compared with about 50 ppb; Slavin, 1992). As shown
in Figure 2.5, the alkali and alkaline earth elements (Na, K, Mg, Ca) are almost
as well-detected by flame AAS as they are by ICP-MS. It is, therefore, not
always the case that the more modern techniques automatically produce more
sensitive analyses (although they may have other advantages). Equally, if not
more, important than the sensitivity are the analytical precisions of the various
techniques – a measure of the repeatability of the experimental method. In

routine applications, the precisions are usually quoted as around 0.5% for flame AAS, 1.5% for ICP-AES, 3% for furnace AAS and 2–3% for ICP-MS, although individual laboratories may be able to improve substantially on these figures for particular applications (Slavin, 1992). The various forms of instrumentation for ICP are substantially more expensive to buy than are AAS instruments, and are generally regarded as requiring more skilled operation. Nevertheless, over the past few years it has gradually become the technique of choice for large-scale analytical programmes where the samples can be brought into solution without too much difficulty. Laser ablation for direct vaporization of solid samples makes ICP analysis even more attractive.

An important consideration in archaeological chemistry is the extent to which the 'legacy' data sets – data collected by OES or AAS – can still be used. A related question is how comparable are data collected by NAA and ICP-MS. In one study, concerning the intercomparability of silicate analyses on archaeological ceramics carried out by ICP-AES and AAS (Hatcher *et al.*, 1995), it was concluded that the results from both methods were sufficiently close that common data banks could be established, providing adequate care had been taken to include certificated standards within each run to monitor performance on the more difficult elements. It has to be said that this is likely to be the exception. In most cases, the differences are likely to be so great that only the most generic of comparisons can be made. With 'legacy' data, it is often the case that quality control data are not published in sufficient detail to allow good comparisons to be made. With current data (such as comparisons between ICP-MS and NAA), it is clearly possible to ensure that good analytical protocol maximizes the chances of having comparable data (Pollard *et al.*, 2007; 208).

2.4 TECHNIQUES USING X-RAYS

The X-ray region of the electromagnetic spectrum consists of wavelengths between 10^{-9} and 10^{-15} m (see Figure A1.2 in Appendix 1). X-ray spectroscopists still use the non-SI standard unit of the Ångstrom (Å) which is defined as 10^{-10} m (so that 10 Å = 1 nm). This is simply because in these units X-rays used in analytical work range between 1 and 10 Å in wavelength. Electronic energy levels deep within the orbital electron structure of the heavier elements have such high energy differences that transitions occurring between these levels give rise to quanta whose energy (or wavelength) cause them to lie within the X-ray region of the spectrum. The situation is exactly analogous to that described above for optical spectra, except for the higher energy of the transitions involved. It is, however, complicated by the fact that X-ray spectroscopists, for historical reasons, insist on labelling the energy levels differently to those used by chemists! They designate the innermost orbitals K, L, M, N, *etc.*, corresponding to the principal energy levels $n = 1, 2, 3, 4$ as described in Appendix 1, and use a different notation for defining the sub-shells of each energy level. Thus, $2s$ is designated L_I, but the $2p$ orbital is split into two levels, labelled L_{II} and L_{III}. The correspondence between these different systems is set out in most books on X-ray spectroscopy, such as Jenkins (1988; Chapter 2), and is incidentally illustrated in Figures 2.7a and 2.8a.

If an electron is removed (by a means discussed below) from one of the inner energy levels of one of the heavier elements – in practice, generally those above sodium in the periodic table – a *vacancy* or a *hole* is created in the electronic structure. Two competing processes can occur to rectify this unstable arrangement, one resulting in the emission of an electron (the *Auger process*), and one in the emission of an X-ray. Figure 2.6 illustrates these two processes – in both cases the hole is in the K shell, and an L electron drops down to fill the vacancy. In the X-ray process, internal re-arrangement of the outer electrons results in one of the electrons from a higher energy level dropping down to fill the vacancy. The energy difference between the two levels is carried away as an X-ray of energy E, as defined by the usual equation:

$$E = E_K - E_L = hc/\lambda$$

In the Auger process, an outer electron drops down to fill the vacancy as before, but instead of emitting a photon, a third electron is ejected, whose kinetic energy is approximately given by the difference between the energy levels involved:

$$E \approx E_K - E_L - E_M$$

In this case, an M electron is emitted as an Auger electron. The Auger process is termed a *radiationless transition*. The probability that an inner shell vacancy

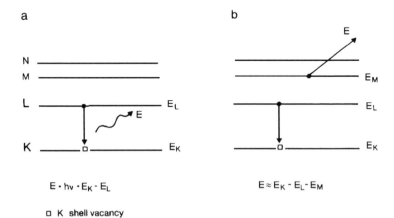

a   b

$E \cdot h\nu \cdot E_K - E_L$ $E \approx E_K - E_L - E_M$

□ K shell vacancy

Figure 2.6   The X-ray emission and Auger processes. An inner vacancy in the K shell de-excites via one of two competing processes – (a) X-ray emission, in which an L electron drops down and the excess energy is carried away by an X-ray photon, or (b) the Auger process, in which an L electron drops down, but the excess energy is carried away by a third electron – in this case from the M shell.

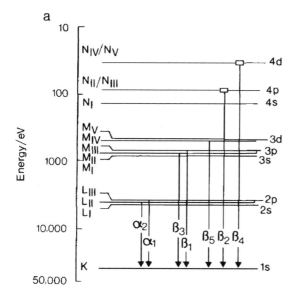

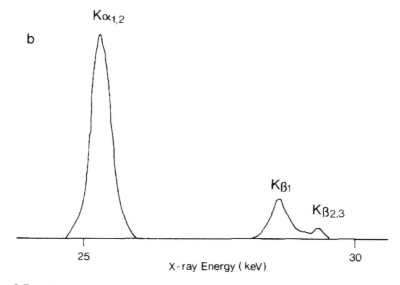

Figure 2.7   Electronic transitions giving rise to the K spectrum of tin. The K spectrum is normally resolved into two lines, K_α and K_β (as shown), which are composed of unresolved X-rays from two and five transitions respectively. (After Jenkins, 1974; Fig. 2–4. © John Wiley & Sons Limited. Reproduced with permission.)

will de-excite by one or the other of these processes depends on the energy level of the initial vacancy, and on the atomic weight of the atom. The fluorescent yield ω is defined as the number of X-ray photons emitted per unit vacancy, and is a measure of the probability (value between 0 and 1) of a particular vacancy

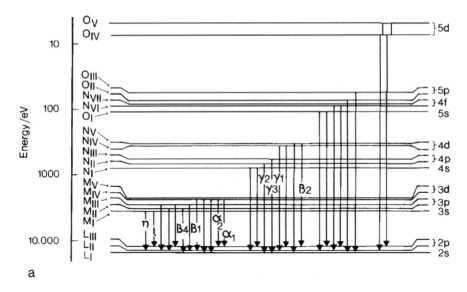

a

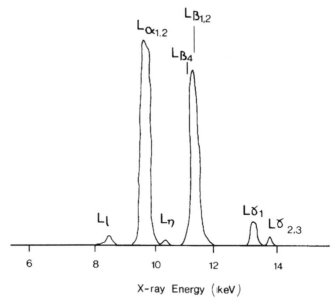

b

Figure 2.8 Electronic transitions giving rise to the L spectrum of gold. The L
spectrum is considerably more complicated with three main lines normally resolved as shown in the accompanying spectrum – L_α (arising from
two transitions), L_β (with up to 17 contributing transitions) and L_γ (up to
8 transitions), plus a number of 'forbidden' transitions. (After Jenkins,
1974; Fig. 2–11. © John Wiley & Sons Limited. Reproduced with
permission.)

resulting in an emitted X-ray. Fluorescent yields are defined for each energy level (ω_K, ω_L, *etc.*), but in practice the K shell Auger processes are only really significant for the lighter elements. L and higher level vacancies, however, are more likely to result in Auger electrons than in X-ray photons.

As with optical emission lines, selection rules apply, defining those electron transitions which are allowed. The details are available in standard texts (*e.g.*, Jenkins, 1988), but the net result for all elements is that vacancies created in the K shell give rise to two separable emission lines – a stronger one, labelled the K_α line, which results from $2p-1s$ transitions, and a weaker one labelled the K_β, resulting from $3p-1s$, $4p-1s$, $3d-1s$ and $4d-1s$ transitions. Figure 2.7 shows the detailed attributions of the K transitions in tin and the resulting appearance of the X-ray emission lines at normal analytical resolution. Although several transitions may contribute to these α and β lines, the resolution of most detection systems is insufficient to separate the fine detail within them, and most tabulations of emission line energies list only the two average values. The separation in energy between the K_α and K_β lines varies from element to element, increasing with atomic weight, and the intensity of the K_β line is typically only 10% of the K_α. The L spectra arise from vacancies created in the second ($n = 2$, or L) shell, and are considerably more compli-cated, but are usually only resolved into three lines, labelled L_α, L_β and L_γ. The L_α line is the strongest, resulting from some $3d$–$2p$ transitions (specifically M_{IV}–L_{III} and M_V–L_{III}). The L_β line, principally due to another $3d$–$2p$ transition (M_{IV}–L_{II}), but including many others (up to 17 separate transitions may contribute), is normally only slightly weaker in intensity (typically 70% or more of the L_α). The L_γ is considerably weaker (typically 10% of L_α), and due largely to a $4d$–$2p$ transition (N_{IV}–L_{II}). Figure 2.8 shows the L transitions in gold and the resulting spectrum as seen at normal resolution. The details of the relative intensities of each transition in X-ray emission depend on the quantum-mechanical transition probabilities. Some, however (just as in optical transitions), are theoretically 'forbidden' by transition rules, such as $3d-1s$, but they can occur and appear as very weak lines. Other lines can appear in the spectrum, such as *satellite lines*, resulting from transitions in doubly-ionized atoms – the Auger process, for example, leaves the atom in such a state – but these are usually very weak, and are not normally used for bulk chemical analysis.

The precise energy of an Auger electron (as given approximately by the equation above) is particularly sensitive to the chemical state of the atom from which it is ejected, since the outer orbitals from whence the Auger electron originated are often involved in chemical bonding. Hence *Auger electron spectroscopy* (AES – the study of such electrons) is extremely valuable for looking at the chemical state of the surfaces of solids. It is possible, for example, to differentiate between clean, oxidized and carbon-covered surfaces of metals (Briggs and Seah, 1990). The extreme surface sensitivity of Auger spectroscopy arises from the fact that Auger electrons have very low kinetic energies (usually less than 1500 eV), and so only emerge with useful information if they originate in the top 25 Å of the solid (*i.e.*, from the top two or three atomic layers). This is a particularly valuable attribute when studying the extreme surfaces of materials, but has restricted its applications in archaeology. One of the few examples is a detailed study of the corroded surfaces of Medieval glass (Dawson *et al.*, 1978). In marked contrast to this, because they arise from inner shell transitions, the chemical environment has virtually no influence on the energy of the emitted X-rays resulting from the other de-excitation process [apart from specialized studies using *extended X-ray absorption fine studies* (EXAFS) and related techniques, which use synchrotron radiation]. This makes the X-ray emission spectra uniquely and quantitatively characteristic of the parent atom, which is why techniques using such X-rays are very powerful tools for chemical analysis.

Three analytical techniques which differ in how the primary vacancies are created share the use of such X-rays to identify the elements present. In X-ray fluorescence, the solid sample is irradiated by an X-ray beam (called the *primary beam*), which interacts with the atoms in the solid to create inner shell vacancies, which then de-excite via the emission of *secondary* or *fluorescent* X-rays – hence the name of the technique. The second uses a beam of electrons to create the initial vacancies, giving rise to the family of techniques known collectively as *electron microscopy*. The third and most recently developed instrumentation uses (usually) a proton beam to cause the initial vacancies, and is known as *particle-* (or *proton-*) *induced X-ray emission* (PIXE).

2.4.1   X-Ray Fluorescence Spectrometry

In XRF, the primary X-rays (those used to irradiate the sample) are most commonly produced by an X-ray tube, although in certain circumstances other sources can be used. An X-ray tube has an anode made from a metal (often tungsten or molybdenum) which emits X-rays efficiently when bombarded with electrons. Thus a tungsten anode will emit the characteristic X-ray lines of tungsten, but in addition to line spectra, a solid will emit a continuous X-ray spectrum when bombarded by electrons. This is because high-energy electrons impacting the target material can give up their energy in a stepwise series of processes, each one resulting in the emission of an X-ray, and thus the output contains a continuous range of X-ray energies (termed *brehmsstrahlung*), up to a maximum limit set by the accelerating voltage applied to the electrons in the X-ray tube. The output of such a tube therefore consists of a continuum up to this maximum energy, superimposed upon which is the line spectrum of the target material. It is important to know what the target material is when using an X-ray tube, because its characteristic lines will almost certainly be seen in the secondary X-ray spectrum of the sample, and must be discounted.

Alternative sources of primary X-rays now include synchrotron radiation (Pollard *et al.*, 2007; 290). The synchrotron is a large electron accelerator which produces electromagnetic radiation across the entire spectrum, with high spectral purity and very high beam intensity. At specific stations around the storage ring, particular sections of the electromagnetic spectrum are selected (such as X-rays), and the extracted beam can be filtered and focused down to sub-micron spot sizes, whilst retaining very high beam intensity. This beam can therefore be used as the primary beam in XRF analysis, with the advantage of spectral purity, combined with high intensity and small beam size. One disadvantage is that because of the physical size of the synchrotron, samples must be taken to it for analysis. There are also restrictions of access – in most countries, synchrotrons are national or international facilities, and access is regulated, but is usually free for approved projects. At the other end of the size scale, powerful radioactive sources have been used to provide the primary X-ray excitation, giving rise to portable XRF machines. In the past few years these have been replaced with systems using miniature X-ray tubes (Thomsen and Schatzlein, 2002). Portable XRF machines have many advantages, the most obvious being that they allow the analyst to go to the object rather than *vice-versa*, but also a number of disadvantages. In general, even with tube-powered systems, the irradiation intensity is low and therefore the analytical precision is worse than that of a conventional instrument (this is not necessarily a critical factor if elemental identification alone is desired). Additionally, in most commercial systems, the primary beam has a diameter of several millimetres at the sample, limiting the spatial resolution. There is also concern expressed in some quarters about the health and safety compliance of some of these instruments. With the development of low-power side-window 50 kV air-cooled X-ray tubes, a new generation of portable machines has been developed (Desnica and Schreiner, 2006) which may overcome many of these limitations.

When the primary X-rays from whatever source is used strike the solid sample, two processes take place – *scattering* and *absorption*. Scattering may be elastic (*coherent*, or *Rayleigh scattering*), in which case the scattered ray has the same wavelength as the primary beam, or inelastic (*incoherent*, or *Compton scattering*), which results in longer wavelength (lower energy) X-rays. Coherent scattering results in the primary spectrum from the X-ray tube being 'reflected' into the detector (hence the appearance of the characteristic tube lines in the resulting spectrum), and can sometimes also result in diffraction phenomena. Incoherent scattering sometimes gives rise to a broadened inelastic peak at the lower energy side of the coherently scattered characteristic tube lines, as well as contributing to the general background. Vacancies are created in the orbital shells of atoms in the sample as a result of energy absorption from the primary beam, when part of the primary energy is transferred to the atom, resulting in the ejection of an orbital electron. When an electron is ejected from an atom as the result of the impact of an X-ray photon, *photoelectric absorption* is said to have occurred, and the ejected electron is termed a *photoelectron*. Study of these photoelectrons is the basis of another surface-sensitive chemical analytical technique called *X-ray photoelectron spectroscopy* [XPS, also referred to as *electron spectroscopy for chemical analysis* (ESCA)], which has been used sparingly in archaeology [*e.g.*, to study the black coatings applied to ceramic vessels (Gillies and Urch, 1983)]. Thus, as a result of all of these processes, on passage through matter the incident X-ray beam is attenuated. According to

Beer's Law, the intensity of the beam $[I(\lambda)]$ after travelling a distance x through the solid is given by:

$$I(\lambda) = I_o \exp(-\mu\rho x)$$

where μ is the *mass absorption coefficient* of the material of density ρ, and I_o is the intensity of the primary beam. The mass absorption coefficient (absorption per unit mass) is dependent on the atomic number of the material, and varies with wavelength (*i.e.*, it is a function of λ). It can be calculated at any wavelength for a complex material by simply summing the mass absorption coefficient of all the elements present, weighted by their fractional abundance. Several tabulations of mass absorption coefficients are available over the normal range of X-ray wavelengths (*e.g.*, Jenkins, 1988). These need to be calculated for the sample concerned in order to produce fully quantitative analytical data by XRF, but this is now normally done automatically by the software controlling the analysis.

Figure 2.9 shows a schematic diagram of the interaction of the primary X-ray beam of intensity $I_o(\lambda)$ incident at an angle of ψ_1 with a flat sample. At some depth d_s the primary beam, whose intensity has been attenuated according to Beer's Law (the *primary absorption*), creates a vacancy which gives rise to a secondary (fluorescent) X-ray with the characteristic wavelength λ_i, which travels towards the detector in the direction characterized by angle ψ_2. On its way out of the solid, it has to travel a distance x_s through the absorbing medium, and its intensity is therefore also reduced, again according to Beer's Law (the *secondary absorption*). The amount of attenuation experienced by the secondary X-ray beam depends on the absorbance of the matrix, and the distance travelled, which depends on the angle ψ_2. Eventually attenuation is so severe that X-rays generated at or greater than the depth d_s cannot escape from the solid – this is termed the *escape depth* and is an important factor in XRF – it

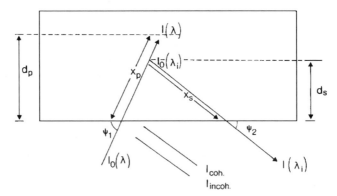

Figure 2.9 Interaction of the primary X-ray beam with a solid sample. See text for discussion. (After Jenkins, 1974; Fig. 3–3. © John Wiley & Sons Limited. Reproduced with permission.)

limits the depth from which analytical information can be obtained. It clearly depends on the nature of the solid matrix – absorption will be much greater in metallic lead, for example, with a very high average atomic number (and therefore high absorption) compared with something like glass, with a lower average atomic number. Calculations show that for the lighter elements such as sodium in silicate glass, for example, the intensity of the secondary radiation is reduced by 90% after passing through only 14 μm of glass. The same percentage reduction is observed with the characteristic radiation of calcium after passage through 122 μm of glass (Cox and Pollard, 1977). For radiation from the heavier elements in such a matrix, such as lead, the escape depth is in excess of half a millimetre. Although not as surface sensitive as AES and XPS, therefore, XRF is still essentially a surface analytical technique, but the exact depth from which information can be obtained depends on the elements of interest and the nature of the solid matrix.

The secondary X-radiation coming from the surface of a solid sample which is being irradiated by X-rays contains a number of components. Of greatest interest to the analyst are the line spectra of the elements contained in the sample – this is the basis of the identification and quantification of the sample chemistry. These line spectra are superimposed on an elastic and inelastic scattered version of the primary irradiation from the X-ray tube, including the characteristic lines of the tube target material, plus a continuous background arising from unspecific processes within the sample. Typical partial XRF spectra from ceramic samples are shown in Figure 4.12 of Chapter 4. The prime requirement of an XRF spectrometer is therefore that it can resolve the separate peaks, identify them (either by measuring their wavelength or energy) and measure their area in order to quantify the data. Two approaches are possible, which mirror the particle-wave duality of electromagnetic radiation – *energy-dispersive X-ray fluorescence* (EDXRF, or EDAX) and *wavelength-dispersive XRF* (WDXRF). Figure 2.10 shows a schematic comparison of the

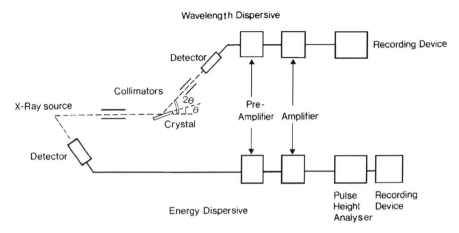

Figure 2.10 Comparison of EDXRF and WDXRF.

two techniques. In the former, the secondary X-ray emitted by the excited atom within the sample is considered to be a particle (an X-ray photon), whose energy is characteristic of the atom from whence it came. An EDXRF system consists of a solid state device which measures the energy of the photon, and counts the number of photons with known energies. This is (conventionally) achieved using a single crystal of silicon, doped (or 'drifted') with lithium to reduce the electronic impurities in the crystal to an absolute minimum (although modern technology allows single silicon crystals to be produced of sufficient purity not to require doping). The crystal is kept at liquid nitrogen temperatures to prevent the lithium diffusing out, and to reduce the electronic noise in the device (in more recent instruments Peltier cooling is used, obviating the need for liquid nitrogen). When an X-ray photon strikes the crystal, the whole energy of the photon is dissipated into the crystal by the creation of a large number of *electron–hole* pairs in the semiconductor – for a Si(Li) detector, each electron–hole pair requires 3.8 eV to form, and therefore the number of pairs created is given by the energy of the incident photon divided by 3.8. A voltage is applied across the crystal via (usually) gold contacts, and the electrons created in the crystal move towards the positive terminal. This constitutes an electric current, the magnitude of which is proportional to the energy carried by the incoming photon. Through a series of electronic devices, the current is measured and a count of one added to the relevant channel in a multi-channel analyser, which therefore records the arrival of a photon within a specific energy band (usually 20 eV). The spectrum is recorded over the range of either 0–20 keV or 0–40 keV, which includes the K and L emission lines of most of the elements of interest. The K and L energies corresponding to all the elements are tabulated in standard texts [*e.g.*, Jenkins (1974; appendices I–III) although this tabulation is by wavelength, energies can be simply obtained by the formula given below]. EDXRF is therefore a simultaneous detection technique, in that information from all elements is recorded at the same time. Software in the controlling computer usually allows peaks to be identified and quantified automatically, and often calibration programmes are included which perform absorption corrections using equations of the type given above. The detector has to be maintained under vacuum to ensure its cleanliness, and it is normally separated from the rest of the spectrometer (which may or may not be capable of being evacuated) by a thin beryllium window, which limits the performance of the system at the very light element end of the periodic table. In a fully evacuated system (*i.e.*, one in which the whole sample chamber can be evacuated), an EDXRF spectrometer should be able to detect elements as light as sodium, although the sensitivity here is usually relatively poor. If an air-path is used (*i.e.*, the sample is in air), then the lightest element quantifiable is usually potassium or calcium. Performance at the heavier end of the table is limited by the fact that the more energetic X-rays from the heavy elements may pass straight through the thin detector crystal without being absorbed, thus also reducing analytical sensitivity.

In an EDXRF system, the two tasks of energy measurement and detection are carried out simultaneously. In a WDXRF system, on the other hand, the two processes are separated. The secondary X-rays from the sample are regarded as being electromagnetic waves, whose wavelength is characteristic of the atom from whence they came. The WDXRF system operates in a manner analogous to the optical systems described in Section 2.3 – a dispersion device is used to separate the radiation into its component wavelengths, and a separate detection system records the intensity of the radiation as a function of wavelength. In this case, however, because of the extremely short wavelength of X-rays (typically 0.1–10 Å for the characteristic lines), a conventional prism or diffraction grating would not work, since the spacing of the diffracting medium has to be similar to the wavelengths to be separated in order for diffraction to occur. Conveniently, however, nature has provided suitable diffraction gratings for X-rays in the form of mineral crystals, whose atomic spacing is similar to the wavelength of X-rays. Early work with X-rays used crystals of calcite or rock salt to disperse the beam, but modern spectrometers tend to use lithium fluoride for general work. Occasionally, more specialist crystals (such as ammonium dihydrogen phosphate) or different crystal geometries are used for particular applications (Jenkins, 1974; 88). Detection of the dispersed X-rays is achieved either with a scintillation counter (comprising a phosphor crystal, which emits light on the impact of an X-ray, and a photomultiplier to record the burst of light) for shorter wavelengths, or at longer wavelengths a gas-flow proportional counter (a sealed tube containing a gas, which ionizes as the X-ray passes through it, and the resulting electron flow measured via a wire anode passing through the gas). Often these detectors are used in tandem, to cover the entire range of X-ray wavelengths. Both of these can now be replaced with a single CCD device, of course. In order to help in relating the energy-dispersive and wavelength-dispersive approaches to XRF, it is helpful to remember the following simple equation, relating the energy E of an X-ray photon (in keV) with its equivalent wavelength λ (in Å):

$$E = 12.4/\lambda$$

As with the ICP spectrometers described above, two modes of WDXRF operation are possible – sequential or simultaneous. In the simultaneous mode, a bank of X-ray detectors is aligned with the dispersive crystal, each making a specified angle with the crystal, so as to detect the characteristic wavelength of a predetermined element. Up to 20 detectors may be employed, giving information on up to 20 elements. This mode of operation is suited to a situation where a large number of identical samples have to be analysed as quickly as possible, such as in industrial quality control. In the sequential mode, a single detector is used. Traditionally, the detector was linked to the crystal via a goniometer, allowing X-ray intensity to be recorded on a chart recorder as a function of diffraction angle. Modern instruments use a computer-controlled detector, which can be programmed to record as many elements as required by moving to the position corresponding to the diffraction angle of the characteristic wavelength of the element of interest, counting, and then moving on to the next without operator intervention.

XRF spectrometers are designed to accommodate solid samples – preferably prepared into a standard shape (usually a disk), and mounted flat in a sample holder. For bulk commercial metal samples, simply cutting to shape and polishing gives a good sample. In routine geological applications, the samples can either be cut, sectioned, mounted and polished, or be powdered and pressed into a disk with a suitable binding medium, or converted into a glass bead by fluxing with excess borax. Solid rock samples are likely to be unsatisfactory, however, if there is significant inhomogeneity, because in commercial XRF spectrometers the primary X-ray beam usually has a diameter of several millimetres, going up to more than a centimetre. Preparation of pressed disks or glass beads can be time consuming, and a certain amount of care has to go into the choice of the binding medium or flux, but this approach is usually the preferred method for geological analysis. If the object to be analysed cannot be converted into a convenient sample form, as is often the case with archaeological material, specially modified spectrometers can be used which can accommodate large irregular-shaped objects, either in a specially designed sample chamber or simply by holding the sample in some fixed geometry in front of the spectrometer in air. The inability to evacuate the sample chamber means that information on elements lighter than potassium is usually lost due to air absorption of the characteristic X-rays, but the advantages of having a transportable machine which can produce qualitative analyses of museum objects have been well-documented (*e.g.*, Hall *et al.*, 1973).

Using commercially available systems, it is generally assumed that WDXRF has lower (*i.e.*, better) limits of detection than EDXRF, and is capable of higher precision. Detailed comparisons, however, of the trace element analysis of geological material using both methods has shown that this is not necessarily the case (*e.g.*, Potts *et al.*, 1985). This work compared several parameters of interest – the limits of detection of major rock-forming elements as determined by EDXRF and WDXRF on fused glass beads; the limits of detection for trace elements in pelletized powders as determined by both approaches; the analytical precision of routine EDXRF and WDXRF; and finally a 'blind test' of several WDXRF laboratories against their own EDXRF determinations. The general conclusions were that EDXRF has poorer limits of detection than WDXRF for the major light elements [typically 0.2–1 weight percent (wt %) oxide for Na_2O to SiO_2, compared to better than 0.1% by WDXRF], but that for the trace elements the limits of detection are comparable (3–20 ppm for Ni, Cu, Zn, Ga, Rb, Sr, Y, Zr, Nb, Pb, Th and U). Analytical precisions were found to be comparable for most elements at the levels found in silicate rocks (coefficients of variation in the range 0.5–6% for major element oxides, 1.5–16% for most trace elements), and the 'blind test' revealed that the EDXRF determinations were statistically indistinguishable from the data produced by a number of laboratories routinely employing WDXRF. The final point of note was that the limit on the accuracy of the data produced was not the analytical technique employed, but the quality of the data on the international rock standards used for calibration.

It must be emphasized that these figures for the performance of EDXRF systems are based on carefully prepared glass beads or pressed pellets, and cannot be applied directly to analyses performed by instruments adapted to work on unprepared samples, where degradation in detection levels and precision by a factor of two at least might be expected. EDXRF is generally much faster than WDXRF systems, accumulating a usable spectrum in around 100 seconds, and the instrumentation is usually much cheaper to buy. For archaeological and museum purposes, many laboratories have equipped themselves with EDXRF systems for rapid identification and semi-quantitative analysis of a wide range of archaeological materials, including metals, ceramics, glasses, jet, *etc.*, whereas WDXRF has had relatively little use on archaeological materials apart from some studies of ceramics, where effectively the material can be treated as a rock sample.

One major factor in the archaeological use of EDXRF on unprepared archaeological samples such as metals and glass has been a recognition of the problems of surface sensitivity. Although by no means the most surface sensitive of the analytical techniques (*e.g.*, by comparison with AES or ESCA), EDXRF can be regarded as only giving an analysis of the top fraction of a millimetre of the sample. In the case of metals, where the phenomenon of surface enrichment (either deliberately during manufacture or naturally as a result of the burial environment) has long been known (*e.g.*, Hall, 1961; Cowell and La Niece, 1991), this can be a critical restriction. Similar problems, due to the selective leaching of the alkali elements, have been noted in glass (Cox and Pollard, 1977). Although careful sample preparation can minimize these difficulties and provide good quantitative data, it is usual to regard EDXRF analyses of unprepared archaeological materials as qualitative, or semi-quantitative at best. This does not necessarily compromise its usefulness in areas such as conservation, where a rapid identification of the material may be all that is required, or in other areas as a preliminary analytical survey technique.

2.4.2 Analytical Electron Microscopy

Although electron microscopy is approached in this chapter as an analytical technique (a variant of XRF), it is essential to state at the outset that electron microscopy is far more versatile than this. Many standard descriptions of electron microscopy approach the subject from the microscopy end, regarding it as a higher resolution version of optical microscopy. Several texts, such as Goodhew *et al.* (2001), Reed (1993) and Joy *et al.* (1986), are devoted to the broad spectrum of analytical electron microscopy, but the emphasis here on the analytical capacity is justified in the context of a book on archaeological chemistry.

As noted above, there are several ways of creating an inner shell vacancy which may de-excite via the emission of a characteristic X-ray. XRF uses a primary beam of X-rays, but suffers from the fact that the characteristic X-ray spectrum recorded from a solid sample contains a scattered version of the primary spectrum, increasing the background signal and therefore degrading analytical sensitivity. The use of an electron beam to create inner shell vacancies, and thus stimulate X-ray emission, dates back to the early 1950s and offers a number of major advantages over X-ray stimulation:

(i) electrons, being charged particles, can be focused and steered using relatively simple electrostatic lenses (X-rays, being electromagnetic radiation, cannot be focused other than by using curved crystals as focusing mirrors). This means that the primary electron beam can be focused down to a spot diameter of around 1 μm, compared to the millimetre size of X-ray beams, allowing small features such as individual mineral inclusions in a rock or ceramic matrix to be chemically analysed. Additional benefits, such as being able to scan the beam across the surface of a sample (described below) allow spatially resolved analyses to be carried out.

(ii) the optical imaging capability of a microscope using an electron beam can be combined with this analytical facility to allow the operator to observe clearly the components of the sample which are being analysed.

(iii) the use of electrons to stimulate the characteristic X-rays reduces the X-ray background in the detector, improving the analytical detection levels.

(iv) the various phenomena associated with electron scattering from solid surfaces can be used to estimate the average atomic weight of different regions within the sample, assisting the characterization of the phase structure of the sample.

Figure 2.11 shows a simplified diagram of the operation of an electron microscope, fitted with a wavelength-dispersive X-ray detector. The primary beam of electrons is produced in a conventional electron gun, where a heated cathode maintained at ground potential emits electrons which are extracted via a positive potential into the focusing elements of the microscope. This beam is focused into a small cross-sectional area and can be steered to any point on the sample by a series of magnetic lenses. Once the electron beam strikes the sample, a number of processes take place, as illustrated in Figure 2.12, which result in the following:

(i) *secondary electrons*, which are very low-energy electrons (less than 50 eV) knocked out of the loosely bound outer electronic orbitals of surface atoms. Their low energy means they can only escape from atoms in the top few atomic layers, and are very sensitive to surface topography – protruding surface features are more likely to produce secondary electrons which can escape and be detected than are depressed features. The intensity of secondary electrons across the sample surface therefore accurately reflects topography, and is the basis of the image formation process in electron microscopy.

(ii) *backscattered electrons*, of higher energy, which result from interactions of the incident beam with the nuclei of the atoms in the sample. Their higher energy means that they can escape from deeper within the sample

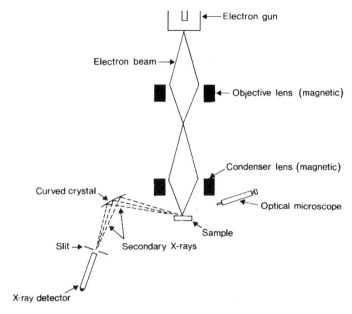

Figure 2.11 Schematic diagram of an electron microscope with a WD-X-ray detector. [From Willard *et al.*, 1988) Fig. 13–18. © 1988 Wadsworth Publishing Company. By permission of Brooks/Cole Publishing Company, a division of International Thomson Publishing Inc., Pacific Grove, CA 93950.]

than can secondary electrons, so they do not reflect surface topography. Their intensity is, however, proportional to the atomic weight (and therefore atomic number) of the interacting nuclei, and the intensity variation across a surface is therefore proportional to the average atomic number of the surface. This gives what is known as a '*backscattered electron (bse) image*', and contains useful structural information.

(iii) some incident electrons will create inner shell vacancies, in a manner similar to that of X-rays described above. The electrons ejected by the primary beam are called *photoelectrons*, and could be used analytically, but are generally neglected. The inner shell vacancy can de-excite via the Auger process (*Auger electrons* are also generally neglected) or via the emission of characteristic X-rays, which are the basis of the analytical operation of the electron microscope.

(iv) if the sample is thin enough, incident electrons may go straight through and be detected, as well as elastically and inelastically scattered electrons which are scattered in a forward direction. These form the basis of *transmission electron microscopy* (TEM), which is beyond the scope of this chapter, and has as yet had limited application in archaeology – an exception is the work of Barber and Freestone (1990), who used TEM to identify the nature of the tiny particles involved in the phenomenon of dichroism in the magnificent example of Roman glass known as the Lycurgus Cup, illustrated on the cover.

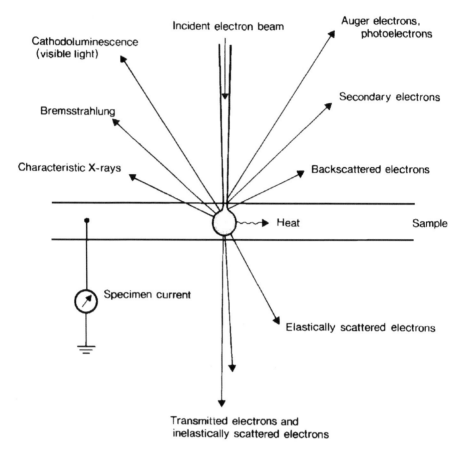

Figure 2.12 Interaction of primary electrons with a thin solid sample, showing the various processes which can take place. (After Woldseth, 1973; Fig. 4–1.)

As an analytical tool, the electron microscope can be operated with either a wavelength-dispersive X-ray detector (in which case it is often called an *electron microprobe*), or an energy-dispersive detector. Both are essentially as described in the previous sections, with the same advantages and disadvantages, and operational characteristics. Some more-powerful instruments are fitted with several detectors, sometimes of both types. Comparative studies have been made of the performance of wavelength-dispersive and energy-dispersive detectors in electron microprobe analysis of silicates (Dunham and Wilkinson, 1978) and, more recently, vitreous material (Verità *et al.*, 1994). Both of these studies agree that the accuracy and precision yielded by both techniques are comparable over the normal compositional ranges found in geological and archaeological material, but energy-dispersive has considerably poorer limits of detection for all elements, typically by one or two orders of magnitude. The exact figures depend on the matrix and counting time, but representative figures for the limit of detection by energy-dispersive detection are 0.05–0.26 wt % of the element.

These figures are comparable to, or slightly poorer than, the equivalent figures for XRF instruments, but one advantage of electron beam stimulation is the 'steerability' of the primary beam. Not only does this allow the analysis to be carried out on small regions of the sample, identified by either optical or electron microscopy, but also it allows line scans and area scans to be carried out, giving spatially resolved chemical information from the surface of the sample. Line scans (produced by monitoring the elemental composition as the electron beam is moved slowly across the surface) are particularly valuable in the study of the changes caused by corrosion and artificial patination of metal surfaces if the sample has been mounted to present a cross-section to the beam. Area scans give rise to '*elemental maps*' of the sample surface, by *rastering* the electron beam across the surface and monitoring the characteristic wavelength of the element of interest. Thus, for example, the distribution of lead can be looked at across the surface of a section through a copper alloy object. This facility is particularly valuable in archaeology, where the samples tend to be inhomogeneous and corroded. Several examples of the applications of this type of work to the study of archaeological metalwork can be found in the work of Cowell and La Niece (1991). Most routine analyses carried out by electron microscopy are performed on prepared samples – cut and mounted to fit standard sample holders, which typically accommodate samples several millimetres across. All samples have to be electrically conducting for electron microscopy, otherwise charge build-up will prevent the beam focusing on the sample. Metals are no problem, providing they are mounted in a conducting resin, but electrical insulators such as glass and ceramics normally need to be coated with a thin layer of carbon or gold to ensure good analysis. Some modern instruments have bigger sample chambers to allow the analysis of large, unprepared samples, and some also allow the analysis and imaging to be carried out under low vacuum (*i.e.*, at a higher ambient pressure than a conventional machine), which is useful for wet samples and biological tissue. These *environmental chambers* (Thiel, 2004) have the additional advantage that the samples do not need to be coated, because the higher pressure in the chamber ensures that there are enough ions around in the proximity of the sample to neutralize the charge build-up.

2.4.3 Proton-Induced X-Ray Emission

During the 1980s an alternative approach to X-ray analyses of inorganic materials was developed, utilizing the existence of large Van de Graaff accelerators (developed for particle beam research), and their ability to produce high intensity, highly focused beams of particles. These beams can be 'tapped off' from the accelerator and focused onto a sample outside the accelerator (*i.e.*, not in a chamber under high vacuum). This is ideal for archaeological material, since it removes the need for sampling (although prepared sections can, of course, also be analysed). The particle beam most usually used for analytical purposes consists of protons, which can be focused and steered just like electrons. The first generation of PIXE machines had beam diameters of the order of half a millimetre (Fleming and Swann, 1986), but more recent machines have micrometre diameter beams (termed μ-PIXE: Johansson and Campbell, 1988), giving spatial resolution similar to that obtainable by electron microprobe analysis. The proton beam strikes the sample, producing inner shell vacancies, which, as before, may de-excite via the emission of characteristic X-rays, which are detected using an energy-dispersive detector as described above. The major advantage offered by this instrument over an electron microprobe is that the use of a primary beam of protons does not give rise to such a high X-ray background as experienced by electron stimulation, thus giving improved analytical sensitivity. The reason is that when electrons interact with solids, a relatively high background of X-rays is produced (the *brehmsstrahlung*), as described above. Protons, being much heavier and accelerated to a higher energy, tend to suffer less energy loss on their passage through the sample, producing less *brehmsstrahlung*. Detection levels in conventional PIXE may be as low as 0.5–5 ppm for a wide range of elements in thin organic specimens such as biological tissue samples (Johansson and Campbell, 1988).

Archaeological work tends to use the external beam arrangement, whereby the whole object, or a sample removed from it, is situated outside the accelerator. The passage of the primary beam and (more importantly) the characteristic X-rays through air (or sometimes helium to minimize absorption) therefore limit the sensitivity of the method, particularly for the light elements. Even so, limits of detection of better than 100 ppm have been reported for elements above calcium in the periodic table on archaeological material. Analyses of archaeological and art historical samples ranging from ceramics, metals, paintings and even postage stamps have been reviewed by Johansson and Campbell (1988; Chapter 14).

2.5 NEUTRON ACTIVATION ANALYSIS

Until the advent of ICP and PIXE during the 1980s, the standard analytical method for producing multi-element analyses with detection limits at the ppm level or better was NAA, and the archaeological literature has many examples of its application. NAA has been used on archaeological material from the inception of the technique in 1950s, particularly for coinage [Kraay (1958, using a technique described by Emeleus (1958)] and ceramics (Sayre and Dodson, 1957). It has been a major technique in archaeological chemistry for these and other materials for the past 50 years (*e.g.*, Hughes *et al.*, 1991a), but increasing difficulties associated with obtaining irradiation facilities, and increasing competition from ICP-MS, is seriously challenging this position. In celebration of 50 years of NAA in archaeology, a whole issue of the journal *Archaeometry* (**49**, part 2, 2007) has been devoted to it.

NAA at its simplest is a technique whereby some of the elements in the sample are converted into artificial radioactive elements by irradiation with neutrons. Figure 2.13 shows a schematic diagram of this process. These artificial nuclei decay by one or more of the standard pathways for radioactive

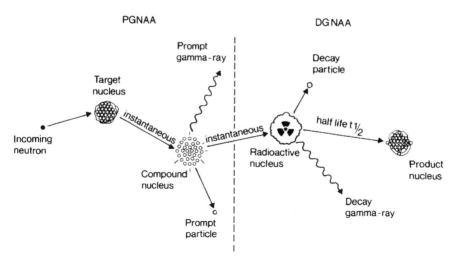

Figure 2.13 Schematic diagram of the nuclear processes involved in neutron activation analysis. Prompt gamma neutron activation analysis (PGNAA) occurs within the reactor; delayed gamma NAA (DGNAA) occurs at some remote site. (After Glascock, 1994; Fig. 1. © John Wiley & Sons Limited. Reproduced with permission.)

decay (normally involving the emission of an α, β or γ particle). By suitable instrumentation the radioactive decay can be detected, and, by measuring the intensity of the emission, can be related back to the original concentration of the parent element in the irradiated sample. Most often the radiation detected is γ emission, although β emission can also be used.

A nuclear reactor of the type used for activation analysis is essentially a source of high fluxes of neutrons – typically of the order of 10^{12} neutrons cm^{-2} s^{-1}. Most irradiations are carried out using relatively low-energy neutrons, known as *thermal neutrons*, with kinetic energies less than 0.2 eV. Other applications can use higher energy neutrons, which is known as *fast neutron activation analysis* (FNAA), but for most elements slow neutrons are the most effective at causing nuclear transformations. At these energies, the thermal neutrons can be 'captured' by the nuclei of elements in the sample, causing an instantaneous nuclear transformation to take place. There are several possible types of reaction, but the two most relevant in activation analysis are known as (n,p) or, more commonly, (n,γ) reactions. Detailed discussion of all the nuclear reactions relevant to the common elements in archaeological ceramics is given in the appendix to the paper by Perlman and Asaro (1969). A brief explanation of this notation is given here. As noted in Appendix 2, a nucleus can be thought of as being composed of Z protons (the atomic number, which uniquely defines the element, *e.g.*, $Z=11$ is the element sodium) and N neutrons, where the neutron number can vary giving rise to different isotopes of the same element. The symbol A is the total number of nucleons (protons plus neutrons) in the nucleus: clearly $A=Z+N$. Isotopes of the same element differ in their A

number, and the average atomic weight of the element is the weighted average of the natural abundances of the different isotopes (*e.g.*, chlorine, $Z = 17$, has stable two isotopes of $A = 35$ and 37, with natural abundances 75% and 25%. respectively, giving an average atomic weight of 35.5). Symbolically, the notation $_Z^A X$ is used, where X stands for the chemical symbol of the element defined by atomic number Z. The superscript refers to the number of atomic mass units in the nucleus (with the proton and neutron regarded for these purposes as having the same mass of one unit) and the subscript as the number of positive charges in the nucleus.

When a slow neutron is captured by the nucleus of element X, another isotope of the same element is instantaneously formed, in an excited state because of the impact (labelled *compound nucleus* in Figure 2.13), which then de-excites by the emission of a gamma particle (and possibly other particles) from the nucleus to produce a radioactive nucleus. For example, when ^{23}Na captures a neutron (signified by $_0^1 n$, since neutrons have a mass of one unit, but no electrical charge), it becomes the radioactive nucleus ^{24}Na, as follows:

$$^{23}_{11}\text{Na} + {}^1_0\text{n} = {}^{24}_{11}\text{Na} + \gamma$$

The γ particle is emitted virtually instantaneously on the capture of the neutron, and is known as a *prompt γ* – it can be used analytically, in a technique known as *prompt gamma neutron activation analysis* (PGNAA), but only if such γ's can be measured in the reactor during irradiation. Under the conditions normally used it would be lost within the nuclear reactor. In this reaction, no other prompt particle is emitted. The isotope of sodium formed (^{24}Na) is radioactively unstable, decaying by beta emission to the element magnesium (the *product nucleus* in Figure 2.13), as follows:

$$^{24}_{11}\text{Na} = {}^{24}_{12}\text{Mg} + {}^{\ 0}_{-1}\beta + \gamma$$

The beta particle emitted by this reaction has a mass of (conventionally) zero mass units, and a charge of –1; hence the notation $_{-1}^{\ 0}\beta$. Equations of this type should be balanced top and bottom – *i.e.*, the sum of the superscripts on the right-hand side should equal the sum of the superscripts on the left-hand side, and the same for the subscripts. The half life of ^{24}Na is approximately 0.623 days, and so the radioisotope formed decays away relatively quickly following irradiation. The gamma particle has a characteristic energy of 1369 keV, and so the decay of ^{24}Na to ^{24}Mg can be monitored by either measuring the β particle or the γ particle. It is important to appreciate that the two γ's in the above equations are quite different, and for the purposes of ordinary (delayed) NAA only that given in the second equation is of any value. In the notation used by radiochemists, all of the above information can be summarized in the following formulation:

$$^{23}\text{Na}(n,\gamma)^{24}\text{Na}; \ 0.623 \text{ d}, \ 1369 \text{ keV}$$

which means that the atom of ^{23}Na is converted into ^{24}Na via the process called neutron capture, accompanied by the emission of a prompt gamma. The nucleus formed (^{24}Na) is radioactive, decaying by the emission of a gamma particle of energy 1369 keV, and having a half life of 0.623 days. This statement contains all the information required by a radiochemist to understand the process, although, as can be seen, some of the detail is omitted.

A slightly more complicated nuclear reaction, but one which is also utilized in the analysis of archaeological material, is the so-called *transmutation*, or (n,p) reaction. In this case, the nucleus captures a neutron, but internal re-arrangements occur, and a proton is immediately ejected from the nucleus (the 'prompt particle'), changing the Z number and hence the chemical identity of the nucleus. For example, neutron irradiation of titanium does not result in any isotopes of titanium which have half lives suitable for measurement, but the transmutation reaction ^{47}Ti(n,p)^{47}Sc yields the radioactive isotope ^{47}Sc, which decays by β emission, with the emission of a γ particle with an energy of 159 keV and a half life of 3.43 days. Written out in full, this is:

$$^{47}_{22}\text{Ti} + {}^{1}_{0}\text{n} = {}^{47}_{21}\text{Sc} + {}^{1}_{1}\text{p}$$

and:

$$^{47}_{21}\text{Sc} = {}^{47}_{22}\text{Ti} + {}^{0}_{-1}\beta + \gamma$$

or, more concisely:

$$^{47}\text{Ti}(n, p)^{47}\text{Sc}; 3.43 \text{ d}, 159 \text{ keV}$$

Although many other types of nuclear reaction are possible as a result of high neutron fluxes, these two are the ones of prime importance in radioanalytical chemistry. The two principal requirements for a reaction to be useful analytically are that the element of interest must be capable of undergoing a nuclear reaction of some sort, and the product of that reaction (the *daughter*) must itself be radioactively unstable. Ideally, the daughter nucleus should have a half life which is in the range of a few days to a few months, and should emit a particle which has a characteristic energy, and is free from interference from other particles which may be produced by other elements within the sample.

The gamma detectors which are used to monitor the γ particles produced by the decay of the irradiated samples are essentially identical to the instruments used in EDXRF (described above), with the exception that the solid-state detector is usually a single crystal of germanium rather than silicon. This is because γ rays have energies which are typically 100 times greater than those of X-rays, and germanium is more efficient at measuring energies in this region. Originally lithium-drifted germanium detectors where used in order to get sufficiently pure crystals, but improvements in manufacturing technology have meant that pure (*intrinsic*) germanium detectors can now be obtained of sufficient quality. They are still, however, maintained at liquid nitrogen temperatures in order to reduce the electronic noise in the system. The spectra produced are very similar to those shown for EDXRF, but have a higher energy range.

The standard procedure in use at the British Museum for ceramic analysis by NAA has been described in detail by Hughes *et al.* (1991b), and is typical of the methods used on a range of materials. Powdered samples (40–80 mg) removed from the ceramic to be analysed (drilled or abraded using a material chosen to minimize contamination) are sealed into silica glass tubes (2 mm internal diameter, length 33 mm). Bundles of five to seven tubes are wrapped in aluminium foil, and then several bundles are packed into an irradiation canister, which can contain up to 70 samples. The canisters are then delivered to the nuclear reactor, where they are subjected to a known neutron flux for a fixed period of time, depending on the elements to be determined. They are then returned to the laboratory four days after irradiation, for a programme of γ counting using a germanium solid-state detector. The British Museum programme allows 23 elements to be determined from the same irradiation, making the whole analytical procedure extremely economical in terms of information obtained for the cost and time involved.

The artificially produced radioisotopes of interest can have a wide range of half lives, so it is normal to have a measuring strategy which involves a series of distinctive steps. Some isotopes, such as ^{28}Al [produced by the reaction ^{27}Al(n,γ)^{28}Al] have half lives so short (in this case 2.3 minutes) that they can only be measured if they are placed into the gamma counter immediately after removal from the irradiation source. They can, in effect, only be measured 'on site' at the reactor. Others, with slightly longer half lives (such as ^{24}Na, with a half life of 0.623 days) can be measured if placed into the counter within a few days of irradiation. Those with relatively long half lives (*e.g.*, ^{46}Sc at 80 days), are better left until the more short-lived isotopes have decayed away, reducing the radiation background and therefore the possible interferences. Measuring laboratories with an on-site reactor clearly have an advantage, in that they can measure the very short-lived isotopes immediately. Those laboratories remote from the reactor have devised appropriate measurement programmes, the details of which depend on the nature of the material being analysed. The British Museum, for example, measures pottery samples for a period of 3000 seconds immediately after receipt of the irradiated samples (four days after irradiation) to measure Na, K, Ca, As, Sb, La, Sm, Yb, Lu and Np (the latter to quantify U). A second measurement run is then carried out 18 days after irradiation, for a period of 6000 seconds, to measure Sc, Cr, Fe, Co, Rb, Cs, Ba, Ce, Eu, Tb, Hf, Ta and Pa (to measure Th; Hughes *et al.*, 1991b). Other laboratories use different strategies, and clearly the measurement of other materials such as metals requires a different programme.

Detection levels for NAA can be as low as 1.5×10^{-5} ppb for very sensitive elements in a suitable matrix (Glascock, 1994). More routinely, one would expect to see figures of the order of 10 ppb or better, up to perhaps 10 ppm, for trace elements in geological or biological material. It is important to realize that not all elements can be analysed by 'normal' NAA. As with all analytical techniques, there is a variation in the sensitivity and detection levels from element to element, but additionally with NAA there are some elements which cannot be 'seen' at all, perhaps because neutron irradiation as described does not produce suitable radioactive nuclei, or possibly because the spectrum has severe spectral interference. Most important of these archaeologically are elements such as lead and silicon, which means that NAA cannot produce 'total analyses' for some metals, ceramics and glasses. Some of these problems can be overcome using variations on NAA, such as PGNAA, but routinely they lead the analyst to resort to other methods for some elements.

There have been a limited number of comparisons of NAA (regarded as the established technique) and ICP (either using emission or mass spectrometric detection) on a range of sample materials. Bettinelli *et al.* (1992) compared ICP-AES, graphite furnace AAS, XRF and NAA for the inorganic analysis of coal fly ash. No simple conclusions were obtained – certain elements (*e.g.*, As, Ca, Cr, Mn, Na and V) gave good results by NAA, ICP and XRF, whereas other elements (such as Zn) gave uniformly poor results. This study confirmed the conclusion that the principal limitation on the accuracy of those techniques which require solid reference samples (in this case XRF and NAA) is the quality of the certified Standard Reference Materials. Other, more limited, comparisons have been carried out, such as the study by Rouchaud *et al.* (1993) on impurities in aluminium metal at the 0.1 to 10 ppm level, which concluded that both methods were accurate in this case (apart, obviously, from those elements such as Si, Ca and Mg which are not easily detected by NAA). Attention has also been paid to comparisons of these techniques in biological materials, such as Awadallah *et al.* (1986), who compared NAA, ICP-AES and flameless AAS for the analysis of a range of Egyptian crops and associated soil samples. Again, the general conclusion was that all methods provided reliable results, or at least results in line with those on the certified standards.

More recently, attention has switched to comparing NAA with ICP-MS – acknowledged to be generally more sensitive than ICP-AES, and with LA-ICP-MS, which has the distinct advantage of operating on solid samples. Ward *et al.* (1990) compared NAA and ICP-MS (solution and laser ablation) for the trace element analysis of biological reference materials. They found that both techniques gave good agreement with certified or published values for 18 elements down to the level of 30 ppb. The comparison between solid and solution ICP-MS work is described as 'fair for most elements'. This comparison has been taken further by Durrant and Ward (1993), who compared LA-ICP-MS with NAA on seven Chinese reference soils. Thirty elements were analysed, and it is stated that 80% of the LA-ICP-MS measurements were within a factor of two of the NAA determinations, although many were considerably closer. Precisions of 2–10% were obtained for most elements by LA-ICP-MS, although this figure deteriorated at lower concentrations. More recent work on archaeological material by Gratuze *et al.* (2001) and James *et al.* (2005) have clearly demonstrated the utility of LA-ICP-MS as a tool for the compositional analysis of a wide range of materials, including obsidian, glass, glazes and flint. In particular, James *et al.* (2005; 697) conclude that '*while absolute accuracy and precision for the ICP data are inferior to INAA, multivariate statistical analysis of data resulting from the two methods demonstrates a high degree of compatability.*' This does suggest that, with good QA procedures, the comparability of NAA and ICP-MS data is likely to be reasonable and that the large historical data banks of NAA results are comparable with ICP-MS.

2.6 MASS SPECTROMETRIC TECHNIQUES

Mass spectrometry is becoming increasingly important in archaeology, both as a technique in its own right and also as a sensitive detection system in the so-called *hyphenated techniques* (*e.g.*, ICP-MS, GC-MS, *etc.*). It is used in its own right for the determination of heavy stable isotope ratios such as lead (Chapter 9), or for light stable isotopes as used in dietary reconstruction (Chapter 10). It also forms the basis of several dating techniques involving radioactive isotope determinations (*e.g.*, K-Ar dating) or in accelerator-based methods for radiocarbon dating purposes (Aitken, 1990). In the past 20 years, however, it has played an expanding role as a 'bolt on' detection system offering much greater sensitivity (as well as isotopically resolved measurements, if required) in a wide range of analytical applications. A good example of this is its use as a detector in ICP.

Mass spectrometry is based on the principle that electrically charged atoms and molecules can be separated on the basis of their different atomic masses (strictly, mass-to-charge ratio) by controlling their motion through externally imposed electrical and/or magnetic fields. A simple MS (of the type first constructed at the beginning of the twentieth century) consists of a source of positively charged ions of the same energy, a magnetic and/or electrostatic deflection system for separating the charged ions and an ion collector to measure the current flowing in the selected beam. The pressure within the system must be kept low by pumping to ensure that the ions are not absorbed or deflected by passage through residual air. These components are indicated schematically in Figure 2.14. Modern spectrometers are designed to handle

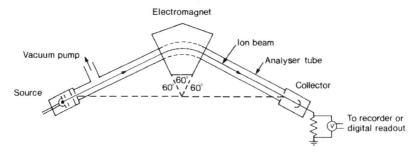

Figure 2.14   Schematic diagram of a 60° sector mass spectrometer. (Faure, 1986; Fig. 5–2. © 1986 John Wiley & Sons, Inc. Reprinted by permission of the publisher.)

either a gaseous sample, such as CO_2, or a solid sample deposited on a wire (usually made of Ta, Re or W). In the first case, ionization is achieved by injecting the gas into the vacuum system and bombarding the stream of gas with a beam of electrons, which causes ionization. The positively charged particles are then extracted from the ion source and given a fixed energy by acceleration through fine holes in a series of negatively charged plates, which also gives a degree of ion beam focusing. This type of ion source is suitable for the light elements which can be conveniently converted into gases, or for relatively volatile organic compounds. For heavier elements, a solid source is needed, and this is the basis of the *thermal ionization mass spectrometer* (TIMS) which has become of prime importance in isotope geochemistry. Here the sample is deposited as one of its salts on a refractory metal wire, which is then loaded into the MS and electrically heated to cause volatilization. The ions so produced are extracted and accelerated as before. Other types of ion source are available for specialist applications, and modern instrumentation uses more sophisticated systems than those described here, such as the use of a triple filament in TIMS, which allows samples as small as 10^{-9} g to be measured satisfactorily (Duckworth *et al.*, 1986; 43).

Once the positive ions have been injected into the body of the MS, the beam is separated into its component masses by one or more deflection devices. In the simpler system, such as that shown in Figure 2.14, this can be an electromagnet designed so that the magnetic field is uniform and perpendicular to the ion beam at the point where it passes the magnet. Under such conditions, the ions experience a force which deflects them in a circular path, the degree of deflection depending on their mass – the heavier ions are deflected less. If the ion has a mass m and carries a charge of e and it is extracted from the ion source by an applied voltage V, then it is given a kinetic energy of E, where:

$$E = eV = 1/2mv^2$$

All ions leaving the ion source have the same kinetic energy E, but the velocity v of a particular ion will depend on its mass. The velocity of an ion is given by rearranging the above:

$$v = (2eV/m)^{1/2}$$

The equation of motion of a charged particle in a magnetic field of strength B is given by the following, where r is the radius of the circular track taken by the ion:

$$Bev = mv^2/r$$

Combining these last two equations to eliminate v gives the following:

$$Be = mv/r = m(2eV/m)^{1/2}/r$$

$$m/e = B^2r^2/2V$$

Giving:

$$r = 1.414(V \cdot m/e)^{1/2}/B$$

Calculations using this equation can be simplified if it is assumed that m is in atomic mass units, e in units of atomic charge (*i.e.*, for a singly ionized atom $e = 1$), B in gauss and r in centimetres. It can be seen that the radius taken by an ion is dependent on the square root of the accelerating voltage V and inversely proportional to the magnetic field strength B. For a fixed geometry of MS, therefore, the value of these two parameters governs which ion will pass through the magnet and into the detector. If B and V are known, the value of m can be calculated for those ions which are detected. Alternatively, if either B or V is varied systematically, ions of different m will be selected to pass through sequentially. Thus, if one or other is scanned, the mass spectrum of the sample can be obtained. At its simplest, the detector is a Faraday cup-type charge detector which monitors the current of the beam as it is passed to earth – the magnitude of the current flowing is directly proportional to the number of ions being received at the detector.

For mass spectrometry in organic chemistry, where the aim is to identify the mass of the molecular ions present (and therefore identify the chemical nature of the ionized molecules and fragments), the important factor to be able to measure is the precise value of the mass, and a spectrometer of the type shown in Figure 2.14 is adequate. In biochemistry, newer *soft ionization* techniques have been developed to look at larger biomolecules (Pollard *et al.*, 2007; 162). For isotope work in geology and archaeology, what is often more important is the abundance of the ions, or the abundance ratio of two ions, such as $^{13}C/^{12}C$ in dietary reconstruction (Chapter 10). For this work, it is better to have a dual collector system, so that both ion beams can be monitored at the same time, thus eliminating any fluctuation in the intensity of the beam leaving the ion source and hence giving a more precise ratio estimate. Such systems and their applications are reviewed by Platzner (1997). More complex systems have been evolved for specific types of measurement, such as the use of the 'double focusing' spectrometer in which enhanced mass-resolving power is given by the addition of an electrostatic field via electric sector plates in the path of the ion beam before the electromagnet (Figure 2.15). This vastly improves the mass resolution of the system (Duckworth *et al.*, 1986; 92).

Systems based on the principles outlined above are those most regularly used for 'straight' mass spectrometry. The increased use of mass spectrometry as a detector in 'hyphenated' techniques has relied principally on a different approach, using a system known as the *quadrupole mass spectrometer*. This is much more compact than the magnetic sector devices shown in Figures 2.14 and 2.15, and allows the mass spectrum to be scanned through very rapidly, but at the expense of poorer mass resolution. It does not require the ions to be accelerated to such high energies as the magnetic sector machines, making it suitable to receive ions from sources other than the conventional ion source, and therefore ideal for a 'bolt on' detector. A quadrupole mass spectrometer

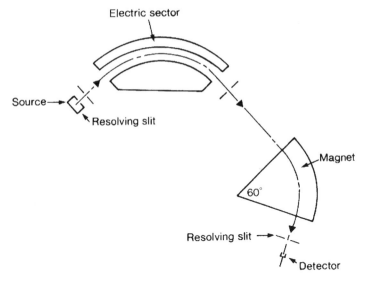

Figure 2.15 A double-focusing mass spectrometer. (After Beynon and Brenton, 1982; Fig. 4–9, by permission of University of Wales Press.)

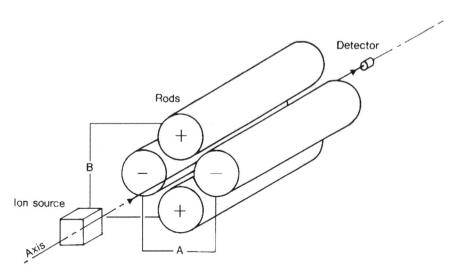

Figure 2.16 A quadrupole mass spectrometer. (Adapted from Beynon and Brenton, 1982; Figs 4–6 and 4–7, by permission of University of Wales Press.)

consists of four metal rods arranged in parallel to each other, as shown in Figure 2.16. The ion beam passes down the gap at the centre of the rods. Typically these rods might have a length of around 20 cm, and a diameter of 1 cm, and they are electrically connected together in opposite pairs, as shown in the figure. If one pair carries a small positive potential and the other pair an equal negative potential then the path down the centre of the rods is at zero potential. The ion beam can pass down the centre without deflection, but it is unstable with respect to the negative charges, since a positive beam is attracted to this polarity. In practice, the polarity of the rods is set to be a combination of a DC (steady) voltage and an oscillating voltage of frequency ω. The equations of motion of an ion beam under such conditions are complicated, and involve the solution of differential equations which are beyond the scope of this volume (see, for example, Duckworth *et al.*, 1986; 124). The mass spectrum can be scanned by varying systematically the magnitude of the DC and oscillating voltages in such a way that their ratio to each other does not change. At any particular value of these voltages, only ions of a certain mass-to-charge ratio will pass through the quadrupole and be detected (again usually by a simple Faraday cup) at the other end – all other ions will be deflected out of the system and lost. The mass spectrum can be scanned very rapidly – 0 to 800 mass units in a matter of seconds – and the normal mode of operation is to scan many times and accumulate an average spectrum in an attached computer. The mass resolution of the machine depends on the length of the rods and the frequency of oscillation, but is generally less than that achievable by instruments of the magnetic sector type. Nevertheless, it is a rapid, relatively cheap and sensitive detector, and has been increasingly used in ICP and chromatography applications, as well as for applications such as secondary ion mass spectrometry (SIMS) and other surface analytical techniques.

ICP-MS instrumentation of this type has been used in environmental applications to measure isotope ratios of heavy elements such as lead in body fluids, plant material and dust samples to determine the source of metal contamination (*e.g.*, Hamester *et al.*, 1994). It is generally accepted that quadrupole ICP-MS measurements of lead isotope ratios are sufficiently precise to detect differences between sources of pollution, but interlaboratory studies of quadrupole ICP-MS measurements on a range of environmental samples suggest that realistic precisions are 0.3% for $^{206}Pb/^{207}Pb$, 0.8% for $^{206}Pb/^{204}Pb$ and 1.4% for $^{208}Pb/^{204}Pb$ (Furuta, 1991), which are roughly an order of magnitude worse than can be achieved with 'conventional' lead isotope measurements made by TIMS (see Chapter 9). More recently, a new generation of ICP-MS machines has become available, which have an ICP source linked to a double-focusing multiple collector 'conventional' MS, giving precisions even better than those achievable by TIMS (Walder and Freedman, 1992; Walder and Furuta, 1993; Halliday *et al.*, 1998). The net result of this is isotopic ratio measurements of sensitivity equal to (if not better than) that of TIMS, but considerably faster, largely because the use of ICP-MS reduces the need for complex sample preparation and purification. This has become known as *high-resolution* ICP-MS.

Laser ablation is another important development, allowing solid samples to be directly analysed by ICP, since it offers the possibility of spatially-resolved microanalysis of solid samples, in a manner similar to electron microscopy, but with greater sensitivity and the potential for isotopic analysis. In this technique, a high-energy pulsed laser is directed onto a solid sample, with a beam diameter of less than 25 µm. The pulse vaporizes about 1 µg of material, to leave a crater

50 µm deep. The vaporized sample is swept into the ICP torch via a carrier gas. Current equipment allows solid samples, approximately flat, of at least 35 mm diameter to be accommodated in the laser ablation unit. The analytical sensitivity in terms of minimum detection limits is generally poorer for laser ablation than for solution analysis, and the detection limit for a particular element (or isotope) depends on the solid matrix which is being analysed. These developments have led LA-ICP-MS to rapidly overtake all other analytical techniques in the earth and environmental sciences, as reviewed by Halliday *et al.* (1998).

2.7 CHROMATOGRAPHIC TECHNIQUES

The term chromatography is used to encompass a wide range of related techniques which enable the separation of the components in a mixture as a result of their distribution between two phases – one *stationary* and one *mobile*. The most widely used chromatographic technique in archaeological applications during the past three decades has been *gas chromatography* (GC), although examples of *liquid chromatography* (LC) are becoming more common. When coupled to a MS, combined GC-MS and LC-MS offer powerful tools for analysis of a wide range of biomarkers remnant in the archaeological record. Complementary to this approach is *compound-specific isotope ratio mass spectrometry* [see Evershed *et al.* (2001) for applications to archaeology, and also Chapter 10].

Superficially, chromatography can be described in terms of placing blotting paper into a bath of ink. The liquid moves up the paper by means of capillary action. After a time, closer observation shows that a separation has occurred – there is a 'front' which marks how far a clear component has moved, and a dark front, lagging behind it, showing the lesser movement of the coloured matter in the ink. This shows that the clear liquid – water – has moved slightly faster through the paper than the coloured component. This is the basis of a separation technique called *paper chromatography*, which still has its uses today. In this simple example, the *mobile phase* is the aqueous ink solution and the *stationary phase* is the paper. The separation is a result of the differing interaction between the various components in the mobile phase and the stationary phase.

The mobile phase can be either a gas or a liquid, and the stationary phase can be either a solid or a liquid, giving rise to the two broad families of instrumental chromatographic techniques, namely GC and LC respectively. In GC, a gaseous mobile phase containing the volatilized mixture to be separated is passed through a column. GC is suitable for the separation of thermally stable and volatile organic and inorganic compounds. The volatility of solutes can be promoted by various chemical derivatization procedures. Although GC has wide application and is very sensitive, only about 20% of known compounds can be analysed – the rest are either insufficiently volatile and cannot be made to travel in the gaseous state, or are thermally unstable and decompose under the conditions employed (Willard *et al.*, 1988; Chapters 17–20). LC is not limited by these considerations and can be used for a wide range of materials. The most common form of LC is *high performance* or *high pressure liquid chromatography* (HPLC) whereby the liquid mobile phase is forced under pressure through a column containing a solid packing (*liquid–solid* or *adsorption chromatography*) or a solid packing coated with a stationary liquid phase (*liquid–liquid chromatography*).

Both GC and LC operate on essentially the same principle, which can be simply described (for further detail on the theory of chromatographic separation see Braithwaite and Smith, 1985; 11–23). The mobile phase is made up of a *solvent* (either a gas or a liquid) which carries the mixture through the system, and a *solute*, which is the mixture to be separated. The molecules of the solute at any particular time are distributed between those carried by the mobile phase and those in the stationary phase. At any stage there is an equilibrium established between the concentration of the solute distributed in each phase. The competition between the two phases for the solute molecules depends on their physical properties and affinity for the stationary phase [see Braithwaite and Smith (1985; 13–14) for a discussion of the molecular interactions influencing retention]. In column chromatography it is usual to define a *partition* or *capacity ratio* (k′) which is a measure of how long a time the molecules of a given species spend in the stationary phase relative to their time in the mobile phase. If a component has a low partition ratio it will pass through the column relatively quickly. At the extreme, if k′ is zero (*i.e.*, the compound spends no time in the stationary phase) it will pass through the column at the same rate as the mobile phase. If k′ is very high, the 'compound' spends a long time in the stationary phase, and may never emerge from the column! The time taken for each molecular species to emerge or elute from a column is termed the *retention time*. The *relative retention* (α) of two compounds is a measure of how well they can be separated (*i.e.*, the resolution of the system), and is the ratio of the two *partition ratios*. For any compound, the value of k′ depends on the combination of stationary–mobile phase chosen, and on the temperature of the system. The retention time depends on these and other factors such as the length of the column and the flow rate of the mobile phase. Complex mixtures can be separated by careful selection of these variables.

A gas chromatograph system requires a means of introducing a sample onto the column where separation is achieved. The column feeds into a detector which acts as a monitor of the column effluent. A basic GC system is shown in Figure 2.17. The mobile phase is an inert gas (also known as the carrier gas) such as helium, hydrogen or nitrogen, the rate of flow of which must be accurately controlled for the purposes of reproducibility. A number of different injection systems are available (Tipler, 1993; 20–30), including automated samplers which give more reproducible injections than can be achieved by a human operator. The column is housed in an oven to allow the column temperature to be controlled. In GC two basic types of column are employed; the *packed column* and the *wall-coated open tubular* (WCOT) or *capillary column*. In the majority of cases the latter have replaced the former. Packed columns are generally made of stainless steel or glass tubing, with an internal

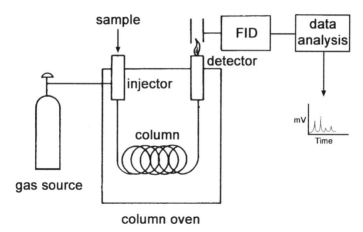

Figure 2.17 Schematic diagram of a gas chromatography system. (Adapted from Bartle, 1993; Fig. 1, by permission of Oxford University Press.)

diameter of 2–6 mm, and a length of up to 5 m. They are packed with a fine-grained inert support material which is coated with the stationary liquid phase. A capillary column has an internal diameter of less than 1 mm, and may have a length of 10–30 m although longer columns are possible. Normally the stationary phase is coated and bonded to the internal wall of the column. The column is housed in an oven which, through the use of incrementally increasing temperatures, means that the least volatile components in a mixture can be eluted more rapidly. The majority of archaeological applications have made use of capillary columns employing low polarity, bonded stationary phases.

The most common detector used in GC is the *flame ionization detector* (FID), in which solutes separated during the chromatographic process are burned in a hydrogen–air flame. Designs vary, but, in the simplest form, just above the tip of the flame are a pair of metal plates which carry a potential difference of around 400 V. The current flowing across the flame between the electrodes is measured. Normally it is very low, but as charged particles enter the gap between the plates from the flame, the resistance between the plates drops and the current flow increases markedly. If only the carrier gas is present in the flame, the current flow will be a fixed baseline value, but as the separated components elute from the column and into the flame, the current will rise and fall as each separated component passes through. The peaks are recorded on a chart recorder, or digitally in a computer, and the output takes the form of a plot of peak height as a function of time, with the signal produced proportional to the amount of each component in the mixture. The FID is very sensitive, responds to the vast majority of organic compounds and retains good linearity over a wide sample concentration range. Other detectors (such as *thermionic emission* or *thermal conductivity*) can be used for specific applications. A more complete description of GC can be found in Baugh (1993).

Equipment for HPLC is slightly more complicated (see Figure 2.18), due to the need to handle a liquid mobile phase. This requires the presence of one or

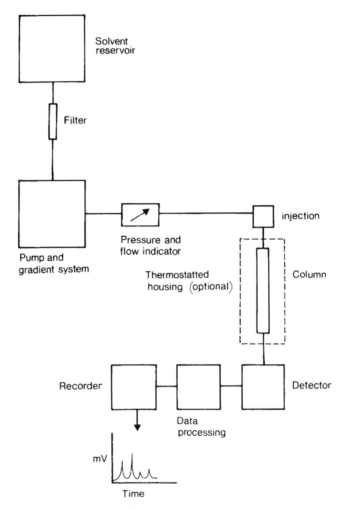

Figure 2.18   Schematic diagram of an HPLC system. (From Lindsay, 1992; Fig. 1–2a, © John Wiley & Sons Limited. Reproduced with permission.)

more reservoirs for the solvent which forms the mobile phase – many systems have the facility for changing the composition of the mobile phase as the run progresses, by mixing solvents from two or more reservoirs (*gradient elution*). The mobile phase is delivered to the column by a pumping system, which can give a high-pressure liquid and control the flow rate without undue variation in the pressure or flow. The most common type of pump is a reciprocating piston, often coupled with a pulse damper to ensure smooth flow. The sample (dissolved, if possible, in the solvent which comprises the mobile phase) is injected into a micro-sampling injector valve, which holds the sample in an injection loop, and releases a fixed volume (usually either 10 or 20 μL) into the liquid flow at the top of the column. Again, auto-samplers are used to relieve the tedium for the operator, and ensure better reproducibility of injection. Columns for HPLC need to be of sturdy construction to withstand the high pressures involved (up to several hundred atmospheres). Typically they are made of stainless steel, or glass-lined metal for chemical inertness, and are straight, with a length of 10–30 cm depending on application. They are usually fitted with a guard column or an in-line filter to prevent unwanted particles entering the column, and are housed in an oven to allow temperature control.

Once the liquid sample has passed through the column (and hopefully been separated) it is fed directly into a detector which detects and quantifies the eluting compounds. Several types of detector are available, depending on the application, but the two most common types are *ultraviolet-visible spectro-photometers* or *fluorescence* detectors. The UV/visible (UV/vis) type of detector comes in a variety of forms, but again the most common type is a fixed wavelength detector. The mobile phase plus separated compound passes through a detector cell (typically of 2–8 μL capacity) through which is focused radiation of the selected wavelength, and the mobile phase only passes through a reference cell. The absorbance at this wavelength is measured as the difference between the absorbance of the two cells, as determined by a photometer. Many compounds absorb around the 254 nm wavelength, which is one of the wavelengths emitted from a medium-pressure mercury vapour lamp, and so this is usually the wavelength chosen. A fluorescence detector is similar to a UV/vis detector, but the irradiating light illuminates the sample cell and the resulting fluorescence is detected at right angles to this. Since not all compounds exhibit fluorescence to UV/vis radiation, a fluorophor may have to be added to the sample by derivatization prior to injection. The fluorescence detector is generally considerably more sensitive (perhaps by two or three orders of magnitude), but the need to sensitize the sample to the fluorescing radiation may make the sample preparation more complex.

Chromatographic techniques have developed rapidly over the past few years, resulting in a wide range of 'hyphenated techniques'. For GC, one of the main drawbacks is the requirement that the compounds to be separated should be volatile but stable. This can be overcome in certain cases by using a *pyrolysis injection system* (Willard *et al.*, 1988; 453), in which a solid sample (*e.g.*, rubber, resin, soil or coal, and so on) is heated rapidly (by Curie point or ohmic heating) to cleave thermally certain bonds producing fragments of lower molecular weight which then enter the chromatographic system. This gives a technique called *pyrolysis-GC* (Py-GC). Commonly, the column of either a GC (or less frequently an HPLC) is interfaced to a quadrupole mass spectrometer. The combination of a separation technique coupled with a versatile detection system, which results in the generation of a mass spectrum of each component separated, allows much greater opportunities of structure identification. Increasingly, combined GC-MS is finding applications in archaeology.

The combination of GC and isotope ratio mass spectrometry was first demonstrated in 1978 (Matthews and Hayes, 1978). This has opened up a field with huge potential in archaeology and many other disciplines – that is, the isotopic analysis of single compounds separated by chromatographic processes.

Applications of compound-specific isotope ratio mass spectrometry in archaeology are demonstrated in Chapters 10 and 11. The first application of GC–combustion–isotope ratio mass spectrometry (GC-C-IRMS) to archaeology appeared in 1994 (Evershed *et al.*, 1994). In this study, carbon isotope measurements were obtained on long-chain alkanes and alcohols from suspected *Brassica* species leaf waxes preserved in late Saxon to early Medieval pottery sherds from West Cotton (Northamptonshire, UK) and compared with these compounds in soil extracts from the site and in contemporary reference samples. This successfully demonstrated the consumption of leafy vegetables, which would normally be archaeologically invisible. Over the past decade a wide range of applications has been reported. Most studies to date have measured carbon isotope ratios in lipid biomarkers and in amino acids but increasingly single-compound analyses of other isotope systems, including hydrogen ($^2H/^1H$), nitrogen ($^{15}N/^{14}N$) and oxygen ($^{18}O/^{16}O$), are being reported.

As noted earlier, the compounds separated by GC must be suitably volatile and thermally stable. Derivatization of polar compounds can be performed but corrections to account for the addition to the compound of carbon atoms with their own isotopic ratios needs to be undertaken. Compounds eluting from the chromatographic column pass through a combustion reactor (normally an alumina tube containing Cu, Ni and Pt wires maintained at high temperatures) where they are combusted. This is followed by a reduction reactor (an alumina tube containing three Cu wires maintained at 600 °C) to reduce any nitrogen oxides to nitrogen. For hydrogen and oxygen a high temperature thermal conversion reactor is required. In order to prevent interference caused by the formation of HCO_2^+, water is removed by a water trap and the sample is introduced into the ion source of the MS. Ionization of the analyte gas is achieved using electron ionization (EI). The ionized gases are resolved into beams of different mass in a magnetic sector analyser and are detected by Faraday cups, the output from which is used to calculate the final stable isotope ratio. For carbon the relative abundances of the ions at m/z 44, 45 and 46 are determined. This is calculated relative to a standard of known isotopic composition and expressed using the familiar 'per mil' notation. Applications of GC-C-IRMS to the study of fatty acids can be found in Meier-Augenstein (2002).

2.8   INFRARED AND RAMAN SPECTROSCOPY

Infrared radiation (IR) is basically heat, and arises from relatively low-energy transitions between molecular vibrational and rotational energy states. It is part of the electromagnetic spectrum (see Figure A1.2 in Appendix 1), but longer in wavelength than the optical region (400–750 nm), and is measured in units known as *wave number*, rather than wavelength. Wave number ($\bar{\nu}$) is the number of waves per centimetre, and is the reciprocal of wavelength. The most analytically useful region of the IR spectrum is 2.5–15 μm (wave numbers 4000–650 cm^{-1}). In this region, called the 'fingerprint region', many organic

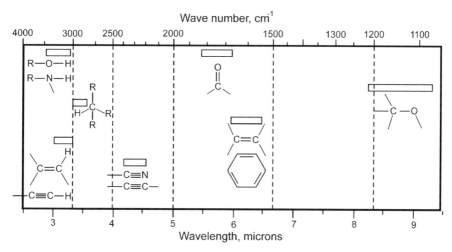

Figure 2.19 Infrared correlation chart, showing approximate wavenumber ranges of common bond vibrations in organic molecules. More detailed information can be found in, for example, Ewing (1985; 95). (From Pollard, *et al.*, 2007 Fig. 4–4, by permission of Cambridge University Press.)

compounds produce a characteristic absorption spectrum. Even simple molecules, however, have a large number of possible vibrational modes, corresponding to the stretching and bending of the various bonds. Thus, peaks in IR spectra are labelled 'C–H stretch', 'C–H bend', *etc.*, and give information on the types of chemical bonds present in a sample. IR should, therefore, be regarded as a 'chemical fingerprinting' technique: it does not provide precise chemical characterization of a sample, and different compounds may have very similar IR spectra. For this reason, infrared spectra are usually interpreted using a correlation chart (Figure 2.19), or a computerized database derived from such a chart. Although useful for identifying functional groups in simple molecules, for more complex molecules and mixtures, or degraded archaeological organic material, IR is of limited use, and is often simply the starting point for a more detailed analysis, such as GC (see above). The most common form of infrared machine for research purposes is now the Fourier transform infrared (FTIR) spectrometer, based on the Michelson interferometer. IR spectroscopy has been extensively used to characterize the sources of European amber (Beck, 1986; 1995; Beck and Shennan, 1991), and has also been applied to the study of the differential preservation of bone (Weiner *et al.*, 1993). *Infrared microspectroscopy* [the use of a microscope with IR capability (Kempfert *et al.*, 2001)] has found application for the identification of mineralized fibres from archaeological contexts (Gillard *et al.*, 1994). Undoubtedly, because of its relative cheapness and versatility, IR (especially FTIR and microspectroscopy) will become even more widely used in archaeology.

Closely related to IR spectroscopy is a technique called Raman spectroscopy. As radiation passes through a transparent medium, a small proportion of the incident beam is scattered in all directions. Most of the incident radiation is scattered at exactly the same wavelength as the incident radiation (*Rayleigh* or elastic scattering). In 1928, however, the physicist Chandrasekhara Raman noticed that a small amount of radiation was scattered at wavelengths different from the incident radiation, as a result of *Compton* or inelastic scattering. More significantly, he noticed that the difference in wavelength between the incident and scattered radiation is characteristic of the material responsible for the scattering. This effect is very weak, and was later named after Raman.

The discussion of infrared absorption above implies that all molecular vibrational modes are capable of absorbing IR. This is not true – infrared absorption only occurs if the vibrational motion produces a change in the dipole moment of the molecule. Dipole moment arises because the charge distribution associated with the valency electrons in a heteronuclear molecule (one containing more than one type of atom) is not uniform. The overall charge on the molecule is still neutral, but the charge distribution is such that one part is more negative, whilst another is more positive. Water (H_2O, but structurally H–O–H), for example, has a strong dipole because the electrons are more strongly attracted to the oxygen than to the hydrogen, which gives rise to a small excess positive charge on each of the hydrogen atoms and a small negative charge on the oxygen. Water is a non-linear molecule, so all the vibrational modes give rise to changes in the dipole moment (see Pollard *et al.*, 2007; Figure 4.3), and therefore water is a strong infrared absorber. In contrast, in CO_2, which is also dipolar but is a linear molecule, the symmetric stretching mode does not give rise to a change in dipole moment, and consequently it is infrared inactive, and can therefore be seen in Raman spectroscopy. The asymmetric stretching mode does change the dipole moment, however, and this vibration is infrared active. Raman and IR spectroscopies are therefore complementary, although rarely do they share the same instrumentation.

Raman spectroscopy, because of its versatility, has been used to characterize a wide range of materials in art historical and conservation science (Edwards, 2000; Vandenabeele *et al.*, 2007), and in archaeological applications (Smith and Clark, 2004). Fourier transform Raman spectroscopy (FTRS), in particular, has the advantage of being a reflective method which allows direct, non-destructive analysis. It can also be used through a microscope to allow the characterization of small samples. Consequently, it has found a great deal of application in the study of paint pigments on works of art (*e.g.*, Burgio *et al.*, 2000) and wall paintings (*e.g.*, Vandenabeele *et al.*, 2005).

2.9 OTHER TECHNIQUES

A wide range of other methods from analytical chemistry have been applied to archaeological samples, but space precludes detailed descriptions of them all. Some, such as XPS, have only been employed sporadically because of the specialized nature of the technique. Others are increasing in application as their archaeological potential is explored. One class of methods which has had some application are *resonance techniques* (*e.g.*, Ewing, 1985; Chapter 13). These are based on another aspect of the interaction between matter and electromagnetic radiation, in which a sample is subjected simultaneously to two magnetic fields – one stationary and the other varying at a known frequency [usually a radiofrequency (RF)]. Under these conditions, sub-atomic particles in the sample are aligned in a particular orientation by the steady field, and, as the frequency of the RF is varied, a resonance condition is met when the 'spin' of the particle is caused to 'flip' causing an absorption in the frequency spectrum. ESR is now best-known archaeologically for its application as a dating technique (Aitken, 1990; Schwarcz and Grün, 1992; Rink *et al.*, 2002), but it has been used for a variety of other applications, such as the elucidation of the thermal history of flint (Robins *et al.*, 1978) and charred organic materials (*e.g.*, Schurr *et al.*, 2001). NMR has also been used for the characterization of amber (Lambert *et al.*, 1988) and coal-like material (Lambert *et al.*, 1992). Although the instrumentation for resonance measurement is relatively expensive, it is likely that selected archaeological problems, particularly those involving the solution of structural organic problems, may well benefit from further applications of such approaches. Further details on NMR can be found in Willard *et al.* (1988; Chapter 15), and Ewing (1985; 255).

One final area worthy of brief mention because of its great potential in archaeology is the thermal analysis of solid materials (*e.g.*, Ewing, 1985; Chapter 23) A range of techniques are included under this heading, but the basic principle is the quantitative measurement of the weight change in a solid sample under controlled heating conditions. In its simplest form [*thermogravimetry* (TGA)] the weight of the sample is recorded as the temperature is raised, thus recording the loss of components such as water and CO_2, allowing the composition to be elucidated. In differential thermal analysis (DTA) the heat evolved or absorbed by the sample as it is heated is measured by comparison with an inert reference material, which allows phase changes to be monitored as well as chemical changes. The related technique of differential scanning calorimetry (DSC) allows more quantitative estimates to be made of the thermodynamic heats of transition. Thermal analytical techniques are widely used in materials science to identify and characterize mineral species, polymers, *etc.*, and to study the high-temperature thermodynamic properties of compounds. A review of the application of DTA to archaeological ceramics is given by Kingery (1974). An example of the use of TGA on other archaeological material can be found in the work of Hunter *et al.* (1993), in which ESR, TGA, Raman spectroscopy, IR and X-radiography were used in combination with the more usual methods to characterize the various black lithic materials (jet, shale, lignite, *etc.*) used in antiquity. Clearly, archaeological chemists are becoming conversant with a much wider range of techniques than has been the case hitherto to characterize and study the enormous range of materials from the past.

REFERENCES

Aitken, M.J. (1990). *Science-based Dating in Archaeology*. Longman, London.

Awadallah, R.M., Sherif, M.K., Amrallah, A.H. and Grass, F. (1986). Determination of trace elements of some Egyptian crops by instrumental neutron activation analysis, inductively coupled-plasma atomic emission spectro-metric and flameless atomic absorption spectrophotometric analysis. *Journal of Radioanalytical and Nuclear Chemistry Articles* **98** 235–246.

Barber, D.J. and Freestone, I.C. (1990). An investigation of the origin of the colour of the Lycurgus Cup by analytical transmission electron microscopy. *Archaeometry* **32** 33–45.

Bartle, K.D. (1993). Introduction to the theory of chromatographic separa-tions. In *Gas Chromatography: A Practical Approach*, ed. Baugh, P.J., Oxford University Press, Oxford, pp. 1–14.

Baugh, P.J. (ed.) (1993). *Gas Chromatography: A Practical Approach*. Oxford University Press, Oxford.

Beck, C.W. (1986). Spectroscopic investigations of amber. *Applied Spectro-scopy Reviews* **22** 57–110.

Beck, C.W. (1995). The provenance analysis of amber. *American Journal of Archaeology* **99** 125–127.

Beck, C.W. and Shennan, S. (1991). *Amber in Prehistoric Britain*. Oxbow, Oxford.

Bettinelli, M., Baroni, U., Pastorelli, N. and Bizzarri, G. (1992). ICP-AES, GFAAS, XRF and NAA coal fly-ash analysis – comparison of different analytical techniques. In *Elemental Analysis of Coal and Its By-Products*, ed. Vourvopoulos, G., World Scientific, Kentucky, pp. 372–394.

Beynon, J.H. and Brenton, A.G. (1982). *An Introduction to Mass Spectrometry*. University of Wales Press, Cardiff.

Brady, J.E. (1990). *General Chemistry*. John Wiley, New York, 5th edn.

Braithwaite, A. and Smith, F.J. (1985). *Chromatographic Methods*. Chapman and Hall, London.

Brenner, I.B. and Zander, A.T. (2000). Axially and radially viewed inductively coupled plasmas – a critical review. *Spectrochimica Acta* **B55** 1195–1240.

Briggs, D. and Seah, M.P. (ed.) (1990). *Practical Surface Analysis. Vol. 1. Auger and X-Ray Photoelectron Spectroscopy*. John Wiley, Chichester.

Britton, D. and Richards, E.E. (1969). Optical emission spectroscopy and the study of metallurgy in the European Bronze Age. In *Science in Archaeology*, ed. Brothwell, D. and Higgs, E., Thames and Hudson, London, 2nd edn, pp. 603–613.

Burgio, L., Clark, R.J.H., Stratoudaki, T., Doulgeridis, M. and Anglos, D. (2000). Pigment identification in painted artworks: a dual analytical approach employing laser-induced breakdown spectroscopy and Raman microscopy. *Applied Spectroscopy* **54** 463–469.

Cazes, J. (ed.) (2005). *Ewing's Analytical Instrumentation Handbook*. Marcel Dekker, New York, 3rd edn.

Christian, G. (1994). *Analytical Chemistry*. John Wiley, New York, 5th edn.

Cowell, M. and La Niece, S. (1991). Metalwork: artifice and artistry. In *Science and the Past*, ed. Bowman, S., British Museum Publications, London, pp. 74–98.

Cox, G.A. and Pollard, A.M. (1977). X-ray fluorescence of ancient glass: the importance of sample preparation. *Archaeometry* **19** 45–54.

Dawson, P.T., Heavens, O.S. and Pollard, A.M. (1978). Glass surface analysis by Auger electron spectroscopy. *Journal of Physics C* **11** 2183–2193.

Desnica, V. and Schreiner, M. (2006). A LabVIEW-controlled portable X-ray fluorescence spectrometer for the analysis of art objects. *X-Ray Spectrometry* **35** 280–286.

Duckworth, H.E., Barber, R.C. and Venkatasubramanian, V.S. (1986). *Mass Spectroscopy*. Cambridge University Press, Cambridge, 2nd edn.

Dunham, A.C. and Wilkinson, F.C.F. (1978). Accuracy, precision and detection limits of energy-dispersive electron-microprobe analyses of silicates. *X-Ray Spectrometry* **7** 50–56.

Durrant, S.F. and Ward, N.I. (1993). Rapid multielemental analysis of Chinese reference soils by laser ablation inductively coupled plasma mass spectrometry. *Fresenius Journal of Analytical Chemistry* **345** 512–517.

Edwards, H.G.M. (2000). Art works studied using IR and Raman spectroscopy. In *Encyclopaedia of Spectroscopy and Spectrometry*, ed. Lindon, J.C., Tranter, G.E. and Holmes, J.L., Academic Press, London, pp. 2–17.

Emeleus, V.M. (1958). The technique of neutron activation analysis as applied to trace element determination in pottery and coins. *Archaeometry* **1** 6–15.

Evershed, R.P., Arnot, K.I., Collister, J., Eglinton, G. and Charters, S. (1994). Application of isotope ratio monitoring gas-chromatography mass-spectrometry to the analysis of organic residues of archaeological origin. *Analyst* **119** 909–914.

Evershed, R.P., Dudd, S.M., Lockheart, M.J. and Jim, S. (2001). Lipids in archaeology. In *Handbook of Archaeological Sciences*, ed. Brothwell, D.R. and Pollard, A.M., Wiley, Chichester, pp. 331–349.

Ewing, G.W. (1985). *Instrumental Methods of Chemical Analysis*. McGraw-Hill, New York, 5th edn.

Faure, G. (1986). *Principles of Isotope Geology*. John Wiley, Chichester, 2nd edn.

Fleming, S.J. and Swann, C.P. (1986). PIXE spectrometry as an archaeometric tool. *Nuclear Instruments and Methods in Physics Research* **A242** 626–631.

Furuta, N. (1991). Interlaboratory comparison study on lead isotope ratios determined by inductively coupled plasma mass spectrometry. *Analytical Sciences* **7** 823–826.

Gillard, R.D., Hardman, S.M., Thomas, R.G. and Watkinson, D.E. (1994). The mineralization of fibres in burial environments. *Studies in Conservation* **39** 132–140.

Gillies, K.J.S. and Urch, D.S. (1983). Spectroscopic studies of iron and carbon in black surfaced wares. *Archaeometry* **25** 29–44.

Glascock, M.D. (1994). Nuclear reaction chemical analysis: prompt and delayed measurements. In *Chemical Analysis by Nuclear Methods*, ed. Alfassi, Z.B., John Wiley, Chichester, pp. 75–99.

Goodhew, P.J., Humphreys, J. and Beanland, R. (2001). *Electron Microscopy and Analysis*. Taylor and Francis, London, 3rd edn.

Gratuze, B., Blet-Lemarquand, M. and Barrandon, J.N. (2001). Mass spectrometry with laser sampling: a new tool to characterize archaeological materials. *Journal of Radioanalytical and Nuclear Chemistry* **247** 645–656.

Hall, E.T. (1961). Surface enrichment of buried metals. *Archaeometry* **4** 62–66.

Hall, E.T., Schweizer, F. and Toller, P.A. (1973). X-ray fluorescence analysis of museum objects: a new instrument. *Archaeometry* **15** 53–78.

Halliday, A., Lee, D.-C., Christensen, J. N., Rehkamper, M., Yi, W., Luo, X., Hall, C.M., Ballentine, C.J., Pettke, T. and Stirling, C. (1998). Applications of multiple collector-ICPMS to cosmochemistry, geochemistry and palaeoceanography. *Geochimica et Cosmochimica Acta* **62** 919–940.

Hamester, M., Stechmann, H., Steiger, M. and Dannecker, W. (1994). The origin of lead in urban aerosols – a lead isotopic ratio study. *Science of the Total Environment* **146/7** 321–323.

Hatcher, H., Tite, M.S. and Walsh, J.N. (1995). A comparison of inductively-coupled plasma emission spectrometry and atomic absorption spectrometry analysis on standard reference silicate materials and ceramics. *Archaeometry* **37** 83–94.

Hughes, M.J., Cowell, M.R. and Craddock, P.T. (1976). Atomic absorption techniques in archaeology. *Archaeometry* **18** 19–37.

Hughes, M.J., Cowell, M.R. and Hook, D.R. (1991a). *Neutron Activation and Plasma Emission Spectrometric Analysis in Archaeology*. British Museum Occasional Paper 82, London.

Hughes, M.J., Cowell, M.R. and Hook, D.R. (1991b). NAA procedure at the BM Research Laboratory. In *Neutron Activation and Plasma Emission Spectrometric Analysis in Archaeology*, ed. Hughes, M.J., Cowell, M.R. and Hook, D.R., British Museum Occasional Paper 82, London, pp. 29–46.

Hunter, F.J., McDonnell, J.G., Pollard, A.M., Morris, C.R. and Rowlands, C.C. (1993). The scientific identification of archaeological jet-like artefacts. *Archaeometry* **35** 69–89.

James, W.D., Dahlin, E.S. and Carlson, D.L. (2005). Chemical compositional studies of archaeological artifacts: comparison of LA-ICP-MS to INAA measurements. *Journal of Radioanalytical and Nuclear Chemistry* **263** 697–702.

Jarvis, K.E., Gray, A.L. and Houk, R.S. (1992). *Handbook of Inductively Coupled Plasma Mass Spectrometry*. Blackie, Glasgow.

Jenkins, R. (1974). *An Introduction to X-Ray Spectrometry*. John Wiley, Chichester.

Jenkins, R. (1988). *X-Ray Fluorescence Spectrometry*. Wiley-Interscience, Chichester.

Johansson, S.A.E. and Campbell, J.L. (1988). *PIXE: A Novel Technique for Elemental Analysis*. John Wiley, Chichester.

Jones, R.E. (1986). *Greek and Cypriot Pottery*. British School at Athens Fitch Laboratory Occasional Paper 1, Athens.

Joy, D.C., Romig, A.D. Jr and Goldstein, J.I. (1986). *Principles of Analytical Electron Microscopy*. Plenum, New York.

Kempfert, K.D., Coel, B., Troost, P. and Lavery, D.S. (2001). Advancements in FTIR microscope design for faster and easier microanalysis. *American Laboratory* **33** (Nov) 22–27.

Kingery, W.D. (1974). Differential thermal analysis of archaeological ceramics. *Archaeometry* **16** 109–112.

Kraay, C.M. (1958). Gold and copper traces in early Greek silver. *Archaeometry* **1** 1–5.

Lambert, J.B., Beck, C.W. and Frye, J.S. (1988). Analysis of European amber by carbon-13 nuclear magnetic resonance spectroscopy. *Archaeometry* **30** 248–263.

Lambert, J.B., Frye, J.S. and Jurkiewicz, A. (1992). The provenance and coal rank of jet by carbon-13 nuclear magnetic resonance spectroscopy. *Archaeometry* **34** 121–128.

Lindsay, S. (1992). *High Performance Liquid Chromatography*. Wiley, Chichester, 2nd edn.

Matthews, D.E. and Hayes, J.M. (1978). Isotope-ratio-monitoring gas chromatography–mass spectrometry. *Analytical Chemistry* **50** 1465–1473.

Meier-Augenstein, W. (2002). Stable isotope analysis of fatty acids by gas chromatography–isotope ratio mass spectrometry. *Analytica Chimica Acta* **465** 63–79.

Perlman, I. and Asaro, F. (1969). Pottery analysis by neutron activation analysis. *Archaeometry* **11** 21–52.

Platzner, I.T. (1997). *Modern Isotope Ratio Mass Spectrometry*. Wiley, Chichester.

Pollard, A.M., Batt, C.M., Stern, B. and Young, S.M.M. (2007*). Analytical Chemistry in Archaeology*. Cambridge University Press, Cambridge.

Potts, P.J., Webb, P.C. and Watson, J.S. (1985). Energy-dispersive X-ray fluorescence analysis of silicate rocks: comparisons with wavelength-dispersive performance. *Analyst* **110** 507–513.

Reed, S.J.B. (1993). *Electron Microprobe Analysis*. Cambridge University Press, Cambridge, 2nd edn.

Rink, W.J., Kandel, A.W. and Conrad, N.J. (2002). The ESR geochronology and geology of the open-air Paleolithic deposits in Bollschweil, Germany. *Archaeometry* **44** 635–650.

Robins, G.V., Seeley, N.J., McNeil, D.A.C. and Symons, M.R.C. (1978). Identification of ancient heat treatment in flint artifacts by ESR spectroscopy. *Nature* **276** 703–704.

Rouchaud, J.D., Boisseau, N. and Federoff, M. (1993). Multielement analysis of aluminium by NAA and ICP/AES. *Journal of Radioanalytical and Nuclear Chemistry Letters* **175** 25–31.

Sayre, E.V. and Dodson, R.W. (1957). Neutron activation study of Mediterranean potsherds. *American Journal of Archaeology* **61** 35–41.

Schurr, M.R., Hayes, R. and Bush, L.L. (2001). The thermal history of maize kernels determined by electron spin resonance. *Archaeometry* **43** 407–419.

Schwarcz, H.P. and Grün, R. (1992). Electron spin resonance (ESR) dating of the origin of modern man. *Philosophical Transactions of the Royal Society of London B* **337** 145–148.

Sibilia, J.P. (1988). *A Guide to Materials Characterization and Chemical Analysis*. VCH, New York.

Skoog, D.A. and Leary, J.J. (1992). *Principles of Instrumental Analysis*. Saunders College Publishing, Fort Worth, 4th edn.

Slavin, W. (1992). A comparison of atomic spectroscopic analytical techniques. *Spectroscopy International* **4** 22–27.

Smith, G.D. and Clark, R.J.H. (2004). Raman microscopy in archaeological science. *Journal of Archaeological Science* **31** 1137–1160.

Thiel, B.L. (2004). Imaging and analysis in materials science by low vacuum scanning electron microscopy. *International Materials Reviews* **49** 109–122.

Thompson, M. and Walsh, J.N. (1989). *A Handbook of ICP Spectrometry*. Blackie, Glasgow, 2nd edn.

Thomsen, V. and Schatzlein, D. (2002). Advances in field-portable XRF. *Spectroscopy* **17-7** 14–21.

Tipler, A. (1993). Gas chromatography instrumentation, operation, and experimental considerations. In *Gas Chromatography: A Practical Approach*, ed. Baugh, P.J., Oxford University Press, Oxford, pp. 15–70.

van Loon, J.C. (1980). *Analytical Atomic Absorption Spectroscopy: Selected Methods*. Academic Press, New York.

van Loon, J.C. (1985). *Selected Methods of Trace Metal Analysis: Biological and Environmental Samples*. John Wiley, Chichester.

Vandenabeele, P., Bode, S., Alonso, A. and Moens, L. (2005). Raman spectroscopic analysis of the Maya wall paintings in Ek'Balam, Mexico. *Spectrochimica Acta Part A* **61** 2349–2356.

Vandenabeele, P., Edwards, H.G.M. and Moens, L. (2007). A decade of Raman spectroscopy in art and archaeology. *Chemical Reviews* **107** 675–686.

Verità, M., Basso, R., Wypyski, M.T. and Koestler, R.J. (1994). X-ray microanalysis of ancient glassy materials: a comparative study of wavelength dispersive and energy dispersive techniques. *Archaeometry* **36** 241–251.

Walder, A.J. and Freedman, P.A. (1992). Isotopic ratio measurement using a double focusing magnetic sector mass analyser with an inductively coupled plasma as an ion source. *Journal of Analytical Atomic Spectrometry* **7** 571–575.

Walder, A.J. and Furuta, N. (1993). High-precision lead isotope ratio measurements by inductively coupled plasma multiple collector mass spectrometry. *Analytical Sciences* **9** 675–680.

Ward, N.I., Abou-Shakra, F.R. and Durrant, S.F. (1990). Trace element content of biological materials – a comparison of NAA and ICP-MS analysis. *Biological Trace Element Research* **26-7** 177–187.

Weiner, S., Goldberg, P. and Bar-Yosef, O. (1993). Bone preservation in Kebara Cave, Israel, using on-site Fourier transform infrared spectrometry. *Journal of Archaeological Science* **20** 613–627.

Willard, H.H., Merritt, L.L. Jr, Dean, J.A. and Settle, F.A. Jr (1988). *Instrumental Methods of Analysis*. Wadsworth, Belmont, California, 7th edn.

Woldseth, R. (1973). *X-Ray Energy Spectrometry*. Kevex Corporation, Burlingame, California.

CHAPTER 3

Obsidian Characterization in the Eastern Mediterranean

3.1 INTRODUCTION

The most well-known explosive volcanic eruption in the Aegean took place on the island of Santorini (Thera) sometime between 1627 and 1600 BC, according to Friedrich *et al.* (2006). It is also one of the youngest. Volcanic activity of much greater antiquity in other areas of the Aegean and beyond gave rise to a valuable artefactual material known as *obsidian*. Obsidian is a volcanic glass, formed when lava is cooled rapidly, often at the margins of a flow. It is normally shiny in appearance, and dark in colour (black or grey), but may be colourless, red, green or brown, depending on the composition and circumstances of formation. The characteristic conchoidal fracture of obsidian ensured its early use in the manufacture of stone tools, particularly for sharp flakes and blades. Obsidian was also valued as a decorative item – small vessels, statuettes and mirrors could be formed from nodules of the glassy rock. Obsidian has been used in the Aegean for more than 10 000 years. When seashells modelled from a form of obsidian with distinctive white inclusions, or *spherulites*, were found at the Minoan palace of Knossos on the island of Crete, Arthur Evans, excavator of the site, considered the obsidian to have come from the western Mediterranean island of Lipari where natural deposits of similar spotted obsidian are known to occur (Evans, 1921; 87; 412). He referred to these examples as 'Liparite'. Scientific study carried out several decades later demonstrated that the obsidian came not from Lipari but from a source in the Aegean. This chapter examines how suggestions regarding source attribution of archaeological materials might be confirmed on the basis of chemical composition. Furthermore, models developed to explain the patterns underlying the distribution of obsidian away from the natural source are reviewed briefly.

Chapter 1 identified source attribution of archaeological materials as one of the most important areas of scientific analysis. This chapter focuses on one of

Archaeological Chemistry, Second Edition
By A. Mark Pollard and Carl Heron
© The Royal Society of Chemistry 2008

the most successful applications of archaeological chemistry: the provenance of obsidian. With hindsight, it is possible to conclude that it was successful for two reasons – the relative ease with which compositional data can be used to characterize obsidian, and the fact that the vast majority of studies have been embedded within the clear archaeological objective of establishing exchange mechanisms. The ability to source this durable volcanic glass has provided archaeologists with a material record of cultural contact over wide areas, and, indirectly, with the earliest evidence for seafaring (and therefore the existence of boats), since the exploitation of island sources would have necessitated marine travel. Williams-Thorpe (1995) has summarized obsidian research undertaken in the Mediterranean and the Near East. More widely, Shackley's (1998) edited volume reviewed the analysis of obsidian through a number of case studies from around the world.

Obsidian became widely used in parts of the Aegean in the Neolithic (beginning around 7000 BC) and Early Bronze Age (beginning around 2500 BC). By the 6th Millennium BC, obsidian is found at sites from Crete to Macedonia. Every Early Neolithic site so far excavated in southern Greece has surrendered at least some obsidian. Systematic and intensive field surveys of the modern land surface in these regions, a powerful tool for assessing the development of settlement patterns and land use, have recovered obsidian (in addition to large quantities of pottery) scattered across the landscape. These finds are indicators of prehistoric activity which enable the location of settlements and other activity areas from the surface record alone. The earliest use of obsidian, identified chemically as coming from a source on the Cycladic island of Melos, has been traced back to the Upper Palaeolithic levels at Franchthi Cave in the Argolid, Greece, around 160 km from Melos by sea (Perlès, 1987), prompting the discussions noted above concerning the earliest seafaring and deep sea fishing. In the Upper Mesolithic, significant quantities of large bluefin tunny (*Thunnus thynnus*) bones appear in the faunal assemblage. According to Perlès (1990a; 46–47), Melian obsidian becomes abundant soon after. The link between obsidian procurement and deep-sea fishing has been disputed by some (*e.g.*, Bloedow, 1987; Pickard and Bonsall, 2004). It is generally assumed that the use of obsidian declined (around 2300 BC) with the increasing availability of metals, although there is evidence for continued use of obsidian throughout the Bronze Age (*e.g.*, Carter and Kilikoglu, 2007). As a footnote to the history of human use of obsidian, some surgeons have used obsidian blades instead of steel for operations – it is said that obsidian gives a cleaner wound which heals faster (Disa *et al.*, 1993). At his own request, the archaeologist and lithic specialist François Bordes underwent major surgery using obsidian blades.

3.2 ORIGIN AND FORMATION OF OBSIDIAN

Igneous rocks (*i.e.*, rocks of magmatic origin) are classified chemically according to their percentage silica (SiO_2). Those with more than 66% are generally termed *acidic*; between 66% and 52% *intermediate*, *basic* between 52% and 45% and *ultrabasic* less than 45% (Read, 1970; 204). They may additionally be

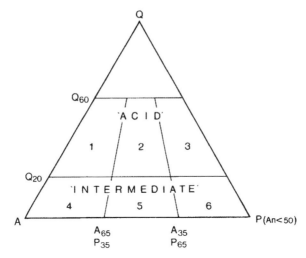

Figure 3.1 A QAP triangular diagram for the classification of quartz-bearing rocks, showing the fields of composition of the six major families of coarse-grained rocks and their fine-grained equivalents. (Redrawn from Hatch *et al.*, 1972; Figure 76.)

classified according to their grain size, loosely falling into two categories – the *coarse grained* (which includes granites) and *fine grained*. Since most igneous rocks are rich in quartz (crystalline SiO_2), they are also traditionally classified according to their relative proportions of three families of minerals, quartz (Q), alkali-feldspar (A) and plagioclase (P). These can be plotted on a so-called *QAP triangular diagram* (Figure 3.1), which shows how the relative abundances of these three minerals relate approximately to the chemical classifications quoted above. These QAP diagrams are divided into six sub-regions within the acidic and intermediate categories. In coarse-grained rocks, these include *granite* (1), *adamellite* (2), *granodiorite* (3), *syenite* (4), *diorite* (5) and *gabbro* (6) according to their chemical composition. Similar categories for fine-grained rocks include *rhyolite* (1), *rhyodacite* (2), *dacite* (3), *trachyte* (4), *andesite* (5) and *basalt* (6). Thus, *rhyolite* is the fine-grained equivalent of *granite*, and so on.

Obsidian is an acidic volcanic glass formed when silica-rich (65–75% SiO_2) magma cools quickly so that little or no crystallization occurs. The abundance of silica confers a high viscosity on the lava. This combination of high viscosity and rapid chilling, either at the thin margins of a flow or as a result of extrusion under water, causes the molten siliceous material to vitrify into a glass rather than crystallize into a rock. A detailed description of the process of vitrification, and the structure of silicate glasses, is given in Chapter 5. If the magma which solidifies in this way were to crystallize completely, it would almost always be classified on the above definitions as a *rhyolite*, and hence obsidians are often described as being *rhyolitic*. Figure 3.2 shows a schematic cross-section through a typical obsidian zone of artefact-quality obsidian in a rhyolite flow-dome structure. The outer shell of the structure is highly porous (*pumiceous*) glass

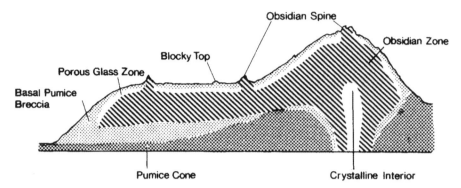

Figure 3.2   A zone of obsidian in a rhyolite flow-dome structure. (Redrawn with permission from Hughes and Smith, 1993; Figure 2A.)

except where spines of dense obsidian protrude through the porous top. The interior of the structure is crystalline. In some rare cases, artefact-quality obsidian may not form. In reality a combination of artefact-quality and lesser quality obsidian will normally be present in the same structure [see Hughes and Smith (1993; 81–82) for a more detailed discussion of the formation and geology of obsidian]. Obsidian exhibits an excellent conchoidal fracture and a vitreous lustre, and hence its value as a raw material for tools. It has a hardness of 6 on the Moh's scale, but is brittle, and has a specific gravity of around 2.4 and a refractive index of about 1.5 (Williams, 1978; 93). True obsidians contain little or no original crystalline inclusions, but variations in the magma chemistry and cooling conditions can give rise to a range of vitreous materials, some of which contain appreciable levels of crystalline material. Vitreous material which contains more crystallites is often called *pitchstone*, which can usually be recognized in hand specimens from its dull resinous lustre. Chemically, pitchstone is also characterized by having a much higher water content than obsidian – levels of up to 10% by weight are not uncommon, whereas obsidian usually contains less than 4%, and often below 1%. During the extrusion of the magma a fraction may become highly aerated as a result of the rapid expansion (*exsolution*) of dissolved gasses. The resulting solidified material is called *pumice*. This is a highly vesicular version of obsidian, which is geologically unusual in that, because of the porous structure, it is less dense than water. Thus, it may float for long periods of time, to be washed up on a beach far distant from the source of the magma (Whitham and Sparks, 1986).

Obsidians formed during the Quaternary (within the last two million years) possess excellent conchoidal fracture, as do certain obsidians which formed during the Tertiary (up to 65 million years old). However, many pre-Tertiary obsidians have lost their ability to fracture conchoidally and have undergone spontaneous recrystallization, forming a rhyolitic acid rock (Cann and Renfrew, 1964). This is because obsidian formed at high-quench temperatures is such that it is not stable over geological periods. Generally, vitreous silicates are thermodynamically less stable than their crystalline counterparts, and most glasses will crystallize spontaneously if sufficient energy is available to the atoms within the structure to enable them to reorganize themselves into a crystalline form. In areas of volcanic activity, such energy is often available in the form of heat as a result of further lava flows, or from hot water percolating through the surface rocks. Therefore, on a geological scale obsidians have a short lifetime, and few workable deposits are older than 10 million years (Cann, 1983). This restricts the number of potential sources for obsidian in antiquity, and can give a good indication as to the geographical location of sources on the basis of their geological age.

Another type of alteration which obsidian can undergo forms the basis of a dating technique for tools, called *obsidian hydration dating*. Once a fresh surface of obsidian is formed, such as by knapping to make a tool, the surface starts to take up water from its surroundings to form what is known as a *hydration layer* on the surface. As noted above, obsidians are characterized by a low original water content, largely because the high temperatures of formation drive off most of the water at the time of solidification. Typical values for water content are 0.1–0.3% by weight, depending in detail on the water pressure in the parent magma and the temperature of extrusion. This water is thermally stable within the obsidian up to temperatures in the region of 800–1000 °C, again because of the high initial temperatures (Ericson *et al.*, 1976). As the magma flow cools (and after cooling) the surface of the obsidian reacts with water vapour in the atmosphere to form a hydration product termed *perlite*, which has a characteristic appearance under the microscope as a result of strain birefringence. The reaction is essentially one of ion exchange, whereby ions of the alkali metals are removed and replaced by hydroxyl ions, in exactly the same manner as described in Chapter 5 for the corrosion of glass. The birefringence seen optically is a result of changing the chemistry of the layer without a change in volume to accommodate the hydroxyl ion. The result is a clear hydrated layer which can be seen under a polarizing microscope. If a fresh surface is created as a result of tool manufacture, then any geologically accumulated hydration layer is removed, and the rate of build up of the new layer as a result of atmospheric exposure can be used to estimate the archaeological age of the tool. For archaeological dating purposes, this hydration 'rim' is typically of the order of 1 to 50 μm thick. It is assumed that hydration approximately follows *Fick's Law* of diffusion:

$$x = Dt^{1/2}$$

where x is the thickness of the layer, D is the hydration rate and t is the time since exposure of the fresh surface. Complications arise due to the influence of the variations in chemistry between different obsidians and the original water content, as well as expected variations due to moisture content in the environment and temperature. The result is that calibration calculations have to be done to accommodate different obsidians in different burial environments. Despite these problems, obsidian hydration dating is still claimed to be capable of producing accurate dates for the manufacture of these artefacts, in the range of 200 to 100 000 years ago (Aitken, 1990; 217). A good recent example is the use of secondary ion mass spectrometry (SIMS) to give high-precision depth profiling of the obsidian hydration layer as one of the chronological indicators for the Hopewell Culture earthworks (*ca.* 200 BC to AD 500) in central Ohio, USA (Stevenson *et al.*, 2004).

Chemically, obsidian is composed principally of silica (SiO_2, typically around 74%), together with a number of other major elements, such as aluminium (Al_2O_3, average 13%), sodium (4% as oxide), potassium (4% as oxide) and calcium (1.5% as oxide). Iron oxides vary from around 0.5% to several percent, and the ratio of Fe^{2+}/Fe^{3+} can vary considerably (Longworth and Warren, 1979). Obsidians may be classified on the basis of their major element composition into two groups: *peralkaline*, with higher combined levels of sodium and potassium than aluminium, and *subalkaline*, in which the reverse is true. They may be further divided into the following sub-groups – *calcic*, with high levels of calcium, and typically low alkalis; *calc-alkaline*, with high levels of both calcium and alkalis, and *alkaline*, with high alkali but low calcium (Williams-Thorpe, 1995; 219). A wide range of trace elements is also present, either dispersed in the glassy matrix or associated with the crystalline phases present, which are normally either quartz or feldspar. Geochemical associations between major and trace elements mean that, for example, calc-alkaline obsidians are also likely to be richer in other alkaline earth trace elements, such as barium and strontium.

The partitioning of trace elements between crystalline and glassy phases gives rise to inhomogeneity on a fine scale, but since solidification from the parent magma is, by definition, rapid, obsidian is normally said to be chemically relatively homogeneous on a large scale within a single flow. Different flows in the same geographical area may well be different, reflecting changes in composition within the magma chamber with time. Extensive geochemical prospecting of obsidian flows is, however, needed to substantiate this hypothesis, since it is crucial to the chemical characterization of obsidian sources, but it has not always been done adequately in the past. For example, one of the major early publications relating to the sources of artefact grade obsidians in California characterized most of the major obsidian sources with less than five analyses, on the assumption of large scale homogeneity (Jack, 1976). Hughes (1994) has investigated the geochemical variability between quarry areas within an outcrop of obsidian in California previously believed to represent one chemically defined 'source'. By analysing 200 samples from 20 distinct geographical locations within the Casa Diablo region, he demonstrated that two or possibly three chemically distinct sources of obsidian exist within this area. It has been appreciated for some time that certain flows, such as the Borax Lake flow in California's North Coast Ranges, exhibit systematic variations in composition, as evidenced by a coherent linear variation between six trace elements and iron content (Bowman *et al.*, 1973), perhaps reflecting the mixing of two different magmas prior to eruption. The implications of the work by Hughes (1994) and others in California are that more detailed geochemical mapping of obsidian sources than has previously been the norm may be necessary before homogeneity can be accepted, even for single flows within the same volcanic field or dome complex. More recently, however, Eerkens and Rosenthal (2004) have questioned the archaeological value of pinning down obsidian sources to the level of sub-flow, arguing that this information has little anthropological value. This echoes a similar debate in the use of lead isotope data to identify the exact mine from which a particular metal ore was extracted (see Chapter 9) – is this information archaeologically useful, or even useable?

3.3 SOURCES OF OBSIDIAN IN THE EASTERN MEDITERRANEAN AND NEIGHBOURING REGIONS

For the reasons outlined above, outcrops of workable obsidian are relatively few in number and are restricted to areas of geologically recent lava flows. Most sources are therefore reasonably well known, and, because of these constraints, identification of new sources in the eastern Mediterranean region becomes ever more unlikely. This makes the exercise of characterizing archaeological obsidians an attractive proposition, since, unlike potential clay sources for pottery provenance, the existence of completely unknown sources can be (cautiously) ignored. This is, of course, subject to the requirement noted above for more detailed geochemical characterization of existing sources.

Systematic exploration of the natural sources of obsidian in the eastern Mediterranean began towards the end of the 19th Century. In 1897, Duncan Mackenzie published the results of a survey of the sources of obsidian on the Cycladic island of Melos which lies around 160 km north of Crete (see Figure 3.3). This was followed shortly after in 1904 by Bosanquet's study of the prehistoric supply of obsidian from Melos to the Aegean. Two sources were identified on the island at Sta Nychia or Adhamas and Demenghaki (Shelford *et al.*, 1982). Other sources in the eastern Mediterranean include the tiny island of Giali (Yali) in the Dodecanese some 240 km to the east of Melos (see Figure 3.3). Here, the obsidian contains white spherulites and was used to make the small stone vases and other items found on Minoan Crete, which the excavator (Evans) originally ascribed to a source on the island of Lipari on the basis of visual appearance. There have been suggestions that 'pure' obsidian may also have been available from this source (Torrence and Cherry 1976, quoted in Shelford *et al.*, 1982; 190). Confirmation of this has recently been reported (Bassiakos *et al.*, 2005). Small nodules of obsidian on the island of Antiparos, 60 km northeast of Melos are thought to be of negligible significance, although two unworked pieces were found nearby at the Neolithic site of Saliagos near Antiparos (Cann *et al.*, 1968; 106).

Any provenance investigation of obsidians used in the eastern Mediterranean obviously must also take account of sources of workable obsidian outside the region, although a judgement has to be made about how far this consideration must extend. A number of sources are known in the western Mediterranean, namely the islands of Sardinia, Lipari, Pantelleria and the Pontine. These sources have been investigated in detail (*e.g.*, Hallam *et al.*, 1976; Tykot, 1998), enabling valuable discussions to take place regarding the nature of Neolithic

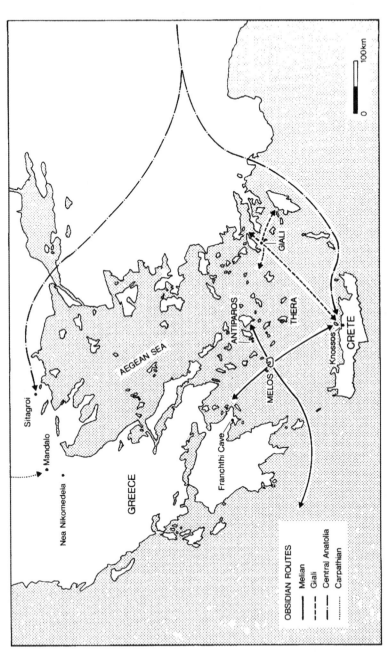

Figure 3.3   Map of the Aegean showing the location of obsidian sources (Melos, Antiparos, and Giali) and sites referred to in the text. Melian obsidian is found in abundance on many Aegean islands and on the southern Greek mainland. It is also found in northern Greece, western Anatolia, and, apparently, mainland Italy in much smaller quantities. Other arrows represent the most southerly location of Carpathian obsidian (Mandalo) and the presence of central Anatolian obsidian at Sitagroi and Knossos (see text for details).

exchange (Ammerman *et al.*, 1990; Tykot *et al.*, 2005). The rich sources in the Carpathian mountains in southeast Slovakia (Zemplén mountains) and the Tokaj mountains of northeast Hungary some 1200 km to the north-north-west of Melos have been studied in detail (Williams, 1978). Some 900 km to the east of Melos, major sources are known in central Anatolia (Turkey) and in Eastern Anatolia. Further afield, sources in Armenia, the Caucasus, Ethiopia and Kenya are also known, but are generally discounted from any discussion of Neolithic/Bronze Age supply to the eastern Mediterranean.

3.4 REVIEW OF ANALYTICAL WORK

Early scientific investigations of obsidian examined the utility of a number of properties based on appearance. Variations in colour (in transmitted and reflected light), fracture and translucency did not provide a reliable method of characterization, nor did variations in specific gravity, density and refractive index. However, some of these properties have proved useful when used in conjunction with the results of chemical analysis. The largely glassy nature of the material means that thin (petrographic) sections of obsidian are too isotropic to be used as a means of characterization. Analysts subsequently turned to methods of chemical characterization. Only a restricted range of magma compositions can satisfy the conditions for vitrification, so analyses of the major elements (*e.g.*, Si, Al, Na, K and Ca) are only helpful in characterizing whether obsidian belongs to peralkaline, alkaline and calc-alkaline types, so trace element analysis was pursued as a geographical characterization technique. Trace elements are those which occur at the level of parts per million (ppm, or 1 in 10^6) by weight in the sample, and this restricts the choice of analytical techniques available.

The results of a preliminary investigation of obsidian from the western Mediterranean using wet chemical determinations of elements present in the glass were published by Cornaggia Castiglioni and co-workers in 1963. This was followed shortly after by the first application of optical emission spectro-scopy (OES: Cann and Renfrew, 1964; Renfrew *et al.*, 1966; 1968 – see Chapter 2 for details of the techniques described in this section). Sample preparation involved crushing small samples of obsidian (*ca.* 60 mg). The technique proved able to determine the amounts of trace elements present in obsidian in proportions between 1 ppm and around 1%. The number of samples of obsidian studied was initially rather low, particularly in authentic samples collected at the obsidian source. Quantitative values for 16 elements were determined, although Ba, Zr, Nb and Y appeared best to differentiate many of the Mediterranean sources. Simple bivariate displays of element concentrations plotted on a logarithmic scale (for example, Ba versus Zr) underlined the potential value of trace element analysis for separating the relevant sources (Figure 3.4). Other pairs of elements could be used to improve definition. However, overlap between certain sources was evident: for example, the so-called Group I included obsidians from Giali, Acigöl in central Anatolia, both sources on Melos and two sources in the Carpathian mountains. It was

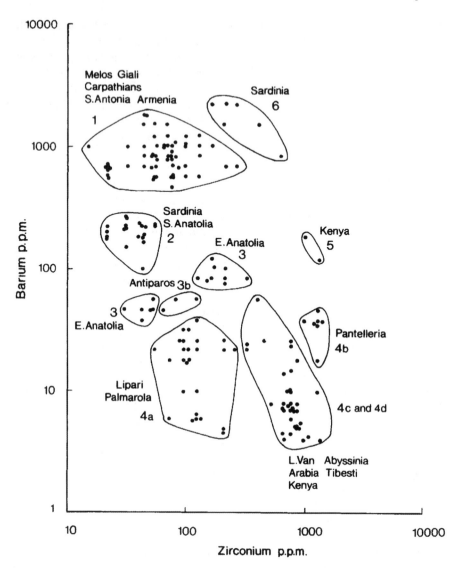

Figure 3.4 The content of Ba and Zr in a number of Old World obsidians. Each dot represents a single specimen and divisions into discrete groups are indicated. (Redrawn with permission from Cann *et al.*, 1969; Figure 106.)

evident that the technique of OES did not have sufficient precision or the ability to analyse a large enough range of elements to discriminate between these sources.

It was not until the application of neutron activation analysis (NAA) that the problem of overlapping sources could be tackled. NAA is a highly sensitive and essentially non-destructive technique, although samples have to be taken which remain radioactive for some time after analysis. The use of NAA in characterizing obsidian was first demonstrated in the early 1970s (Aspinall *et al.*, 1972). Sample preparation for NAA is straightforward: samples of obsidian weighing from around 300 to 1000 mg could be analysed intact after washing the tool surface to remove surface impurities. By suitable choice of analytical strategy, a wide range of elements can be detected, in some cases at the sub-ppm level. However, the determination of some elements is problematic. For example, although the aluminium nucleus captures a neutron, the unstable isotope formed has a very short half-life of 2.3 minutes and can easily be missed if counting takes place some days after irradiation. It is also not possible to determine light elements such as silicon and oxygen, since they present a very low cross-section for neutron capture. In the case of obsidian, the initial gamma ray spectrum is dominated by ^{24}Na (with a half-life of 15 hours), so first examination normally takes place after several days have elapsed. In NAA, direct comparison is made between the unknown sample and a set of standards of known composition which is irradiated along with the samples under consideration.

The original technique employed allowed quantitative measurement of 16 elements with a much greater precision than that of OES. In contrast to the OES data, plots of simple elemental ratios allowed a clear distinction to be made between sources. A procedure of internal normalization of a number of elements (*e.g.*, Cs, Ta, Rb, Th, Tb, Ce and Fe) to Sc (an element determined in obsidian with high precision) was devised to mitigate the effects of precision and accuracy loss arising from neutron-flux variations during sample irradiation and variations in geometry during counting due to the widely varying shape and mass of the samples. The normalization procedure led to an improvement in intra-source homogeneity, although at the cost of a loss of transparency in the initial data. Although the number of samples investigated in the preliminary study was small, these NAA data served to separate all the obsidians in Group I as defined by the OES data, including the two sources on Melos only ten km apart.

Although historically NAA has been the method of choice for this work, more recently, because of the increasing difficulties associated with obtaining irradiation facilities for NAA, an increasing number of studies have been undertaken using X-ray fluorescence (XRF; *e.g.*, Giaque *et al.*, 1993). In such a situation, the question of intercomparability between analytical techniques is bound to become significant – otherwise, all the data obtained by one method becomes inaccessible to studies using an alternative approach. Shelford *et al.* (1982) report a study of compositional variation within the two sources on Melos using both XRF and NAA. Although the analyses were carried out on different specimens, values could be compared between six elements common to both analyses, and were found to be in good agreement. In this case, the combination of the XRF and NAA data underlined the homogeneity of the flows on Melos. For most chemical elements each source is homogeneous to better than 7% and there are significant differences between the sources for major, minor and trace element compositions.

NAA and XRF remain the most widely used techniques for obsidian analysis. However, a wide array of other analytical, geochemical, dating and magnetic approaches have been utilized during the past 30 years. These include fission track dating, thermoluminescence dating, potassium–argon dating, electron spin resonance, strontium isotope ratios, inductively coupled plasma mass spectrometry, back-scattered electron imaging, electron microprobe analysis, Mössbauer spectroscopy, magnetic properties and so on (see Williams-Thorpe, 1995; Torrence, 1986; 95; Table 4). These techniques have been applied variously in order to verify conclusions based on earlier analyses, to explore rapid, non-destructive or minimally destructive approaches and, perhaps more cynically, simply to parade new techniques to an analytical audience. The past decade has seen a greater use of inductively coupled plasma techniques. Kilikoglou *et al.* (1997) compared NAA with inductively coupled plasma and atomic emission (ICP-AES) in a study of Aegean and Carpathian sources. Both approaches worked successfully, although NAA was considered to show greater discrimination of neighbouring sources.

One of the most interesting complementary techniques has been the use of *fission track dating* to obtain the geological age of the obsidian flows, originally proposed by Durrani *et al.* (1971). Fission track dating is conceptually simple, in that a freshly polished surface is etched with acid to reveal the structural damage caused by the fission of a uranium nucleus. The 'tracks' are the result of the energetic recoil of the two halves of the uranium nucleus. If the number of tracks can be counted, and the uranium content of the obsidian measured, it is possible, in principle, from a knowledge of the half life for fission of the uranium nucleus to calculate the time elapsed since the vitrification of the obsidian (Aitken, 1990; 132). The date obtained is generally the geological age of the flow, unless re-heating of the sample has taken place to a temperature sufficient to anneal out the tracks accumulated over geological time, in which case the age may be the time since the last heating event. The obsidian sources in central Anatolia, Giali and Melos appeared to have very similar volcanic histories (two general periods of tectonic activity) on the basis of the fission track dating of the eruptions in these areas (see also Dixon, 1976; 299–300). Keller *et al.* (1990; 24) report much younger dates by fission track methods. The earliest dates for Melos are 3.15–2.7 million years. By contrast, Giali (at 24000 BP) is amongst the youngest events in eastern Aegean volcanism. In the western Mediterranean, although obsidian from Sardinia, Lipari and Pantelleria can be distinguished by fission track dating, the combination of this method and potassium–argon dating appears to give similar dates to the various flows on Sardinia [see the review by Tykot (1998)]. Fission track dating has been used in a combined investigation with NAA to examine obsidian from Anatolian sites near Istanbul in an effort to evaluate the possibility of Aegean or Carpathian obsidian in the Bosphorus region (Bigazzi *et al.*, 1993).

Using another approach, Gale (1981) investigated the use of strontium isotope ratios ($^{87}Sr/^{86}Sr$) to characterize the different sources (see Chapter 10 for a discussion of strontium isotope systematics). This ratio depends on the strontium isotope ratio in the initial magma melt and the time elapsed since formation, since the decay of the radioactively unstable isotope ^{87}Rb into ^{87}Sr contributes to the abundance of ^{87}Sr. Strontium-86 is stable so elevated ratios of $^{87}Sr/^{86}Sr$ should correlate with older rocks. Whilst the measurement of this parameter proved insufficient by itself to discriminate between all sources, plotting the strontium isotopic composition against the rubidium concentration (as determined by XRF) enabled separation of the major groups, as did plotting the actual concentrations of Sr versus Rb.

The usefulness of the magnetic properties of obsidian for characterization purposes has been considered in a study by McDougall *et al.* (1983). On cooling, the magma can be regarded as acquiring a permanent record of the strength and direction of the Earth's magnetic field appropriate to the place and time. If subsequent alteration to these parameters can be ruled out, then this too should provide a 'fingerprint' of the parent obsidian flow. However, although the original magnetic direction can be measured in the parent flow, this cannot subsequently be determined from pieces of obsidian removed from the flow, since the original orientation of the piece will be unknown. Nevertheless, measurements of the intensity of magnetization, saturation magnetization and low field susceptibility (properties which are independent of the original orientation) have proved to be a rapid and useful means for discriminating between many Mediterranean, central European and Near Eastern sources. Somewhat surprisingly, this promising line of research does not seem to have been followed up.

3.5   ARCHAEOLOGICAL IMPLICATIONS

The principal aim of any archaeological provenance study is an assessment of the economic and social factors which underlie the movement of materials. It was pointed out in 1991 by Cherry and Knapp that few provenance programmes had actually gone as far as contributing directly to quantitative models of exchange. The early success of NAA and XRF in providing an elemental 'fingerprint' for each known source in Europe and the Near East has enabled the definition of a series of obsidian '*interaction zones*' in the western Mediterranean, Aegean, central Europe and the Near East. An interaction zone has been defined as '*the area within which sites, within the time-range considered, derived thirty per cent or more of their obsidian from the same specific source*' (Renfrew and Dixon, 1976; 147). Since the figure of 30% was arbitrary, Torrence (1986; 13) suggests that all sites comprising obsidian should be included in any study of exchange mechanisms.

Both sources on Melos have proved to be the origin for the great majority of the obsidian used in the Aegean during the Neolithic and Early Bronze Age (*i.e.*, in the above terminology, much of the Aegean falls into the single interaction zone of obsidian from Melos). Melian obsidian served much of the Greek mainland as far north as the sites at Nea Nikomedeia and Servia in Macedonia and Sitagroi in Thrace, although very few pieces are found in this region. At Sitagroi, Melian obsidian occurs with Anatolian obsidian from Çiftlik (one piece). The latter piece was not well stratified but probably dates from the Early Bronze Age. Renfrew *et al.* (1966) found that three pieces from later Neolithic levels at Knossos on Crete are also from Çiftlik. This suggests relatively long-range contacts at a very early date. The characteristic obsidian from Giali in the Dodecanese is found in Crete, Kos, Rhodes, Kalymnos and Saliagos (Cherry, 1985; 15).

The obsidian in the Upper Palaeolithic and Mesolithic levels at Franchthi Cave in the Argolid was also found to be Melian (Renfrew and Aspinall, 1990). Obsidian exploitation at this time pre-dates settlement evidence on Melos, which may not have been inhabited until the Late Neolithic. Melian obsidian is also found in Western Anatolia, albeit in low abundance. For example, Renfrew *et al.* (1965; 238), characterized two pieces of Melian obsidian at the Late Neolithic/Early Chalcolithic site of Morali. In the study by Gale (1981), one find from Haçilar in Turkey, previously examined by Renfrew *et al.* (1966) and assigned to the Acigöl source in Central Anatolia on the basis of trace element composition was considered to originate from the Demenghaki source on Melos using the isotopic and chemical data. However, this reinterpretation has subsequently been regarded as unlikely, on the grounds that insufficient samples were used to define the characteristics of each source on the basis of the $^{87}Sr/^{86}Sr$ ratio and Sr/Rb concentrations (Renfrew and Aspinall, 1990; 270). Adequate characterization of sources is discussed in detail by Hughes and Smith (1993). In their study of obsidian in the Bosphorus region, Bigazzi *et al.* (1993) concluded that most of the obsidian derived from central and northern Anatolia, as well as some minor local occurrences and/or eastern Anatolian sources. Recent evidence has come to light of Carpathian obsidian at the Neolithic/Bronze Age site at Mandalo in Macedonia, Greece, where the unusually large obsidian assemblage for this area also includes obsidian from Melos (Kilikoglou *et al.*, 1996). NAA was carried out on 12 samples; ten from the Late Neolithic Phase II (5th Millennium BC) and two from the Bronze Age Phase III (4th- to 3rd Millennium BC). One of the latter samples corresponded with the Demenghaki source on Melos, whilst the other 11 fell within the expected range for the Carpathian I source in southeastern Slovakia. This finding extends the known distribution of Carpathian obsidian for another 400 km to the south into modern Greece where it also occurs with obsidian originating from Melos.

Obsidian from the Carpathian I source (a grey transparent obsidian from Szöllöske and Malá Torona in Slovakia) was widely used and distributed. A single piece of central European obsidian from this source has been found at the Neolithic cave site of Grotta Tartaruga in northeast Italy, where obsidian from the island of Lipari predominates (Williams-Thorpe *et al.*, 1984). Lipari lies off the north coast of Sicily and supplied large quantities of obsidian, mainly to the southern Italy peninsula. Randle *et al.* (1993) identified a single piece of Carpathian obsidian at the Early Neolithic site at Sammardenchia in northeast Italy. Thus the western Mediterranean source shows at least some overlap with the Central European (Carpathian I) source, which in turn overlaps with Melian obsidian at Mandalo in Macedonia.

Suggestions made in the early 20th Century that obsidian found on the island of Malta (with no natural sources) might have been brought there by Minoan traders were supported by the findings of Cornaggia Castiglioni *et al.* (1963).

However, the analytical programme of trace element analysis carried out by Renfrew, Cann and Dixon suggested strongly that this obsidian derived from Lipari and Pantelleria, a tiny island 150 miles northwest of Malta. More recently, the case for Melian obsidian in mainland Italy has been made on the basis of fission track dates on obsidian from Grotta del Leone and Capraia Island (both in Tuscany) and Ponte Peschio in Abruzzi (Bigazzi and Radi, 1981; Arias *et al.*, 1986). Obsidian from Grotta del Leone was subsequently analysed by NAA (Bigazzi *et al.*, 1986; 1992). The data appear to reinforce the suggestion that Melos is the source of the obsidian. Unfortunately the obsidian is from archaeologically undatable contexts.

The underlying aim of obsidian provenance has always been to understand the mechanisms which account for its distribution over large areas. A quantitative approach to obsidian distribution in the Neolithic of the Near East was proposed by Renfrew and co-workers (Renfrew *et al.*, 1968). This study proceeded by noting the 'fall-off' with distance of obsidian as a proportion of the obsidian from that source in the total worked stone assemblage. Up to around 300 km from each source, the proportion of obsidian in the worked stone assemblage remained close to 100% (*i.e.*, the slope of the line is nearly flat). At greater distances, the proportion of obsidian declines dramatically. The concept of a '*supply zone*' and a '*contact zone*' was developed to explain this pattern of obsidian distribution around the Near Eastern sources (Renfrew *et al.*, 1968). The supply zone encompassed the area around each source where groups travelled to the source directly in order to procure their own supplies of obsidian. Quantitatively the supply zone incorporated those sites where the lithic assemblage contained more than 80% obsidian. The contact zone is the region where the proportion of obsidian in the assemblage is declining. Within this zone settlements obtained obsidian through regular reciprocal exchanges with neighbours located nearer the source rather than through direct access to the source itself. The system of communities passing on some of their obsidian to adjacent villages was termed '*down-the-line exchange*'. Since each village would retain a proportion of the obsidian, the fall-off with distance within the contact zone approximates an exponential curve. Organized trade in obsidian involving profit motives and fixed rates of exchange has been rejected (see Renfrew, 1973; 180) in favour of these anthropological models of so-called '*primitive*' exchange (see Torrence, 1986; 14).

In the Aegean, the application of such models has been made difficult by the absence of quantitative data on the proportion of obsidian in stone tool assemblages and the fact that in some areas obsidian is the only stone type in the worked stone assemblage (see Torrence, 1986; 97–98). However, in the Neolithic (and earlier periods), Renfrew suggested a *direct-access model*, whereby most consumers travelled to Melos and obtained their own supplies from the quarries (Renfrew, 1972; 442–443). On the Greek mainland, down-the-line exchange could account for an expansive distribution network. However, the sea precludes the establishing of straightforward supply and contact zones with plots of the fall-off in obsidian with distance from Melos. The direct-access model may have continued in the Bronze Age (Torrence, 1986; 105).

A comprehensive review and analysis of the theories of obsidian exchange applied in the eastern Mediterranean has been carried out (Torrence, 1986). Many of Renfrew's suggestions have been supported (and augmented) by detailed study of the quarry sites on Melos. For example, Torrence suggests that direct access could have included skilled knappers visiting the island intentionally to collect suitable obsidian as well as visits made as part of some other activity (such as fishing) when unmodified nodules were collected for use back home. In contrast, Perlès (1990b) argues that in the Early to Middle Neolithic the movement of obsidian was the result of specialized trading, whereas in the Late Neolithic this system becomes 'despecialized', with nearer groups obtaining their obsidian by direct access [see Phillips (1992) for a review]. This debate requires a knowledge of the exact nature of the obsidian assemblages recovered from both occupation sites and obsidian sources, including the presence or absence of unworked nodules, preformed cores, debitage and so on. More generally, Perlès (1992) has considered the evidence for the movement of a number of rock types in Greece, including andesite, emery and honey flint, and emphasized the importance of viewing a distribution network of a specific material in the context of the production and distribution of other materials.

Moving away from the Aegean, it is important to recognize that these models might not be universally applicable. In the western Mediterranean, obsidian could be obtained from the islands of Sardinia, Lipari, Palmarola (Pontine Islands) and Pantelleria. The early work of Hallam *et al.* (1976) and others has been refined and developed by Rob Tykot through resolving multiple obsidian flows at specific sources. ICP-MS analysis of geological specimens undertaken by Tykot has clarified five separate chemical subgroups of obsidian occurring in the Monte Arci region of Sardinia (Tykot, 1997; 1998). Work by Ammerman *et al.* (1990) suggests that in northern Italy direct access to the sources is precluded by the large distances. In this region sites where obsidian is found in high quantities may have served as nodes of exchange. Certain types of obsidian (such as the translucent blades from Lipari) may have been more valued for use as prestige items, compared to the Sardinian obsidian which was predominantly used to fabricate utilitarian tools. The implication of this is that a number of different exchange mechanisms (both synchronic and diachronic) may be needed to account for obsidian originating from different sources. Further analysis of 94 pieces of obsidian from the Middle Neolithic site of Gaione in northern Italy has been undertaken (Tykot *et al.*, 2005). The earlier investigation had proposed that Lipari obsidian may have been transported as finished blades, although cores have now been found suggesting that blade tools could have been fabricated from these at Gaione. Around two-thirds of the obsidian artefacts by number are from Lipari, although Palmarola (31 artefacts) and Sardinia C (six pieces) are also represented. A new programme of analysis on Neolithic obsidian artefacts from southeastern France has shown [using proton-induced X-ray emission (PIXE)] that the majority of finds here are from the Monte Arci source on Sardinia (Poupeau *et al.*, 2000).

Research undertaken by Marie-Claire Cauvin and colleagues (*e.g.*, Cauvin *et al.*, 1998) has extended the initial conclusions of the work of Renfrew, Dixon and Cann in the Middle East, both through a more comprehensive characterization of the geological sources and by analysing larger numbers of well-dated archaeological finds. The movement of obsidian from Anatolian sources to the Levant and Iraqi Kurdistan took place before or around the time of the first evidence of sedentary communities around 12 500 BC and continued through the Pre-Pottery Neolithic A and B (Cauvin *et al.*, 1998; Renfrew, 2006; Sherratt, 2006). Obsidian came from the Cappadocian source of Göllü Dağ-east (originally termed Çiftlik) and the sources of Bingöl and Nemrut Dağ in eastern Anatolia. In Cappadocia, there is as yet no evidence of settlement until around 8000 BC so the relationships between populations in this region and the Levant in terms of obsidian exchange patterns remain to be addressed (Renfrew, 2006).

Obsidian is found at the well-known Neolithic settlement at Çatalhöyük in central Anatolia. The nearest sources are around 120 miles to the northeast in Cappadocia. Tristan Carter and colleagues (Carter *et al.*, 2006) have recently published the first set of results on 135 obsidian artefacts recovered from the site, of which 128 are from the Neolithic east mound and seven from the early Chalcolithic west mound. Good correspondence in trace element data was found using ICP-MS, ICP-AES and laser ablation ICP-MS (LA-ICP-MS). The obsidian was assigned to the Cappadocian sources Göllü Dağ-east and Nenezi Dağ, with one piece possibly from the Acigöl west source in northern Cappadocia. There is no evidence for obsidian coming from the nearest source at Hasan Dağ. Here the obsidian is both hard to access and '*possesses a poor conchoidal fracture habit*' (Carter *et al.*, 2006; 905). This paper aims to progress beyond source identification to explore more fully the relationship through time between lithic technology and source. Seven discrete obsidian 'industries' have been recognized and, although not all of the industries were sampled in the analytical programme, it is possible to chart the domination of the Göllü Dağ-east source in the earlier phases of the site's history. Nenezi Dağ obsidian then becomes more important during the Early Neolithic, and this correlates with the much greater manufacture of unipolar pressure-flaked blades. This contribution is a good example of the way in which obsidian sourcing has moved beyond the documentation of source towards a closer understanding of lithic technology and ultimately of human behaviour. '*Traditionally obsidian sourcing has been employed to gain a clearer image of cultural interaction and the socio-economic complexity of prehistoric societies. Yet more often than not, interpretations have been forwarded on the basis of a handful of samples, whose descriptions rarely extend beyond an object/lab number, largely ignoring their context, form, or technological characteristics*' (Carter *et al.*, 2006; 894–895).

3.6 SUMMARY

Since completing this chapter in 1995, a 'new wave' of obsidian studies has appeared and the revised chapter seeks to capture some of these developments.

The new programmes have built on the firm foundations of the analytical investigations since the 1960s, but have augmented them in several important ways (see also Williams-Thorpe, 1995; Shackley, 1998). Greater thought is being directed towards the social processes which influenced the physical processes of obsidian transport over long distances. In her consideration of the circulation of obsidian from the island of Lipari, Farr (2006) has explored the role of seafaring as social action, with terms such as '*journey*' and '*travel*' preferred to '*mobility*' and '*raw material circulation*'. According to Farr (2006; 96), '*It may be possible that the real value of obsidian was its symbolism of the journey, the knowledge, skill and risk which had been undertaken.*'

As long ago as 1982, it was possible to state that '*we can now be reasonably confident that the vast majority of the obsidians found on other Cycladic Islands and on the Greek mainland are Melian in origin*' (Shelford *et al.*, 1982; 120). For Cherry and Knapp (1991; 95), the attribution to source of obsidian in the Aegean represents '*an unusually simple provenance problem.*' Without doubt they are correct. For some, early successes with obsidian may have raised 'unrealistic expectations' about the ability of analytical techniques to locate the source of other materials, particularly ceramics (*ibid.*; 95). Whether further large-scale analytical programmes on obsidian remain to be undertaken in the Aegean is questionable, although the quantities of Melian obsidian reaching western Anatolia and the source and availability of obsidian reaching the fringes of the Carpathian and Melian 'spheres of influence' are still to be elucidated. Furthermore, the relationship between obsidian and the metals trade in the Bronze Age is receiving wider consideration (Carter and Kilikoglou, 2007).

In contrast to the Aegean, large-scale obsidian research programmes have continued in the western Mediterranean (Tykot, 1992; 1998). Further east, the volcanic sources of eastern Anatolia and adjacent areas have not been surveyed fully. In addition to the known sources in central Anatolia (Acigöl and Çiftlik) and around Lake Van in eastern Anatolia, new important sources may yet come to light (Keller and Seifried, 1992). In Anatolia and Armenia a large number of obsidian occurrences (at least 25) have been characterized, stretching over very substantial areas. Some of these sources comprise a number of flows where obsidian has been extruded from several centres at different times. Many of these can be distinguished geochemically, with sufficient sampling. In other areas of Anatolia, some sources may have been covered by alluvium several metres thick. Although many issues (such as the relative importance of Aegean and Anatolian sources in providing obsidian for western Turkey, and the geography of the interaction zones for the various sources) will continue to be debated, the characterization of obsidian exemplifies the successful integration of chemistry and physics within archaeological goals.

Not all rocks can be characterized as successfully as obsidian. The success of obsidian provenance is due to the limited number of workable sources and the fact that each source, while being relatively homogeneous, is sufficiently different from other sources to enable elemental 'fingerprinting'. This picture contrasts markedly with that of, for example, flint – a stone not restricted to a few sources, and which is inhomogeneous in composition. Given the relative suitability of obsidian for analysis, and the potential for reconstructing such diverse factors as the social context of trade and exchange, and the history of seafaring, it is not surprising that a number of researchers around the world continue to study this beautiful material.

REFERENCES

Aitken, M.J. (1990). *Science-based Dating in Archaeology*. Longman, London.

Ammerman, A., Cesana, A., Polglase, C. and Terrani, M. (1990). Neutron activation analysis of obsidian from two Neolithic sites in Italy. *Journal of Archaeological Science* **17** 209–220.

Arias, C., Bigazzi, G., Bonadonna, F.P., Cipolloni, M., Hadler, J.C., Lattes, C.M.G. and Radi, G. (1986). Fission track dating in archaeology: a useful application. In *Scientific Methodologies Applied to Works of Art*, ed. Parrini, P.L., Montedison Progetto Cultura, Milan, pp. 151–159.

Aspinall, A., Feather, S.W. and Renfrew, C. (1972). Neutron activation analysis of Aegean obsidians. *Nature* **237** 333–334.

Bassiakos, Y., Kilikoglou, V. and Sampson, A. (2005). Yali island: Geological and analytical evidence for a new source of workable obsidian. International association of obsidian studies bulletin **33** 18.

Bigazzi, G. and Radi, G. (1981). Datazione con le tracce di fissione per l'identificazione della provenienza dei manufatti di ossidiana. *Rivista di Scienza Preistoriche* **36** 223–250.

Bigazzi, G., Meloni, S., Oddone, M. and Radi, G. (1986). Provenance studies of obsidian artefacts: trace elements and data reduction. *Journal of Radio-analytical and Nuclear Chemistry Articles* **98** 353–363.

Bigazzi, G., Meloni, S., Oddone, M. and Radi, G. (1992). Study on the diffusion of Italian obsidian in the Neolithic settlements. In *Atti del VIII Convegno Nazionale sulla Attività di Ricerca nei Settori della Radio-chimcia e della Chimica Nucleare, delle Radiazioni e dei Radioelementi, Torino, 16–19 Giugno 1992*, CNR and Università degli Studi di Torino, pp. 243–247.

Bigazzi, G., Ercan, T., Oddone, M., Ozdogan, M. and Yegingil, Z. (1993). Application of fission track dating to archaeometry: provenance studies of prehistoric obsidian artifacts. *Nuclear Tracks and Radiation Measurements* **22** 757–762.

Bloedow, E.F. (1987). Aspects of ancient trade in the Mediterranean. *Studi Micenei ed Egeo-Anatolici* **26** 59–124.

Bowman, H.R., Asaro, F. and Perlman, I. (1973). Composition variations in obsidian sources and the archaeological implications. *Archaeometry* **15** 123–127.

Cann, J.R. (1983). Petrology of obsidian artefacts. In *The Petrology of Archaeological Artefacts*, ed. Kempe, D.R.C. and Harvey, A.P., Clarendon Press, Oxford, pp. 227–255.

Cann, J.R. and Renfrew, C. (1964). The characterisation of obsidian and its application to the Mediterranean region. *Proceedings of the Prehistoric Society* **30** 111–133.

Cann, J.R., Dixon, J.E. and Renfrew, C. (1968). Appendix IV: the sources of the Saliagos obsidian. In *Excavations at Saliagos near Antiparos*, ed. Evans, J.D. and Renfrew C., Supplementary Volume 5, British School of Archaeology at Athens, Thames and Hudson, London, pp. 105–107.

Cann, J.R., Dixon, J.E. and Renfrew, C. (1969). Obsidian analysis and the obsidian trade: In *Science and Archaeology*, ed. Brothwell, D. and Higgs, E., Thames and Hudson, London, pp. 578–591.

Carter, T. and Kilikoglou, V. (2007). From reactor to loyalty? Aegean and Anatolian obsidians from Quartier Mu, Malia (Crete). *Journal of Mediterranean Archaeology* **20** 115–143.

Carter, T., Poupeau, G., Bressy, C. and Pearce, N.J.G. (2006). A new programme of obsidian characterization at Çatalhöyük, Turkey. *Journal of Archaeological Science* **33** 893–909.

Cauvin, M.-C., Gourgaud, A., Gratuze, B., Arnaud, N., Poupeau, G., Poidevin, J.-L. and Chataigner, C. (1998). *L'obsidienne au Proche-Orient: du volcan à l'outil*. International Series 738, British Archaeological Reports, Oxford.

Cherry, J.F. (1985). Islands out of the stream. In *Prehistoric Production and Exchange*, ed. Knapp, A.B. and Stech, T., Monograph 25, Institute of Archaeology, UCLA, Los Angeles, pp. 12–29.

Cherry, J.F. and Knapp, A.B. (1991). Quantitative provenance studies and Bronze Age trade in the Mediterranean: some preliminary reflections. In *Bronze Age Trade in the Mediterranean*, ed. Gale, N.H., Studies in Mediterranean Archaeology XC, Paul Åström's Förlag, Jønsered, pp. 92–119.

Cornaggia Castiglioni, C.O., Fussi, F. and D'Agnolo, M. (1963). Indagini sulla provenienza dell'ossidiana utilizzata nelle industrie preistoriche del Mediterraneo occidentale. *Atti della Societa Italiana di Scienz Naturali e del Museo Civico di Stroia Naturale in Milano* **102** 310–322.

Disa, J.J., Vossoughi, J. and Goldberg, N.H. (1993). A comparison of obsidian and surgical steel scalpel wound-healing in rats. *Plastic and Reconstructive Surgery* **92** 884–887.

Dixon, J.E. (1976). Obsidian characterization studies in the Mediterranean and Near East. In *Advances in Obsidian Glass Studies*, ed. Taylor, R.E., Noyes Press, Park Ridge, New Jersey, pp. 288–333.

Durrani, S.A., Khan, H.A., Taj, M. and Renfrew, C. (1971). Obsidian source identification by fission track analysis. *Nature* **233** 242–245.

Eerkens, J.W. and Rosenthal, J.S. (2004). Are obsidian subsources meaningful units of analysis?: temporal and spatial patterning of subsources in the Coso Volcanic Field, southeastern California. *Journal of Archaeological Science* **31** 21–29.

Ericson, J.E., Mackenzie, J.D. and Berger, R. (1976). Physics and chemistry of the hydration process in obsidians. I. Theoretical implications. In *Advances in Obsidian Glass Studies*, ed. Taylor, R.E., Noyes Press, Park Ridge, New Jersey, pp. 25–45.

Evans, A. (1921). *The Palace of Minos, Volume 1*. Macmillan, London.

Farr, H. (2006) Seafaring as social action. *Journal of Maritime Archaeology* **1** 85–99.

Friedrich, W.L., Kromer, B., Friedrich, M., Heinemeier, J., Pfeiffer, T. and Talamo, S. (2006). Santorini eruption radiocarbon dated to 1627–1600 B.C. *Science* **312** 548.

Gale, N.H. (1981). Mediterranean obsidian source characterisation by strontium isotope analysis. *Archaeometry* **23** 41–52.

Giaque, R.D., Asaro, F., Stross, F.H. and Hester, T.R. (1993). High-precision nondestructive X-ray fluorescence method applicable to establishing the provenance of obsidian artefacts. *X-Ray Spectrometry* **22** 44–53.

Hallam, B.R., Warren, S.E. and Renfrew, C. (1976). Obsidian in the western Mediterranean. *Proceedings of the Prehistoric Society* **42** 85–110.

Hatch, F.H., Wells, A.K. and Wells, M.K. (1972). *Petrology of the Igneous Rocks*. Thomas Murby, London, 13th edn.

Hughes, R.E. (1994). Intrasource chemical variability of artefact-quality obsidians from the Casa Diablo area, California. *Journal of Archaeological Science* **21** 263–271.

Hughes, R.E. and Smith, R.L. (1993). Archaeology, geology and geochemistry in obsidian provenance studies. In *Effects of Scale on Archaeological and Geoscientific Perspectives*, ed. Stein, J.K. and Linse, A.R., Special Papers 283, Geological Society of America, Boulder, pp. 79–91.

Jack, R.N. (1976). Prehistoric obsidian in California. I. Geochemical aspects. In *Advances in Obsidian Glass Studies*, ed. Taylor, R.E., Noyes Press, Park Ridge, New Jersey, pp. 183–217.

Keller, J. and Seifried, C. (1992). The present status of obsidian identification in Anatolia and the Near East. In *Volcanologie et Archeologie*, ed. Livadie, C.A. and Widemann, F., PACT 25, Council of Europe, Strasbourg, pp. 57–87.

Keller, J., Rehren, T. and Stadlbauer, E. (1990). Explosive volcanism in the Hellenic arc. In *Thera and the Aegean World III, Volume 2, Earth Sciences*, ed. Hardy, D.A., Thera Foundation, London, pp. 13–26.

Kilikoglou, V., Bassiakos, Y., Grimanis, A.P., Souvatzis, K., Pilali-Papasteriou, A. and Papanthimou-Papaefthimiou, A. (1996). Carpathian obsidian in Macedonia, Greece. *Journal of Archaeological Science* **23** 343–349.

Kilikoglou, V., Bassiakos, Y., Doonan, R.C. and Stratis, J. (1997). NAA and ICP analysis of obsidian from Central Europe and the Aegean: source characterisation and provenance determination. *Journal of Radioanalytical and Nuclear Chemistry* **216** 87–93.

Longworth, G. and Warren, S.E. (1979). The application of Mössbauer spectroscopy to the characterisation of Western Mediterranean obsidian. *Journal of Archaeological Science* **6** 179–193.

McDougall, J.M., Tarling, D.H. and Warren, S.E. (1983). The magnetic sourcing of obsidian samples from Mediterranean and Near Eastern sources. *Journal of Archaeological Science* **10** 441–452.

Perlès, C. (ed.) (1987). *Les Industries Lithiques Taillées de Franchthi (Argolide, Grèce). Tome 1, Présentation générale et industries paléolithiques,*

Excavations at Franchthi Greece, Fasicule 3. Indiana University Press, Bloomington.

Perlès, C. (ed.) (1990a). *Les Industries Lithiques Taillées de Franchthi (Argolide, Grèce). Tome II: Les Industries de Mésolithique et du Néolithique Initial*, Fasicule 5. Indiana University Press, Bloomington.

Perlès, C. (ed.) (1990b). L'outillage de pierre taillée néolithique en Grèce: approvisionnement et exploitation des premieres materiéres. *Bulletin de Correspondance Hellenique* **114** 1–42.

Perlès, C. (1992). Systems of exchange and organization of production in Neolithic Greece. *Journal of Mediterranean Archaeology* **5** 115–164.

Phillips, P. (1992). Western Mediterranean obsidian distribution and the European Neolithic. In *Sardinia in the Mediterranean: A Footprint in the Sea*, ed. Tykot, R.H. and Andrews, T.K., Sheffield Academic Press, Sheffield, pp. 71–82.

Pickard, C. and Bonsall, C. (2004). Deep-sea fishing in the European Mesolithic: fact or fantasy? *European Journal of Archaeology* **7** 273–290.

Poupeau, G., Bellot-Gurlet, L., Brisotto, V. and Dorighel, O. (2000). Nouvelles données sur la provenance de l'obsidienne des sites néolithiques du Sud-Est de la France. *Comptes Rendus de l'Académie des Sciences de la Terre et des Planètes* **330** 291–303.

Randle, K., Barfield, L.H. and Bagolini, B. (1993). Recent Italian obsidian analyses. *Journal of Archaeological Science* **20** 503–509.

Read, H.H. (1970). *Rutley's Elements of Mineralogy*. Thomas Murby and Co., London.

Renfrew, C. (1972). *The Emergence of Civilisation: the Cyclades and the Aegean in the Third Millennium BC*. Methuen, London.

Renfrew, C. (1973). Trade and craft specialisation. In *Neolithic Greece*, ed. Theocharis, D.R., National Bank of Greece, Athens, pp. 179–200.

Renfrew, C. (2006). Inception of agriculture and rearing in the Middle East. *Comptes Rendus de l'Académie des Sciences – Series IIA– Earth and Planetary Science* **5** 395–404.

Renfrew, C. and Aspinall, A. (1990). Aegean obsidian and Franchthi Cave. In *Les Industries Lithiques Taillées de Franchthi (Argolide, Grèce). Tome II Les Industries de Mésolithique et du Néolithique Initial*, Fasicule 5, ed. Perlès, C., Indiana University Press, Bloomington, pp. 258–270.

Renfrew, C. and Dixon, J.E. (1976). Obsidian in Western Asia: a review. In *Problems in Economic and Social Archaeology*, ed. Longworth, I.H. and Sieveking, G., Duckworth, London, pp. 137–150.

Renfrew, C., Cann, J.R. and Dixon, J.E. (1965). Obsidian in the Aegean. *Annual of the British School of Archaeology at Athens* **60** 225–247.

Renfrew, C., Dixon, J.E. and Cann, J.R. (1966). Obsidian and early culture contact in the Near East. *Proceedings of the Prehistoric Society* **32** 30–72.

Renfrew, C., Dixon, J.E. and Cann, J.R. (1968). Further analysis of Near Eastern obsidians. *Proceedings of the Prehistoric Society* **34** 319–331.

Shackley, M.S. (ed.) (1998). *Archaeological Obsidian Studies: Method and Theory*. Advances in Archaeological and Musuem Science Volume 3, Plenum Press, New York.

Shelford, P., Hodson, F., Cosgrove, M.E., Warren, S.E. and Renfrew, C. (1982). The sources and characterisation of Melian obsidian. In *An Island Polity: The Archaeology of Exploitation in Melos*, ed. Renfrew, C. and Wagstaff, J.M., Cambridge University Press, Cambridge, pp. 182–192.

Sherratt, A. (2006). *The Obsidian Trade in the Near East, 14,000 to 6500 BC.* Archtlas, 2nd edn, http://www.archatlas.org/ObsidianRoutes/Obsidian-Routes.php. Accessed: 19 January 2007.

Stevenson, C.M., Abdel-Rehim, L. and Novak, S.W. (2004). High precision measurement of obsidian hydration layers on artifacts from the Hopewell site using secondary ion mass spectrometry. *American Antiquity* **69** 555–567.

Torrence, R. (1986). *Production and Exchange of Stone Tools.* Cambridge University Press, Cambridge.

Tykot, R.H. (1992). The sources and distribution of Sardinian obsidian. In *Sardinia in the Mediterranean: A Footprint in the Sea*, ed. Tykot, R.H. and Andrews, T.K., Sheffield Academic Press, Sheffield, pp. 57–70.

Tykot, R.H. (1997). Characterization of the Monte Arci (Sardinia) obsidian sources. *Journal of Archaeological Science* **24** 467–479.

Tykot, R.H. (1998). Mediterranean islands and multiple flows: the sources and exploitation of Sardinian obsidian. In *Archaeological Obsidian Studies: Method and Theory.*, ed. Shackley, M.S., Advances in Archaeology and Museum Science Series, Plenum Press, New York, pp. 67–82.

Tykot, R.H., Ammerman, A.J., Bernabò Brea, M., Glascock, M.D. and Speakman, R.J. (2005). Source analysis and the socioeconomic role of obsidian trade in Northern Italy: new data from the Middle Neolithic site of Gaione. *Geoarchaeological and Bioarchaeological Studies* **3** 103–106.

Whitham, A.G. and Sparks, R.S.J. (1986). Pumice. *Bulletin of Volcanology* **48** 209–223.

Williams, O. (1978). *A Study of Obsidian in Prehistoric Central and Eastern Europe, and Its Trace Element Characterization.* Unpublished PhD Thesis, Postgraduate School of Studies in Physics, University of Bradford.

Williams-Thorpe, O. (1995). Obsidian in the Mediterranean and the Near East: a provenancing success story. *Archaeometry* **37** 217–248.

Williams-Thorpe, O., Warren, S.E. and Nandris, J.G. (1984). The distribution and provenance of archaeological obsidian in Central and Eastern Europe. *Journal of Archaeological Science* **11** 183–212.

CHAPTER 4

The Geochemistry of Clays and the Provenance of Ceramics

4.1 INTRODUCTION

If clay is heated to a sufficient temperature, an irreversible chemical change takes place. The product, pottery or ceramic, is in most instances a durable material although there are examples of the comminution of low fired, friable ceramics leading to complete mechanical destruction as a result of repeated freeze–thaw cycles or by cryoturbation. The earliest 'ceramics' derive from the assemblage of fired loess figurines (including the famous 10 cm Venus figure) and other objects from the Palaeolithic site cluster at Dolni Věstonice and Pavlov in the Czech Republic (Vandiver *et al.*, 1989), dated by radiocarbon to *ca.* 26 000 BP (uncalibrated). The earliest pottery vessels date from around 10,000 years later in three different regions of East Asia (Japan, Southern China and the Russian Far East). It has been suggested that, on the basis of the variation in the pottery found in these regions, pottery making originated in these different places independently (Keally *et al.*, 2004). Thus the use of pottery vessels pre-dates the onset of farming and sedentism (the 'Neolithic Revolution'), traditionally seen as being the impetus for the development of ceramic technology, and therefore the reasons for the adoption of pottery by hunter–gatherers are not as straightforward as once thought. Rice (1999) argues that pottery-making represented a 'prestige technology' amongst Upper Palaeolithic communities, and in the complex hunter–gatherer societies of the Late Pleistocene was indicative of social intensification in regions of abundant food supplies. Whatever the cause, equating pottery with the first farming communities is anachronistic since in some regions hunter–gatherer populations used pottery, whilst in others early farming communities were aceramic (*e.g.*, the 'pre-pottery Neolithic' in the Near East). In other cultures, such as the Viking period on Orkney, manufacture ceased because pottery became redundant. The causes of adoption and use (or disuse) of ceramics therefore varies

Archaeological Chemistry, Second Edition
By A. Mark Pollard and Carl Heron
© The Royal Society of Chemistry 2008

according to regional social, economic and environmental factors. Nevertheless, pottery forms one of the major archaeological find materials in most parts of the world from the Neolithic onwards. It has been used for both functional purposes (*e.g.*, food storage and cooking vessels, transport vessels) and as decorative items (*e.g.*, high-quality table ware). It combines functionality with ubiquity, and has therefore received a great deal of archaeological attention. The links between technology, aesthetics and the role of pottery in society have been explored extensively by David Kingery in a number of publications (Kingery, 1986; 1990; Kingery and Vandiver, 1986).

Some years ago, Renfrew (1977; 3) considered that the study of pottery had '*won itself a bad name in some archaeological circles.*' Archaeologists had become obsessed with typologies of pottery styles. For many years, sites and cultures could only be dated through comparison of characteristic pottery types with an established pottery typology, based on the evolution of form or decoration to construct a sequence. In some cultures (*e.g.*, the Classical Mediterranean world), dating by pottery is thought to be far more precise than any scientific dating technique, and is only surpassed if identifiable coins are found in stratified contexts. In the Greek and Roman periods, for example, it is not unusual to see pottery typologically dated to within 25 years. Renfrew identified a number of other possible research areas for ceramic studies, including the study of the origins of pottery, the wider implications for form and decoration of pottery, manufacturing technology, provenance, function and the influence of formation processes on ceramic evidence. The considerable literature focusing on at least some of these topics and the deployment of a battery of scientific, experimental and ethnoarchaeological approaches attests to a healthy future for ceramic studies in general, although problems remain in the lack of interdisciplinary overlap (Bishop and Lange, 1991). General texts covering pottery analysis and interpretation include Rice (1987), Orton *et al.* (1993), Freestone and Gaimster (1997) and Velde and Druc (1999).

Identification of traded pottery was originally important for dating purposes, since the presence of imported goods can be used to construct temporal linkages between undated cultures and those with either a calendrical chronology (such as Egypt) or with a well-established ceramic chronology, such as in the prehistoric Aegean. There is less emphasis on this in much of modern archaeology, since chronologies can now be constructed in more rigorous ways. However, the movement of pottery is of significance in its own right for examining cultural relationships and the economic and social factors responsible for promoting trade. In the ancient world, the significance of pottery must however be kept in perspective. It was almost always only the container for the item of real trade significance (*e.g.*, the use of *amphorae* to transport wine around the Roman empire), and it only represents a fraction of the trade economy – much of this would have been perishable, and therefore is largely invisible today. The problem of ceramic provenance is one of immense subtlety, since even today the supply of local (or cheaper imported) imitations of recognized 'brand names' can be of major economic significance. So it was in antiquity – for example, large pottery workshops were set up in Britain after the Roman conquest to produce vessels similar to those made on the Continent. The identification of the place of manufacture of some ceramic vessels is therefore not always straightforward. Consequently, one of the major applications of analytical chemistry to archaeology has for some time been to measure the major, minor or trace element composition of archaeological objects for the purposes of assigning *provenance* – defined here as the geographical origin of the raw materials used. Given the importance of traded pottery, it is not surprising that one of the biggest single classes of material to be studied has been pottery. A good example of the use of chemical techniques of ceramic provenance is the work carried out on pottery from the prehistoric Greek and Cypriot Mediterranean, summarized and synthesized by Jones (1986).

The theory behind chemical provenancing of ceramics is extremely simple – perhaps to the point of being naive. It is assumed that the chemical composition of the fired ceramic is indicative of the chemical composition of the principal raw material – clay. It has always been acknowledged that the transformation of raw clay into fired pottery is potentially a complex process, which might involve a number of factors, all of which could influence the final composition of the product. These include:

(i) the natural variability of the clay beds themselves;
(ii) selection and mixing of clays from different sources to give the correct colour and working properties;
(iii) *levigation* and/or processing of the raw clay to remove unwanted material and give the desired texture;
(iv) addition of *temper* (non-clay filler), usually to modify the thermal properties of the body;
(v) the firing cycle itself, which might affect the composition via the volatility of some components.

An additional problem in the study of archaeological ceramics is the possibility of post-depositional chemical alteration of the fabric.

In order to compensate for these problems, the normal procedure has been to compare chemically the finished pottery not with the raw clay from which it was produced, but with fired pottery of certain or assumed provenance. Very few chemical studies have attempted to relate a fired piece of pottery to the actual clay bed from which it was made [for an example of one such successful attempt from Guatemala see Neff (2001), which shows what can be done with extensive sampling of regional clay resources]. Most commonly, however, 'control groups' are established from either kiln '*wasters*' (distorted or broken vessels found by archaeological investigation in the vicinity of the kilns themselves), or by comparison with material of the best possible provenance – examples certified by experts as being 'type specimens', or, occasionally, bearing a maker's stamp (although even this can sometimes be deceptive). A brief review of the theory and application of chemical provenancing applied to ceramics is given in Pollard *et al.* (2007; 12).

There has been endless (and ultimately rather pointless) debate in the archaeological literature over the past 40 years or more about the 'best' procedure for carrying out ceramic provenance work. Much of it has centred around the 'best' instrumentation to use, which relates to deciding which elements are 'best'. Neutron activation analysis (NAA), which is capable of measuring more than 20 trace (parts per million) or ultra-trace (parts per billion) elements, has been widely used, although the various types of inductively-coupled plasma (ICP) spectroscopies with similar analytical capabilities have tended to replace it in recent years because of the increased difficulty in accessing nuclear irradiation facilities. A strong theoretical case has been made for using these trace and ultra-trace elements for provenance, since they are extremely unlikely to represent deliberate additions to the fabric, and therefore give a true reflection of the geochemical 'fingerprint' of the clay sources. It is also apparent that the serious efforts made in the past to standardize NAA measurements between laboratories has ensured that databases from different sources are comparable, which is not always the case using other techniques (Pollard *et al.*, 2007; 133). Other instrumentation which can produce a 'total' analysis of major and minor constituents in silicates, such as atomic absorption spectrometry (AAS) or X-ray fluorescence (XRF), has been used for some studies, particularly those with an eye on the relationship between original clay mineralogy and the composition of the finished product (see Section 4.3). There has even been expressed a 'reductionist' (but to our mind misleading) view which believes that chemical provenancing of ceramics can be reduced to a standardized procedure whereby the same limited number of elements are measured by the same method, regardless of the mineralogy or geochemistry of the clays involved, or the manufacturing process used. Although standardization of methods is generally desirable, the inherent variability of clay, temper and processing, and the complex and subtle nature of the questions to be addressed demands a greater degree of flexibility of approach. In the context of understanding the processes of manufacturing pottery, the contribution of *ceramic petrology* has been of paramount importance in enabling the reconstruction of production techniques, and in relating the ethnographic and historical record of pottery production with that observed in archaeological examples – see, for example, Middleton and Freestone (1991).

Much of the debate about the 'best' analytical method, and also that which ensues about the 'best' multivariate analysis techniques to be used on the data, is largely fruitless. A much greater understanding of the analytical data is obtained if it is considered in the light of the known geochemical behaviour of the elements in clays, and in the context of the archaeological question to be answered. This chapter therefore reviews the basic structure of silicate minerals, leading to an overview of the chemistry of the aluminosilicate (clay) minerals, followed by a discussion of the relevant trace element geochemistry of clays and the effect of firing and post-depositional alteration on the composition of clay. It concludes with a case study of the provenance of a particular class of Roman fineware known as '*Rhenish*' ware.

4.2   THE STRUCTURE OF CLAY MINERALS

The term 'clay' is a somewhat ambiguous one. To a soil scientist, it implies a particular size fraction of the fine portion of the soil – one whose size is such that its properties can be regarded as those of a *colloid* [a large surface area per unit weight, with the consequent presence of surface electrical charges which attract ionic species and water (Brady, 1974)]. The term 'clay' is usually used for those particles with a diameter of less than 0.002 mm, with 'silt' referring to diameters between 0.002 and 0.05 mm, and 'sand' up to 2 mm. Soil classifications depend on the exact particle size distribution, with regions defined on a triangular diagram with axes labelled 'clay', 'silt' and 'sand' (Figure 4.1). To a chemical mineralogist, clay minerals are part of the large family of silicates, which form the majority of the rocks of the Earth's surface. Specifically, clay minerals are related to the *micas*, which are *sheet silicates* (see below), and chemically they are generally hydrated silicates of aluminium. The structural chemistry of silicates in general and clay minerals in particular is discussed below. To a geologist, clays are the weathering products of rocks. They have been classified by Ries (1927) according to factors such as the nature of their parent rock (*e.g.*, granite, gneiss, *etc.*), the weathering process involved and their position with respect to their parent rock (*e.g.*, residual, colluvial or

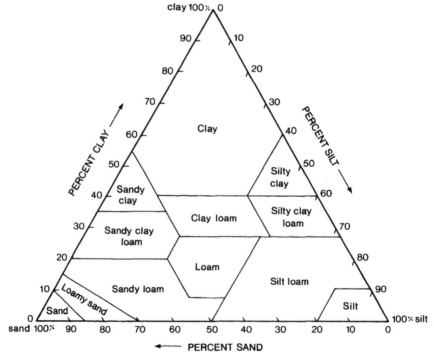

Figure 4.1   Relationship between class name of a soil and its particle size distribution. (From Brady and Buckman, 1990, reproduced by permission of Prentice Hall, Upper Saddle River, New Jersey.)

Table 4.1   Classification of clays by their uses and properties. (From Singer and Singer, 1963; 26. Copyright Chapman and Hall, with kind permission of Springer Science and Business Media.)

White-burning clays (used in whitewares)
(1) Kaolins: (a) residual; (b) sedimentary.
(2) Ball clays.

Refractory clays (fusion point above 1600 °C but not necessarily white-burning)
(1) Kaolins (sedimentary).
(2) Fire clays: (a) flint; (b) plastic.
(3) High alumina clays: (a) gibbsite; (b) diaspore.

Heavy clay products (low plasticity but containing fluxes)
(1) Paving brick clay and shales.
(2) Sewer-pipe clays and shales.
(3) Brick and hollow tile clays and shales.

Stoneware clays (plastic, containing fluxes)

Brick clays (plastic, containing iron oxide)
(1) Terra-cotta clays.
(2) Face and common brick.

Slip clays (containing more iron oxide)

---

transported). To a ceramic technologist, clays have been classified according to the system devised by Norton (1952), which is based on their properties, such as white-burning, refractory, brick clays, *etc.* The geological and ceramic classifications of clays are discussed in more detail by Singer and Singer (1963). Table 4.1 reproduces the latter classification, since it is of relevance to archaeological ceramicists. Of the clays important in ceramic production, kaolins or china clays are white-burning, low plasticity and highly refractory (*ca.* 1750 °C) clays, formed from feldspars, often in primary deposits. Ball clays are secondary deposits, often containing organic material, but which will fire white or cream. Refractory clays are low in fluxes, having a fusion point above 1600 °C, but are not necessarily white-burning. Strictly, the term fire clay refers to refractory clays and shales which are not plastic and do not take up water until ground to a powder, but it has also been used to refer to clay deposits associated with coal measures. Stonewares will vitrify at lower temperatures (*ca.* 1100 °C), but not necessarily to a white body. Brick clay refers to mixed clay mineral deposits which fire to coloured bodies at low temperatures, without vitrifying to stoneware and thus retaining a high porosity. Other plate-like minerals are often incorporated, such as micas and chlorites.

Chemical mineralogy has a long history in western science, stretching as far back as the observation by Libavius in 1597 that certain salts crystallized during alchemical processes could be identified by their *crystal habit* (Evans, 1966). Habit is a crystallographic term which refers to the characteristic shape of crystals, which often reflects the internal structure. Observations were made throughout the 17th and 18th Centuries of the crystal habits of many common inorganic minerals, including precise measurements of crystallographic angles after the invention of the contact goniometer in 1780. The classification of minerals as stoichiometric compounds is attributed to Berzelius in 1824 (von Meyer, 1891), but it was not until the development of X-ray crystallography in 1912 that serious structural analysis became possible (Bragg, 1937).

Despite the fact that 95% of the Earth's crust is made up of only five silicate mineral groups – *feldspars, quartz, amphiboles, pyroxenes* and *micas* (Putnis, 1992) – the silicates proved particularly intractable to classify before the advent of structural analysis by X-ray crystallography. Initially, classification was attempted on the basis of regarding them as salts of fictitious silicic acids, but this was unsuccessful. There are many reasons for this failure, largely relating to the wide variety of chemical compositions of the silicate minerals and the resulting lack of any apparent simple stoichiometric formulae. When this is combined with the fundamental difficulty of studying silicates – they are virtually insoluble in anything other than hydrofluoric acid, and their thermal behaviour lacks any sharply defined features – it is hardly surprising that virtually nothing was known about their classification until X-ray analysis became routine during the first half of the 20th Century. It is only when the full complexity of the possible structures of silicates, and their associated propensity for substitution and the formation of solid solution series, has been understood, that some insight can be gained into the properties and behaviour of these minerals.

In order to study in more detail the clay minerals, it is first helpful to review briefly the basic structural classification of the silicates in general. Although ultimately complicated, the general progression is logical, and is based on the degree of 'polymerization' of the basic structural unit which is the SiO_4 tetrahedron (see below). The sequence runs as follows:

    (i) minerals containing *isolated* SiO_4 tetrahedra (or isolated groups of SiO_4 tetrahedra, including SiO_4 rings), which are bonded only to other cations;

   (ii) minerals consisting of *single infinite chains* of linked SiO_4 tetrahedra;

  (iii) minerals consisting of *double infinite chains* of linked SiO_4 tetrahedra;

  (iv) minerals consisting of *infinite sheets* of SiO_4 tetrahedra (which includes most of the clay minerals);

   (v) minerals made up of infinite *three dimensional frameworks* of SiO_4 tetrahedra, the simplest of which is crystalline quartz.

Table 4.2 summarizes this basic classification and terminology of silicates. It is worth remembering that the term 'infinite' refers here to an atomic scale of measurement, and does not imply that the chains extend to infinity in the mathematical sense of the word! The terminology surrounding the silicates and clay minerals, in common with other mineralogical terms, has never been fully systematized, and so the names given are often ill-defined, or not unique to a

Table 4.2 Summary of silicate mineral classification.

Structure	Formula of Silicate Group	Common Names	Si:O Ratio	Examples of Mineral Groups
(i) Separate SiO_4 terahedra	$[SiO_4]^{4-}$	orthosilicates or nesosilicates	1:4 (0.25)	Olivines Garnets
Separate Si_2O_7 groups Separate silicate rings	$[Si_2O_7]^{6-}$ $[(SiO_3)_n]^{2n-}$	sorosilicates metasilicates or cyclosilicates	2:7 (0.29) 1:3 (0.33)	Melelite $Ca_2MgSi_2O_7$ Beryls $(Si_6O_{18})^{12-}$
(ii) Single chain silicates	$[(SiO_3)_n]^{2n-}$	inosilicates	1:3 (0.33)	Pyroxenes
(iii) Double chain silicates	$[(Si_4O_{11})_n]^{6n-}$	inosilicates	4:11 (0.36)	Amphiboles
(iv) Layer (sheet) silicates	$[(Si_2O_5)_n]^{2n-}$	phyollosilicates	2:5 (0.4)	Micas Clay minerals
(v) Framework silicates	SiO_2	tectosilicates	1:2 (0.5)	Quartz (feldspars)

structural class, and should be regarded as the mineralogical equivalent of common names in chemistry.

The basic structural unit of all silicate rocks is the tetrahedron formed by the tetrahedral co-ordination of four oxygen atoms around a single silicon ion. The Si–O bond is the most stable bond formed between any other element and oxygen, and it is therefore unsurprising that elemental silicon is not found naturally in the presence of oxygen. It is conventional to assume that the tetrahedron is held together by purely electrostatic forces as a result of the attraction between the Si^{4+} ion and the O^{2-} ions. Whilst this is not entirely true – the Si–O bond is thought to have about a 50% covalent character (Putnis, 1992), otherwise the tetrahedron would not have the shape it does – it is adequate to consider the bonding in silicates as an ionic phenomenon, and to regard the SiO_4 tetrahedron as being $[SiO_4]^{4-}$. The huge difference in ionic radii between the O^{2-} (1.40 Å or 0.14 nm) and the Si^{4+} (0.41 Å, 0.041 nm) means that on the atomic scale the central silicon ion is virtually lost in the centre of the four co-ordinating oxygens – in fact, it sits comfortably in the tetrahedral hole formed in the centre of the four oxygen spheres (Figure 4.2). The sequence of silicate structures given above can therefore be interpreted as an increasing number of direct linkages between adjacent tetrahedra via a 'bridging oxygen' (Si–O–Si) bond. Isolated tetrahedra have no shared corners, single chains share two, double chains alternately share two or three, sheets share three and frameworks share four.

Minerals containing isolated single tetrahedra are referred to as *orthosilicates* or *nesosilicates*. Simple examples are the *olivine* and *garnet* minerals. The general formula for the olivines is M_2SiO_4, where M can either be Mg^{2+}, Fe^{2+}, Ca^{2+} or Mn^{2+}. The structure of the olivine group of minerals is shown in Figure 4.3, where the SiO_4 tetrahedra are shown as triangles – those with full internal lines represent tetrahedra with the apex pointing up, those with dashed lines point down. Even with this simple silicate the actual unit cell structure of the mineral is complicated, but it is easy to see that the SiO_4 tetrahedra are not linked together other than by the intervening M^{2+} cations. The pure mineral Mg_2SiO_4 is called *forsterite*, whilst Fe_2SiO_4 is called *fayalite*, which is an extremely important mineral archaeologically since it is one of the major constituents of iron-smelting slag. *Forsterite* and *fayalite* form the end members of a complete solid substitutional series, with the stoichiometry $(Mg,Fe)_2SiO_4$, which implies that the cationic site can be randomly occupied by either Mg^{2+} or Fe^{2+}. This is the formula for the natural mineral *olivine*, after which the family takes its name. It should not be surprising, therefore, to find that the fayalite phase of an ironworking slag can contain significant quantities of magnesium. This illustrates the facility with which silicate minerals can form solid substitution series, which is one of the chief difficulties to be faced if they are studied using only chemical information, and not structural data. Many other orthosilicates are known, but all have a silicon:oxygen (Si:O) ratio of 1:4 by number. Thus garnets, with the general formula $A_3^{2+}B_2^{3+}Si_3O_{12}$ (where A^{2+} can be any divalent ion such as Ca^{2+}, Mg^{2+} or Fe^{2+} and B^{3+} is any trivalent ion, such as Al^{3+}, Fe^{3+}, *etc.*) is still classified as an orthosilicate despite the

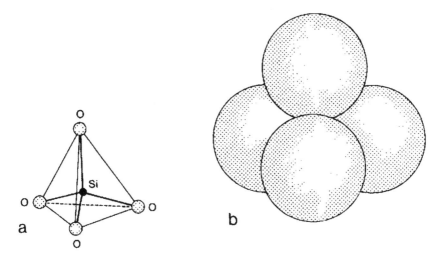

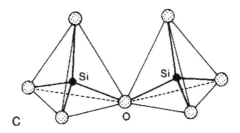

Figure 4.2   Tetrahedral co-ordination of silica, (a) conventional model of [SiO$_4$] tetrahedra, (b) space-filling model of [SiO$_4$] tetrahedra; (c) model of linked tetrahedra [Si$_2$O$_7$]. (From Putnis 1992; Figure 6.1, by permission of Cambridge University Press.)

fact that there are three silicons in the stoichiometric formula (and hence in the unit cell).

More complex isolated units can be formed by two tetrahedra linked together (*i.e.*, one corner shared, conventionally written as [Si$_2$O$_7$]$^{6-}$ and referred to mineralogically as *sorosilicates*). This dimeric tetrahedral structure is also shown in Figure 4.2. An example of this is the mineral *melilite* (Ca$_2$MgSi$_2$O$_7$), which is also a common constituent of vitreous slag. Minerals containing three-membered rings are known, with the structural unit [Si$_3$O$_9$]$^{6-}$, and the general formula for isolated rings of linked SiO$_4$ tetrahedra is [(SiO$_3$)$_n$]$^{2n-}$. These ring structures are known as *metasilicates* or *cyclosilicates*, and are characterized by a Si:O ratio of 1:3. An example of this type of mineral is *beryl*, Be$_3$Al$_2$Si$_6$O$_{18}$, where the large isolated six-membered Si$_6$O$_{18}$ rings are only linked together by either Be or Al cations, thus defining the mineral as a metasilicate.

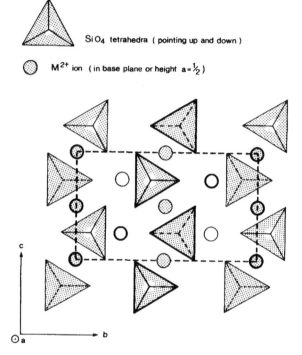

SiO₄ tetrahedra (pointing up and down)

M²⁺ ion (in base plane or height a = ½)

Figure 4.3   Structure of the olivine family of minerals down the a axis of the unit cell.
(After Putnis, 1992; Figure 6.4a, by permission of Cambridge University
Press.)

All of the above therefore, despite the increasing size of the silicate group, are
examples of silicate minerals containing isolated SiO_4 tetrahedra or groups of
tetrahedra. The general rules observed in the polymerization of silicates are
that linking of the tetrahedra only occurs by sharing corners (*i.e.*, via a bridging
Si–O–Si bond), and that only two tetrahedra may share a corner (this follows
from the divalency of the oxygen). Figure 4.4 shows how individual SiO_4
tetrahedra can link together to form infinite chains if these rules are followed. If
the tetrahedra share two corners (Figure 4.4a), the result is an infinite single
chain, termed an *inosilicate*. Each tetrahedron has two bridging and two non-
bridging corners (oxygens), giving an Si:O ratio of 1:3 since each shared oxygen
contributes half towards the count for each tetrahedron. The resulting general
formula is $[(SiO_3)_n]^{2n-}$, the same as for metasilicates (ring silicates), and the net
charge imbalance per tetrahedron is 2^- (equivalent to two unsatisfied valencies,
one on each of the non-bridging oxygen ions). On average, therefore, in order
to maintain electrical neutrality (which is a requirement for all crystal structures) each tetrahedron should be associated with two positive charges (*i.e.*, a
single divalent cation such as Ca^{2+}, or two monovalent cations such as Na^+).
This leads to minerals of the *pyroxene* group, such as *enstatite* ($MgSiO_3$), in
which the single silicate chains stack together lengthwise with, alternately, the

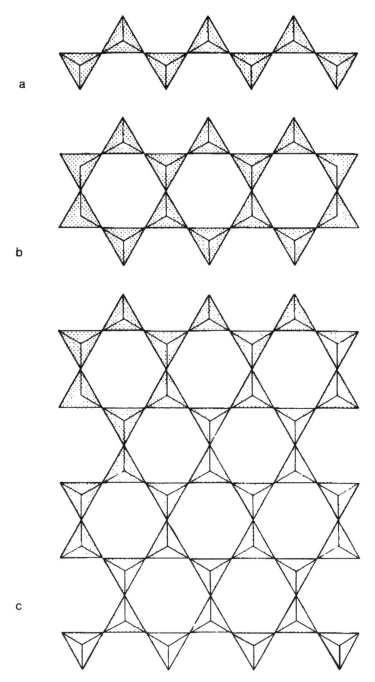

Figure 4.4   Infinite chain silicates (single, double, and sheet): (a) infinite single chain silicate with two corners shared per tetrahedron (pyroxene structure); (b) infinite double chain, with alternate two and three corners shared (amphibole structure); (c) infinite sheet structure, with each tetrahedron sharing three corners (sheet silicates). (From Putnis, 1992; Figure 6.3, by permission of Cambridge University Press.)

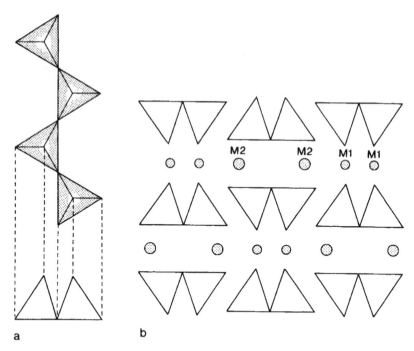

Figure 4.5 Structure of pyroxene minerals: (a) demonstration of the end view of the single silicate chain; (b) end view of the stacking arrangement of single chains, showing the position of the metal cations. There are two different cationic environments, M1 and M2. (After Putnis, 1992; Figure 6.11, by permission of Cambridge University Press.)

apices and the bases pointing towards each other, and the magnesium cations joining pairs of chains together (Figure 4.5). The resulting crystal habit, as might be expected from minerals with an infinite chain length in one direction, is often *acicular* or *needle-like* – the mineralogical term *inosilicate* is derived from the Greek *inos*, meaning fibre.

The double-chain silicates (Figure 4.4b) are also termed *inosilicates* because of their fibrous or needle-like qualities. The double chains can be thought of as two cross-linked single chains, with the link being made between the basal plain of each chain – in other words, as with the single chain, the apices of each individual tetrahedron are all pointing in the same direction. The net result is that alternate tetrahedra have two or three bridging corners, depending on whether they are on the 'outside' of the double chain (with two links) or on the 'inside' (with three). The general formula is $[(Si_4O_{11})_n]^{6n-}$, and is the structure found in the *amphibole* group of minerals, which makes up a large group of rock-forming minerals. The chains are stacked together in a manner very similar to that described for the pyroxenes, with two double chains bridged apex to apex by cations and hydroxyl (OH^-) groups, and then the pairs linked together base-to-base by more cations.

The next step in this logical progression (Figure 4.4c) is to link together an infinite number of these double chains to produce a continuous sheet of SiO_4 tetrahedra, all linked in the basal plane and all sharing three corners, and all with their apices pointing in the same direction. These minerals are known as layer silicates, or *phyllosilicates* (from the Greek *phylon*, meaning leaf), and they have the general formula $[(Si_2O_5)_n]^{2n-}$. The sheets are stacked one on top of the other, alternately 'up' and 'down' (*i.e.*, with the apices pointing up or down). The two layers facing each other are held together by cations and hydroxyl groups, and the resulting 'sandwiches' are held together by other cations. In general, the phyllosilicates are *laminar* minerals, and are characterized by being flaky or plate-like; they are analogous to graphite in carbon chemistry. Aluminium substitution for silicon is common, and these minerals are the starting point for a study of the aluminosilicate clay minerals, discussed below.

The final group of silicate structures are the framework silicates, in which all four corners of the tetrahedra are linked to another tetrahedron, to form an infinite three-dimensional network, analogous to the diamond structure in carbon chemistry. With no substitution, the general formula for these minerals is SiO_2, which is polymorphic (*i.e.*, has several crystalline forms), but is generally known as *quartz*. Aluminium substitution for silicon within the framework (which demands charge compensation by interstitial cations to counteract the resulting charge deficit) gives the large family of *feldspar* minerals, which are also very important rock-forming minerals.

As alluded to above, the chemistry of the silicate minerals is made infinitely more complex by the ability of certain elements to substitute for silicon within the structure. The most important substitution is Al^{3+} for the Si^{4+}; since Al^{3+} is only slightly larger than Si^{4+} ($0.51\,\text{Å}$, compared to $0.42\,\text{Å}$) and the tetrahedral co-ordination of the four oxygens is maintained (with some distortion of the bond angles), but obviously for each ion substituted there is a net charge deficit equivalent to one additional negative charge per substituted tetrahedron. This has to be compensated by an additional positive charge, supplied by a cation which has to be accommodated *interstitially* (*i.e.*, in the holes in the structure). This is most easily illustrated by considering the framework silicates, as indicated above. A pure quartz structure has an infinite framework of SiO_4 tetrahedra, each linked to each other via a corner (Si–O–Si) bond. All bonds are bridging, and the Si:O ratio is 1:2. If one silicon ion in four is substituted by an aluminium ion, the framework structure can be symbolized on average as being $MAlSi_3O_8$, where M is a monovalent cation such as K^+ (in which case the mineral becomes *orthoclase*). If two silicons are substituted, the general formula becomes $NAl_2Si_2O_8$, where N could be a divalent cation such as Ca^{2+}, in which case the mineral becomes *anorthite*. These aluminium-substituted framework silicates are called *feldspars*. Most naturally occurring feldspars have between 25 and 50% aluminium substitution, and their composition can be usually plotted on a ternary $KAlSi_3O_8$–$NaAlSi_3O_8$–$CaAl_2Si_2O_8$ (*orthoclase–albite–anorthite*) diagram. These general rules for silicon substitution by aluminium (and other small highly charged cations) apply to all the structural categories of silicate minerals described above, and give rise to the full complexity of the subject. The standard text on the structure and properties of the rock-forming minerals (in five volumes) is that of Deer *et al.* (1962; 1963). One point to note is that the Al^{3+} ion can also enter the mineral as an interstitial cation as well as a substituent for silicon in the network. The aluminium ions in these two environments are not equivalent in terms of oxygen co-ordination, *etc.*, and it is important to distinguish between the two. For example, the mineral *muscovite* is one of the mica family of phyllosilicates, with aluminium substituted for about one atom in four of the silicons, but it also includes potassium and aluminium as cations between the aluminosilicate sheets. It is given the general formula $KAl_2(AlSi_3O_{10})(OH,F)_2$ to signify that some of the aluminium (that within the brackets) is substituted for one in four of the silicon ions in the sheets, and also that some of the aluminium exists between the sheets, together with potassium. The existence of both OH and F in the second set of brackets indicates that the hydroxyl group between the silicate layers can also be substituted at random by the fluoride ion. This example indicates the difficulty of classifying the aluminosilicate minerals from chemical analytical data alone, since a traditional chemical analysis of the mineral would give only the total amount of aluminium present, but would not identify the extent of the partitioning of the aluminium between the two sites, and it would therefore be impossible to deduce the stoichiometry of the compound from the analytical data without the additional structural information.

The clay minerals can now be discussed in terms of their relationship with the phyllosilicates (sheet silicates). It is important to keep clearly in mind here the difference between 'clay' – the material which is dug out of the ground, and which may be a mixture of different clay minerals, together with various non-clay minerals (such as quartz, pyrite, *etc.*), as well as unaltered rock fragments and incorporated organic material (Grim, 1968) – and the clay minerals themselves, which are crystalline compounds of specified stoichiometry and structure. At this stage, we are only considering the structure of the clay minerals.

It is instructive to begin with the mineral *kaolinite*, $Al_2(Si_2O_5)(OH)_4$, which does not have aluminium substituted into the silicon sheets, and is therefore relatively simple. It used to be thought that kaolinite was the only true clay mineral, and that natural clay deposits differed in having different proportions of kaolinite mixed with other finely divided minerals, but this is not the case. Figure 4.4c shows part of the infinite sheet of SiO_4 tetrahedra which form the basis of the phyllosilicates and clay minerals. Since all the tetrahedra point in the same direction, the result is a sheet with 'mountain peaks' (the non-bridging oxygens) rising above the basal plane, and these peaks make hexagonal patterns of oxygens when seen from above. A hydroxyl ion sits in this hexagonal cradle like an egg in an egg box, forming a close-packed raft of oxygen (from the silica) and hydroxyl ions. On top of this layer lies an aluminium ion, with a further close-packed layer of hydroxyls above this (Figure 4.6). The aluminium ion, being small, sits in the hollow formed between three touching spheres (two oxygens from the SiO_4 tetrahedra, and a hydroxyl) in the middle of the sandwich, with the top layer of hydroxyls sitting above it. Its nearest

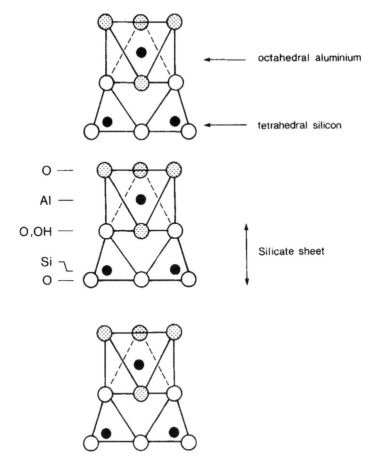

Figure 4.6 Layer structure of kaolinite. Edge view showing the aluminosilicate sheets, and the stacking arrangement of these sheets. (After Evans, 1966; Figure 11.11, by permission of Cambridge University Press.)

neighbours are four hydroxyls and two oxygens, and it is therefore in *octahedral co-ordination*. The simplest way of thinking about this structure is to imagine it as a tetrahedral layer of SiO_4 bonded to an octahedral layer of aluminium co-ordinated by (O + OH). The complete structure is made up of stacks of these double layers but (unusually for phyllosilicates) all oriented in the same way, with the double layers held together by weak van der Waals (London) forces – hence the plasticity of such minerals.

In phyllosilicate terminology, kaolinite is called a 1:1 structure type, referring to the ratio of tetrahedral to octahedral layers (Brown, 1984). Most phyllosilicates and clay minerals are 2:1 structures, with the sequence tetrahedral–octahedral–inverted tetrahedral (*i.e.*, like kaolinite but with an additional inverted layer of SiO_4 tetrahedra on top). This is illustrated by Figure 4.7, which shows an edge view of the structure of one of the *smectite* group of clay minerals, specifically *montmorillonite*, $Al_2Si_4O_{10}(OH)_2.nH_2O$ (which again, for

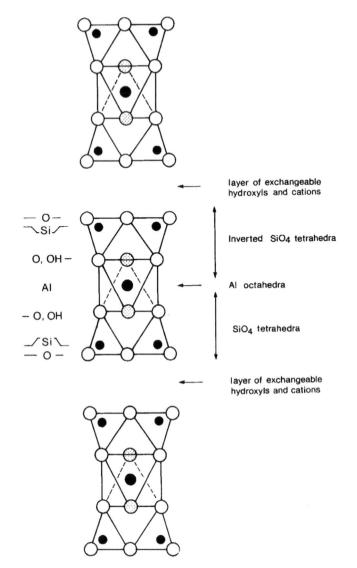

Figure 4.7   Structure of one of the smectite group of clay minerals: montmorillonite, $Al_2Si_4O_{10}(OH)_2.nH_2O$. (After Evans, 1966; Figure 11.08b, by permission of Cambridge University Press.)

simplicity, has no Al substitution into the SiO_4 tetrahedra). The Al ion sits in the octahedral hole in the (O + OH) layer, as described above, but the upper layer, instead of being simply a layer of hydroxyls, is now an inverted version of the basal layer of SiO_4 tetrahedra and associated OH groups. The Al is still in octahedral co-ordination, this time by four oxygens (two from each of the SiO_4 tetrahedral layers) and two hydroxyls (one associated with each layer). These triple layers are stacked upon each other, but now a variable number of water molecules and loosely-bound (*exchangeable*) cations may occupy this interlayer plane. This water is easily lost on heating, but can be regained by absorbing water, thus explaining one of the familiar properties of clays – shrinkage on drying, and swelling on rehydration. In montmorillonite, as in other clay minerals, the full complexity of the family is developed by allowing Al to replace Si in the lattice (with the attendant necessary charge-balancing cations occupying holes in the structure, or the interlayer gap), and by substitution occurring for the Al in the octahedral layer – commonly Fe^{2+} or Mg^{2+}, again with the necessity for an additional cation somewhere (usually the interplanar layer) to maintain electrical neutrality. It can be seen that the detailed structure becomes very complicated as more and more substitutions occur because of the demands of electrical neutrality, but the overall systematic structural classification of the clay minerals remains valid. Variations on this 2:1 structures account for the majority of phyllosilicate and clay mineral structures, including *talcs, vermiculites, micas* and *chlorites*. For a full description of these structures see Deer *et al.* (particularly volume 3, 1962), Grim (1968) or Brown (1984).

4.3 THE FIRING OF CLAYS AND THE MINERALOGICAL COMPOSITION OF CERAMICS

The discovery of the working properties of clays must have resulted in one of humankind's first expressions of representational art, roughly contemporaneous with the discovery of the colouring properties of natural pigments and their use in cave art. The additional discovery that the result of the manipulation of this art form could be rendered permanent by the use of fire must indeed have been a source of wonder. The earliest fired ceramic so far known is a small moulded figurine from Dolni Věstonice in what was Czechoslovakia, dated to approximately 26 000 years BP (Vandiver *et al.*, 1989). By approximately 10 000 years ago, simple utilitarian vessels were being produced in the Near and Far East.

The value of clay as a raw material for the manufacture of vessels stems from two properties – its plasticity when wet, and its hardness when heated to a moderately high temperature. Most clays suitable for pottery making will fire at a temperature achievable in a small bonfire (very approximately 800 °C), although specialist products such as porcelain require temperatures in excess of 1200 °C to mature. Plasticity is the property whereby sufficiently wet clay can be deformed by modest pressure, and the clay will retain the given form when the pressure is removed. It stems largely from the sheet-like structure of the clay minerals, as described above. Most clay minerals incorporate water molecules as part of their structure between the composite layers of aluminosilicates, and it is the presence of these which allows the individual crystals to slide over each other and renders clay plastic. Plasticity varies from clay mineral to clay mineral – montmorillonite and kaolinite are highly plastic, whereas illite and dickite are considerably less so. The plasticity of a lump of clay therefore depends partially on the exact mixture and particle size distribution of the clay minerals within it, but also on the proportion of non-plastic minerals present, and the pH of the water (Singer and Singer, 1963; 62).

The firing behaviour of clay minerals and commercially valuable clay deposits has of course been extensively studied (Singer and Singer, 1963; Kingery *et al.*, 1976; Rice, 1987). The key factor is the irreversible removal of the interplanar water molecules, but the study of this process is complicated by the multiplicity of sites which water may occupy within the structure – *adsorbed water*, weakly held on the surface of the crystals themselves, which is probably lost below 100 °C, *interplanar water* between the layers, relatively weakly held, and lost at a few hundred degrees, and *structural OH* groups which may require up to 1000 °C to remove them. The phenomenon of water loss is best studied using the various techniques of thermal analysis (Grim, 1968; Rice, 1987; see also Chapter 2), but is complicated by the fact that results may vary as a function of the degree of crystallinity of the mineral under study, and also depending on the rate of heating employed. The most commercially important thermal decomposition of a clay mineral is that of *kaolinite* [$Al_2(Si_2O_5)(OH)_4$] to *mullite* ($3Al_2O_3.2SiO_2$), and consequently this is the best studied. In well-crystallized samples of kaolinite little happens up to 400 °C, but above that differential thermal analysis shows a sharp endothermic peak beginning at about 400 °C and usually reaching a maximum at about 600 °C, which corresponds to the onset of the loss of OH groups from the lattice. By about 800 °C, virtually all of the water has gone. The resulting mineral has been called *metakaolin*, and there is some discussion about its true nature, since it appears to have a slightly disordered structure (Murad and Wagner, 1991), but it is generally thought to have a similar structure to kaolinite, with the loss of OH causing the aluminium octahedral layer to become re-ordered into a tetrahedral (oxygen) co-ordination (Grim, 1968). Rehydration studies have shown that this phase can revert to kaolinite, depending to some extent on the degree of crystallinity of the kaolinite from which it was formed, so that this reaction cannot be regarded as truly irreversible. At higher temperatures, irreversible changes do occur. Differential thermal analysis shows an exothermic reaction occurring between 900 and 1000 °C, attributed to the formation of high-temperature phases of silica (SiO_2) and alumina (γ-Al_2O_3), although there has been considerable debate about the nature of these phases, with some suggestion that minerals with a *spinel* structure can form (Brindley and Lemaitre, 1987). Above 1200 °C *mullite* ($3Al_2O_3.2SiO_2$) forms, together with *cristobalite*, which is the high-temperature form of SiO_2 (stable above 1470 °C).

All the various clay minerals undergo a similar chain of dehydration reactions, ultimately ending in a number of high-temperature aluminosilicate phases, usually with excess silica in the form of cristobalite (or some other high-temperature silica phase) The complication arises from the presence of other cations in most clay minerals, which can result, for example, in phases such as *enstatite* ($MgSiO_3$) if large proportions of magnesium are present. Other substitutions will result in other mineral phases being formed, as documented by Grim (1968) or Brindley and Lemaitre (1987). From the point of view of a ceramicist in general (and an archaeological ceramicist in particular), what becomes important for practical purposes is not the high-temperature thermal behaviour of individual clay minerals, but the interactions at high temperatures between the various phases present. These phases can be the result of the decomposition of the clay minerals themselves, and also of any non-plastic inclusions which might be present, such as carbonates or even organic material. In particular, the melting point of various phases becomes critical in high temperature reactions, since impurities in the clay usually mean that sufficient fluxes are present to melt (or sinter) at least some of the phases. In order to consider these, it is usual to use triangular phase diagrams, the most important of which for archaeologists is either the $CaO–Al_2O_3–SiO_2$ system for earthenware, or the $K_2O–Al_2O_3–SiO_2$ system for higher fired stonewares and porcelains (Heimann, 1989). Examples of these diagrams are shown in Figure 4.8. Although in reality the composition of ancient ceramics is always more complex than the three oxides specified, it is usually sufficiently accurate to sum elements which behave similarly into one of the terms, in order to reduce the dimensionality of the problem to three, which can then be displayed graphically. For example, with earthenware (essentially any archaeological ceramic which is not stoneware or porcelain can be regarded as earthenware), it is usual to construct the 'CaO' value from the total 'fluxes' present $(CaO + MgO + Fe_2O_3 + K_2O + Na_2O)$, which will then allow the composition to be plotted on the 'CaO'–Al_2O_3–SiO_2 ternary diagram, with any other unaccounted oxides contributing little to the total. The resulting plots allow the composition and proportion of phases present to be predicted with reasonable confidence, given that there is already an inbuilt assumption which in some circumstances can limit the usefulness of such an approach – the assumption of equilibrium conditions, which is inherent in the use of all phase diagrams. Nevertheless, using the usual techniques for interpreting phase diagrams such as the *Phase Rule* and the *Lever Principle* (described by Heimann, 1989), it is often possible to obtain a reasonably realistic picture of the phase structure and composition of fired ceramics.

It is always important when working with phase diagrams to be certain of the identity of the units used in the axes of the diagram, and the temperature at which it is applicable. Standard multi-component phase diagrams used in ceramic technology usually deal with concentrations expressed as *weight percent oxide*, in which case routine analytical data (which is normally quoted as 'weight percent oxide') can be used directly to calculate the co-ordinates of a point. Occasionally axes are labelled as *molar percent* (or *molar fraction*) of oxides – conversion between weight and molar units is relatively straight-forward, involving dividing the weight percentage by the stoichiometric formula molecular weight, and re-normalizing to 100%. Slightly more compli-cated are those diagrams labelled not as weight percent oxide, but as weight (and sometimes molar) percentages of other stoichiometric compounds, usually using theoretical mineralogical formulae. This can be particularly useful if the mineralogy of the clay is known (or can reasonably be assumed). For example, in a study of the chemical compositions of the bodies of Chinese porcelains, Pollard and Wood (1986) demonstrated the value of converting the analytical data (expressed as weight percent oxides) into ternary co-ordinates using the assumed mineralogy of the raw materials – in this case quartz (SiO_2), kaolinite

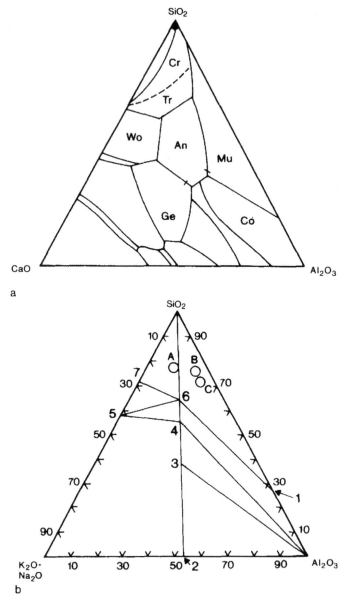

Figure 4.8 Triangular diagrams of the systems CaO–Al$_2$O$_3$–SiO$_2$ and K$_2$O–Al$_2$O$_3$–SiO$_2$. (a) Simplified liquidus surface in the system CaO–Al$_2$O$_3$–SiO$_2$ (the 'ceramic triangle'). Symbols: Cr = cristobalite; Tr = tridymite; Wo = wollastonite; An = anorthite: Mu = mullite; Ge = gehlenite; Co = corundum. (b) Phase diagram K$_2$O–Al$_2$O$_3$–SiO$_2$. Symbols: 1 = 3Al$_2$O$_3$.2SiO$_2$; 2 = K$_2$O.Al$_2$O$_3$; 3 = K$_2$O.Al$_2$O$_3$.2SiO$_2$; 4 = K$_2$O.Al$_2$O$_3$.4SiO$_2$; 5 = K$_2$O.SiO$_2$; 6 = K$_2$O.Al$_2$O$_3$.6SiO$_2$; 7 = K$_2$O.4SiO$_2$; A = Medici 'porcelain'; B = Thai stoneware; C = 'Rhenish' stoneware. (After Heimann, 1989; Figures 8 and 12, by permission of the author.)

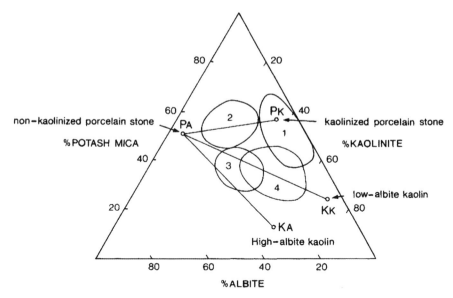

Figure 4.9 Possible mineralogical constitution of Jingdezhen porcelain bodies. Average composition of four types of raw material are marked (P_A, P_K, K_A, K_K), and the approximate compositions of Chinese porcelains from four periods: 1 = predominantly 10th to 12th Century AD; 2 = 12th to 13th Century AD; 3 = predominantly 14th Century AD; 4 = post-17th Century AD. (From Pollard and Wood, 1986; Figure 10.7, with permission of Smithsonian Institution Scholarly Press.)

[$Al_2(Si_2O_5)(OH)_4$], potash mica [*muscovite* or *sericite*, $KAl_2(AlSi_3)O_{10}(OH)_2$] and soda feldspar (*albite*, $NaAlSi_3O_8$). In order to plot the data onto a triangular diagram, it is necessary to recalculate the weight percentages into these four components and then to normalize the three non-quartz components to 100% which then gives the triangular co-ordinates (see Figure 4.9). A sample calculation, using the average body composition of 18 samples of Tianqi porcelain, is given in Table 4.3. Tianqi porcelain is a 17th Century AD oriental porcelain of disputed origin, but thought on the basis of these analyses to have been made at Jingdezhen in China rather than in Japan, as was originally postulated (Impey *et al.*, 1983). In this way, it proved possible not only to predict the raw material mixtures used to produce the porcelain by comparison with the known mineralogies of the raw material sources (and changes in this pattern with time), but also to suggest a provenance using the major and minor element analyses, rather than the more usual trace element patterns (Section 4.4).

4.4 TRACE ELEMENT GEOCHEMISTRY IN CLAYS

Section 4.3 sets out the principles underlying the structure of the silicate mineral family. Natural clay deposits are formed by the *chemical weathering* of rocks – largely as a result of the attack by slightly acidic surface waters. Rainwater,

Table 4.3 Example of calculation of theoretical mineralogy of Tianqi porcelain.

Theoretical composition of minerals involved:
Sericite $KAl_2(AlSi_3)O_{10}(OH)_2$ or $K_2O.3Al_2O_3.6SiO_2.2H_2O$
Albite $NaAlSi_3O_8$ or $Na_2O.Al_2O_3.6SiO_2$
Kaolinite $Al_2(Si_2O_5)(OH)_4$ or $Al_2O_3.2SiO_2.2H_2O$
Quartz SiO_2
Calculate wt % of oxides in minerals:

	$\%SiO_2$	$\%Al_2O_3$	$\%K_2O$	$\%MgO$	$\%Na_2O$	$\%CaO$
Sericite	45.2	38.4	11.8			
Albite	68.8	19.4			11.8	
Kaolinite	46.6	39.5				
Quartz	100.0					

Average composition of Tianqi Porcelain body:

	$\%SiO_2$	$\%Al_2O_3$	$\%K_2O$	$\%MgO$	$\%Na_2O$	$\%CaO$
	67.2	25.5	3.47	0.25	1.7	0.94

Step 1. Calculate proportion of sericite in porcelain:
Assuming all K_2O comes from sericite: **% sericite** $= 3.47/11.8 \times 100 = $ **29.4%**
Calculate how much SiO_2 and Al_2O_3 are contributed by sericite and subtract from original analysis:

	$\%SiO_2$	$\%Al_2O_3$	$\%K_2O$	$\%MgO$	$\%Na_2O$	$\%CaO$
Sericite	13.3	11.3	3.47			
Residual	53.9	14.2		0.25	1.7	0.94

Step 2. Calculate proportion of albite in porcelain:
Assuming all Na_2O comes from albite: **% albite** $= 1.7/11.8 \times 100 = $ **14.4%**
Calculate how much SiO_2 and Al_2O_3 are contributed by albite and subtract from this residual analysis:

	$\%SiO_2$	$\%Al_2O_3$	$\%K_2O$	$\%MgO$	$\%Na_2O$	$\%CaO$
Albite	9.9	2.8			1.7	
Residual	44.0	11.4		0.25		0.94

Step 3. Repeat for kaolinite:
Assuming all residual Al_2O_3 comes from kaolinite: **% kaolinite** $= 11.4/39.8 \times 100 = $ **28.6%**

	$\%SiO_2$	$\%Al_2O_3$	$\%K_2O$	$\%MgO$	$\%Na_2O$	$\%CaO$
Kaolinite	13.3	11.4				
Residual	30.7			0.25		0.94

Step 4. Quartz: **% quartz** $= 30.7/100 \times 100 =$ **30.7%**

	$\%SiO_2$	$\%Al_2O_3$	$\%K_2O$	$\%MgO$	$\%Na_2O$	$\%CaO$
Quartz	30.7					
Residual	—	—		0.25		0.94

Predicted mineralogical composition of average Tianqi porcelain body:

	As calculated	*Normalized*	*Normalized without quartz*
Sericite	29.4	28.2	40.6
Albite	14.4	13.8	19.9
Kaolinite	28.6	27.4	39.5
Quartz	30.7	29.4	
Others	1.2	1.2	
Total	104.3	100.0	100.0

The predicted raw material composition is therefore:
Sericite = 28.2%
Albite = 13.8%
Kaolinite = 27.4%
Quartz = 29.4%

Predicted coordinates on triangular diagram are obtained by eliminating quartz and re-normalizing:
Sericite = 40.6%
Albite = 19.9%
Kaolinite = 39.5%

after absorbing small amounts of atmospheric carbon dioxide, becomes slightly acidic and interacts with the rock-forming minerals via a cation-exchange reaction, in which the more soluble cations (Na^+, K^+, Mg^{2+}) are removed leaving a new hydrated mineral, which, for most rock types, is a clay mineral. Ultimately the parent rock may be completely decomposed to leave a mixture of clay minerals and silica-rich sand grains, which, if found where it was formed, is termed a *primary* clay deposit. If transported in some way (mostly by water, but sometimes by wind or glacial action) it becomes a *secondary* clay deposit. Different rock minerals decompose at different rates, depending primarily on the temperature of formation and chemistry – high temperature minerals (*e.g.*, volcanics) are less chemically stable than low temperature minerals, and those with high magnesium and iron (Fe^{2+}) more susceptible to decay than those with high aluminium.

Clay deposits, then, are generally made up of a mixture of clay minerals, together with various non-clay inclusions, such as fragments of unweathered rock, or other siliceous materials which may have become mixed in with the clay deposit. The exact composition of the deposit will therefore depend on the composition, age and weathering history of the parent rock, and, for secondary deposits, on the nature of the transport phenomena involved (in terms of what has been 'winnowed' out and what may have been 'picked up' during transport). Most clay deposits are also likely to undergo further weathering and biological modification as part of the normal soil-formation processes.

Before turning to study the use of trace element data in ceramics for provenance studies, it is important to consider briefly some of the general principles of trace element geochemistry, so that we may look at the data from a more informed viewpoint. The literature on the geochemistry of the elements is vast, and much of our knowledge stems directly from the pioneering work of Goldschmidt (1954). Our theoretical understanding of the behaviour of trace elements in the geochemical environment has progressed considerably since then, but the majority of this effort has gone into understanding the processes involved in the partitioning of trace elements between fluid and rock during the crystallization of the rock minerals. Only relatively recently has the behaviour of the trace elements in the sedimentary environment become of central interest, with the increasing importance of environmental geochemistry. It is nevertheless useful to review briefly those factors which control the distribution of trace elements in geochemistry, since the general principles will apply to the sedimentary environment of clays, and allow us to predict some of the features we would expect in the chemistry of clays. Of particular interest in the context of clay is the *cation exchange capacity* (CEC) of clay, which governs the degree to which clays can pick up soluble elements from the surrounding aqueous environment. Finally, some attention must be given to the vastly under-researched area of post-depositional alteration to fired archaeological ceramics.

The factors which control the distribution of trace elements [defined arbitrarily in geochemistry as those elements present at less than 0.1 weight percent (wt %)] can be discussed under a number of headings – *structural*, *thermodynamic*, *kinetic* and, in the sedimentary environment, *solubility* and *speciation*.

At the crudest level, the chemical elements can be divided into three categories, from a consideration of the free energies of formation of the relevant oxides and sulfides. It is generally assumed that as it condensed the Earth separated into three liquid phases – metallic iron, molten iron sulfide and a molten silicate phase – which rapidly solidified. The distribution of the elements between these phases is largely governed first of all by the stability of the appropriate oxides relative to that of iron oxide (FeO). All those elements which form more stable oxides were oxidized preferentially, whilst those less stable remained in the reduced metallic state and were concentrated in the molten iron core of the earth – the so-called *siderophil* elements (principally Ni, Co and the platinum and palladium group metals, as well as P, but this for other reasons). All of the other elements concentrated in the sulfidic or silicate regions of the Earth's crust, and are known as *chalcophil* elements. Some of these, essentially those with a greater affinity for oxygen than that of sulfur, concentrated in the *lithosphere*, and became the rock-forming (*lithophil*) elements – largely Mg, Al, Na and Si. The remainder – Pb, Zn, Cd, Hg, Cu and Ag – concentrated in the sulfidic ores and have become the familiar primary ore metals from near-surface deposits.

Beyond this crude level of partitioning of the elements during the formation of the Earth, it is possible to consider the factors which control the partitioning of the elements during mineralization processes, such as those which occur when rock-forming minerals crystallize from magma. The principal factors which determine the distribution of trace elements between different minerals, as first demonstrated by Goldschmidt, are those relating to the size and charge of the ionic species. By studying the tendencies of various pairs of elements to substitute for each other in different silicate minerals, it is possible to observe that similarity of ionic radius is the principal factor. This gives a whole series of pairwise associations, some of which are unsurprising, such as Mg^{2+} and Ni^{2+} (Goldschmidt originally thought both had radii of 0.78 Å, but they are now tabulated as 0.66 and 0.69 Å, respectively; *e.g.* Aylward and Findlay, 1974; Table 4). This simply reflects the ability of one ion to physically replace another in the crystal lattice without causing undue distortion. Other pairs are a little less obvious, such as Na^+ and Ca^{2+} (radii now given as 0.97 and 0.99 Å, respectively) or K^+ and Ba^{2+} (1.33 and 1.34 Å), where substitution can occur despite a difference in valency (which of course has to be balanced up elsewhere in the structure). Goldschmidt introduced the concept of '*ionic potential*' – the ratio of the ionic charge to the ionic radius – into the theoretical discussion of the distribution of the elements, and his ideas have since been termed '*Goldschmidt's Rules*' for the relationship between the composition of a liquid and that of the crystals forming from it (*e.g.*, Henderson, 1982; 125). These can be summarized as:

(i) ions of similar radii and the same charge will be incorporated into the crystal in the same ratio as their concentration in the liquid;

(ii) an ion of smaller radius but the same charge will be incorporated preferentially into a growing crystal;

(iii) an ion of similar radius but with a higher charge will be incorporated preferentially into a growing crystal.

These rules form the basis of our understanding of the distribution of the elements in rocks and ores, which, at least in igneous rocks, has long been held to conform to a log-normal law (Ahrens, 1954). When combined with some knowledge of the kinetics of crystal growth, they can also be used for calculations of the temperature (and/or pressure) at which crystallization took place (*geothermometry* and *geobarometry*: Henderson, 1982; 175). This procedure may well have applications in some areas of archaeological chemistry (*e.g.*, in considering the partitioning of trace elements between slag and metal during metallurgical processes), but it has long been realized that these factors are of limited value in considerations of surface phenomena such as weathering, and in aqueous environments in general. Weathering takes place under a very limited range of temperatures, pressures, acidity (pH), redox potential (Eh) and ionic concentrations, and is overwhelmingly conditioned by the behaviour of the various ions in aqueous media. The composition of surface waters is defined by the oxidation state of the ions in solution, the speciation of the ions and the stability of various solid phases present with respect to the solution – all of which are controlled to a greater or lesser extent by the redox conditions of the environment. Furthermore, one cannot simply use the concept of the ionic radius as an indicator of the behaviour of a particular element, as is possible in higher temperature systems. A classic example of this is the chemical composition of seawater and the sediments which precipitate from it (Millot, 1970; 49). It is well known that the major dissolved species in typical seawater are Na^+ and Cl^- – effectively, seawater can usually be considered as a half molar solution of sodium chloride. In fact, the molar concentrations of sodium and potassium in seawaters are given as approximately 0.486 M and 0.010 M respectively (Henderson, 1982; Table 11.2), giving an approximate ratio of K/Na by weight of 1/28.5. The equivalent figure for fresh water is approximately 1/10, indicating that sodium is by far the dominant cationic species in surface waters. In contrast, the ratio of K/Na by weight in surface rocks ranges from about 1/2.8 for schists to 1/7.7 for limestones, with an average of 1/3 for sedimentary deposits. It is clear that sediments are significantly enriched in potassium compared to surface waters, which runs counter to Goldschmidt's Rules. The explanation lies in the fact that in an aqueous medium the sodium ion is normally surrounded by a *hydration sphere* of H_2O molecules, whereas potassium ions are not (see Chapter 5). This gives an *effective radius* for these ions in water as nearer 1.83 Å for the sodium ion (compared to 0.97 Å without hydration), whereas potassium is calculated as being around 1.24 Å, compared to its 'normal' radius of 1.33 Å (Millot, 1970; 52). The sodium ion is therefore effectively considerably larger than the potassium ion, and is consequently discriminated against when crystallization and sedimentation occur (as would be expected from Goldschmidt's Rules, providing the solvation effect is taken into account). Rubidium and caesium behave similarly to potassium, and are therefore similarly concentrated in aqueous sediments, whereas most of the other elements behave like sodium.

The geochemistry of a clay deposit, therefore, depends on its parent rock type, the degree of chemical weathering and the nature of the transport mechanism involved, if any. Factors which might affect the chemical composition of a secondary clay deposit include the distance moved, since either aeolian or fluvial transport will cause sorting of the grain sizes, with the smaller minerals remaining longer in suspension and being transported further. Deposition in a marine environment is likely to produce clays enriched in a whole suite of elements, since clay mineral particles in the marine environment have been seen to be responsible for the removal of major and trace elements from seawater (Goldschmidt, 1954; 64). For the reason described above, 35% of the sodium dissolved in the oceans is removed by the sediment, whereas 98% of the potassium is removed, with the result that marine clays are enriched in potassium compared to sodium. Similar considerations based on mass balances and the redox behaviour of the elements predicts that oxides of magnesium, scandium, gallium, trivalent vanadium and chromium, titanium, zirconium, cerium and hafnium, thorium, tin, niobium and tantalum should also concentrate in marine sediments (Goldschmidt, 1954; 67). Thus, pottery made using clay from marine sediments should be relatively easily distinguished from that made from terrestrial or fluvial deposits.

It is clear that the environment with which we are mostly concerned in archaeological chemistry is one in which the aqueous behaviour of the elements is of prime importance. Here the geochemical behaviour of the elements is subject to the thermodynamic and kinetic controls first postulated by Goldschmidt, and the association between elements in clays, sediments and ultimately in archaeological ceramics is thus in principle predictable. We should not, therefore, be surprised to find certain inter-element associations in the analysis of archaeological ceramics – in fact, we should be more surprised if we do not! Certain common behaviour has been noted by some authors, *e.g.*, Glascock (1992), who observed that Ca is often correlated with elements such as Sr and Ba. Fe, Sc and other transition elements are well known for exhibiting high correlations, and the rare earths (the *lanthanides*) exhibit strong correlations with each other. This latter point has been commented on geochemically in the archaeological literature (Allen *et al.*, 1975), bringing attention to the well-known '*europium anomaly*' in rare earth geochemistry (arising from the fact that most lanthanides exist primarily in the 3^+ oxidation state, but europium can be reduced to 2^+ under natural conditions, giving it a slightly different pattern of behaviour). In general however, discussion of these correlations is usually restricted to simple observations of their occurrence, and tends to focus on the undoubted influence of correlation on the multivariate data analysis techniques commonly employed (*e.g.*, Bishop and Neff, 1989). These studies usually conclude that ceramic data cannot be analysed without taking correlation effects into account, and that what often distinguishes different clay sources is not the absolute concentrations of the various trace elements, but the degree of correlation between pairs of elements. The point is

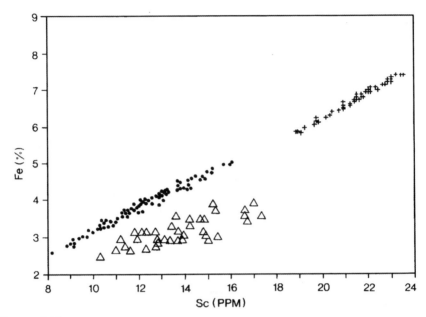

Figure 4.10   Scatterplot of Fe and Sc values for three different pottery groups, showing the effect of correlation on the data. (Redrawn with permission from Bishop and Neff, 1989; Figure 2. Copyright 1989, American Chemical Society.)

illustrated in Figure 4.10 (from Bishop and Neff, 1989), showing the correlation between Fe and Sc in three distinct clay sources. Two sources have approximately the same degree of correlation, but differ in absolute concentrations, but the third shows a different relationship between the two elements. This is simply a reflection of the different geochemical forces acting on a clay deposit as a result of deposition and weathering, and is a very good argument for attempting to understand these processes as part of the scientific approach towards the study of archaeological ceramics.

If the geochemical correlations between elements in clays are little discussed in the archaeological literature, then this paucity is nothing compared with the lack of consideration of post-depositional geochemistry on the composition of archaeological ceramics. Almost every discussion of ceramic provenance studies assumes without question that the composition of the ceramic vessel as recovered from the ground is unchanged from the composition of the newly fired vessel in antiquity. This is unlikely to be so if there is any water around in the burial environment – a fact widely acknowledged in archaeological bone studies (*e.g.*, Hancock *et al.*, 1989), but largely unconsidered in pottery work. There are a few exceptions. In 1967, Freeth reported a study of some Bronze Age sherds from Lincolnshire, in which he compared the major and minor chemical element analysis (as determined by optical emission spectrometry) of several sherds from the same vessel, some of which were recovered from below the current water table, and some from above. He concluded that two of the oxides measured (CaO and MnO) showed conclusive evidence of post-depositional change, with the CaO being three times greater in the sherds found above the water table. This vessel was originally highly calcareous ('shell tempered'), implying that calcium has been removed from the sherds below the water table. It is not clear to what extent these data can be taken as general, but it obviously provides some evidence for a potential problem. More recent work on the post-depositional alteration of glass and ceramics has been reviewed by Freestone (2001).

Much of this more recent interest has focused on the question of the possible uptake of phosphorus into archaeological ceramics, the evidence for which was summarized by Freestone *et al.* (1985). This shows that the evidence from archaeological material is contradictory, as is the outcome of laboratory simulation experiments. Freestone *et al.* (1985) report the results of electron microprobe analyses of cross-sections from a range of archaeological samples, all of which showed evidence for enhanced P_2O_5 concentrations at the surfaces of the sherds, which is mirrored by the behaviour of CaO and FeO. They interpret this as evidence of extensive post-depositional alteration in the surface layers of these sherds. A subsequent (but recurring) suggestion that enhanced levels of phosphate in pottery could be taken as a marker for the presence of residues of blood and fats (*e.g.*, Bollong *et al.*, 1993) has been vigorously (and, in our view, correctly) rebuffed by Freestone *et al.* (1994). The more common view that phosphate accumulation is the result of post-depositional uptake and is an indicator of contamination during burial has recently been challenged by Maritan and Mazzoli (2004), who argue that the phosphorus is endogenous to the sample, and is therefore mainly just concentrated into aggregate phosphate minerals such as vivianite by post-depositional processes.

Very little consideration has been given to the actual nature of mineralogical change experienced by ceramic material during burial. Maggetti (1982) notes that low-fired ceramics (below 700 °C) in particular are susceptible to mineralogical change, with the secondary products being carbonates (particularly calcite, $CaCO_3$), hydrates [hematite (Fe_2O_3) hydrating to goethite, $Fe_2O_3.H_2O$], hydrosilicates (zeolites and various clay minerals such as montmorillonites) and hydrated sulfates such as gypsum ($CaSO_4.2H_2O$). Some consideration has also been given to the more aggressive chemical environment of seawater (Bearat *et al.*, 1992), in which long-term experiments have suggested that certain ceramics will lose calcium and strontium, and gain magnesium. These and the other studies summarized here clearly have important implications for the interpretation of analytical data from archaeological ceramics, and highlight the importance of petrological and mineralogical analysis to support the chemical data. They clearly show that, under some circumstances at least, archaeological ceramics may undergo considerable alteration.

One profitable line of enquiry has been the application of the concept of CEC, familiar in the study of clays, but only sporadically applied to fired ceramics (*e.g.*, Hedges and McLellan, 1976). This is an experimental measure of the proportion of exchangeable cations held by the clay (or ceramic), and may have implications for radiometric dating techniques, as well as for provenance applications using trace elements which might be gained or lost from groundwaters. A further approach is to use the techniques of geochemical groundwater computer modelling, particularly those capable of simulating the reaction between rock minerals and groundwater, since this is analogous to the events occurring in the post-depositional alteration of ceramics (Wilson and Pollard, 2002). Geochemical modelling has rarely been used in archaeological research to date. Very simple thermodynamic equilibrium data have been applied to understand the post-depositional stability of certain archaeological materials, usually via the generation and interpretation of Pourbaix (stability) diagrams. Of these, studies of metal corrosion, including copper (Thomas, 1990; McNeil and Selwyn, 2001) and lead minerals (Edwards *et al.*, 1992; Edwards, 1998), have been the most successful. The potential use of geochemical models in the study of bone diagenesis has been actively promoted (Pollard, 1995) but no substantial studies on the inorganic phase have been published, with the exception of the wide-ranging pilot work carried out by Wilson (2004) using The Geochemist's Workbench^{TM} (Bethke, 1996). In her thesis, Wilson simulated the alteration of a wide range of different archaeological materials, including metal, bone, glass and ceramic. The latter included modelling the alteration of three different types of ceramics [high-fired Tianqi porcelain as described above, mid-temperature Roman terra sigillata (see Section 4.5) and low-fired Early Neolithic Knossos earthenware]. The aim of this was to simulate published experimental studies, to allow a comparison of predicted and observed mineral alteration. One model successfully simulated the alteration of ceramic in seawater, as carried out experimentally by Bearat *et al.* (1992; 153). In their experimental work, a synthetic mixture of 50% calcite and 50% kaolinite was fired at 1000 °C and submerged in simulated seawater for 2 months. After this time XRD revealed the disappearance of portlandite [$Ca(OH)_2$] and larnite (β-Ca_2SiO_4), and the appearance of brucite [$Mg(OH)_2$]. For the modelling (Wilson, 2004), the assumption was made that this ceramic would degrade at a rate similar to Neolithic Knossos earthenware, since both are simple clay and calcite composites. The geochemical model does not allow diagenesis to be traced over the same short time period as that in Bearat *et al.* (1992), but it does permit mineralogical dissolution and precipitation to be investigated over the course of 10 years (Figure 4.11). Over this time period, the model predicts that larnite and portlandite will disappear as stable crystalline phases. Quartz survives for the first year, but thereafter dissolves and re-precipitates in secondary mineral form (antigorite [$Mg_{48}Si_{24}O_{85}(OH)_{62}$], a kaolinite polymorph). After 10 years, brucite and antigorite are the dominant phases, followed by magnesite ($MgCO_3$). Bearat *et al.* (1992; 159) also observed equimolar loss of calcium and gain of magnesium in their ceramic, which is predicted by the model (Wilson, 2004). The model therefore replicates all the main features of Bearat *et al.*'s experimental observations. Extension of the reaction path over 150 years predicts monticellite ($CaMgSiO_4$) precipitation will also occur as larnite dissolution proceeds. Crude and relatively unrealistic though this pilot study may be, it does indicate that this is a fertile area for

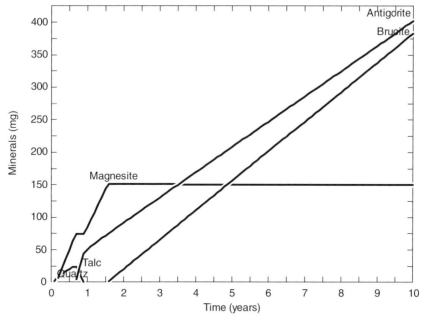

Figure 4.11 Simulated reaction pathway between $25\,cm^3$ ceramic and 1 kg seawater, at STP, showing the predicted alteration of the minerals in the ceramic. Total corrosion rate = 104.62 mg/year. [From Wilson, 2004; 199, generated using The Geochemist's Workbench™ (Bethke, 1996), with permission of the author.]

further analytical and theoretical study, and that the post-depositional alteration of ceramics may, in some circumstances, be significant.

4.5   THE PROVENANCE OF ARCHAEOLOGICAL CERAMICS: ROMAN FINEWARES

The majority of chemical provenance studies carried out since the 1970s have utilized, for good theoretical reasons, trace element analysis by NAA, and more recently ICP spectrometry (either ICP-AES, using an optical emission detector, or ICP-MS, where the plasma source is coupled to an even more sensitive mass spectrometer; see Chapter 2). A review of the chemical characterization of archaeological ceramics has been published by Neff (1992). The example illustrated here, which was carried out at the Research Laboratory for Archaeology and the History of Art, University of Oxford, has been specifically chosen to illustrate the possibility of using major and minor element analysis (via AAS or even XRF). Although inherently less powerful than the trace element techniques for provenancing, this type of approach does give more insight into the nature of the clays used and, occasionally (as described above in connection with the Japanese porcelain study), allows something to be said about the nature of the raw materials being used. Rarely can this be deduced from trace element data.

'*Rhenish*' *wares* are a type of fine Roman pottery made in the Rhineland and central Gaul (around the Lezoux area), and imported into Britain from the early 2nd to the late 3rd Centuries AD (Symonds, 1992). Their manufacture is closely related to that of *terra sigillata*, or *samian* – the most famous of the Roman fine wares, typically with a dark red glossy surface slip. Both wares are characterized by a fine red fabric. Rhenish wares often show a light or dark grey sandwich-like core in a cross-section of the paste, and have a very glossy polished black, or dark red or dark green colour coating. They usually occur in the forms of beakers and cups, and, less frequently, carafes and flagons. Typical decorations are rouletting, indenting, folding and decorations '*en barbotine*', either under the colour coating or painted on it in a white paint. Barbotine decoration under the colour coating is more typical of central Gaulish vessels, and sometimes includes figure types commonly found on central Gaulish *terra sigillata*. The white-painted decoration is more common to vessels from the Rhineland, where they are referred to as *spruchbechers*, or 'motto' beakers, as the decoration often consists of a repeating scroll underneath a 'motto', usually an exhortation to drink.

Chemical research on 'Rhenish' wares was initiated in order to study their distribution in Roman Britain, since it had been found to be very difficult to distinguish visually between vessels from the Rhineland and those from central Gaul, especially with material from Roman–British occupation sites, which tends to produce small sherds, as opposed to whole vessels from graves. A comparison of published vessels from the Rhineland, France and Britain suggested the existence of several visual source indicators, but especially in the limited British literature these seemed speculative and perhaps inadequate. These indicators included: straight, pedestal bases on central Gaulish vessels, versus curved, beaded bases on east Gaulish vessels; the absence of a grey, sandwich-like core in the fabric of the former, versus its presence with the latter vessels; and the decorations '*en barbotine*', which are under the colour coating on the former vessels, and above it and in white paint on the latter vessels. Therefore a pilot programme of analysis of 'Rhenish' wares by AAS and XRF was initiated in Oxford. This consisted of 20 samples from several British excavations, divided into two groups thought to correspond to the two production areas according to the visual criteria described above. The results confirmed these two groups as chemically distinct and relatively internally homogeneous, as shown in Figure 4.12, where sections of two typical qualitative XRF spectra are shown. Most striking is the presence of As in the central Gaulish vessel (estimated average value expressed as As_2O_3 around 0.01%), along with higher values of Zn, Rb and Sr. The geochemistry of As is complicated by the fact that it exhibits variable valency (existing as $3+$ and $5+$), but it is usually associated with oxidizing primary mineralization, but can also be concentrated in reduced marine sediments. The general enhancement of trace elements in the central Gaulish vessels may suggest a reducing marine environment for the deposition of the parent clay beds, but this is speculation.

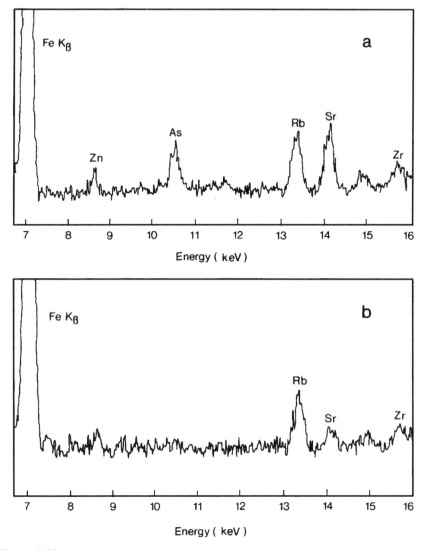

Figure 4.12   Part of XRF spectra from two 'Rhenish' wares of different suspected
origin (a) Alcester 76, attributed to Central Gaul; (b) Verulamium 58,
attributed to Eastern Gaul. (From Pollard *et al.*, 1981; Figure 1, with
permission from *Revue d'Archéométrie*.)

Following this successful pilot, attention turned to the more critical question
of where these vessels were actually made. It had long been suspected that
'Rhenish' wares are closely related to *terra sigillata*. Typologically, the relationship is closer with central Gaulish wares, where identical applied figure types
are found on central Gaulish *sigillata* and on central Gaulish 'Rhenish' vessels.
East Gaulish 'Rhenish' wares are less closely related to *sigillata*: similar

decoration is found on red vessels only rarely. 'Rhenish' wares are nonetheless the only pottery type made in the Rhineland of a quality similar to that of east Gaulish *sigillata* (or better) during the same period. *Terra sigillata* has been extensively studied, both typologically and chemically (*e.g.*, Picon *et al.*, 1971; 1975). Velde and Druc (1999; 211) have published a synthesis of the chemical work carried out on French and Italian samian ware ('sigillate'). They demonstrate that the major French and Italian workshops can be distinguished on the basis of major element chemistry (although they point out that this is '*largely unnecessary*', since '*decoration ... is usually sufficient to identify samples and give a production age*' – a point with which we do not entirely agree!).

If it could be shown, therefore, that 'Rhenish' wares from either production area matched chemically with *sigillata* from any particular known workshop in that area, then it could reasonably be surmised that the 'Rhenish' wares and *sigillata* came from the same clay source and were therefore probably made in the same workshop. As stamps or any obvious typological evolution are absent, 'Rhenish' wares have always been rather broadly dated: if there is a direct link with *sigillata*, which is generally well-dated, there ought to be a better chance of dating 'Rhenish' wares more closely. Further chemical analysis was therefore undertaken on 50 more samples of 'Rhenish' wares (from several British excavations) along with 50 samples of *sigillata* identified on typological grounds as being from Trier, Rheinzabern, Blickweiler, Chemery, La Madeleine and Lavoye (all east Gaulish sites), Lezoux and Les Martres de Veyre (central Gaulish sites), and Colchester (the only then known British kiln making *sigillata*: the chemical composition of material from this site has been considered in detail by Hart *et al.*, 1987). These results were discussed in Pollard *et al.* (1981). Subsequent work supplemented these sites with *sigillata* samples from the central Gaulish sites of Clermont Ferrand and Toulon-sur-Allier, the 'mid-Gaulish' sites of Gueugnon, Jaulges-Villiers-Vineux and Domecy-sur-Cure (see map, Figure 4.13). These data were discussed in detail in Pollard *et al.* (1982).

The major element chemical data provided by AAS analysis were analysed using the multivariate techniques developed at the time in Oxford (Pollard, 1986). In the first study, principal components analysis (PCA) and cluster analysis showed that the Lezoux *terra sigillata* and the supposed central Gaulish 'Rhenish' wares grouped together, well-separated from the east Gaulish material – principally distinguished by higher values of SrO, CaO, As_2O_3 and lower values of NiO, K_2O and Cr_2O_3 (Figure 4.14). The east Gaulish vessels occupy a different region of the plot, with the Trier and Rheinzabern groups distinguished from each other, and with all of the east Gaulish 'Rhenish' vessels associated with the Trier material. It was concluded, therefore, that not only were Trier and Lezoux the principal production centres for the 'Rhenish' wares imported into England, but also that the same clay was being used for black and red gloss vessels. What was also important archaeologically was that it also supported the separation of the 'Rhenish' vessels into east and central Gaulish groups on the grounds of the macroscopic criteria described above, meaning that pottery specialists could continue to use these

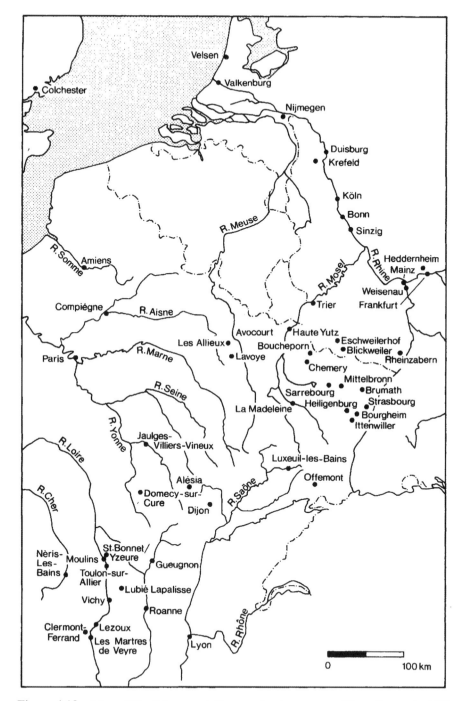

Figure 4.13 Map of Gaulish *terra sigillata* production centers. (From Symonds, 1992, by permission of Oxford University Committee for Archaeology.)

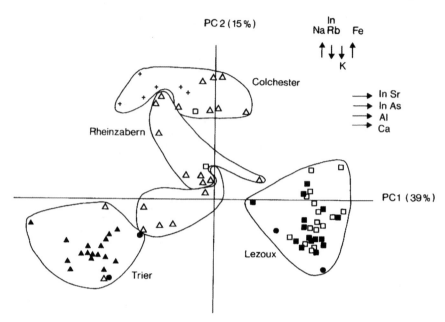

Figure 4.14   Cluster analysis of Rhenish and *terra sigillata* analyses, showing association between Lezoux *sigillata* and Central Gaulish 'Rhenish', and Trier
sigillata and East Gaulish 'Rhenish'. Key: ■ Central Gaulish *sigillata*; □
Central Gaulish 'Rhenish'; △ East Gaulish *sigillata*; ▲ East Gaulish
'Rhenish'; + Colchester *sigillata*; ● uncertain 'Rhenish'. (From Pollard
et al., 1981; Figure 4, with permission from *Revue d'Archéométrie*.)

criteria with a greater degree of confidence. Unfortunately, however, a large
number of sherds are found which cannot be classified by these visual means.
In these cases, chemical analysis appears to be the only reliable guide to
provenance. Only a handful of the 85 British 'Rhenish' vessels analysed did
not appear to originate from these two centres – these may be associated with
minor production centres (minor, that is, in terms of supply to Britain), and this
encouraged us to extend the study a little further, in order to resolve this
problem.

Another notable result from this work was the somewhat surprising success
of the XRF technique, which appeared in this case to provide a rapid, non-
destructive method for distinguishing sherds from east and central Gaul, both
'Rhenish' and *terra sigillata*. A 60 second irradiation of the cleaned edge of
a sherd was sufficient to produce a spectrum of the type reproduced in
Figure 4.12, and on the basis of this an opinion could be given as to the source
of the vessel. It is not, of course, possible to distinguish between the various east
Gaulish production sites on the basis of such a quick test, but with a small plain
body sherd such information has on several occasions proved useful. Ancillary
work indicated that the same trace element characteristics shown by the body
fabric of vessels were possessed by the surface gloss from both sources (as might
be expected if the slip was made from the finest fraction of the same clay as that

used for the body fabric). This therefore offers a useful source indicator when even the minimal damage caused by cleaning the edge of a sherd is not permissible, or on whole vessels. A similar suggestion for the discrimination of *sigillata* from Arezzo, La Graufesenque, Lezoux and Lyon using a plot of the intensity ratios of Ti/Mn versus Fe/K measured on the surface has subsequently been made by Baillié and Stern (1984).

This preliminary work left six of the 85 fragments of 'Rhenish' ware found in Britain unaccounted for. Could any of these have come from any of the smaller Gaulish production centres known to be producing vessels of similar type and quality? It had already been shown that it is possible to distinguish chemically between *sigillata* from closely adjacent kilns, such as Lezoux and Les Martres de Veyre (Picon, unpublished; Velde and Druc, 1999), and so it should be possible to discriminate between the other comparable products from these sources. The problem was approached by analysing separately the analytical data from the three regions under consideration – central Gaul, 'mid Gaul' and east Gaul, or the Rhineland. Simplified dendrograms from the cluster analysis of these three data sets are reproduced in Figure 4.15. The first area to be considered in more detail was the central Gaulish area around Lezoux, including Les Martres de Veyre, Clermont-Ferrand and Toulon-sur-Allier (see map, Figure 4.13). Unfortunately, no samples were obtainable from the suspected workshops at Neris-les-Bains, St Bonnet/Yzeure, Lubié-Lapalisse or Vichy, but a selection of vessels were analysed from the museum at Moulins, which might be expected to contain examples of the work of most of the local kilns. The data set contained 96 samples, including the 'Rhenish' wares found in England and attributed to central Gaul. Cluster analysis of the analytical data identified four distinct groups, each one attributable to one of the four kilns represented, confirming that the products are, indeed, chemically distinct, and suggesting a secondary attribution for some of the vessels imported into England from this area (Figure 4.15a). The distinction between the groups attributed to Lezoux and Les Martres de Veyre is based on the Cr, Mg and Ca content, but several of the Les Martres 'Rhenish' wares were found in the Lezoux group. It was originally suspected that some of the sherds from Les Martres might, in reality, be products of the Lezoux kilns, and this appears to be the case. The group labelled Clermont-Ferrand consists mainly of vessels from Moulins museum, but the presence of three Clermont-Ferrand sherds suggests this to be the source, despite the fact that the Lezoux group also contains two (possibly three) of the vessels also thought to be manufactured in Clermont-Ferrand. This group is interesting, however, because it contains the only British finds not attributed to Lezoux – fragments of three vessels, all found at Milton Keynes.

The second group was attributed to an area loosely described as north central Gaul or Mid Gaul – the region around Alésia (Figure 4.15b). From a study of the material in French museums, it is obvious that several groups of black colour-coated vessels of similar quality to 'Rhenish' wares were being produced in regions other than central and east Gaul. A good example of this was found at the Musée Alésia, which contains a set of high-quality vessels characterized

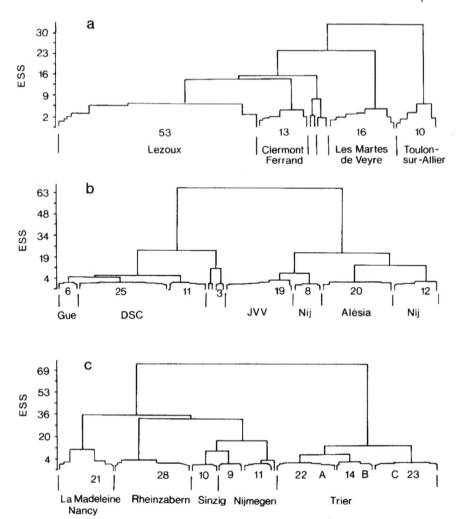

Figure 4.15 Dendrograms of chemical data from 'Rhenish' wares and *terra sigillata* from three regions of Gaul: (a) Central Gaul, (b) Mid Gaul, (c) East Gaul. (From Pollard *et al.*, 1982; Figures 2–4, with permission.)

by tall, everted rims and excised decoration. The nearest known workshops producing *sigillata* are Gueugnon and Jaulges-Villiers-Vineux, and these, together with Domecy-sur-Cure, are also thought to have been making black colour-coated vessels. Samples were obtained from these three workshops and from the museum to determine whether (a) the museum samples were made at Lezoux and (b) if not, whether they were from one of the neighbouring *sigillata* production centres. Subsequently, results from the analyses of fourth century colour-coated vessels from Nijmegen, originally obtained for comparison with Trier, were added to the data set, because preliminary clustering of all the data suggested that they were more closely related to the mid-Gaulish material.

The dendrogram is more difficult to interpret than that for Central Gaul, but nine clusters appear to be significant, seven of which are identifiable as associated with Gueugnon (Gue), Domecy-sur-Cure (DSC), Jaulges-Villiers-Vineux (JVV), Alésia, and two groups containing material only found at Nijmegen (Nij). There is some exchange of samples between the Jaulges-Villiers-Vineux group and that attributed to Gueugnon, but the chemical differences are quite clear, and we may suggest on the basis of chemical analysis that the original attributions are incorrect.

The final area to be studied was east Gaul (Figure 4.15c). This study had two purposes, one of which related to the Nijmegen vessels, with their implications for 4th Century production at Trier after the decline around AD 270. This has been discussed above, and a Trier origin for these vessels has been discounted on chemical grounds. The second purpose, by analogy with the work on central Gaul, was to confirm that all the vessels labelled as east Gaulish 'Rhenish' in Britain did originate in Trier, and not from any other east Gaulish *sigillata* centre. Unfortunately, most of the minor *sigillata* kilns are only represented by two or three sherds, which is not enough to define a reliable chemical signature for these kilns – they appear to have been classified with either Trier or Rheinzabern. This is likely to be an artefact of the numerical analysis, resulting from the much larger numbers of Trier and Rheinzabern samples included. The classification suggested nine significant clusters, identified as La Madeleine A, La Madeleine B, Rheinzabern, Sinzig, Nijmegen A, Nijmegen B and three groups from Trier, labelled A, B and C. The archaeological interpretation of these results gives rise to several difficulties. There is nothing associated with the group labelled 'Trier A' to confirm this attribution, although none of the *sigillata* associated with this group give convincing proof of another source. The close chemical proximity of this group to the other two clusters identified with Trier on the dendrogram is the sole evidence for a Trier origin. Two scenarios are possible – (a) the group labelled Trier A does not originate in Trier, but from some other chemically very similar kiln site, or (b) the production of 'Rhenish' and *sigillata* at Trier was not as closely related at Trier as at Lezoux, resulting in slight differences in the fabric. There could also be a temporal difference in the production of the two types of vessel, with some associated slight change of clay source. A third explanation is also possible, relating to the clustering procedures used. Ward's method (used to produce Figure 4.15) can sometimes yield unacceptable divisions in dense data, arbitrarily dividing up large groups into a number of clusters which are approximately spherical in multivariate space (Pollard, 1982). Other clustering algorithms, such as the average linkage cluster algorithm (ALCA), when applied to these data sets do not divide the Trier material in this way, but neither do they separate clearly the groups attributed to Rheinzabern, Sinzig and Nijmegen, indicating the sensitivity of such interpretations to the exact data analysis procedures used. A time variation in the composition of the ceramics produced at such long-lived production centres, as in the *sigillata* production at Lezoux (Picon *et al.*, 1971), has been postulated, but there is no firm evidence for this at Trier, since none of the *sigillata* analysed from Trier show any tendency to split into different compositional groups. Subsequent work on Lezoux *sigillata* (Argyropoulos, 1992) has suggested that once a satisfactory raw material recipe had been obtained for *sigillata*, the clay sources used by at least six workshops in Lezoux remained constant throughout the 2nd Century AD. The resolution of this problem is important to our understanding of imported Roman finewares in Britain, since half of the east Gaulish 'Rhenish' wares from England fall into this 'Trier A' group (the other half being securely associated with Trier *sigillata* in the 'Trier C' group). It is clearly a matter of some concern to decide whether we are dealing with an artefact of the data analysis employed, the products of a minor *sigillata* kiln in the Trier region or a time distinction in the composition of the products from Trier itself. This is a matter which can only be resolved by further analysis of well-provenanced material, and serves to illustrate the usual outcome of chemical provenance work – some questions can be confidently answered, but others are inevitably left undecided.

4.6   SUMMARY

The example of chemical provenancing of Roman finewares described above is atypical of the wider field of analytical studies of archaeological ceramics, but does illustrate both the power of the technique and some of the potential problems, particularly in the interpretation of the data. In terms of the problem of identifying the Continental sources of 'Rhenish' wares found on British sites, the work has been highly successful. With very few exceptions, all the samples analysed came from either Lezoux or, in all probability, Trier, and the clear chemical discrimination allowed the typological criteria to be examined from a firm base. The clear compatibility of 'Rhenish' wares with *sigillata* (at least at Lezoux) has allowed the archaeological study of 'Rhenish' vessels to proceed at a much faster pace by comparison with the much better-known *sigillata*, and led us to believe that many of the smaller 'Rhenish' and *sigillata* production sites in France could be chemically distinguished. The chemical characteristics of some of the samples found at Milton Keynes compared very closely with those of samples from Clermont-Ferrand, and these are, so far, the only vessels of this type found in Britain not to be attributed to either Lezoux or Trier. This is a remarkable tribute to the value of chemical studies of archaeological ceramics, when combined with an archaeologically coherent research design.

Methodologically, inevitably some questions must always be considered. Do the 'control sherds' adequately represent the output of all of the relevant vessels from all of the relevant kiln sites? There is always potentially the problem of 'kiln X' – an unknown (or at least, uncharacterized) kiln which is producing wares similar to those under consideration, but not included either because it has not been identified archaeologically, or because it was not known to produce the type of ware under question. This problem should be at least considered since a geochemical knowledge of the type of clay deposits being exploited may restrict the geographical extent of the possible clay sources. The question of the representativeness of the samples is more difficult. Most kiln sources are characterized by at best 100 analyses, probably covering a range of 'qualities' of vessels produced, a range of vessel types and, possibly, a time span of tens or even hundreds of years. Essentially the archaeological chemist is relying on the quality-control procedures applied in antiquity to ensure that the sample is representative of the range of compositions produced! It is undoubtedly essential to match as closely as possible the fabric, shape, quality and date of the control sample with the unknown material. Again, this emphasizes the need for extremely good archaeological definition of a project to give any chance of success. Finally, of course, the data analysis needs to be handled with extreme care, from a knowledge of the propensities of the different methods which might be applied. It is all too easy to rely on one technique, just because it gives an easily interpreted picture.

More generally, we hope to have shown in the preceding discussion that there is scope for further investigation even in this rather well-developed technique in archaeological chemistry – the chemical provenancing of ceramics. There is room for a more systematic application of knowledge of the geochemical behaviour of trace elements to aid in understanding the relationship of secondary clays and provenance, perhaps even to develop a predictive model of the type: 'the clay used for these ceramics must have come from a particular sort of deposit, and these are known to occur only in these areas'. Perhaps more fundamentally, it is becoming increasingly apparent that there are observable effects due to post-depositional alteration which could in some cases have a significant effect on the interpretations made by archaeologists from the chemical analysis of ceramics. Ironically, it could be that these effects have only become significant now that the analytical precisions delivered by the latest generations of instrumentation are such that small changes matter! The price of progress?

REFERENCES

Ahrens, L.H. (1954). The lognormal distribution of the elements 1 (A fundamental law of geochemistry and its subsidiary). *Geochimica et Cosmochimica Acta* **5** 49–75.

Allen, R.O., Luckenbach, A.H. and Holland, C.G. (1975). The application of instrumental neutron activation analysis to a study of prehistoric steatite artifacts and source material. *Archaeometry* **17** 69–83.

Argyropoulos, V. (1992). *Chemical Studies of the Roman Samian Pottery Industry of Central Gaul*. Unpublished PhD Thesis, Department of Archaeological Sciences, University of Bradford.

Aylward, G.H. and Findlay, T.J.V. (1974). *S.I. Chemical Data*. John Wiley, Chichester, 2nd edn.

Baillié, P.J. and Stern, W.B. (1984). Non-destructive surface analysis of Roman terra sigillata: a possible tool in provenance studies? *Archaeometry* **26** 62–68.

Bearat, H., Dufournier, D. and Nouet, Y. (1992). Alteration of ceramics due to contact with sea water. *Archaeologia Polona* **30** 151–162.

Bethke, C.M. (1996). *Geochemical Reaction Modeling: Concepts and Applications*. Oxford University Press, Oxford.

Bishop, R.L. and Lange, F.W. (eds.) (1991). *The Ceramic Legacy of Anna O. Shepard*. University Press of Colorado, Boulder, Colorado.

Bishop, R.L. and Neff, H. (1989). Compositional data analysis in archaeology. In *Archaeological Chemistry IV*, ed. Allen, R.O., American Chemical Society Advances in Chemistry Series 220, Washington D.C., pp. 57–86.

Bollong, C.A., Vogel, J.C., Jacobson, L., van der Westhuizen, W.A. and Sampson, C.G. (1993). Direct dating and identity of fibre temper in pre-contact Bushman (Basarwa) pottery. *Journal of Archaeological Science* **20** 41–55.

Brady, N.C. (1974). *The Nature and Properties of Soils*. Macmillan, New York, 8th edn.

Brady, N.C. and Buckman H.O. (1990). *The Nature and Property of Soils*. Prentice Hall, Upper Saddle River, New Jersey, 10th edn.

Bragg, W.L. (1937). *The Atomic Structure of Minerals*. Cornell University Press, New York.

Brindley, G.W. and Lemaitre, J. (1987). Thermal, oxidation and reduction reactions of clay minerals. In *Chemistry of Clay and Clay Minerals*, ed. Newman, A.C.D., Mineralogical Society Monograph No. 6, Longmans, London, pp. 319–370.

Brown, G. (1984). Crystal structure of clay minerals and related phyllosilicates. In *Clay Minerals: Their Structure, Behaviour and Uses*, ed. Fowden, L., Barrer, R.M. and Tinker, P.B., Royal Society, London, pp. 1–20.

Deer, W.A., Howie, R.A. and Zussman, J. (1962, 1963). *Rock-Forming Minerals*. Longmans, London (5 vols).

Edwards, R. (1998). The effects of changes in groundwater geochemistry on the survival of buried metal artefacts. In *Preserving Archaeological Remains In-Situ*, ed. Corfield, M., Hinton, P., Nixon, T., Pollard, A.M., Museum of London Archaeology Service, London, pp. 86–92.

Edwards, R. Gillard, R.D., Williams, P.A. and Pollard, A.M. (1992). Studies of secondary mineral formation in the PbO–H₂O–HCl system. *Mineralogical Magazine* **56** 51–63.

Evans, R.C. (1966). *An Introduction to Crystal Chemistry*. Cambridge University Press, Cambridge, 2nd edn.

Freestone, I.C. (2001). Post-depositional changes in archaeological ceramics and glass. In *Handbook of Archaeological Sciences*, ed. Brothwell, D.R. and Pollard, A.M., Wiley, Chichester, pp. 615–625.

Freestone, I. and Gaimster, D. (eds) (1997). *Pottery in the Making*. British Museum Press, London.

Freestone, I.C., Meeks, N.D. and Middleton, A.P. (1985). Retention of phosphate in buried ceramics: an electron microbeam approach. *Archaeometry* **27** 161–177.

Freestone, I.C., Middleton, A.P. and Meeks, N.D. (1994). Significance of phosphate in ceramic bodies: discussion of paper by Bollong *et al. Journal of Archaeological Science* **21** 425–426.

Freeth, S.J. (1967). A chemical study of some bronze age sherds. *Archaeometry* **10** 104–119.

Glascock, M.D. (1992). Characterization of archaeological ceramics at MURR by neutron activation analysis and multivariate statistics. In *Chemical Characterization of Ceramic Pastes in Archaeology*, ed. Neff, H., Prehistory Press, Madison, Wisconsin, pp. 11–26.

Goldschmidt, V.M. (1954). *Geochemistry*. Clarendon, Oxford.

Grim, R.E. (1968). *Clay Mineralogy*. McGraw-Hill, New York, 2nd edn.

Hancock, R.G.V., Grynpas, M.D. and Pritzker, K.P.H. (1989). The abuse of bone analyses for archaeological dietary studies. *Archaeometry* **31** 169–179.

Hart, F.A., Storey, J.M.V., Adams, S.J., Symonds, R.P. and Walsh, J.N. (1987). An analytical study, using inductively coupled plasma (ICP) spectrometry, of samian and colour-coated wares from the Roman town at Colchester together with related Continental samian wares. *Journal of Archaeological Science* **14** 577–598.

Hedges, R.E.M. and McClennan, M. (1976). On the cation exchange capacity of fired clays and its effect on the chemical and radiometric analysis of pottery. *Archaeometry* **18** 203–207.

Heimann, R.B. (1989). Assessing the technology of ancient pottery: the use of ceramic phase diagrams. *Archeomaterials* **3** 123–148.

Henderson, P. (1982). *Inorganic Geochemistry*. Pergamon Press, Oxford.

Impey, O.R., Pollard, A.M., Wood, N. and Tregear, M. (1983). An investigation into the provenance and technical properties of Tianqi porcelain. *Trade Ceramic Studies* **3** 102–158.

Jones, R.E. (1986). *Greek and Cypriot Pottery: A Review of Scientific Studies*. British School at Athens Fitch Laboratory Occasional Paper 1, Athens.

Keally, C.T., Taniguchi, Y., Kuzmin, Y.V. and Shewkomud, I.Y. (2004). Chronology of the beginning of pottery manufacture in East Asia. *Radiocarbon* **46** 345–351.

Kingery. W.D. (ed.) (1986). *Technology and Style: Symposium on Ceramic History and Archaeology*. American Ceramic Society, Columbus, Ohio.

Kingery. W.D. (ed.) (1990). *The Changing Roles of Ceramics in Society: 26,000 B.P. to the Present*. American Ceramic Society, Westerville, Ohio.

Kingery W.D and Vandiver P.B. (1986). *Ceramic Masterpieces: Art, Structure, and Technology*. Free Press, New York.

Kingery, W.D., Bowen, H.K. and Uhlmann, D.R. (1976). *Introduction to Ceramics*. John Wiley, New York, 2nd edn.

Maggetti, M. (1982). Phase analysis and its significance for technology and origin. In *Archaeological Ceramics*, ed. Olin, J.S. and Franklin, J.D., Smithsonian Institution Press, Washington D.C., pp. 121–133.

Maritan, L. and Mazzoli, C. (2004). Phosphates in archaeological finds: implications for environmental conditions of burial. *Archaeometry* **46** 673–683.

McNeil, M. and Selwyn, L.S. (2001). Electrochemical processes in metallic corrosion. In *Handbook of Archaeological Sciences*, ed. Brothwell, D.R. and Pollard, A.M., Wiley, Chichester, pp. 605–614.

Middleton, A. and Freestone, I. (eds) (1991). *Recent Developments in Ceramic Petrology*. Occasional Paper 81, British Museum, London.

Millot, G. (1970). *Geology of Clays: Weathering, Sedimentology, Geochemistry*. Chapman and Hall, London.

Murad, E. and Wagner, U. (1991). Mössbauer spectra of kaolinite, halloysite and the firing products of kaolinite – new results and a reappraisal of published work. *Neues Jahrbuch für Mineralogie-Abhandlungen* **162** 281–309.

Neff, H. (ed.) (1992). *Chemical Characterization of Ceramic Pastes in Archaeology*. Prehistory Press, Madison, Wisconsin.

Neff, H. (2001). Synthesising analytical data – spatial results from pottery provenance investigation. In *Handook of Archaeological Sciences*, ed. Brothwell, D.R. and Pollard, A.M., Wiley, Chichester, pp. 733–747.

Norton, F.H. (1952). *Elements of Ceramics*. Addison-Wesley, Cambridge, Massachussets.

Orton, C., Tyers, P. and Vince, A. (1993). *Pottery in Archaeology*. Cambridge University Press, Cambridge.

Picon, M., Vichy, M. and Meille, E. (1971). Composition of the Lezoux, Lyon and Arezzo samian ware. *Archaeometry* **13** 191–208.

Picon, M., Carre, C., Cordoliani, M.L., Vichy, M., Hernandez, J.A. and Mignard, J.L. (1975). Composition of the La Graufesenque, Banassac and Montans terra sigillata. *Archaeometry* **17** 191–199.

Pollard, A.M. (1982). A critical study of multivariate methods as applied to provenance data. In *Proceedings of the 22nd Symposium on Archaeometry, University of Bradford, 30th March–3rd April 1982*, ed. Aspinall, A. and Warren, S.E., University of Bradford Press, Bradford, pp. 56–66.

Pollard, A.M. (1986). Multivariate methods of data analysis. In *Greek and Cypriot Pottery: A Review of Scientific Studies*, ed. Jones, R.E., British School at Athens Fitch Laboratory Occasional Paper 1, Athens, pp. 56–83.

Pollard, A.M. (1995). Groundwater modeling in archaeology – the need and the potential. In *Science and Site*, ed. Beavis J. and Barker K., Bournemouth University School of Conservation Sciences Occasional Paper 1, pp. 93–98.

Pollard, A.M. and Wood, N. (1986). Development of Chinese porcelain technology at Jingdezhen. In *Proceedings of the 24th International Archaeometry Symposium*, ed. Olin, J.S. and Blackman, M.J., Smithsonian Institution, Washington, pp. 105–114.

Pollard, A.M., Hatcher, H. and Symonds, R.P. (1981). Provenance studies of 'Rhenish' pottery by comparison with *terra sigillata*. *Revue d'Archéométrie, Actes due XX Symposium International d'Archéométrie, Paris 26–29 Mars 1980*, Vol II, 177–185.

Pollard, A.M., Hatcher, H. and Symonds, R.P. (1982). Provenance studies of 'Rhenish' wares – a concluding report. In *Proceedings of the 22nd Symposium on Archaeometry, University of Bradford, 30th March–3rd April 1982*, ed. Aspinall, A. and Warren, S.E., University of Bradford Press, Bradford, pp. 343–354.

Pollard, A.M., Batt, C.M., Stern, B. and Young, S.M.M. (2007). *Analytical Chemistry in Archaeology*, Cambridge University Press, Cambridge.

Putnis, A. (1992). *Introduction to Mineral Sciences*. Cambridge University Press, Cambridge.

Renfrew, C. (1977). Introduction: production and exchange in early state societies, the evidence of pottery. In *Pottery and Early Commerce: Characterization and Trade in Roman and Later Ceramics*, ed. Peacock, D.P.S., Academic Press, London, pp. 1–20.

Rice, P.M. (1987). *Pottery Analysis: A Sourcebook*. University of Chicago Press, Chicago.

Rice, P.M. (1999). On the origins of pottery. *Journal of Archaeological Method and Theory* **6** 1–54.

Ries, H. (1927). *Clays. Their Occurence, Properties and Uses*. Chapman and Hall, London.

Singer, F. and Singer, S.S. (1963). *Industrial Ceramics*. Chapman and Hall, London.

Symonds, R.P. (1992). *Rhenish Wares. Fine Dark Coloured Pottery from Gaul and Germany*. Monograph No. 23, Oxford University Committee for Archaeology, Oxford.

Thomas, R.G. (1990). *Mineralogy of Copper Corrosion Products*. Unpublished PhD Thesis, Department of Chemistry, University of Wales, Cardiff.

Vandiver, P.B., Soffer, O., Klima, B. and Svoboda, J. (1989). The origins of ceramic technology at Dolni Vstonice, Czechoslovakia. *Science* **246** 1002–1008.

Velde, B. and Druc, I.C. (1999). *Archaeological Ceramic Materials*. Springer-Verlag, Berlin.

von Meyer, E. (1891). *A History of Chemistry*. Macmillan, London.

Wilson, L. (2004). *Geochemical Approaches to Understanding In Situ Archaeological Diagenesis*. Unpublished PhD Dissertation, Department of Archaeological Sciences, University of Bradford.

Wilson, L. and Pollard, A.M. (2002). Here today, gone tomorrow? Integrated experimention and geochemical modeling in studies of archaeological diagenetic change. *Accounts of Chemical Research* **35** 644–651.

CHAPTER 5

The Chemistry, Corrosion and Provenance of Archaeological Glass

5.1 INTRODUCTION

In the modern world, we are accustomed to taking the chemical stability of glass very much for granted – we rely on the durability of glass for so many things, such as windows and (until the widespread availability of plastics) bottles, as well as its use in the chemical laboratory as an extremely inert and unreactive container. In addition to its apparent inertness, glass has a number of other beneficial properties, such as its transparency or the ability to take on virtually any colour as the result of the addition of a small amount of transition metals.

Even with modern glass, however, this apparent chemical stability is largely an illusion: for example, Bacon (1968) observed that an ordinary soda-lime–silica glass bottle full of water will lose approximately 30 mg of material per year into the solution. The situation with archaeological glass is considerably more complex. During the Roman period and through the early Medieval period (to about AD 1100) in Europe, glass was made from either soda-rich plant ash and pure sand, or *natron* (a hydrated sodium carbonate mineral, $Na_2CO_3.10H_2O$, or, more likely, the mineral *trona*, $Na_3HCO_3CO_3.2H_2O$, obtained from evaporite lake deposits) mixed with a calcareous sand (Newton, 1982). Both result in a glass with a remarkably modern composition. Like modern glass, it is relatively durable, although the majority of excavated examples show evidence of surface iridescence due to deterioration (see the following sections). During the European Medieval period (*ca.* AD 1100–1600), however, the 'building boom' in great Cathedrals required the supply of large quantities of coloured glass to provide the so-called 'stained glass windows' (technically, the majority from the earlier period are painted glass, not stained glass – see below), which was satisfied by glass manufactured using beech ash as the source of the alkali. The manufacture of such glass is described in the

Archaeological Chemistry, Second Edition
By A. Mark Pollard and Carl Heron
© The Royal Society of Chemistry 2008

contemporary writing of Theophilus (Hawthorne and Smith, 1979). This *'forest glass'* has proved to be a major problem for modern-day conservators, since its composition has rendered it almost water soluble in extreme cases, and the maintenance of the surviving glass in a large Cathedral has been likened to the legendary painting of the Forth Road Bridge in Scotland – as soon as the job is completed, it is time to start again! The cause of this poor durability in Medieval glass is to be found largely in the composition of the glass itself, although storage in damp conditions during the Second World War as protection against bomb damage has exacerbated the problem in some cases. Contrary to popular opinion, the increased atmospheric pollution during the 20th Century, whilst clearly not beneficial, has probably not been the primary cause of this deterioration.

This chapter reviews the chemistry and structure of silicate glasses (Section 5.2), and looks at the chemistry of the colouring process (Section 5.3). It then summarizes some work on the relationship between the composition and durability of Medieval European window glass (Section 5.4). Section 5.5 looks briefly at the much more complex decay mechanisms of buried glass, and extends this concept to the chemical stability of vitreous tephra (natural volcanic ash) in the depositional environment, which has implications for tephrochronology (the use of this material as a dating or stratigraphic tool). Section 5.6 reviews some recent developments in the chemical provenancing of glass, using strontium and neodymium isotopes.

5.2 THE STRUCTURE AND CHEMISTRY OF ARCHAEOLOGICAL GLASS

Before describing the structure of glass, it is important to first define the term 'glass', and then to specify the particular branch of this large class of materials with which we are concerned. The definition given by the American Society for Testing and Materials is: *'Glass is an inorganic product of fusion which has cooled to a rigid condition without crystallising.'* This by no means covers all substances found in the vitreous state (*e.g.*, toffee made from heated sugar), nor does it describe all the possible current methods of manufacturing glasses, but it is sufficient for archaeological purposes. An alternative structural definition quoted by Holland (1964) is *'the glass-like or vitreous state is believed to be that of a solid with the molecular disorder of a liquid frozen into its structure.'* A more formal statement of this is given by Paul (1990; 4): *"glass is a state of matter which maintains the energy, volume and atomic arrangement of a liquid, but for which the changes in energy and volume with temperature and pressure are similar in magnitude to those of a crystalline solid.'* This then may also include organic polymers and resins as well as non-oxide glasses, but it is generally better to consider the vitreous state as an additional state of matter (to be added to solid, liquid and gas) rather than to define a glass by the process used in its manufacture. In archaeological studies we are almost exclusively concerned with oxide glasses based on silica, and with few exceptions belonging to the *alkali–alkaline earth–silica* family of glasses. Much of what is said in this chapter also applies to ceramic glazes, although these also include systems such as lead silicates not explicitly described here. The production of such alkali glasses is still the most important aspect (by volume) of the modern glass industry, although many other oxide glasses of technical importance have been developed during the 20th Century. New manufacturing techniques now allow a much wider range of materials to be made in the vitreous state, including products such as metallic glasses, originally produced by the rapid cooling of liquid metals (*e.g.*, Greer, 1995; see below). Such materials have been developed intensively over the past 20 years because of their interesting physical and electrical properties, which are potentially of great technological importance (*e.g.*, Zhang *et al.*, 2006).

An important distinguishing feature of a glass is that it shows no discontinuous change of any measurable property on cooling from the liquid to the vitreous state. Hence it is necessary to define various reference points, usually in terms of *viscosity*, in order to discuss the glass-forming process. Table 5.1 gives the commonly accepted reference points defined by the viscosity of the melt, a brief definition of each point and the corresponding temperatures for a typical soda–lime–silica glass and 96% vitreous silica. The viscosities are quoted in the old CGS units of *poise*, with dimensions ($g\,cm^{-1}\,s^{-1}$), as is conventional in glass technology (the SI equivalent is the Pascal second, with units Pa s, equal to $kg\,m^{-1}\,s^{-1}$). For comparison, glacier ice has a viscosity of 1.2×10^{14} poise, glycerol $10^{1.2}$ poise and water at 25 °C approximately 10^{-2} poise. Table 5.1 clearly shows the benefit in terms of workability that is achieved by modifying the composition of pure silica glass by the addition of alkali and alkaline earths – roughly a factor-of-two reduction in the reference temperatures. The specific volume (volume per unit mass) of a glass

Table 5.1 Characteristic reference temperatures (°C) in glass working. (From Holloway, 1973; 19, with permission of Thomson Publishing Services.)

Reference Point	Definition	Viscosity (poise)	96% SiO_2 glass (°C)	Soda-lime–silica glass (°C)
Working point	Sufficiently soft for shaping	10^4	–	1000
Softening point	Glass tubes can be bent in a flame	$10^{7.6}$	1500	700
Annealing point	Internal stresses removed in a few minutes	$10^{13.4}$	900	510
Strain point	Highest temperature from which the glass can be rapidly cooled without serious internal stress	$10^{14.5}$	820	470

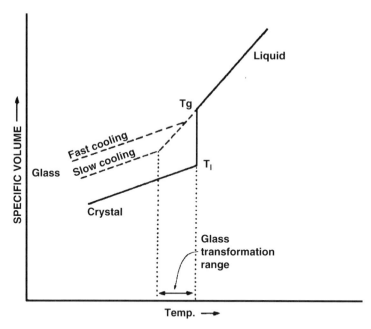

Figure 5.1   Transformation from a liquid to a crystalline or glassy state. (From Paul, 1990; Figure 1.2. With kind permission of Springer Science and Business Media.)

melt is shown in Figure 5.1 as a function of temperature. The solid line shows the result of allowing the melt to cool slowly and crystallize, with a characteristic sharp drop in specific volume at the melting point, corresponding to the phase transition from a liquid to a crystalline solid. This temperature depends on the composition of the melt and is known as the *liquidus temperature* (T_1), defined as the temperature above which no crystalline solid exists. In the glassy state (the broken line), achieved by a slower cooling rate, the curve bends at some temperature below T_1 which depends on the rate of cooling, and attains a gradient similar to that of the crystalline state. The bend defines the *glass formation temperature* (T_g) for that particular melt and cooling rate. Below T_g the melt has become too viscous for molecular movement, and so the system is metastable with respect to the crystalline state, and has the structure of the liquid at T_g 'frozen' into the solid. Glasses of the same composition exhibit a range of temperatures over which vitrification occurs (called the *glass transformation range*) and the exact position of T_g depends on the rate of cooling of the melt. Properties of the glass which depend on the structure (*e.g.*, ionic conductivity, electrical resistance, chemical resistance) depend on T_g and hence on the thermal history of the glass. T_g is also known as the *fictive temperature*.

   The first theory of the structure of glass to become widely accepted was that of Zachariasen (1932), called the *random network theory* [now commonly referred to as the *continuous random network* (CRN) theory]. This arose from the following comparison of the behaviour of crystalline and vitreous silica:

(i)   the mechanical properties of the glass are equivalent to the crystalline form over a large range of temperatures – thus the bonding forces must be similar,

(ii) X-ray diffraction (XRD) studies of the time showed no periodic structure in the vitreous state,

(iii) the vitreous structure cannot be entirely random because the intermolecular distances do not fall below certain minimum values.

From these observations Zachariasen concluded that glass can be thought of as an infinitely large unit cell containing an infinite number of atoms. The difference between the crystalline and the vitreous state is, therefore, found in the presence or absence of periodicity – what we would now describe as '*long range order*', which is a fundamental property of a crystalline structure. This theory explains many of the differences between glasses and crystals, such as:

(i)   The optical isotropy of glasses results from the random atomic arrangement.

(ii) In a random network no two atoms occupy exactly identical sites in terms of their chemical environment, and so the lack of abrupt changes of state can be understood. The bond energies all vary slightly (contrasting with the fixed bond energies in an ideal crystal) and hence the thermal breakdown of the network occurs over a large range of temperatures in a glass – it gradually softens into a liquid rather than melts at a characteristic temperature, as is the case with a crystalline solid. (A glass is conventionally said to have 'melted' when its viscosity is reduced to between 10^8 and 10^2 poise.)

(iii) The composition of glass is not stoichiometric – it is a mixture, not a compound.

(iv) There are no crystal cleavage planes in glass – hence the characteristic conchoidal fracture patterns.

It also allowed him to predict on the basis of molecular geometry those oxides which ought to be capable of forming glasses. Goldschmidt (1926; 137–9) had previously observed that the radius ratio of the metals and oxygen (R_A/R_O) in glass-forming oxides of the formula A_mO_n was between 0.2 and 0.4, and that *tetrahedral coordination* was also necessary (*i.e.*, four oxygen atoms surround the metal atom in a tetrahedral shape – see Figure 5.2). Zachariasen generalized this theory by giving the following rules for the formation of an oxide glass:

(i)   Oxygen atoms are linked to not more than two other atoms.

(ii) The number of oxygen atoms surrounding the cations must be small.

(iii) The oxygen polyhedra share corners with each other, not edges or faces. To give a three-dimensional structure at least three corners in each polyhedron must be shared.

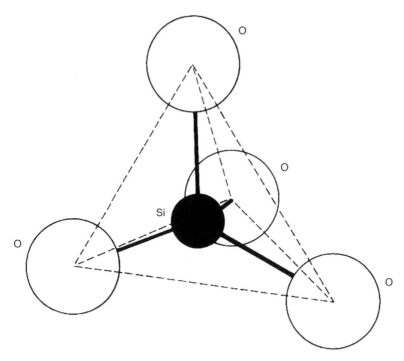

Figure 5.2   Tetrahedral co-ordination of four oxygens around a silicon atom.

On the basis of these rules, he predicted that B_2O_3, SiO_2, GeO_2, P_2O_5, As_2O_5, P_2O_3, As_2O_3, Sb_2O_3, V_2O_5, Sb_2O_5, Cb_2O_5 and Ta_2O_5 may all occur as a single substance in the vitreous state. Of these, B_2O_3, SiO_2, GeO_2, P_2O_5, As_2O_5 and As_2O_3 are known to do so, thus giving good support to Zachariasen's model.

Experimental support for this theory came in a series of papers from Warren and co-workers (Warren, 1937; Warren and Biscoe, 1938). They obtained experimental X-ray scattering curves from vitreous silica and applied the Fourier analysis technique developed for the study of liquid structures (Warren, 1937) to obtain a radial distribution curve. This is a function which gives the number of atoms about any atom and their average distance from it. From peaks in the radial distribution curves, the average atomic separations, up to a distance of about 7–8 Å ($10\,\text{Å} = 1\,\text{nm} = 10^{-9}$ m), can be measured, and from the area beneath each peak an estimate of the co-ordination numbers (number of surrounding atoms) can be made. For vitreous silica (Figure 5.3) Warren found four oxygens tetrahedrally distributed around the silicon, with an average Si–O bond distance of 1.62 Å (compared with 1.60 Å in crystalline silica). The oxygen–oxygen separation in the tetrahedron was found to be 2.65 Å, and further regular interatomic distances were identified out to about 4.5 Å. From this, he concluded that Zachariasen's model was a valid interpretation of the structure of vitreous silica. Warren and Biscoe (1938) studied binary soda–silica glasses in a similar manner and found an extra peak due to

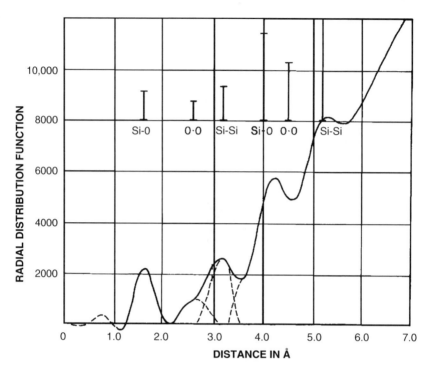

Figure 5.3   Radial distribution curve for vitreous SiO_2 from X-ray studies. Peaks
correspond to average atomic separations of Si–O, O–O, *etc.* (Warren,
1937; Fig. 2. Copyright 1937, American Institute of Physics.)

average the Na–O separation at 2.35 Å. They concluded that the average
structure of this binary SiO_2–Na_2O glass is a silicon atom tetrahedrally
surrounded by four oxygen atoms at a distance of 1.62 Å. The sodium ions
are situated interstitially in the network, surrounded on average by six oxygens
at 2.35 Å. Arguments based on co-ordination numbers rule out the existence of
discrete molecules with the formula Na_2O, $Na_2Si_2O_5$, Na_2SiO_3, *etc.*, but the
structure determined by this method is an average, and the presence of small
regions showing crystalline form cannot be ruled out on this evidence (see
below). Warren calculated the maximum size allowable for crystalline structures in the glass on the basis of the measured diffraction patterns, and obtained
a figure of 7 Å. This is approximately the size of the crystalline silica unit cell, so
the term 'crystal' cannot meaningfully be applied to these minuscule aggrega-
tions. Subsequent studies using more sophisticated techniques such as nuclear
magnetic resonance (NMR; see Chapter 2) to study the distribution of the
oxygen atoms, and extended X-ray absorption fine structure (EXAFS) to look
at the oxygen distributions around the sodium ions in binary soda–silica glasses
(Greaves *et al.*, 1981) have emphasized the importance of this short-range order
in glass, but have substantially supported the views formed in the 1930s.
   Figure 5.4 shows this classical representation of glass structure in two
dimensions, following the work of Zachariasen. The compound R_2O_3 is shown

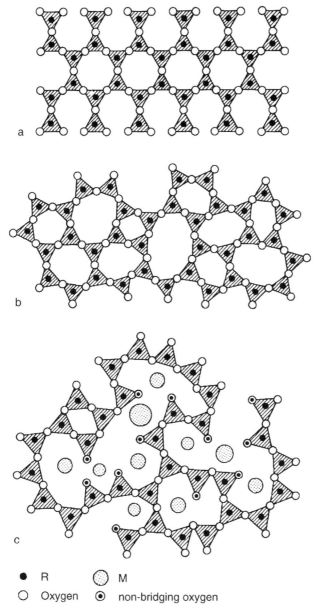

• R ◉ M
○ Oxygen ⊙ non-bridging oxygen

Figure 5.4 Schematic representation of the vitreous state of a compound of stoichio-metry R_2O_3, and as modified by a compound M_2O, according to the random network theory: (a) R_2O_3 in the crystalline state; (b) R_2O_3 in the vitreous state; (c) vitreous R_2O_3 modified by M_2O, showing bridging and non-bridging oxygen sites. (Reprinted with permission from Zachariasen, 1932; Figures 1a and 1b (copyright 1932, American Chemical Society), and Warren and Biscoe, 1938; Figure 4, by permission of Blackwell Publishing Ltd.)

in its crystalline and vitreous forms (R in this model is a trivalent cation so that the structure can be represented in two dimensions), and also the vitreous structure when modified by the addition of a substance of the form M_2O. In order to picture the structure of a silicate glass, Figure 5.4 should be considered as a 'slice' through the full three-dimensional network, remembering, however, that the tetrahedral co-ordination of the oxygen around the silicon means that in reality the silicon and oxygen atoms are not in the same plane, and therefore the 'slice' should not be regarded as flat – more as a crinkly sheet, with the oxygens slightly lower, below the plane of the silicon atoms.

As a result of the random network theory, glass-forming oxides have been classified as *network formers, network modifiers* and *intermediates.* Network formers are those which may be found in the vitreous state as pure substances, and include SiO_2, B_2O_3, GeO_2, P_2O_5 and As_2O_5. The network modifiers disrupt the continuity of the network, changing the physical and chemical properties, as illustrated in Figure 5.4c. This category includes the alkali metal oxides (*e.g.*, Na_2O, K_2O) and the alkaline earth oxides (*e.g.*, CaO, MgO). Intermediates are oxides which can either enter the network as a network former or occupy interstitial holes (*i.e.*, act as a network modifier), but are unable to form glasses themselves. Included in this category are Al_2O_3, TiO_2 and ZrO_2. In archaeological glasses, the principal network former is silica (SiO_2), with the alkalis (soda, Na_2O, and potash or potassia, K_2O) and alkaline earths (calcia or lime, CaO, and magnesia, MgO) as the network modifiers. Alumina (Al_2O_3) plays an important role in stabilizing the network and is the most important of the glassmaking intermediates. As can be seen from Table 5.1, although pure silica in the vitreous state is a perfectly well-behaved glass, its value as a usable material is limited by its very high working temperatures – for example, not softening until 1500 °C. These temperatures are far too high to have been achieved routinely by ancient glassworkers, and it was only by the discovery of the fluxing effect of the alkalis on these thermal properties that glassmaking became possible. The traditional story of the discovery of the beneficial properties of alkali on glassmaking as first told by Pliny in the 1st Century AD is recounted by Vose (1980). Phoenician merchants camping on the banks of the River Belus (a small river in modern-day Israel, now called Nahr Na'mân) used some cakes of natron (natural sodium carbonate) to support their cooking pots on the fire. In the morning they found the natron had fused the sand to form a glass. As all subsequent commentators have pointed out, the story is unlikely to be true, since the temperature of a cooking fire is unlikely to have reached the 1000 °C or more required to produce glass, but it clearly indicates a knowledge of the use of alkalis to modify the properties of silicate glasses, and that the River Belus was an important source of glass-making sand in antiquity.

From Figure 5.4c, it is apparent that some of the oxygen atoms in the modified structure are bonded to two silicon atoms, and some to only one, with the resulting excess negative charge being balanced by an interstitial ion of the type M^+. Oxygen ions which join two polyhedra are known as *bridging oxygens*, and those linked to only one *non-bridging*. Stevels (1960–1961) characterized the structure of the network with four parameters X, Y, Z and R, defined as:

X = average number of non-bridging oxygen ions per polyhedron;

Y = average number of bridging oxygen ions per polyhedron;

Z = average total number of oxygen ions per polyhedron;

R = ratio of total number of oxygen ions to total number of network formers.

Z is known from the identity of the principal network former; for silicates, Z is 4. R can be deduced from the molar composition of the glass, although in 'real' glasses R is not predictable precisely because of the presence of intermediate oxides. Knowing Z and R allows the values of X and Y to be calculated for a particular glass, using the simple relationships $X + Y = Z$ and $X + Y/2 = R$. Stevels demonstrated that many properties of the glass depend on the value of Y, and showed that glasses of different chemical composition, but with the same Y value, have the same physical properties, such as viscosity and thermal expansion coefficient. From studies of the variations in properties with changing composition, he was able to show that a marked change in behaviour occurs when the value of Y falls below 3. This figure marks the point below which polyhedra can exist that are connected by only two points of contact with neighbouring polyhedra. Thus the rigidity of the network decreases significantly, and properties which depend on ionic transport (such as electrical conductivity, *etc.*) show a significant change. In binary soda–silicate glasses, for example, this point corresponds to the composition $Na_2O.2SiO_2$ (33.3 mole % Na_2O, 66.67 mole % SiO_2) and the gradient of a plot of electrical conductivity against composition shows a marked change at this value (Stevels, 1960–1961; Figure 3). [Mole % is a measure of the abundance of a particular component by number of (theoretical) molecules, rather than by weight.]

A similar result can be obtained by studying the molar ratio of silicon to oxygen atoms in a silicate glass. In vitreous SiO_2 the ratio is 0.5, simply calculated from the stoichiometric ratio of 1 silicon to 2 oxygens, but this figure falls as the number of non-bridging oxygen atoms increases (*i.e.*, as a result of adding extra oxygen *via* the network-modifying oxides). Most chemical analyses of glasses are reported as weight percent (wt %) oxide, but it can be seen that many of the important properties depend not on this but on the molar percentage (*i.e.*, the relative number) of atoms present. The molar ratio Si/O can be calculated for any glass from its chemical (wt % oxide) analysis using the formula due to Huggins and Sun (1943; eqn. 16):

$$Si/O = \frac{f_{Si}}{60.06 \sum \frac{n_M \times f_M}{W_M}}$$

where f_{si} = wt fraction of SiO_2 (= wt % SiO_2/100),
         60.06 = molecular weight of SiO_2
         n_M = number of oxygens in formula for oxide ($M_M O_N$)
         W_M = molecular weight of oxide $M_M O_N$
and f_M = wt fraction of $M_M O_N$ (= wt % $M_M O_N$/100)

and the summation is over all the oxides in the glass including SiO_2. Table 5.2 shows the relationships between the calculated value of the Si/O ratio, the number of bridging and non-bridging oxygen ions and the probable bonding of the silicon in the glass, where the bond shown as a line signifies a bridging (Si–O–Si) bond. As the value of the Si/O ratio increases, the number of bridging oxygens increases and the degree of 'polymerization' of the network increases. These figures have important consequences for the interpretation of the chemical durability of glass, as discussed below.

An alternative theory of glass structure, attributed to Lebedev in the Soviet Union in 1921 (Porai-Koshits, 1977; Vogel, 1977), was the *crystallite theory*. This model maintained that glass is made up of small crystalline regions varying in size from 10 Å (three to four silica units) up to perhaps 300 Å. A critical review of these two competing views on the nature of glass was published by Porai-Koshits (1990). As mentioned above, the experimental work of Warren specifically ruled out the existence of crystalline regions larger than about 10 Å, but it is appreciated that the random network theory presents only an average picture of the structure, and says nothing about the local distribution of ions. Porai-Koshits argues that regions of maximum order in the continuous network may be conveniently termed 'crystallites', and, using this

Table 5.2 Relationship between the Si/O ratio in a glass, the number of non-bridging oxygens per tetrahedra, and the structure. (Adapted from Volf, 1961; 20.)

Si/O Molar ratio	Number of bridging oxygens per tetrahedron	Number of non-bridging oxygens per tetrahedron	Pattern of structure
<0.286	1	3	— Si
0.286–0.333	1–2	2–3	— Si to — Si (with vertical bond)
0.334–0.400	2–3	1–2	— Si to — Si —
0.401–0.444	3–4	0–1	— Si— to —Si—
0.500	4	0	— Si —

definition of a crystallite, the two theories are seen to blend into one another. In 1971, Soviet academicians concluded that modern methods of analysis made it impossible to hold on to the original concept of crystallites in glass (Porai-Koshits, 1985), and yet direct evidence for structural inhomogeneity on the sub-micron to micron scale has subsequently been obtained using transmission electron microscopy and small angle X-ray scattering (SAXS). Application of this technique to glasses, together with direct observation by electron microscopy, has conclusively demonstrated the phenomenon of phase separation in many vitreous systems (Vogel, 1977). The composition of the various phases in multi-component glasses has been confirmed using electron microprobe analysis. The current understanding of the structure of glass may be summarized by combining the comments of Vogel (1977) and Porai-Koshits (1977):

(i) In one-component glasses (such as pure vitreous silica), no inhomogeneous structure occurs, except for frozen thermal density fluctuations corresponding to the situation in the melt at the glass transformation temperature. These glasses are adequately described by the random network theory of Zachariasen and Warren.

(ii) All multi-component glasses whose composition is intermediate between two stable compounds exhibit phase separation, with the compositions of each phase approximating the stable compositions. Maximum phase separation occurs in compositions which are furthest removed from the stable compositions.

(iii) The addition of further components such as colorants does not lead to a homogeneous distribution of additives. Highly charged cations will concentrate in anion-rich phases.

There is now ample evidence to support the view that glasses cannot simply be regarded as a homogeneous continuous network, as predicted by the random network theory, but nor is it a series of domains of highly ordered crystallites, as originally proposed by Lebedev. For example, detailed studies of sodium silicate glasses using molecular dynamic simulations to model the properties and structure of glasses have shown that the sodium ions are not randomly distributed through the structure, but associate with non-bridging oxygens to produce silica-rich and alkali-rich regions on the atomic scale within the glass, thus supporting the above consensus view (Huang and Cormack, 1990).

The phase structure of glasses has a significant effect on their physical properties, which is discussed below with reference to chemical durability. The magnitude of the phase separation can be altered by heat treatment, and enhanced or reduced by the addition of various oxides to the melt. In particular, the addition of alumina to commercial soda-lime–silica glasses reduces the tendency to phase separation, improving chemical resistance (Doremus, 1973). A detailed study of the microstructure of soda-lime–silica glasses has been published by Burnett and Douglas (1970). The control of phase separation in the melt is now commercially important for processes such as the production of 96% silica glass for laboratory use from phase-separated sodium borosilicate glasses by the Vycor process (Paul, 1990). In this process, a glass with composition approximately 8% Na_2O, 20% B_2O_3 and 72% SiO_2 is heat treated at between 500 and 800 °C, inducing the separation of two phases, one silica rich and the other rich in boric acid. When treated with dilute hydrochloric acid, the boron-rich phase dissolves, leaving a highly porous but almost pure silica glass, which, when heated to 1200 °C, becomes more compact and dense. The advantage of this is that the vitreous silica has been produced at a temperature much below that which would have been necessary to melt a glass of pure silica. Phase-separation phenomena are also vitally important in understanding that other branch of silicate chemistry important in archaeology – the art of glaze manufacture. One spectacular example is the production of Jun ware glazes by Chinese potters during the Song dynasty (AD 960–1263) and later, in which the characteristic milky sky-blue colour is produced not by additives but by the formation of an emulsion of two liquids in the glaze at high temperature, and subsequently retained by being cooled at exactly the correct rate (Kingery and Vandiver, 1986) – a phenomenal achievement for potters working 1000 years ago without the aid of modern technology.

The vitreous state should therefore be seen as an addition to the traditional three states of matter – one that is described by an essentially random network characteristic of a liquid, but which, unlike true liquids, may still retain inhomogeneities. The crucial factor in deciding whether a melt will crystallize or become vitreous is kinetically controlled – the rate of cooling relative to the bond energies involved. The stronger the chemical bonds, the more sluggish will be the rearrangement processes close to the melting point, and the more likely the melt will be to undergo the glass transition process rather than crystallization, which requires a great deal of molecular reorganization. Silica, with a high Si–O bond energy (368 kJ mol^{-1}), is particularly prone to vitrify, but almost any melt can be vitrified if the rate of cooling is fast enough. This is the basis of the '*splat cooling*' process which can produce metallic glasses, in which a jet of molten metal is sprayed onto a rotating cooled drum, achieving a temperature drop of 1000 °C in a millisecond, causing the metal to vitrify (Greer, 1995).

The discussion in this section has used the terms 'atom', 'ion' and 'molecule' virtually indiscriminately, reflecting the usage of these terms in the glass chemistry literature. The Zachariasen model is phrased in terms of atoms, and assumes total covalency of bonding, which gives the required directionality of the bonds, but does not admit much flexibility in the bond lengths, as is required in the model. Ionic or metallic bonds do not have the required directionality. Subsequent workers have modified the theory to account for a 'mixed' (covalent–ionic) bond which is a more realistic picture of the bonding in real glasses, but these modifications have little impact on the structural models as outlined, and are generally ignored in the literature (Paul, 1990). One final point to be emphasized is that although glass analyses are conventionally reported as 'weight percent oxides' or 'molar percent oxides', the presence of compounds such as Na_2O or CaO in a glass is purely fictitious, in that compounds of this stoichiometry do not physically occur in the glass. Remembering that this oxide notation is purely for convenience is particularly important when dealing with transition metal oxides such as those of Fe^{II} or Fe^{III}, since the assumption of a valence state for a particular transition metal is also largely conventional (see below). The assumed valency of the metal could, however, affect the total analysis of the glass depending on which of the 'fictitious oxides' the iron content is reported as (FeO or Fe_2O_3). This is discussed in more detail in Section 5.3.

5.3   THE COLOUR OF GLASS

The colour of a glass is usually (but not always) attributable either to the presence of small amounts of transition metal ions or to metallic atoms within the structure, and the colours resulting from various additives have been studied for well over 100 years (Weyl, 1976). Exceptions to this rule are *opaque glasses*, in which the opacity is often due to the presence of a second finely distributed immiscible phase within the glass, giving rise to the scattering of light which results in a milky or opaque appearance (as in Jun ware glazes, described earlier). Another colour-forming situation is the presence of thin layers on the surface of the glass, which can give rise to an iridescent appearance as the result of diffraction phenomena within the surface layers. Such iridescence often arises as a result of weathering (or corrosion in the case of buried glasses), and is a characteristic feature of many otherwise well-preserved archaeological glasses. Opacity is usually more important in the production of ceramic glazes or enamelled metalwork, apart from certain classes of artefact such as glass beads, which are deliberately made to be opaque. This section concentrates on the chemical explanation of the colours found in transparent glasses as a result of the presence of transition metal ions.

The classic treatise on the subject of the colour of silicate glasses is that of Weyl (1976). He describes the colours produced by metallic ions in glasses (including the various oxidation states of Fe, Mn, Cr, V, Cu, Co, Ni, U, Ti, W, Mo and the rare earths), and also those produced by metal atoms in glass, principally Au, Ag and Cu, as well as describing the effect of replacing the oxygen in the glass with other anions. The results of his synthesis are well-known in archaeology, although the detailed chemistry of the causes of the different colours is often less well-appreciated. Table 5.3 lists the colours produced by the traditional colouring ions, together with the modern raw materials used to impart such colours. As can be seen, the colour produced by a metal ion depends not only on its oxidation state, but also on the position it occupies in the glass structure. The reasons for this are explained below. Even this table of colours is by no means exhaustive – for example, Fe can give a range of colours from pale yellow through to blues and greens, depending on the oxidation state of the ion, the presence of other transition metal ions and the composition of the base glass. Indeed, an understanding of the principles underlying the colour of glasses is far more complex than one might believe, and requires a discussion of factors such as the redox conditions in the furnace

Table 5.3 Some of the chromophores responsible for the colours in ionically coloured glasses. (Compiled from Weyl (1976; 60) and Vogel (1994; 232) with kind permission of Springer Science and Business Media.)

Transition metal	Modern raw material	Colouring ion	Colour in tetrahedral co-ordination (network former)	Colour in octahedral co-ordination (network modifier)
Chromium	Chromium(III) oxide Cr_2O_3	Cr^{III}	–	Green
	Potassium dichromate(VI) $K_2Cr_2O_7$	Cr^{VI}	Yellow	–
Copper	Copper(II) oxide CuO	Cu^{I}	–	Colourless to red, brown fluorescence
	Copper(II) sulfate $CuSO_4.5H_2O$	Cu^{II}	Yellowish brown	Blue
Cobalt	Cobalt(III) oxide Co_2O_3 Cobalt(II) carbonate $CoCO_3$	Co^{II}	Blue	Pink
Nickel	Nickel(II) oxide NiO Nickel(III) oxide Ni_2O_3 Nickel(II) carbonate $NiCO_3$	Ni^{II}	Purple	Yellow
Manganese	Manganese(IV) oxide MnO_2	Mn^{II}	Colourless to faint yellow, green fluorescence	Weak orange, red fluorescence
	Potassium manganate(VII) $KMnO_4$	Mn^{III}	Purple	–
Iron	Iron(II) oxalate $Fe(C_2O_4).2H_2O$	Fe^{II}	–	Absorbs in infrared
	Iron(III) oxide Fe_2O_3	Fe^{III}	Deep brown	Weak yellow to pink
Uranium	Uranium oxide U_3O_8 Sodium uranate $Na_2U_2O_7.3H_2O$	U^{VI}	Yellowish orange	Weak yellow, green fluorescence
Vanadium	Vanadium(V) oxide V_2O_5	V^{III}	–	Green
		V^{IV}	–	Blue
		V^{V}	Colourless to yellow	–

and the nature of the co-ordination sphere around the colouring ion – all concepts familiar to inorganic chemists.

Before attempting to explain the phenomenon of colour in glasses, it is instructive to look at simple aqueous solutions coloured by transition metal ions. A well-known undergraduate chemistry experiment is to observe the difference in colour between an aqueous solution of copper sulfate and a solution of copper sulfate crystals dissolved in a mixture of water and concentrated ammonia solution. The aqueous solution is the familiar pale blue–green of copper solutions, but as ammonia is added the colour deepens to a much more intense blue. The explanation of this apparently simple change requires a detailed investigation of *co-ordination chemistry*. Even the original blue crystals of copper sulfate are not what they seem – in fact they are the pentaquo compound $CuSO_4.5H_2O$ – in other words, the central copper ion is surrounded by six groups (called *ligands*) – one sulfate, and five water molecules. These six ligands are arranged in a distorted octahedron – four of the water groups in a square plane surrounding the copper, plus one water below and the sulfate above the plane (Figure 5.5). The co-ordination is said to be *octahedral*. The term 'distorted' implies that the Cu–O bond lengths to the square planar water groups are equal, but are not the same as the Cu–O bond to the other water group, or the Cu–O bond length to the oxygen in the sulfate group. Anhydrous copper sulfate – without the five waters – is colourless, indicating that the colour of the crystal is strongly related to presence of this *co-ordination sphere* around the copper ion, which is the case.

The colour of the transition metal ions, which is one of their major characteristics, arises directly from the interaction between the outer orbital electrons of the transition metal and the electric field created by the presence of the co-ordinating ligands. The theory of this is called *ligand field theory*, and is well covered in most basic textbooks on inorganic chemistry (*e.g.*, Cotton and Wilkinson, 1976). Ligand field theory is an extremely powerful tool, which

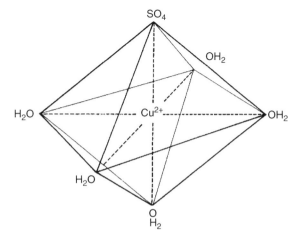

Figure 5.5 Octahedral co-ordination of the Cu^{2+} ion in copper sulfate, $CuSO_4.5H_2O$.

explains all manner of spectroscopic, electronic and magnetic properties of the transition elements. A simplified version, based purely on electrostatic considerations, is called *crystal field theory*, which allows a relatively simple explanation to be given for the causes of colour in transition metal compounds. Transition metals are defined as those elements which have partially filled *d* orbitals (or *f*, but we will restrict this discussion to the *d*-block elements: see the Periodic Table, Appendix 5). As explained in Appendix 1, the *d*-orbitals are made up of five possible energy levels, and because each energy level can accommodate two electrons (one spin 'up' and one spin 'down'), they can therefore hold a maximum of ten electrons before they are complete. In free space (*i.e.*, far removed from any electric or magnetic field), the five energy levels are exactly equivalent in energy, and therefore a transition metal ion with, say, three *d*-electrons (*e.g.*, Cr^{III}, or Mn^{IV}) could put one electron in any three energy levels, leaving the other two empty. (Putting two electrons in the same orbital whilst another orbital of the same energy level is empty is energetically unfavourable.) However, when the transition metal ion is co-ordinated by other ions or molecules, as it invariably is in a crystal or in solution, then there is an interaction between the charges on the neighbouring ions, atoms or molecules and the *d*-orbitals of the transition metal ion. The five *d*-orbitals have different shapes in space around the nucleus (see Pollard *et al.*, 2007; 240), so some are affected more than others by the surrounding charges, and the precise nature of the interaction depends on the relationship between the fixed geometry of the *d*-orbitals and the distribution of nearest neighbours around the ion. In octahedral co-ordination (as shown in Figure 5.5), for example, three of the *d*-orbitals are depressed in energy compared to the *d*-orbital energy in the free ion, and two are increased. In tetrahedral co-ordination (*i.e.*, with four co-ordinating ligands making a tetrahedral shape), the pattern is different, with three orbitals being depressed and two raised. The energy difference between the upper and the lower groups is not the same in these two cases (Δ_o is the energy difference in octahedral splitting, Δ_t the tetrahedral: see Figure 5.6).

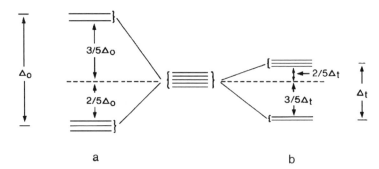

a   b

Figure 5.6   Energy level diagram of the splitting of the *d*-orbitals of a transition metal ion as a result of (a) octahedral co-ordination and (b) tetrahedral co-ordination, according to the crystal field theory. (From Cotton and Wilkinson, 1976; Figure 23–4. Copyright 1976 John Wiley & Sons, Inc. Reprinted by permission of the publisher.)

Now when the outer *d*-electrons are being allocated to orbitals, there is not a free choice for the ion. For our ion with three *d*-electrons to distribute, in the octahedral case all three can occupy one each of three lower orbitals, but in the tetrahedral case there is a choice between putting two in one each of the lower and the third in one of the upper orbitals, or putting the third electron into one of the lower orbitals by pairing two electrons (one 'spin up' and one 'spin down') in one orbital. These two choices are known as 'high spin' and 'low spin', respectively, since the first has three unpaired electrons, and the second only has one. The number of paired electrons in the *d*-orbital is of paramount importance in dictating the magnetic properties of the compound, and is the basis for an atomic explanation of the magnetic properties of materials.

The splitting of the *d*-orbital energies by the presence of other atoms has one further point of central importance here. It now becomes possible for electrons in one of the lower *d*-orbital energy states to be promoted to the upper energy state, and to drop down again, emitting a quantum of electromagnetic energy equal to the energy difference between the split orbitals, as described in Chapter 2. (In fact, this so-called *d*–*d* band transition is forbidden by the laws of quantum mechanics, but it happens anyway because of quantum mechanical tunnelling.) The energy difference between the split orbitals (Δ_o or Δ_t) is small, but when converted to a wavelength using the usual formula [$E = h\nu$, or $E = hc/\lambda$, where E is the energy difference between the orbitals, h is Planck's constant (6.626×10^{-34} J s), c is the speed of light (3×10^8 m s^{-1}) and λ and ν are the wavelength and frequency of the emitted radiation, respectively] it transpires that the electromagnetic radiation produced is in the visible region of the spectrum, unlike much of the other radiation arising from transitions between atomic orbitals. This then is the source of the colour in most inorganic compounds. One corollary is that, because the energy difference between the split orbitals depends on the geometry of the co-ordinating ligands, the colour of a compound can change if there is a change in co-ordination, *i.e.*, if an ion is surrounded by six water molecules in octahedral co-ordination, the colour will be different to that produced by the same ion surrounded tetrahedrally by only four water molecules. Likewise, if one or more of the water molecules is replaced by another ligand, then the strength of the electrostatic interaction will change, the energy splitting of the *d* orbitals will change, and so the colour will again change.

In view of this, let us return to our simple experiment. When the crystals of hydrated copper sulfate dissolve in water, the simple formula:

$$Cu_2SO_4 = Cu^{2+} + SO_4^{2-}$$

is an acceptable version of the truth in most cases, but it implies that there is a 'bare' copper ion free to move through the solvent water molecules, which is not the case. Water molecules develop a dipole moment, with the oxygen becoming slightly negatively charged and the two hydrogens slightly positive, so the slight negative charge on the oxygen is attracted to the positive charge on the Cu^{2+}. Geometrical considerations mean that only six waters can cluster around the copper ion. This means that the copper(II) ion exists in water as the hexaquo ion $[Cu(H_2O)_6]^{2+}$, with the six waters octahedrally distributed around the copper ion. (The notation Cu^{II} implies the divalent copper ion, and is equivalent to Cu^{2+}). The familiar pale blue–green colour is characteristic of *d–d* band transitions within this octahedral copper(II) hexaquo complex. When the same crystals are dissolved in water containing various additions of ammonia, there is a gradual replacement of the solvating water molecules by ammonia molecules, resulting in a series of complexes ranging from $[Cu(NH_3)(H_2O)_5]^{2+}$ up to $[Cu(NH_3)_4(H_2O)_2]^{2+}$ resulting in a deeper blue colour as the number of ammonia groups increase. [It is usually very difficult to entirely replace all the waters in copper(II) complexes.] This is an example of a change in the chemical nature of the co-ordination sphere altering the colour of a transition metal ion.

It is not difficult to translate this knowledge to an explanation of the colour of transition metal ions in glass. It is conventional to talk of the glass as a solid 'solvent' for the colouring ions, which either exist as interstitial network modifiers (like Na^+) or as substitutes for the silicon in the network, and to discuss the co-ordination of the non-bridging oxygens around the metal ion in just the same way as the water was treated above. Interstitial ions usually have at least six or more oxygens surrounding them, and are approximately octahedrally co-ordinated. Certain ions (such as Fe^{3+}) can also substitute for the silicon in the network itself, and thus will have fourfold (approximately tetrahedral) oxygen co-ordination. From the above arguments, it is obvious why the colour of a glass will change if the oxygen co-ordination changes from four to six – the equivalent in a crystal of changing from tetrahedral to octahedral co-ordination, although in a glass the geometry will not be quite so regular. According to Weyl (1976: see also Table 5.3), for example, if cobalt ions replace the sodium ions interstitially in a silicate glass, the resulting colour will be pink. If the same ions substitute for silicon in the network, then the colour will be deep blue. Nickel ions give a yellow colour if they act as network modifiers, but purple if acting as network formers. Normally, the substituting ion will distribute itself between both possible sites, depending on the details of the manufacturing process and the furnace conditions, and the result for nickel would be a mixture of yellow and purple colours, which would normally balance to give a greyish glass. A secondary influence might also be the nature of the other network modifiers present, since these may also have an influence on the crystal field experienced by the colouring ion. It is conceivable, for example, that the colour developed by an ion in a glass with soda as the only alkali might be different from that given by the same ion in the same glass, but with potash substituted as the alkali. This factor has received relatively little attention, in view of the lack of commercial interest in potash glasses, although it may be historically relevant because of the use of 'forest glass' in the great cathedrals. A full treatment of the influence of the co-ordination of transition metal ions on the colour of glasses is given in Bamford (1977) and Paul (1990). An interesting footnote to this debate is the observation, made by Newton (1978), that the use of beech wood ash to make coloured glass as advocated by the Medieval monk Theophilus (Hawthorne and Smith, 1979; Book II,

Chapter 4) might have been influenced by the fact that all the colours required (except red, which used additional copper) could be obtained from glass made in such a way simply by controlling the furnace conditions, *i.e.*, without any explicit addition of colorants. It might also explain the observation that deep blue glasses (coloured by cobalt) in the 12th Century AD were made principally from soda glass, either because potash glass does not give such a deep colour (an idea which could be tested experimentally), or arising from the re-use of Roman glass (Cox and Gillies, 1986). Clearly, whatever the answer to this, Medieval glassmakers were, inadvertently, adept co-ordination chemists!

Colouration by metal atoms as opposed to ions in glasses is less complicated chemically, but the physics of the situation tends to be more involved. The more noble of the transition metals (Cu, Ag and Au) can exist in glasses as metallic clusters; Cu and Au give rise to the highly prized red colours in glasses and glazes, and silver is responsible for the yellow colour in the true Medieval *stained glass*. In red glass, the colour is imparted by a fine dispersion of precipitated metal throughout the glass, and is a result of scattering from the particles rather than absorption. The colour depends on the size of the precipitated particles, which in turn depends on the cooling regime adopted in the manufacturing process (Weyl, 1976). One of the technical marvels of the ancient glassmakers art, the 4th Century AD Lycurgus Cup, which is dichroic – red in transmitted light but green in reflected light – is coloured by a very fine distribution of small (<10 nm) metallic particles, made up on average of 66% Ag and 31% Au, with 3% Cu (Barber and Freestone, 1990). Silver in glass is particularly important in a discussion of Medieval glass, because it is the cause of the yellow colour in later Medieval *stained glass*. Earlier glass is more correctly termed *painted*, because an opaque paint is simply applied to the surface of a transparent piece of coloured glass. The manufacturing process for stained glass involves the diffusion of silver into the surface after the base glass has been made, a process said to have been invented in Ulm in AD 1460 (Weyl, 1976). The process is known as *cementation*, and involves the application of a paste of a silver compound (usually the oxide mixed with a clay binder) to the surface of the glass, and heating to about 600 °C, at which temperature the silver ions diffuse into the glass by exchange with the sodium ions. Once the diffusion has taken place, the silver ions are reduced to the metal by a *redox reaction* with the other transition metal ions present (Fe^{II} or similar), as follows:

$$Ag^+ + Fe^{2+} = Ag + Fe^{3+}$$

The final stage is to control the temperature programme in order to promote the growth of the silver droplets to the size required to produce the yellow colour, in the same way as the production of ruby red glasses.

Clearly then, in glasses coloured by metal ions, the co-ordination chemistry of the transition metal ion has a major influence on the colour. The other major influence is the oxidation state of the metal ion, since variable valency is another characteristic of the transition metals. All other things being equal, for example, iron in the Fe^{II} form will give a pale blue colour, whereas Fe^{III} gives rise to a brown colour in tetrahedral co-ordination. The two principal factors which control the oxidation state of the transition metals are the redox conditions in the furnace (*i.e.*, the partial pressure of oxygen in the furnace atmosphere, or more simply the degree of oxidation or reduction achieved), and the redox reactions between the various transition metal ions present in the melt – this is returned to below.

Control of the furnace conditions, together with the rate of cooling, is the traditional kernel of the glassmaker's art, and the primary means of control over the colour and quality of the product. More specifically, the glassmaker's art consists of matching the production process to the recipe to produce the desired result. Up until recent times, all glasses contained impurities from the raw materials, and the production of truly colourless glass required careful control over the recipe and the furnace conditions. The most important trace elements in this respect are iron and manganese. Most ancient and historical glasses contain significant levels of iron (0.3–1.5 wt % oxide) as an impurity from the sand used, which imparts the 'natural' green tinge in archaeological glass. Under oxidizing furnace conditions, iron forms the Fe^{III} species, which usually imparts darker green or brown colours, whereas under reduction, Fe^{II} predominates, giving paler blues and greens. The reason for this actually relates to the previous discussion – Fe^{III} has an ionic radius of 0.74 Å compared to 0.64 Å for Fe^{II}, and is more likely to occupy the tetrahedral site of silicon (ionic radii of Si^{IV} 0.42 Å) in the network, whereas Fe^{II} is more likely to occupy the larger octahedral site vacated by sodium (Na^{I} ionic radius 0.97 Å).

The exact redox state of the iron in the melt depends on the redox potential of the furnace (strictly, the partial pressure of oxygen in the atmosphere: Paul, 1990; 148), but also on the total amount of iron present, and on the redox reactions between the various possible redox couples in the melt – the most important one being that between manganese and iron. The equilibrium between Fe^{II} and Fe^{III} is affected by the total amount of iron present in the glass – as the total iron increases, the Fe^{III} state predominates over the Fe^{II} (Jackson, 1992), thus darkening the colour. Manganese in a glass can act as both a colorant and as a decolourizer – the latter partly by oxidizing the iron present, and partly by compensating for the resulting green colour by its own purple colour in the oxidized Mn^{III} state – Mn^{II} is a faint yellow or brown, and not a strong colorant. The Mn–Fe redox couple is as follows:

$$Mn^{3+} + Fe^{2+} = Mn^{2+} + Fe^{3+}$$

or, using an oxide formulation:

$$Mn_2O_3 + 2FeO = 2MnO + Fe_2O_3$$

Under the conditions normally found in a glassmaking furnace, the equilibrium lies well over to the right – so much so that Mn_2O_3 does not usually occur until all the iron present is in the Fe^{III} form. Hence it is conventional to report the manganese present in glass in terms of MnO, and the iron as Fe_2O_3, although it can be seen that this may not always be entirely true. The resulting colour is therefore due largely to the Fe^{III}, which, although it gives a dark colour, is a weak chromophore (*i.e.*, a lot of Fe^{III} is needed to give a measurable colour). The well-known decolourizing effect of manganese requires excess manganese to be present in the Mn^{II} state, when the purple colour counteracts the green. More generally, Bamford (1977; 82) gives the following table to help work out which oxidation states will predominate when any two transition metals are present in a melt:

$$
\begin{array}{c}
\Rightarrow \\
Cr^{6+}/Cr^{3+} \\
Mn^{3+}/Mn^{2+} \\
Ce^{4+}/Ce^{3+} \\
As^{5+}/As^{3+} \\
Sb^{5+}/Sb^{3+} \\
Fe^{3+}/Fe^{2+} \\
\Leftarrow
\end{array}
$$

If a particular pair of transition metals is present (such as Cr and Fe), then the lower of the pair in this table (in this case, iron) will tend to exist as the left-hand (oxidized) form of the ion (Fe^{3+}) and the upper of the pair will be in the right hand (reduced) state (Cr^{3+}). This tendency is shown by the direction of the arrows. In most ancient glasses iron is present, so the iron will always tend to be in the Fe^{III} state, and any other transition metal in its lower oxidation state. This is important not just from the point of the colour, but also because in glasses with a high transition metal content (such as vitrified nuclear waste or natural glasses, *i.e.*, obsidian) the redox state of these metals can influence the durability of the glass (Jantzen and Plodinec, 1984). For example, Fe^{III} improves the durability, whereas Fe^{II} reduces it (see below).

It has been repeatedly demonstrated that a straightforward chemical analysis using a technique which simply measures the total iron or manganese present (as is usually the case, even if the data *appear* to be in the form of oxide measurements) does not give an adequate guide to the colour, and that a knowledge of the exact oxidation states of the transition metal ions is required (Sellner *et al.*, 1979). This implies that techniques such as Mössbauer spectroscopy or electron spin resonance (ESR) are useful in this context, or direct measurement of the visible absorption spectrum of the glass compared to the absorption spectra of the various transition metal ions (Green and Hart, 1987).

It is clear that the colour of a glass is the result of a complex interplay between the co-ordination of the transition metal ions, the redox reactions between the various ions present and the redox potential in the furnace. The traditional archaeological view that colour can be simply related to the presence of various 'colouring agents' can only be regarded as a very crude guide.

5.4   THE DECAY OF MEDIEVAL WINDOW GLASS

The *durability* of glass may be defined as its resistance to attack by water, aqueous acid or base solutions, steam or atmospheric agents. The particular case of attack of glass by water combined with atmospheric gases (*e.g.*, SO_2, CO_2) is termed *weathering*. The resistance of glass to attack by these agents is assessed commercially by several standard tests described in detail by Volf (1961). In general these involve grinding and sieving the glass to a specified size, subjecting a known weight to extraction by a fixed volume of liquid at a known pH and temperature for a specified time, then determining the weight of each component extracted chemically. The resistance of a particular glass to any one of the above agents is then classified according to the amount of leached material found.

Over the past 60 years, much literature has appeared on the reactions of glass with water, and comprehensive reviews of the subject can be found in Clark *et al.* (1979), Newton (1985) and Scholze (1988). Most of the early work was on simple systems (usually binary soda–silica or ternary soda–lime–silica glasses), which provided a good framework for the understanding of the behaviour of real glasses (Perera *et al.*, 1991). In the following section we review the evidence from studies on binary glasses, and then consider ternary systems (Na_2O–CaO–SiO_2 and K_2O–CaO–SiO_2) which allow a comparison to be made of the different behaviour of Na^+ and K^+ in such systems. We briefly consider the role of phase separation in the context of ionic diffusion through glasses, and then review the evidence for the durability of synthetic Medieval glasses, which leads to a short discussion of the behaviour of mixed alkali glasses. The section concludes with a summary of some work done on relating durability to bulk composition of Medieval glass from York Minster.

Douglas and Isard (1949) investigated the reaction of water with soda-lime–silica glass, and concluded that a double-diffusion process occurs, with H^+ ions from the water entering the network to establish electrical neutrality as Na^+ ions are removed. Strictly, this statement should be phrased in terms of the *hydronium ion*, H_3O^+, since this is the correct ionic form of the positive ion resulting from the dissociation of water, but it is conventional to refer to it as the H^+ ion, and it makes little difference except when dealing with ionic transport phenomena. If electrical neutrality was not maintained by the above mechanism, an electrical double layer would form, rapidly bringing the process to a halt. The mathematics of this double-diffusion process shows that if one diffusion coefficient is much less than the other, then the process may be regarded as a single-diffusion problem with a diffusion coefficient similar to that of the slower ion. The electrical conductivity of silicate glasses is known to be due to the diffusion of alkali ions (Morey, 1954), and Douglas and Isard deduced that, since the activation energies for electrical conduction and for water leaching were found to be very nearly equal (0.77 and 0.79 eV, respectively), the diffusion of sodium ions was the rate-controlling factor.

The basis for the currently accepted mechanism for the attack of glass by aqueous solutions was proposed by Charles (1958), after observing that vitreous silica is not attacked by water vapour even over long periods, whereas multi-component glasses show signs of leaching into the solution relatively quickly. He concluded that dissolution occurs via the 'terminal structure' of alkali ions (*i.e.*, those associated with the non-bridging oxygen sites):

$$- Si - O - Na + H_2O \rightarrow - Si - OH + Na^+ + OH^- \tag{5.1}$$

Equation (5.1) represents the hydrolysis of the salt of a weak acid. The sodium ion migrates into the solution, and the hydroxyl ion released in the glass can then attack the otherwise stable silica network, converting a bridging oxygen into a non-bridging site, thus disrupting the network:

$$- Si - O - Si - + OH^- = - Si - OH + - Si - O^- \tag{5.2}$$

The non-bridging oxygen ion so produced is now capable of dissociating another water molecule:

$$- Si - O^- + H_2O \rightarrow - Si - OH + OH^- \tag{5.3}$$

These equations may be summarized as follows, but the intermediate stages involving the non-bridging oxygen sites are of crucial importance, as demonstrated by the observed resistance of vitreous silica to aqueous attack:

$$- Si - O - Si - + H_2O = 2 \left[- Si - OH \right] \tag{5.4}$$

From Equations (5.1) and (5.3), excess hydroxyl ions are produced during the reaction, which, depending on the environment of the glass, may accumulate in the corrosion layer, increasing the pH of the attacking solution and therefore accelerating the dissolution of the network. Once the pH has risen to greater than 9, the silica network begins to break up and silicon is removed into solution as $Si(OH)_4$. Ion exchange of H^+ for Na^+ is referred to as Stage I of the process, and the break-up of the network as Stage II (Clark *et al.*, 1979). These equations explain the following observations:

(i) corrosion of glass proceeds faster in basic solutions;
(ii) pure silica is resistant to aqueous attack at around neutral pH;
(iii) glasses are normally resistant to acid attack (except HF, where the fluorine ion behaves like the hydroxyl group), since the hydroxyls produced tend to be neutralized before the network is attacked;
(iv) faster corrosion rates are observed with steam than with water at the same temperature because dilution effects on the hydroxyl concentration are reduced.

Charles also made another important observation, namely that corrosion of an expanded glass structure (*i.e.*, that resulting from a high fictive temperature) proceeds faster than corrosion of a compacted glass structure (annealed) even if conditions and glass composition are the same. Thus the weathering characteristics of a piece of glass also depend on its thermal history.

Douglas and El-Shamy (1967) studied the rate of leaching of alkali and silica from binary glasses as a function of pH, and El-Shamy *et al.* (1975) extended this work to include soda-lime–silica glasses. Results for a binary potash–silica glass are shown in Figure 5.7. Below pH 9 the rate of alkali extraction is approximately constant and that of silica negligible. This corresponds to attack *via* Equation (5.1), with the concentration of OH$^-$ ions being insufficient to break up the silica network as described by Equation (5.2). Above pH 9 the rate of alkali extraction falls and the quantity of silica extracted rises. The rate of extraction of silica depends strongly on the alkali content of the glass and the pH of the attacking solution. Douglas and El-Shamy confirmed that the alkali in solution produced by ion exchange in the glass affected the rate of extraction of silica, as proposed by Charles, by leaching identical glasses in water in two vessels, one containing a cation-exchange resin to remove the alkali ions. The extraction rate of silica in the alkali-free solution was significantly less than that in the alkaline solution. Thus the rate of extraction of silica depends on the rate of removal of alkali. Addition of calcium oxide to the glass reduced the quantity of alkali extracted provided the CaO content of the glass did not exceed 10 mole %. No calcium was detected in the solution. Above 10 mole % calcium was extracted at pH's below four, with a corresponding increase in the alkali extraction. This reaction is described by Equation (5.5):

$$-\overset{\displaystyle |}{\underset{\displaystyle |}{Si}} - O - Ca - O - \overset{\displaystyle |}{\underset{\displaystyle |}{Si}} - + 2H^+ = 2\left[-\overset{\displaystyle |}{\underset{\displaystyle |}{Si}} - OH\right] + Ca^{2+} \qquad (5.5)$$

The terminal network sites so produced then provide a path for the easy movement of alkali ions, explaining the accompanying increase in alkali extraction.

In a comprehensive series of experiments, El-Shamy and co-workers (1972) studied the leaching rates of powdered glasses with differing compositions in aqueous solutions. The apparatus and experimental method used are described by Rana and Douglas (1961a). These authors (Rana and Douglas, 1961a, 1961b) studied the rates of extraction of all components from soda-lime–silica and potash–lime–silica glasses in water, and concluded that two mechanisms were operating:

(i) over short time periods and at low temperatures the rate of alkali extraction varied with the square root of time;

(ii) over long time periods the amount of alkali and silica extracted vary linearly with time.

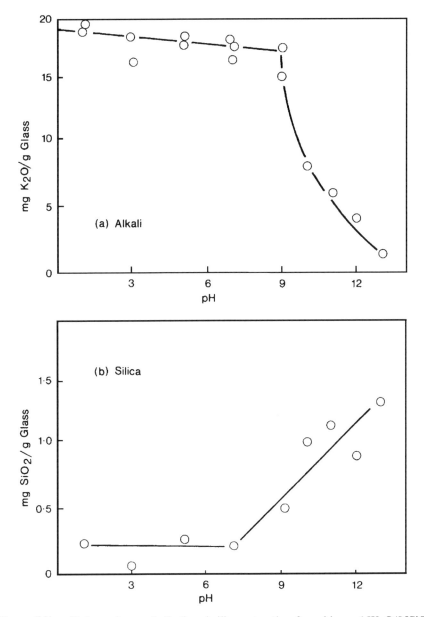

Figure 5.7 pH dependence of alkali and silica extraction from binary $15K_2O/85SiO_2$ glass at 35 °C. The y-axis shows the amount of alkali and silica extracted from a given weight of glass over a fixed time period. (From Douglas and El-Shamy, 1967; Figures 5 and 6, reprinted by permission of Blackwell Publishing Ltd.)

It was found that the rate of extraction of potash from potash glasses was always greater than the extraction rate of soda from the equivalent soda glass, even though the electrical conductivity of the potash glass is lower. Douglas and El-Shamy (1967) also observed that the ratio of alkali to silica in the solution is greater than the ratio in the glass, implying a preferential removal of alkali. This must lead to a leached layer being formed on the glass which is depleted in alkali [sometimes referred to as a *gel layer*, or, on volcanic glass, *palagonite* (Stroncik and Schminke, 2002)], the thickness of which depends on time and the composition of the glass. For glasses of low durability the leached layer may eventually extend through the whole bulk of the glass. The transition from a square-root dependent time coefficient to a linear dependence can be understood in terms of the time taken for a leached layer to reach equilibrium. Initially, the process will be diffusion-controlled as the alkali and hydrogen ion exchange occurs. When equilibrium is reached, there are two reaction planes, the gel–solution interface (*gel* is the term applied to the porous silica network) and the glass–gel boundary. A simple diagram of the situation is given in Figure 5.8, from Doremus (1973). At the gel–solution interface, the porous silica network is being broken down by the action of hydroxyl groups, at a rate which depends on the OH^- concentration at the surface. This concentration depends on the pH of the attacking solution, the composition of the glass and the experimental conditions. The gel layer itself is an open network allowing relatively free ionic movement. At the glass–gel interface, the ion-exchange reaction is proceeding, with H^+ ions entering the network and alkali ions leaving it. Thus at equilibrium there is a leached layer (the gel) of constant thickness, which progresses steadily into the glass as the silica network at the gel–solution interface is destroyed. The reaction between glasses and water can

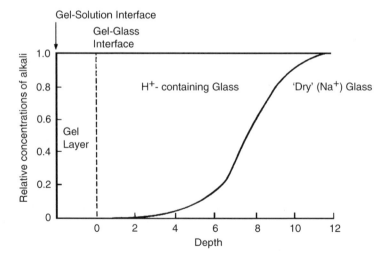

Figure 5.8   Model of hydrated glass surface showing the development of the hydrated layer. (Reprinted from Doremus, 1975; Figure 3. Copyright 1975; with permission from Elsevier.)

be treated mathematically as a double-diffusion problem with a moving boundary, as shown by Frischat in his monograph on ionic diffusion in oxide glasses (1975). Subsequent work by Hench and co-workers has extended this simple picture by defining six types of glass surface resulting from the attack of glass by water (summarized by Newton, 1985), which model more closely the behaviour of real glasses in contact with water of varying pH.

An important advance in the understanding of the chemical behaviour of glasses in aqueous solution was made in 1977, when Paul (1977) published a theoretical model for the various processes based on the calculation of the standard free energy ($\Delta G°$) and equilibrium constants for the reactions of the components with water. This model successfully predicted many of the empirically derived phenomena described above, such as the increased durability resulting from the addition of small amounts of CaO to the glass, and forms the basis for our current understanding of the kinetic and thermodynamic behaviour of glass in aqueous media.

In addition to the composition of the glass, the corrosion resistance is determined by the degree of phase separation present and the nature of the surface before the glass is subjected to corrosion. Other factors which may affect the behaviour are the presence of any crystalline particles, either due to heat treatment or from incomplete melting of the raw materials, and the presence of any stresses, either internal or external. The surface composition of glass can be modified by reactions which occur during the manufacturing process, and on contact with the atmosphere. These include the migration of alkali ions to the surface during manufacture to satisfy the 'dangling' bonds caused by termination of the silica network, partially compensated by the high temperature volatility of the alkalis. Fresh glass surfaces are also de-alkalized very rapidly by contact with atmospheric water vapour. Several techniques are available for determining the composition of these surface layers – refractive index measurements (Pfund, 1946), infrared reflection spectroscopy (IRRS: Sanders *et al.*, 1972), infrared microspectroscopy (Cooper *et al.*, 1993), Auger electron spectroscopy (AES: Dawson *et al.*, 1978) and electron loss spectroscopy (ELS: Pollard *et al.*, 1980). All these methods confirm that the chemical composition of the surface is a complicated function of age and thermal history.

The influence of phase separation on the physical properties of glass has been summarized by Porai-Koshits (1977): '*the viscosity of a two phase glass depends on the character of distribution of the high viscous phase, the electrical conductivity – on the distribution of the conducting phase, the chemical durability – on the distribution of the chemically more unstable phase.*' The effect of microstructure in the sodium silicate system has been studied by Redwine and Field (1968). They observed two types of phase structure after differing heat treatments of glasses with the same composition – two interconnected phases, or a dispersion of particles in a continuous phase. In the second case, the properties of the glass were very similar to those of the continuous phase, but in the first case the properties depended on the amount and composition of the low-soda phase. Frischat (1975) investigated the effect of phase separation on the diffusion of ^{22}Na through a soda–silica glass, and concluded that two diffusion paths are possible, one through the soda-rich phase, and one with a lower diffusion coefficient through the silica-rich phase. Reviews of the phase-separation tendencies of multi-component glasses are given by Doremus (1973) and Vogel (1977). The magnitude of phase separation is controlled by the addition of certain substances to the melt, as well as by heat treatment. One of the most important additives, as already noted, is alumina, which, when added in small quantities ($<2.5\%$), considerably reduces the tendency to phase separation. This is the basis of the manufacture of chemically resistant silicate glasses (Volf, 1961). Of importance to the study of ancient glasses, particularly 'forest glass', is the fact that phosphorus pentoxide (P_2O_5) is known to increase the tendency to phase separation, as is the addition of small quantities of halogens to the melt (Vogel, 1977). Medieval forest glass has high levels of phosphorus, and often significant traces of chlorine.

Modern chemically stable glasses have been developed empirically over the years, well in advance of the theoretical understanding of their chemical properties. It is interesting to note that the composition of Roman glass is close to the modern composition. Many of the glasses manufactured today for technological purposes are not in the soda-lime–silica system (e.g., Pyrex), but the poor durability of Medieval glass has been known for many years. In 1907, Heaton gave a paper to the Society of Arts (Heaton, 1907) in which he discussed the chemistry of Medieval glasses. He identified the main alkali as potash, derived from wood ash instead of seaweed, which produces a soda glass. The wood ash also supplied lime, magnesia and phosphorus pentoxide. The diversity of composition of Medieval glasses is attributable to the variability of composition of burnt wood, not just from species to species but even from different parts of the same tree. El-Shamy (1973) studied the reactions with aqueous solutions of simulated Medieval glasses (quaternary K_2O–CaO–MgO–SiO_2 glasses) with systematically varying composition. Replacement of SiO_2 by CaO (above 10 mole % CaO) resulted in an increased extraction of all oxides except SiO_2 in acid solution, even though the MgO and K_2O contents were unchanged. In acid solution the leaching of K_2O, CaO and MgO was greater, in accordance with Equation (5.5). Replacement of SiO_2 by MgO had the same effect. Replacing SiO_2 by K_2O caused a considerable increase in the extraction of all oxides in water. In glasses with 15 mole % K_2O, complete de-alkalization was observed within 72 hours. Replacement of K_2O by CaO or MgO has the effect of reducing the rate of alkali extraction by one-quarter for a 5% replacement. It seems that CaO and MgO are interchangeable, as replacement of CaO by MgO had no effect.

El-Shamy (1973) noticed a sudden decrease in the chemical resistance of these glasses when the SiO_2 content fell below 66 mole % SiO_2, and related this to the geometry of the structure. Below this point, every silicon atom becomes associated with a basic atom as second neighbour, so there is always an interconnecting path of non-bridging oxygen sites available for the diffusion process. This is the same argument as was used by Stevels (1960–1961: see above) to explain the importance of the point at which the number of bridging oxygen ions (Y) falls below three.

The fact that potash glasses are considerably less durable than equivalent soda glasses is attributable largely to the greater ionic radius of the K^+ ion (1.33 Å compared with 0.95 Å for Na^+). Thus as the melt passes through the glass formation temperature, the network is formed around the much larger K^+ ions, resulting in a more open structure which is more easily attacked by water. The difference in corrosion resistance of sodium and potassium binary silicate glasses with 30 mole % alkali has been studied by Hench (1975). Initially (at $t = 1$ min), a silica-rich film was developed by de-alkalization, penetrating deeper into the bulk in the potash glass. After 100 minutes, the silica network of the potash glass was being dissolved, whilst the soda glass suffered de-alkalization but not dissolution. After a long time (10^3 minutes), both glasses were being destroyed because of the high alkali content. Addition of calcium and aluminium tended to cause a protective film on the glass, preventing dissolution.

When two alkalis are present together in a glass, the glass exhibits non-linearity in many of its physical properties as the ratio of one alkali to the other is changed, the total alkali being kept constant. This is called the *mixed alkali effect* (MAE), and is most pronounced in those properties which depend on ionic transport – electrical conductivity, diffusion, chemical durability. Day (1976) has reviewed the effect, and discussed the many theories which attempt an explanation. Most properties show a pronounced maximum or minimum at a particular alkali ratio, and Day states that mixed alkali glasses have a higher durability than single alkali glasses. He reports a minimum in the durability for a molar potash/soda ratio of 0.7–0.8. Hench (1975), however, claims that there is no minimum in the durability curve, and that previous observations of minima are due to presenting the results in terms of weight rather than molar percentages. He also observed that the addition of calcium to the mixed alkali glasses did not produce a minimum. A more comprehensive experimental study of the MAE in mixed-alkali silicate and mixed-alkali lime silicate glasses by Dilmore (1977) led to the following conclusions about the MAE:

(i) Some glasses do exhibit a minimum in the concentration of leached alkalis – in general these are glasses of high durability;
(ii) the presence of MAE is a function of the pH of the attacking solution. Below pH 9, selective leaching is the dominant mode of decay, and the MAE may be present. Above pH 9, total dissolution takes place and the MAE is unimportant.

Hence it is concluded that in multi-component silicate glasses, the presence or absence of a MAE is a complicated function of composition and environment, and not necessarily entirely dependent on the ratio of one alkali to the other.

In an analytical study of the relationship between the chemical composition and corrosion behaviour of Medieval window glass from York Minster (Pollard, 1979), a total of more than 200 samples were analysed, dating mainly from the 12th, 14th and 15th Centuries. A summary of the findings has been published in Cox *et al.* (1979). The glasses were visually classified according to weathering characteristics into four broad categories: unweathered (u), pitted (p), pitted and crusted (p/c), and crusted (c). This classification is a gross simplification of the true variation in weathering characteristics, since no two corroded samples show exactly the same features. In particular, the pitted-and-crusted group contains a wide variety of specimens: some exhibit pits which contain a white deposit, whilst others show a severe loss of volume caused by total dissolution of the surface. Obviously, these four groups are extremely crude and can only be considered as a rough guide. It is also possible that some of the glasses which now exhibit pure pitting once had a crust in the pits and have subsequently lost it in the course of restoration, since the history of each piece is not precisely known.

The samples were chosen to give as broad a spread as possible of dates and weathering characteristics, but the group is not representative of the true range available in York Minster. No 13th Century glass was available at the time although several large windows of 13th Century work exist in the Minster. Only three glasses from the 16th Century were analysed. One piece of 11th Century glass was added to the group – this was an excavated deep-blue glass which was completely unweathered. Amongst the 12th and 14th Century glasses the most common decay mechanism is pitting, with relatively few cases of crusting, and few unweathered pieces surviving. By the 15th Century the majority of pieces are unweathered, with crusting in particular becoming very rare.

The complete results of the quantitative analysis of the samples by X-ray fluorescence (XRF) are given in Pollard (1979). The results of the XRF analysis were converted from weight to molar percentages, which were then used to calculate triangular co-ordinates according to the empirical formula based on that given by Iliffe and Newton (1976). The three axes are network formers 'SiO$_2$', alkali network modifiers ('R$_2$O') and alkaline earth network modifiers ('RO'). The formulae for calculating these co-ordinates from the molar percentages are:

$$\text{'SiO}_2\text{'} = SiO_2 + P_2O_5 + 2 \times (Al_2O_3 + Fe_2O_3)$$
$$\text{'R}_2\text{O'} = Na_2O + K_2O - (Al_2O_3 + Fe_2O_3)$$
$$\text{'RO'} = CaO + MgO + MnO + CuO + ZnO + PbO$$

The justification for adding twice the molar concentration of the trivalent oxides to the 'SiO$_2$' figure is that they can enter the network and also immobilize the alkali cations, which is why they are also deducted from the 'R$_2$O' value.

The data are most simply summarized in the triangular diagram shown in Figure 5.9. It is immediately apparent that there is a correlation between weathering behaviour and composition: the 60% 'SiO$_2$' value is the lower limit for durable glasses. This diagram also shows another striking feature – that is the existence of a small group of high 'SiO$_2$', low 'RO' glasses, toward the top left-hand side, well removed from any other group in the sample. This group contains five samples, which form a distinctive set in that they are all soda

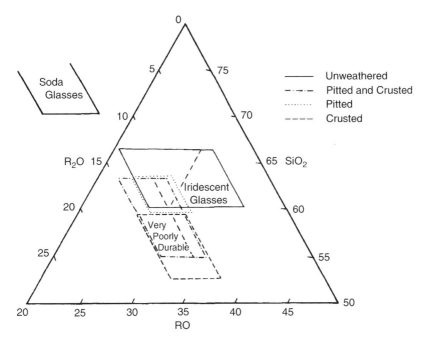

Figure 5.9 Triangular diagram showing the distribution of the various weathering classifications of York Minster glass. (From Pollard, 1979; Figure 46.)

glasses, all deep blue and all date from the 12th Century. Subsequent work (Cox and Gillies, 1986) has shown that such samples (all deep cobalt blue, all soda glasses and all late 12th Century AD in date) are more common in York Minster than was previously thought. The existence of this group is of great historical interest. The high sodium and low magnesium, phosphorus and potassium values of these glasses indicate a different manufacturing process, which may imply a different place of origin from the main body of the glasses analysed. The composition is very similar to that of Saxon glass as reported by Cramp (1970), but the pieces are larger than the common Saxon samples found. Possibly the glass was re-made from Saxon glass, or even, as suggested by Cox and Gillies (1986), from Roman (see below).

It can be seen from Figure 5.9 that the chemical distinction between pitted, pitted-and-crusted and crusted glasses is a relatively arbitrary one. All these glasses are characterized by 'R$_2$O' values between 10 and 15%, and it seems again that 60% 'SiO$_2$' is an important figure. Above this, decay is predominantly by pitting, below it by crusting, but a large number of pitted-and-crusted glasses straddle this line indicating that there is no real distinction between these groups. Above 60% 'SiO$_2$' it is likely that the network formers are linked by enough bridging oxygen atoms to resist the general dissolution of the network, giving rise to localized decay (pits). Below this point the network is more disrupted by modifiers and a uniform removal of the surface is possible. In his study of the durability of synthetic Medieval glasses discussed above,

El-Shamy (1973) observed a sudden drop in the chemical resistance of his glasses as the SiO_2 content dropped below two-thirds of the molar composition. This silica content marks the point at which, on average, each silicon atom has a basic ion as second neighbour, leading to an interconnected path of cation sites, with a corresponding increase in ionic mobility. In the glasses analysed here, aluminium and phosphorus will act as network formers, and this may lower the critical value to the observed 60% 'SiO_2'. In addition to the main regions in Figure 5.9, two further groupings are marked; a very poorly durable group characterized by high alkali content, and a group of otherwise unweathered glasses which show an iridescent surface, characterized by high alkaline earth content. Many glasses in this study were found to have lost a considerable amount of material from the surface, either uniformly over the whole surface or unevenly, as if by severe pitting. These glasses are examples of the poorest durability, although it is quite possible that glasses of even worse durability were manufactured, but have not survived. Despite the fact that great care was taken to produce a representative surface (Cox and Pollard, 1977), it is also possible that glasses of such low resistance to corrosion have moved on the triangular diagram as the alkali has been leached from the whole bulk, resulting in a glass which has an artificially lower 'R_2O' and a higher 'SiO_2' co-ordinate.

The types of weathering behaviour observed in the glasses from York Minster are not necessarily typical of the decay of Medieval glass in general. Several weathering phenomena commonly observed in other cathedrals are rare in York. In particular, glasses forming a thick heavy crust are not at all common in the Minster. As part of an attempt to extend the study to these other types, further analyses were carried out on generally unweathered Medieval glass. When plotted on the triangular diagram, four (two each from Winchester and Chartres) were found to occupy positions typical of the blue 12th Century soda glasses found in York. Again, all were dark blue and 12th century, and all are soda glasses (9.5–17.4 mole % Na_2O), with low MgO (0.6–0.8 mole %), low K_2O (0.4–1.2 mole %) and low CaO (7.4–8.9 mole %). These characteristics agree very well with the York group, and we must conclude that dark-blue soda glasses were fairly widespread during the 12th Century. Subsequently, Cox and Gillies (1986) studied more 12th Century blue soda glass from York, and concluded that it was not rare in the Minster, and postulated that the glass was Roman (or slightly later), re-melted in France, possibly with the addition of some potash glass, and imported into England. This is in agreement with the brief description given around AD 1100 by Theophilus (Hawthorne and Smith, 1979; 59), where he describes '*different kinds of glass . . . found in mosaic work in ancient pagan buildings*' and also '*various small vessels . . . which are collected by the French. . . . They even melt the blue in their furnaces, adding a little of the clear white to it, and they make from it blue glass sheets which are costly and very useful in windows.*' There appears to be good evidence from modern analytical data to substantiate this description, and the widespread distribution of such glass suggests that the re-use of ancient glass may have been more common than was previously thought. Why it should be only the cobalt blue glass which has so far been detected is, however, something of a mystery.

The structural chemistry of the weathering crusts themselves has been studied using a combination of XRD, infrared spectroscopy, atomic absorption analysis and electron microscopy (Gillies and Cox, 1988a, 1988b; Schreiner, 1988). It had previously been thought that the mineralogy of the corrosion products is a function of environmental conditions, such as window orientation. A number of phases were identified, most commonly gypsum ($CaSO_4.2H_2O$), syngenite [$K_2Ca(SO_4)_2.H_2O$], hydrated silica ($SiO_2.xH_2O$) and calcite ($CaCO_3$), but no pattern was found between the mineralogy of the corrosion products and the climatic conditions or orientation of the window. The best relationship appeared to be that between the bulk composition of the glass and the identity of the weathering products.

The main conclusion of these studies is that the weathering behaviour of Medieval glass is dictated primarily by chemical composition, and does not depend on the aspect of the window or the level of pollution in the atmosphere, as has often been thought. This has been confirmed by a study of some excavated Medieval glass from York known to have been from windows destroyed in 1540, which showed that the glass was already suffering from decay before it was buried (Cox and Cooper, 1995), well before the advent of modern levels of atmospheric pollution. Modern observations of synthetic low-durability glasses distributed around various environments in Europe and North America have also shown that the prime cause of decay is aqueous attack (Munier *et al.*, 2002). Local environmental conditions (*e.g.*, degree of exposure, levels of atmospheric pollution) may, however, dictate the nature of any corrosion products which accumulate (Munier *et al.*, 2002; Melcher and Schreiner, 2004).

The distinction between pitted and crusted glasses is shown, in general, to be an arbitrary one, with the two groups coalescing into a band on the triangular diagram, and exhibiting a slow gradation of weathering characteristics. It is here that the environment of the glass may have some influence on the corrosion, governing whether a crust forms or not as the surface is attacked by water. These conclusions agree reasonably well with the data from accelerated corrosion work on simulated medieval glasses, although the figure from this of 66.67 mole % SiO_2 as the lower limit of good durability is higher than the experimental figure. In this study, a figure of 60% 'SiO_2' seems more appropriate, but where the 'SiO_2' co-ordinate includes the intermediate oxides of aluminium and phosphorus, which may explain the difference.

5.5 THE CORROSION OF BURIED GLASS

The composition of Egyptian, Near Eastern and European archaeological glass has been studied for more than 100 years, and a number of well-defined chemical groups have been identified:

(i) high magnesia soda-lime glasses (*ca.* 1500–800 BC);
(ii) low magnesia soda-lime glasses (*ca.* 800 BC to AD 1000);

(iii) high antimony soda-lime glasses (*ca.* 600–200 BC);
(iv) Islamic high-lead glasses (*ca.* AD 1000–1400);
(v) Islamic high-magnesia glasses (*ca.* AD 840–1400).

These groups, first identified by Sayre and Smith (1961), have subsequently been added to by other workers (summarized by Henderson, 2001), including low magnesia high potassium (LMHK, *ca.* 1150–750 BC), high potassium high barium Chinese glasses (206 BC to AD 221), and high alumina glasses in the 1st Millennium AD India. The significance of these groupings, which appear to be remarkably consistent over large intervals of time and geographical regions, is that they are felt to be the result of the systematic use of a specific source of alkali. For example, the switch from high to low magnesia soda-lime–silica glasses around 800 BC in the Near East is interpreted as a change from plant ash (high in magnesium and potassium) to mineral sources of alkali (natron, lower in both), which may or may not also be a reflection of a change in the place of origin of the glass. The vexed question of chemical approaches to the provenance of glass is returned to below. Shortland *et al.* (2006) have recently reviewed the history of the use of natron in glass making, and have suggested that the decline of natron-based glass around the 9th Century AD (replaced once again by soda-rich plant ash in Islamic glasses, and, eventually, by potash-rich 'forest glass' in Europe) was the result of a decline in the ability of Wadi Natrun in Egypt to supply the demand. This, in turn, they suggest was either due to climatic change reducing the rate of production of natron, or, more likely, to political upheaval in the region.

Although buried glass is of more general interest in archaeology than is window glass, the detail of its corrosion behaviour has received considerably less attention in the literature. This is in part due to a general assumption by archaeologists that the glass which has survived is relatively inert, although this is not necessarily the case (Newton and Davison, 1989), but also because of the complex chemistry of the soil environment (Pollard, 1998) as compared to atmospheric attack. Freestone (2001) has summarized our knowledge of the corrosion mechanisms of buried glass [based largely on the competing rates of Type I (de-alkalization) and Type II (network destruction) corrosion as described above], and also the influence of manufacturing technology on the rate of corrosion. The interaction of groundwater with vitreous material is, of course, also crucial to the long-term stability of buried vitrified nuclear waste, and this aspect has been extensively studied using accelerated corrosion techniques (Lutze, 1988; Ewing, 2001). The value of the study of natural (geological) glasses has long been understood in the context of validating these models, but more recently it has come to be appreciated that a study of archaeological material also has a role to play (Petit, 1992; Sterpenich and Libourel, 2001).

Observation of the nature of the corrosion products on buried glass goes back at least as far as Brewster (1863), who reported the characteristic flakiness and surface iridescence. Under the microscope he observed a fine laminar structure with a range of thicknesses between 0.3 and 15 μm, which he deduced was responsible for the iridescence *via* interference phenomena. Subsequent chemical and structural analyses of these multiple layers using electron microscopy, XRD and Raman microspectroscopy (Cox and Ford, 1993) has concluded that these layers are depleted in virtually all oxides except those of Si, Al and Fe, and consist of poorly crystalline hydrated silicates and aluminosilicates, with deposits of calcite, calcium phosphate and manganiferous minerals on the surface, thought to be of external origin. Contrary to the early suggestion that these layers may reflect annual variations in ground conditions, which could be counted to give an age for the glass in the same way as dendrochronology (Brill and Hood, 1961), subsequent studies have shown a number of different morphologies of weathering crusts, including parallel layers, hemispherical layers and zigzag banding (Cox and Ford, 1993). The likely explanation of this is that the dissolution products from the glass (hydrated silica, plus ions of the alkalis and alkaline earths, depending on the bulk composition of the glass) diffuse out into the attacking aqueous solution, exceed the solubility product as a result of pH changes and precipitate as some form of hydrated, poorly crystalline silicate (or aluminosilicate) mineral, possibly incorporating other species which may be present in the attacking solution. The cyclic nature of the events which gives rise to the laminations may be explained by the precipitation event influencing the local pH and ionic concentrations in the solution, and thus temporarily inhibiting the precipitation process. It is clear from this that the process depends very strongly on local conditions, and not on any annual cycles, as was postulated by Brill and Hood (1961).

The precise nature of the chemical interaction between the buried glass and the local groundwater (and, indeed, between any buried archaeological object and its environment) is exceedingly complicated, but is fundamental to our understanding of the long-term alteration and survival of archaeological evidence. For over 30 years, geochemists have used computer modelling in order to provide further understanding of how such systems might behave. Computer models of the chemistry of dilute aqueous solutions such as surface groundwater have been in existence since the mid-1960s, and are now documented (*e.g.*, Bassett and Melchior, 1990) and commercially available. This type of program can be used to predict the stability fields for a large range of material over the range of naturally encountered conditions of pH and Eh. Although widely used in geochemistry, little systematic use has been made of these programs in archaeology, with the exception of the work done by Wilson (2004), who carried out simulations on a wide range of archaeological material, including glass and ceramics, as described in Chapter 4. At the moment the limitation is that for most models, the glass composition has to be converted into an 'equivalent mineralogy', since the thermodynamic parameters are only determined for stoichiometric minerals, which is highly unrealistic. There is a need to develop these tools (together with a knowledge of chemical kinetics), not only to aid our understanding of the interaction between buried glass and its environment, but also to increase our understanding of the nature of the surviving archaeological record in general.

A related problem has recently received some attention – this involves a study of the relative rates of dissolution of vitreous microtephra in the sedimentary environment. Tephra is the fine fraction of volcanic ash which can be ejected high into the stratosphere during explosive eruptions and transported considerable distances. The ash is usually a mixture of volcanic minerals, but contains within it a proportion of vitreous material. Visible deposits associated with well-known volcanic eruptive centres such as Iceland, southern Italy and New Zealand have been studied for many years, and used as stratigraphic markers for correlating geological sections across long distances. If a particular layer can be related to a specific eruption of a known volcano, and the date of that eruption is known, then it is possible to assign a date to this horizon. In some cases, where a historic eruption is identified, this date can be remarkably precise, such as August 24th for the AD 79 eruption of Vesuvius as described by Pliny (Francis and Oppenheimer, 2004; 55). When used for correlation purposes, the technique is known as *tephrostratigraphy*. If a date can be ascribed, then it has become known as *tephrochronology* (Lowe *et al.*, 2001). On the face of it, however, this method looks to be restricted to regions close to volcanic centres. In the past ten years, however, it has been demonstrated that fine particles of volcanic ash, including glass (termed *microtephra*), can be much more widely distributed than was previously thought. It has recently been shown, for example, that vitreous tephra from an Icelandic eruption (the Vedde Ash) can be found in two central European lake cores, *ca.* 2000 km from the source volcano (Blockley *et al.*, in press). At this distance from the volcano, however, they occur, not as visible layers, but as microscopic particles (typically around 25–100 µm) concentrated in particular layers. These minute glass particles have to be laboriously extracted from the surrounding sediment by chemical or flotation techniques. Crucially, identification with source volcano is usually made by comparison of major element analysis of the glass (usually using wavelength dispersive analytical electron microscopy, as described in Chapter 2). It is essential, therefore, to know whether any chemical alteration of the glass might have taken place, either during burial or as a result of chemical extraction procedures in the laboratory, since any such alteration would make it difficult if not impossible to make correct correlations.

Until recently it was not widely accepted by archaeologists and quaternary scientists (*i.e.*, those primarily interested in the past few tens to hundreds of thousands of years) that chemical alteration might be a factor in the identification of microtephra. This view is not entirely unreasonable, given that vitreous tephras have been chemically characterized and successfully identified in deposits dating to several million years (*e.g.*, Peate *et al.*, 2003). However, in the light of the above discussion on the chemical alteration of glass, and bearing in mind the small size of the particles (and hence the high surface-to-volume ratio), it might be suspected that chemical alteration could occur for some glass compositions in some environments. In fact, closer inspection of the literature on several tephras a million years old shows that potential alteration is indeed a prime consideration – any chemical correlations made do not rely on major and minor element composition alone, but are validated by trace element data (especially rare earth profiles) and isotopic ratios of Pb and Nd, precisely because these data are felt to be less susceptible to long-term alteration.

In order to investigate the potential influence of chemical attack on Late Quaternary vitreous tephra chemistry, a theoretical study was carried out by Pollard *et al.* (2003) in which the various parameters used to characterize the chemical durability of anthropogenic glasses (NBO, Si/O ratio and ΔG_H, as described above) were calculated for a range of published Icelandic tephra compositions (Haflidason *et al.*, 2000). The resulting stability plot (Figure 5.10) predicted that there should indeed be a variation in chemical stability, ranging from the silica-rich rhyolitic tephras such as Hekla 5 (dated to 6100 C^{14} years BP) and Hekla 4 (3830 C^{14} years BP), which are predicted to have stabilities similar to modern anthropogenic soda-lime silica glasses (*e.g.*, NIST 612), to the silica-poor basaltic tephras (*e.g.*, Saksunarvatn, *ca.* 9000 C^{14} years BP), which appears to have very poor durability. Subsequent laboratory experiments on a suite of volcanic glasses (Wolff-Boenisch *et al.*, 2004) have shown that major elements are indeed leached from the glass by mineral acid. These authors predicted that the lifetime of a 1 mm basaltic glass sphere (*i.e.*, similar in composition to Saksunarvatn) at pH 4 and 25 °C is 500 years, whereas that of a similar rhyolitic glass sphere (similar to Hekla 4 and 5) is 4500 years. Given that these dimensions are much larger than the typical distal microtephras used in tephrochronology, it is not at all unreasonable to expect that chemical alteration or even total dissolution might be possible in some circumstances, even when lower ambient temperature is allowed for. This suggests that the observed absence of certain more basic tephras in the distal sedimentary record in some cases (*e.g.*, in peat bogs) may be due to chemical factors rather than the poorer dispersive powers of these eruptions, as had been previously thought. It appears, therefore, that the chemistry of glass dissolution could yet have some significance in quaternary science.

5.6 RADIOGENIC ISOTOPES AND THE PROVENANCE OF GLASS

Determining the provenance of glass by chemical means has been one of the great challenges to the archaeological chemist for many years, and, until recently, has proved extremely elusive. In many ancient societies, glass was a relatively common commodity, although it is likely that it had the status of a luxury item. In common with fine pottery and metals, therefore, it is of interest to be able to trace the trade in glass by identification of the source of manufacture, and, therefore, infer the presence of ancient trading routes. This has proved to be a particularly intractable problem, for a number of reasons. One is the necessity to distinguish between primary glass *making* sites (where sand and alkali are fused to form raw glass) and secondary glass *working* sites, where this raw glass is converted into vessels and other artefacts. Glass making and glass working may well occur at very different places, and there is some evidence from shipwrecks that raw glass chunks have been significant items of trade around the Mediterranean since the Bronze Age (Shortland *et al.*, 2007). It is gradually becoming apparent that the number of glass-making sites in the

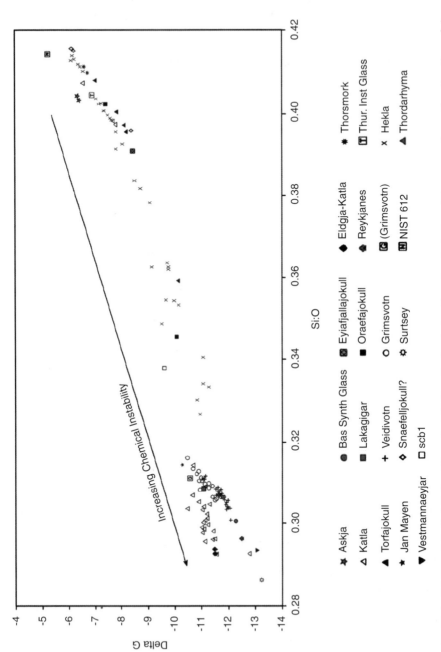

Figure 5.10 Plot of ΔG_{hyd} (labelled Delta G) against Si:O ratio for Icelandic volcanic tephra. The arrow shows the direction of decreasing chemical stability. (From Pollard *et al.*, 2003; Figure 1. Copyright John Wiley and Sons Limited. Reproduced with permission.)

Roman and later period may have been very limited, and confined to places in the Near East where there is a good supply of natron and suitable sand (Freestone *et al.*, 2008). In the absence of any archaeological evidence, however, the actual locations of any major sites of Roman glass making remain unknown – a fact which, given the ubiquity of glass in the Roman empire, is little short of astonishing.

Resolving these issues by chemical means is made difficult by the relative complexity of glass as a material. Chemical methods of characterization have proved generally unhelpful. Minor and trace elements, which in the case of ceramics have proved useful in provenance (see Chapter 4), can enter the glass from the sand, or with the alkali source (which, if plant ash, might be quite variable), or with the source of lime (if the sand does not contain enough calcareous material), or from the colouring pigments added, or even from the crucible. When this uncertainty is compounded by the possibility that glass from different sources may have been mixed during manufacture, and the propensity to use a percentage of cullet (recycled glass) when melting raw glass, it is quite clear that glass is a particularly difficult material to provenance chemically. Although it has generally proved impossible to attribute provenance by chemical analysis, it has sometimes been possible to identify similarities within glass assemblages which may enable archaeological interpretation. For example, Jackson (2005), in her study of colourless Roman glass from three sites in Britain, identified two broad compositional groups which she attributed to differences in the choice of the initial raw materials selected for glass production, in particular the sand. She was unable, however, to say where these raw materials might have come from.

The breakthrough, if such it turns out to be, has come from the recent application of isotopic analysis to glass. The fact that it has taken so long is somewhat surprising, given that glass was one of the first archaeological materials to be studied using lead isotope techniques in the 1960s (see Chapter 9), but even this approach did not solve the problem. This may be due to the fact that lead, like many other trace elements, can enter glass from a number of sources (as listed above). The real breakthrough has come with the measurement, initially, of strontium and neodymium isotopes in glass, and possibly those of oxygen. The isotope systematics of strontium and neodymium are covered in most texts on isotope geochemistry, such as Faure and Mensing (2005). A summary of the strontium system is given in Chapter 10 and the use of Sr isotopes in determining glass provenance has been discussed by Freestone *et al.* (2003). Sr can be regarded as a proxy for the source of calcium in the glass, and in glasses made by mixing a mineral source of alkali such as natron with sand, the ratio of $^{87}Sr/^{86}Sr$ can indicate whether the lime has come from marine shells or from crushed limestone. This is because variation in marine $^{87}Sr/^{86}Sr$ has been extensively studied over geological time (back to 600 million years ago) through measurement in sedimentary rocks, and the value of this ratio in seawater during the Holocene had been very close to 0.7091. This is a uniquely high value in the geological record, with the average being around 0.708, and the lowest known value being *ca.* 0.7067 in the Permian. (These differences, although apparently small, are highly significant and easily measured using modern mass spectrometry.) Thus if the value in the glass is around 0.7091, it is evidence that the source of limestone is modern marine shell. The easiest way for this to enter the glass is as a natural mixture with beach sand. Limestones tend to have $^{87}Sr/^{86}Sr$ between 0.707 and 0.708, and therefore the deliberate addition of limestone can be detected. Sr isotopes are conventionally measured by thermal ionization mass spectrometry (TIMS), but are now capable of measurement by high-resolution inductively-coupled mass spectroscopy (ICP-MS), which is both faster and cheaper (see Chapter 2).

The isotope ratio of neodymium ($^{143}Nd/^{144}Nd$) is variable in the Earth's crust because ^{143}Nd is the daughter of the radioactive decay of ^{147}Sm. It is usually reported as ε_{Nd}, which is a measure of the isotopic ratio in the sample compared to the ratio in the theoretical primordial matter (the CHUR model – chondritic uniform reservoir – see Faure, 1986; 209). ε_{Nd} in the Earth's crust ranges from –45 to +12, with, crudely speaking, the older crustal material having the more negative values. Recent sediments, including glass-making sands, are expected to have a range of –16 to –3. ε_{Nd} is conventionally measured by TIMS. In an early archaeological application of this method in 1999, Wilson, Pollard and Evans measured Sr and Nd isotopes in a small number of Medieval glass samples (seven from York Minster, kindly provided by York Glazier's Trust, and one from Canterbury Cathedral) using the TIMS facilities at the National Isotope Geosciences Laboratories, Keyworth, UK. Figure 5.11 shows a plot of ε_{Nd} against $^{87}Sr/^{86}Sr$ for these samples (unpublished data, listed

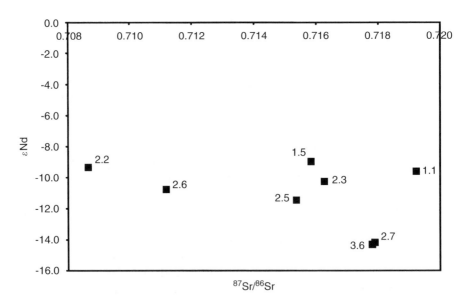

Figure 5.11   Plot of $^{87}Sr/^{86}Sr$ vs ε_{Nd} for some examples of English Medieval cathedral window glass. (Wilson, Pollard and Evans unpublished data; key in Table 5.4)

Table 5.4 Values of ε_{Nd} and $^{87}Sr/^{86}Sr$ for eight samples of Medieval window glass. (Wilson, Pollard and Evans unpublished data.)

Sample	Colour	Century	Nd ppm	$^{143}Nd/^{144}Nd$	Error	ε_{Nd}	$^{87}Sr/^{86}Sr$	Error
YM 1.5	Green	15th	7.148	0.512179	4	−9.0	0.715847	4
YM 3.6	Pink	Medieval	1.723	0.511872	10	−14.3	0.717822	4
YM 1.1	Pink	13th	9.854	0.512160	7	−9.6	0.719240	3
CC 2.7	Pink	12th	1.851	0.511911	14	−14.2	0.717892	4
YM 2.5	Pink	12th	3.211	0.512050	7	−11.5	0.715373	3
YM 2.3	Green	12th	5.535	0.512111	5	−10.3	0.716278	4
YM 2.2	Blue	12th	11.26	0.512158	4	−9.4	0.708658	3
YM 2.6	Ruby	12th	6.093	0.512086	3	−10.8	0.711199	3

in Table 5.4). The two green samples from York, despite being attributed to the 12th and 15th Centuries, are very similar, suggesting either continuity of supply, re-use or an error in the dating of one or the other. Two of the pink samples are also identical – one 12th Century sample from Canterbury, and one, only identified as 'Medieval', from York. This, albeit very tentatively, hints at a common glass supply for the two Cathedrals and highlights the powerful potential of the method. The ruby and the blue samples are very different from the rest, and could possibly be made from re-used Roman glass. Despite the fact that these eight samples were not well-provenanced or dated, the results are sufficient to show that there is a large variation in both ε_{Nd} and $^{87}Sr/^{86}Sr$ in these Medieval glasses, suggesting that further measurements would be extremely worthwhile.

The most systematic and successful isotope provenancing of glass has been carried out by Freestone and co-workers. In a recent study of an assemblage of Late Roman (4th–5th Centuries AD) glass from Billingsgate, London, Freestone *et al.* (2008) have compared these samples with contemporary raw glass chunks from excavations in Carthage, and with some later material from Late Byzantine or Early Islamic glass tank furnaces in Israel. The London glass belongs compositionally to the low magnesium soda–lime glass type (as does most Roman to 9th Century AD glass), taken to indicate natron as the source of the alkali. In fact, it belongs to a sub-group of this called 'HIMT' (high iron, manganese and titanium), which distinguishes later Roman glass from that of the 1st to 3rd Centuries AD. Strontium isotopes in these samples show that the calcium source was modern marine shell, most likely from beach sand. The neodymium isotopes suggest that the sand came from an area in the south-eastern Mediterranean, between the Nile delta and the coast of Israel. For the first time, therefore, isotopic analysis of late Roman glass places its source of production somewhere between Alexandria in Egypt and Gaza in Palestine. This is archaeologically extremely interesting and important, not only because the finding of such glass in London suggests that, even at the end of the Empire, the Roman trading system was still operating, but also because of the ubiquity of HIMT glass. It is not known why HIMT replaced the earlier form of glass, nor, as yet, where this earlier form was manufactured.

Nevertheless, this work shows that the previously intractable question of glass provenance is now amenable to archaeological investigation.

5.7 SUMMARY

A study of the chemistry and corrosion of glass provides ample opportunity for the application of a wide range of inorganic and physical chemical principles. The structure of glass was for many years a challenging area of research for physical chemists, with two competing theories – the CRN model of Zachariasen and co-workers and the 'crystallite' model of Lebedev. As with many scientific debates, the truth lies somewhere in between, and when examined closely the two competing models merge into one another. With improved instrumentation for studying short-range atomic order, the subtleties of the vitreous state, including the influence of phase separation on various properties such as durability, became much clearer, and the debate has now resolved itself. The explanation of the colours observed as a result of the presence of small amounts of transition metals in glasses is a particularly interesting application of inorganic chemistry, since it involves knowledge of co-ordination chemistry, crystal field theory and redox reactions. Some of the most spectacular examples of the glassmakers art, however, also involve physical principles such as the dichroic effect produced by the scattering of light from a colloidal dispersion of gold–silver alloy precipitates in glass, such as in the Lycurgus Cup.

The corrosion and decay of Medieval window glass has received a great deal of attention for many years, and has provided a number of surprises. It has been shown, for example, that the decay of the potash-based coloured glass in the great cathedrals of northern Europe is not primarily caused by increased atmospheric pollution, but by aqueous attack on glasses of inherently low durability. One of the principal factors in this low durability is the composition of the glass itself – partly because of the lower durability of potash glasses compared to equivalent soda glasses, but probably more importantly because the total proportion of 'network formers' in the glass is lower than the critical value required to produce a continuous stable network. This conclusion has been tested by chemical analysis of the glasses themselves, by accelerated corrosion tests on synthetic glasses of Medieval composition and by thermo-dynamic calculations of theoretical stabilities.

The corrosion of buried glass is considerably less well-understood by archaeologists, despite a substantial body of data collected by the nuclear waste disposal industry on the interaction between vitreous materials and groundwater. The basic process of the leaching of glass components and the cyclical precipitation of poorly crystalline silicate minerals can be explained, but the detail of the interaction between buried glass and its groundwater environment requires further modelling in order to become predictive or reconstructive. The value of chemical studies on the corrosion of archaeological glass is now appreciated as being of great value in validating the accelerated testing results on nuclear waste disposal materials, and it suggests that the natural long-term laboratory that is the archaeological record is not yet fully exploited in terms of its relevance to modern-day chemistry.

In common with other areas of archaeological chemistry, glass studies have benefited enormously from the application of isotopic methods of analysis. Although one of the first archaeological materials to be studied isotopically, it is only very recently that the application of strontium and neodymium isotope systematics has revealed the probable eastern Mediterranean provenance of Late Roman vessel glass. This opens up a whole new field of chemical enquiry in archaeology, and one which is all the more exciting because of the previous intractability of glass provenance, and the likely benefits in terms of sample throughput of the application of ICP-MS instrumentation. Perhaps even more significantly, however, is the fact that it allows us to reconstruct for the first time the ancient trading patterns of glass – a common yet valuable material, but one which appears to have a surprisingly small number of primary manufacturing sites – just where the ancient historians told us they should be!

REFERENCES

Bacon, F.R. (1968). The chemical durability of silicate glass Part One. *Glass Industry* **49** 438–446.

Bamford, C.R. (1977). *Colour Generation and Control in Glass.* Glass Science and Technology 2, Elsevier, Amsterdam.

Barber, D.J. and Freestone, I.C. (1990). An investigation of the origin of the colour of the Lycurgus Cup by analytical transmission electron microscopy. *Archaeometry* **32** 33–45.

Bassett, R.L. and Melchior, D.C. (1990). Chemical modeling of aqueous systems: an overview. In *Chemical Modeling of Aqueous Systems II*, ed. Melchior, D.C. and Bassett, R.L., American Chemical Society, Washington, pp. 1–14.

Blockley, S.P.E., Lane, C.S., Lotter, A.F. and Pollard, A.M. (in press). Evidence for the presence of the Vedde Ash in Central Europe. *Quaternary Science Reviews.*

Brewster, D. (1863). On the structure and optical phenomena of ancient decomposed glass. *Philosophical Transactions of the Royal Society of Edinburgh* **23** 193–204.

Brill, R.H. and Hood, H.P. (1961). A new method for dating ancient glass. *Nature* **189** 12–14.

Burnett, D.G. and Douglas, R.W. (1970). Liquid–liquid phase separation in the soda–lime–silica system. *Physics and Chemistry of Glasses* **11** 125–135.

Charles, R.J. (1958). Static fatigue of glass I. *Journal of Applied Physics* **29** 1549–1553.

Clark, D.E., Pantano, C.G. Jr and Hench, L.L. (1979). *Corrosion of Glass.* Books for Industry and the Glass Industry, New York.

Cooper, G.I., Cox, G.A. and Perutz, R.N. (1993). Infra-red microspectroscopy as a complimentary technique to electron-probe microanalysis for the investigation of natural corrosion on potash glasses. *Journal of Microscopy* **170** 111–118.

Cotton, F.A. and Wilkinson, G. (1976). *Basic Inorganic Chemistry*. Wiley, New York.

Cox, G.A. and Cooper, G.I. (1995). Stained glass in York in the mid-16th century – analytical evidence for its decay. *Glass Technology* **36** 129–134.

Cox, G.A. and Ford, B.A. (1993). The long-term corrosion of glass by groundwater. *Journal of Materials Science* **28** 5637–5647.

Cox, G.A. and Gillies, K.J.S. (1986). The X-ray fluorescence analysis of Medieval blue soda glass from York Minster. *Archaeometry* **28** 57–68.

Cox, G.A. and Pollard, A.M. (1977). X-ray fluorescence analysis of ancient glass – the importance of sample preparation. *Archaeometry* **19** 45–54.

Cox, G.A., Heavens, O.S., Newton, R.G. and Pollard, A.M. (1979). A study of the weathering behaviour of Mediaeval glass from York Minster. *Journal of Glass Studies* **21**, 54–75.

Cramp, R. (1970). Decorated window glass and millefiori from Monkwearmouth. *Antiquaries Journal* **50** 327–335.

Dawson, P.T., Heavens, O.S. and Pollard, A.M. (1978). Glass surface analysis by Auger electron spectroscopy. *Journal of Physics C* **11** 2183–2193.

Day, D.E. (1976). Mixed alkali glasses – their properties and uses. *Journal of Non-Crystalline Solids* **21** 343–372.

Dilmore, M.F. (1977). *Chemical Durability of Multicomponent Silicate Glasses*. Unpublished PhD Thesis, University of Florida.

Doremus, R.H. (1973). *Glass Science*. Wiley, New York.

Doremus, R.H. (1975). Interdiffusion of hydrogen and alkali ions in a glass surface. *Journal of Non-Crystalline Solids* **19** 137–144.

Douglas, R.W. and El-Shamy, T.M.M. (1967). Reactions of glasses with aqueous solutions. *Journal of the American Ceramic Society* **50** 1–8.

Douglas, R.W. and Isard, J.O. (1949). The action of water and of sulphur dioxide on glass surfaces. *Journal of the Society of Glass Technologists* **33** 289–335.

El-Shamy, T.M. (1973). The chemical durability of $K_2O–CaO–MgO–SiO_2$ glasses. *Physics and Chemistry of Glasses* **14** 1–5.

El-Shamy, T.M., Lewins, J. and Douglas, R.W. (1972). The dependence on the pH of the decomposition of glasses by aqueous solutions. *Glass Technology* **13** 81–87.

El-Shamy, T.M., Morsi, S.E., Taki-Eldin, H.D. and Ahmed, A.A. (1975). Chemical durability of $Na_2O–CaO–SiO_2$ glasses in acid solutions. *Journal of Non-Crystalline Solids* **19** 241–250.

Ewing, R.C. (2001). Nuclear waste form glasses: the evaluation of very long-term behaviour. *Materials Technology* **16** 30–36.

Faure, G. (1986). *Principles of Isotope Geology*. John Wiley, New York, 2nd edn.

Faure, G. and Mensing, T.M. (2005). *Isotopes: Principles and Applications*. Wiley, Hoboken, NJ.

Francis, P. and Oppenheimer, C. (2004). *Volcanoes*. Oxford University Press, Oxford, 2nd edn.

Frank, S. (1982). *Glass and Archaeology*. Studies in Archaeological Science, Academic Press, London.

Freestone, I.C. (2001). Post-depositional changes in archaeological ceramics and glasses. In Brothwell, D. and Pollard, A.M. (eds) *Handbook of Archaeological Sciences*, Wiley, Chichester, pp. 615–625.

Freestone, I.C., Leslie, K.A., Thirlwall, M. and Gorin-Rosen, Y. (2003). Strontium isotopes in the investigation of early glass production: Byzantine and early Islamic glass from the Near East. *Archaeometry* **45** 19–32.

Freestone, I.C., Degryse, P., Shepherd, J., Gorin-Rosen, Y. and Schneider, J. (2008). Near Eastern origin of Late Roman glass from London using neodymium and strontium isotopes. *Journal of Archaeological Science*.

Frischat, G.H. (1975). *Ionic Diffusion in Oxide Glasses*. Trans Tech Publications, Bay Village, Ohio.

Gillies, K.J.S and Cox, A. (1988a). Decay of medieval stained glass at York, Canterbury and Carlisle. Part 1. Composition of the glass and its weathering crusts. *Glastechnische Berichte* **61** 75–84.

Gillies, K.J.S and Cox, A. (1988b). Decay of medieval stained glass at York, Canterbury and Carlisle. Part 2. Relationship between the composition of the glass, its durability and the weathering products. *Glastechnische Berichte* **61** 101–107.

Goldschmidt, V.M. (1926). Geochemische verteilungsgesetze der elemente VIII. Untersuchungen über bau und eigenschaften von krystallen. *Skrifter utgitt av Det Norske Videnskaps-Akademi I Oslo. I Matematisk-naturvidenskapelig klasse. 2 Bind.* No. 8, pp. 1–156.

Greaves, G.N, Fontaine, A., Lagarde, P., Raoux, D. and Gurman, S.J. (1981). Local structure of silicate glasses. *Nature* **293** 611–616.

Green L.R. and Hart F.A. (1987). Colour and chemical composition of ancient glass: an examination of some Roman and Wealden glass by means of ultraviolet–visible–infra-red spectrometry and electron microprobe analysis. *Journal of Archaeological Science* **14** 271–282.

Greer, A.L. (1995). Metallic glasses. *Science* **267** 1947–1953.

Haflidason, H., Eiriksson, J. and van Kreveld, S. (2000). The tephrochronology of Iceland and the North Atlantic region during the Middle and Late Quaternary: a review. *Journal of Quaternary Science* **15** 3–22.

Hawthorne, J.G. and Smith, C.S. (trans.) (1979). *Theophilus, De Diversis Artibus*. Dover Publications, New York.

Heaton, N. (1907). Mediaeval stained glass: its production and decay. *Journal of the Society of Arts* **55** 468–484.

Hench, L.L. (1975). Characterization of glass corrosion and durability. *Journal of Non-Crystalline Solids* **19** 27–39.

Henderson, J. (2001). Glass and glazes. In *Handbook of Archaeological Sciences*, eds Brothwell, D. and Pollard, A.M., Wiley, Chichester, pp. 471–482.

Holland, L. (1964). *The Properties of Glass Surfaces*. Chapman and Hall, London.

Holloway, D.G. (1973). *The Physical Properties of Glass*. Wykeham, London.

Huang, C. and Cormack, A.N. (1990). The structure of sodium silicate glass. *Journal of Chemical Physics* **93** 8180–8186.

Huggins, M.L. and Sun, K.-H. (1943). Calculations of density and optical constants of a glass from its composition in weight percent. *Journal of the American Ceramic Society.* **26** 4–11.

Iliffe, C.J. and Newton, R.G. (1976). Using triangular diagrams to understand the behaviour of Medieval glass. *Verres et Réfractaires* **30** 30–34.

Jackson, C.M. (1992). *A Compositional Analysis of Roman and Early Post-Roman Glass and Glassworking Waste from Selected British Sites.* Unpublished PhD Thesis, Department of Archaeological Sciences, University of Bradford.

Jackson, C.M. (2005). Making colourless glass in the Roman period. *Archaeometry* **47** 763–780.

Jantzen, C.M. and Plodinec, M.J. (1984). Thermodynamic model of natural, Medieval and nuclear waste glass durability. *Journal of Non-Crystalline Solids* **67** 207–223.

Kingery, W.D. and Vandiver, P.B. (1986). *Ceramic Masterpieces.* Free Press, New York.

Lowe, J.J., Hoek, W.Z. and INTIMATE Group (2001). Inter-regional correlation of palaeoclimatic records for the last Glacial–Interglacial transition: a protocol for improved precision recommended by the INTIMATE project group. *Quaternary Science Reviews* **20** 1175–1187.

Lutze, W. (1988). Silicate glasses. In *Radioactive Waste Forms for the Future*, ed. Lutze, W. and Ewing, R.C., North-Holland, Amsterdam, pp. 3–159.

Melcher, M. and Schreiner, M. (2004). Statistical evaluation of potash–lime–silica glass weathering. *Analytical and Bioanalytical Chemistry* **379** 628–639.

Morey, G.W. (1954). *Properties of Glass.* Reinhold, New York.

Munier, I., Lefevre, R., Geotti-Bianchini, F. and Verità, M. (2002). Influence of polluted urban atmosphere on the weathering of low durability glasses. *Glass Technology* **43** 225–237.

Newton, R.G. (1978). Colouring agents used by Medieval glassmakers. *Glass Technology* **19** 59–60.

Newton, R.G. (1982). *The Deterioration and Conservation of Painted Glass: A Critical Bibliography.* Corpus Vitrearum Medii Aevi Great Britain Occasional Paper II, British Academy, Oxford University Press, Oxford.

Newton, R.G. (1985). The durability of glass – a review. *Glass Technology* **26** 21–38.

Newton, R.G. and Davison, S. (1989). *Conservation of Glass.* Butterworths, London.

Paul, A. (1977). Chemical durability of glasses; a thermodynamic approach. *Journal of Materials Science* **12** 2246–2268.

Paul, A. (1990). *Chemistry of Glasses.* Chapman and Hall, London, 2nd edn.

Peate, I.U., Baker, J.A., Kent, A.J.R., Al-Kadasi, M., Al-Subbary, A., Ayalew, D. and Menzies, M. (2003). Correlation of Indian Ocean tephra to individual Oligocene silicic eruptions from Afro-Arabian flood volcanism. *Earth and Planetary Science Letters* **211** 311–327.

Perera, G., Doremus, R.H. and Lanford, W. (1991). Dissolution rates of silicate glasses in water at pH 7. *Journal of the American Ceramic Society* **74** 1269–1274.

Petit, J.C. (1992). Natural analogs for the design and performance assessment of radioactive-waste forms – a review. *Journal of Geochemical Exploration* **46** 1–33.

Pfund, A.H. (1946). The aging of glass surfaces. *Journal of the Optical Society of America* **36** 95–99.

Pollard, A.M. (1979). *X-Ray Fluorescence and Surface Studies of Glass, with Application to the Durability of Mediaeval Window Glass.* Unpublished D.Phil. Thesis, Department of Physics, University of York.

Pollard A.M. (1998). The chemical nature of the burial environment. In *Preserving Archaeological Remains In Situ*, ed. Corfield, M., Hinton, P. Nixon, T. and Pollard, A.M., Museum of London Archaeology Service, London, pp. 60–65.

Pollard, A.M., Matthew, J.A.D. and Heavens, O.S. (1980). Electron loss spectra of silicate glasses. *Physics and Chemistry of Glasses* **21** 167–170.

Pollard, A.M., Blockley, S.P.E. and Ward, K.R. (2003). Chemical alteration of tephra in the depositional environment: theoretical stability modeling. *Journal of Quaternary Science* **18** 385–394.

Pollard, A.M., Batt, C.M., Stern, B. and Young, S.M.M. (2007). *Analytical Chemistry in Archaeology.* Cambridge University Press, Cambridge.

Porai-Koshits, E.A. (1977). The structure of glass. *Journal of Non-Crystalline Solids* **25** 87–128.

Porai-Koshits, E.A. (1985). Structure of glass: the struggle of ideas and prospects. *Journal of Non-Crystalline Solids* **73** 79–89.

Porai-Koshits, E.A. (1990). Genesis of concepts on structure of inorganic glasses. *Journal of Non-Crystalline Solids* **123** 1–13.

Rana, M.A. and Douglas, R.W. (1961a). The reaction between glass and water. Part I. Experimental methods and observations. *Physics and Chemistry of Glasses* **2** 179–195.

Rana, M.A. and Douglas, R.W. (1961b). The reaction between glass and water. Part 2. Discussion of the results. *Physics and Chemistry of Glasses* **2** 196–205.

Redwine, R.H. and Field, M.B. (1968). The effect of microstructure on the physical properties of glasses in the sodium silicate system. *Journal of Materials Science* **3** 380–388.

Sanders, D.M., Person, W.B. and Hench, L.L. (1972). New methods for studying glass corrosion kinetics. *Applied Spectroscopy* **26** 530–536.

Sayre, E.V. and Smith, R.W. (1961). Compositional categories of ancient glass. *Science* **133** 1824–1826.

Scholze, H. (1988). Glass–water interactions. *Journal of Non-Crystalline Solids* **102** 1–10.

Schreiner, M. (1988). Deterioration of stained medieval glass by atmospheric attack. Part 1. Scanning microscopic investigations of the weathering phenomena. *Glastechnische Berichte* **61** 197–204.

Sellner, C., Oel, H.J. and Camera, B. (1979). Unterscuchung alter gläser (waldglas) auf zusammenhang von zusammenensetzung, farbe und scmelzatmosphäre mit der elektronenspektroskopie und der elektronenspin-resonanz (ESR). *Glastechnische Berichte* **52** 255–264.

Shortland, A., Schachner, L., Freestone, I. and Tite, M. (2006). Natron as a flux in the early vitreous materials industry: sources, beginnings and reasons for decline. *Journal of Archaeological Science* **33** 521–530.

Shortland, A., Rogers, N. and Eremin, K. (2007). Trace element discriminants between Egyptian and Mesopotamian Late Bronze Age glasses. *Journal of Archaeological Science* **34** 781–789.

Sterpenich, J. and Libourel, G. (2001). Using stained glass windows to understand the durability of toxic waste matrices. *Chemical Geology* **174** 181–193.

Stevels, J.M. (1960–1961). New light on the structure of glass. *Philips Technical Review* **22** 300–311.

Stroncik, N.A. and Schminke, H.-U. (2002). Palagonite – a review. *International Journal of Earth Science* **91** 680–697.

Vogel, W. (1977). Phase separation in glass. *Journal of Non-Crystalline Solids* **25** 172–214.

Vogel, W. (1994). *Glass Chemistry*. Springer-Verlag, Berlin, 2nd edn.

Volf, M.B. (1961). *Technical Glasses*. Pitman, London.

Vose, R.H. (1980). *Glass*. Collins, London.

Warren, B.E. (1937). X-ray determination of the structure of liquids and glass. *Journal of Applied Physics* **8** 645–654.

Warren, B.E. and Biscoe, J. (1938). Fourier analysis of X-ray patterns of soda–silica glass. *Journal of the American Ceramic Society* **21** 259–265.

Weyl, W.A. (1976). *Coloured Glasses*. Society of Glass Technology, Sheffield.

Wilson, L. (2004). *Geochemical Approaches to Understanding In Situ Diagenesis*. Unpublished PhD thesis, University of Bradford, UK.

Wolff-Boenisch, D., Gislason, S.R., Oelkers, E.H. and Putnis, C.V. (2004). The dissolution rates of natural glasses as a function of their composition at pH 4 and 10.6, and temperatures from 25 to 74 degrees C. *Geochimica et Cosmochimica Acta* **68** 4843–4858.

Zachariasen, W.H. (1932). The atomic arrangement in glass. *Journal of the American Chemical Society* **54** 3841–3851.

Zhang, Y., Wang, W.H. and Greer, A.L. (2006). Making metallic glasses plastic by control of residual stress. *Nature Materials* **5** 857–860.

CHAPTER 6

The Chemical Study of Metals – the Medieval and Later Brass Industry in Europe

6.1 INTRODUCTION

With the exception of gold and platinum, the great majority of the metal artefacts used in antiquity were obtained by smelting metalliferous ores to give the various metals and alloys required (Craddock, 1995). Historically, one of the first goals of archaeological chemistry was to track these metal objects back to their ore source using trace element analysis, since this may be a direct method of reconstructing ancient trading patterns. In this context, we are using the term 'trace element' to imply a component of the alloy which was not deliberately added, and which may be considered to be characteristic of the ore from which the metal (or one of the metals in an alloy) was extracted. Such 'trace elements' may well be present in amounts up to a few percent, and the issue of deliberate addition or accidental incorporation is contentious, to say the least (*e.g.*, Pollard *et al.*, 1990). Although the principle of trace element provenancing is very similar to that employed successfully with other archaeological materials [*e.g.*, obsidian (Chapter 3) and pottery (Chapter 4)], when applied to metals it has always been a process viewed with considerable scepticism in some quarters, and now it would be fair to say that the majority opinion of archaeometallurgists is that precise chemical provenancing of smelted metal objects is not in general possible. This is because the high temperatures and extreme redox conditions involved in processing the ores and finished metal mean that some, at least, of the control over the trace element composition of the finished product is exerted by thermodynamic and kinetic considerations during processing. In fact, it could be argued that the trace element composition of the finished artefact is governed largely by the chemical equilibria established during these processes, and that this

Archaeological Chemistry, Second Edition
By A. Mark Pollard and Carl Heron
© The Royal Society of Chemistry 2008

information might therefore be more useful in reconstructing the processes used rather than in provenance.

In general, of course, it is probably safest to say that the trace element composition of a finished metal artefact is controlled by a range of factors, including the composition (mineralogical and chemical) of the ore source(s) involved, the thermodynamics and kinetics of the processes used, and deliberate human factors such as the mixing of metals to produce the desired alloy, and the recycling of scrap metal. The influence of these factors is discussed in more detail in Chapter 9. It is no great surprise therefore that the huge but relatively simplistic metal analysis programmes of ancient European copper alloy arte- facts carried out for the purpose of provenance during the 1950s and 1960s simply created confusion (Budd *et al.*, 1996). It would, however, be unwise to take the extreme view that chemical analysis of archaeological metal objects is never worthwhile. The study of the metallographic structure of metal artefacts (using various microscopic techniques on prepared surfaces or sections) is extremely well-established, and is absolutely vital for the understanding of manufacturing processes (*e.g.*, Scott, 1991). Significant additional information can nevertheless be obtained about manufacturing processes from chemical analysis, as is demonstrated by the study of European Medieval and later brasses (Section 6.4).

Brass is an alloy of copper with a minor proportion of zinc – typically in archaeological samples up to about 30%, for reasons explained in Section 6.2. It is an extremely interesting alloy, largely because of the volatility of metallic zinc (boiling point 906 °C), which means that zinc cannot easily be reduced to a metal from its common ore sources without distillation. Consequently, metallic zinc is rare (some would say absent) in the archaeological record before the later Medieval period, although brass first appears in Europe in Roman times. In the absence of metallic zinc, this alloy could not have been made by melting together ingots of copper and zinc, and so some other process is required. The Romans perfected a technique known as the *calamine process* (Section 6.2), which allows brass to be made without recourse to metallic zinc, but which incidentally puts kinetic or thermodynamic controls on the maximum levels of zinc achievable in brass made by this process. Once the distillation process for the manufacture of zinc had been mastered, brass of any composition could be made by a simple mixing of metallic copper and zinc, and thus the upper limit of zinc content imposed by the calamine process became obsolete. Thus a measurement of the level of zinc in a brass alloy can be extremely informative about the manufacturing process involved.

One result of the peculiar history of brassmaking is that since the alloy composition has changed in an explicable way over time as a result of changing manufacturing processes, authenticity studies of supposedly ancient brass artefacts become a real possibility. The so-called 'Drake Plate' (Hedges, 1979) discussed in Section 6.6 is an excellent example of this. Given the high market value of Medieval European brass artefacts, particularly scientific instruments, the ability to give an independent (scientific) opinion on the authenticity of disputed artefacts is extremely important, and one which is now routinely taken advantage of by most large museums.

In this chapter, following a description of what is believed to be the ancient process for the manufacture of brass (Section 6.2), we review the early history of brassmaking (Section 6.3), and the Medieval and later European industry (Section 6.4), largely from literary sources. Following a brief discussion of the problems of the analysis of museum specimens of brass objects, we then summarize some analytical data on European brass tokens and coinage which throws an interesting perspective on the literary evidence (Section 6.6). We conclude with an example of the application of this knowledge to the study of European scientific instruments (Section 6.7), and an archaeological example of the value of studying copper alloy compositions in a north American context (Section 6.9).

6.2 THE PRODUCTION METHODS OF BRASS IN ANTIQUITY

The question of how ancient brass was produced has been discussed extensively. As noted above, the two simplest possibilities are the *direct process* – the mixing of metallic copper with metallic zinc, presumably by adding the zinc to molten copper – and the *calamine* (also known as *cementation*) process, which was in use in Europe by the Roman period. Widespread use of the direct process is thought to be unlikely in Europe before the 18th Century AD because of the difficulty in producing metallic zinc. The calamine process involves heating together solid broken metallic copper and zinc oxide or zinc carbonate ($ZnCO_3$) mixed with charcoal in a closed crucible. Zinc carbonate was traditionally known in Britain as *calamine*, although mineralogically it is now correctly referred to as *smithsonite*, since the term *calamine* used to be used for the silicate mineral *hemimorphite* [$Zn_4Si_2O_7(OH)_2.H_2O$] in the USA. At a temperature of between 906° and around 1000 °C (*i.e.*, hot enough to vaporize the reduced zinc metal, but not hot enough to melt unalloyed copper, which has a melting point of 1083 °C), the zinc vapour is absorbed by the solid copper, and at the end of the process the temperature is raised, the alloy melts and can be poured into an inclined stone mould to form sheet brass. Alternatively, the metal may melt during the process, when the solidus temperature (the temperature below which no liquid phase exists) of the alloy produced falls to the temperature of the furnace – the binary Cu–Zn phase diagram can be used to predict this 'melting point' of the alloy for a given uptake of zinc (Figure 6.1). It is generally stated (but not necessarily true!) that the nature of this solid copper–gaseous zinc reaction is such that the amount of zinc entering the alloy is limited by the reaction equilibrium set up at the temperature of the furnace, and that a maximum zinc uptake of about 28% is observed because the alloy would melt if more zinc was absorbed. This was apparently more or less confirmed in a series of experiments carried out by Werner (1970) and, more comprehensively, by Haedecke (1973). In fact, Werner put an upper figure of 30% on the zinc uptake, and Haedecke gives a graph of zinc uptake as a function of furnace temperature (his Figure 5 is reproduced as Figure 6.2)

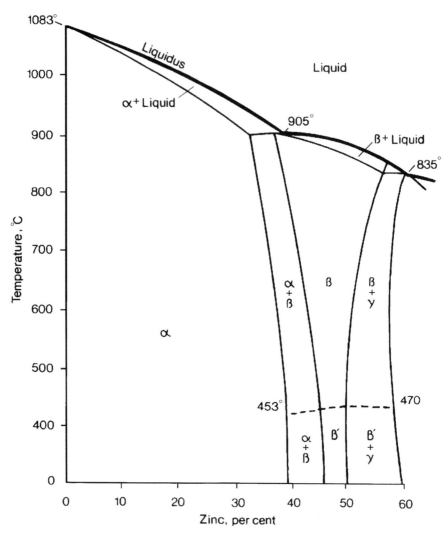

Figure 6.1   Binary Cu–Zn phase diagram. [From Rollason, 1973; Figure 201, Reproduced by permission of Edward Arnold (Publishers) Ltd.]

which shows a theoretical curve, together with some supporting experimental data, which peaks at a zinc uptake of 31% (±1%) at just over 900 °C. His work included starting with solid samples of both copper and brass, and he concluded that the final zinc level in the brass depended only on the temperature of the furnace, providing the temperature was above 970 °C – below this temperature, kinetic factors limit zinc absorption to a thin (<1 mm) surface layer. From these data there can be little doubt that the uptake of zinc is limited by thermodynamic or kinetic considerations during the calamine process. However, in the light of the importance subsequently placed on these figures in the interpretation of the ancient manufacturing techniques (and

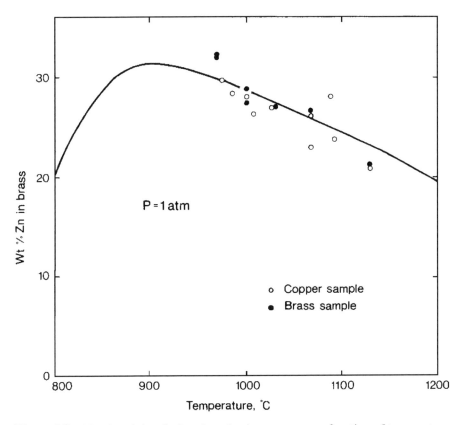

Figure 6.2 Uptake of zinc during the calamine process as a function of temperature. (Adapted from Haedecke, 1973; Figure 6.5, with permission.)

consequently in authenticity judgements), there appears to be scope for further study of the details and, in particular, the mechanism of the limitations on zinc uptake. There can be no doubt that the existing interpretations are inadequate.

6.3 THE EARLY HISTORY OF BRASS AND ZINC

The early history of brass is discussed only briefly here, drawing mainly on the work of Tylecote (1976), Craddock *et al.* (1980) and several chapters in Craddock (1998), all of which summarize the analytical and documentary evidence. The very earliest appearance of zinc in copper alloys cannot necessarily be interpreted as the deliberate production of brass, since the principal primary ore of zinc (ZnS – *sphalerite*, *blende* or *black jack*) commonly occurs in association with sulfides of copper and lead, and many sulfidic copper ores contain traces of zinc. It is not unlikely that the smelting of sulfidic copper ore under suitable conditions might give rise to impure copper containing traces of zinc, although the roasting step normally required for the smelting of sulfidic ores might be expected to limit the uptake because of the volatility of zinc. Despite this, both Craddock and Tylecote accept that arsenical copper or tin bronzes with up to 8% Zn can be explained as the unconscious use of a mixed ore, retaining a variable concentration of zinc. Tylecote asserts that the 23.4% Zn contained in a pin from Gezer, an Early Bronze Age city in Palestine, is the earliest example of a deliberate brass, dating to 1400–1200 BC, but Craddock (1978) is more sceptical, noting that no other Palestinian Bronze Age metal-work has yet been published with a similar composition, that the pin in question was without a reference number and described by the excavator as 'a small corroded worthless pin', and that the site was also occupied until Hellenistic times, throwing some possible doubt on the exact date of the pin. Craddock suggests that the earliest occurrence of brass should be placed in the 8th to 7th Centuries BC in Asia Minor, represented by finds of fibulae ('safety pin' type brooches) from the Gordion Tomb in Phrygia, south-western Anatolia (alloys with 10% Zn, and little tin or lead). Sporadic examples have been published of later 1st Millennium BC brasses with up to 10% Zn, culminating in an Etruscan (Italian) statuette of a naked youth dated to the 3rd or 2nd Centuries BC with around 11% zinc (Craddock, 1978; Figure 2). Brass coinage containing between 13 and 26% Zn has been reported from ancient Anatolia (Bithnyia and Phrygia, in modern-day Turkey), securely dated to the 1st Century BC. One example, dating to between 90 and 75 BC from Amisus in Bithnyia and Pontus (north-eastern Turkey) was found to contain 19% zinc (with 0.83% Pb, 0.31% Sn and 0.5% Fe, the remainder Cu) which is thought to represent the earliest known brass coin definitely produced by the cementation process (Preston, 1980). There can be little doubt that brass was being produced, almost certainly by the cementation process, by the end of the 1st Millennium BC in what is now Turkey, and probably other areas of the eastern Mediterranean.

As always, however, new archaeological evidence has the capacity to cause established views to be overturned. Analyses of copper alloy objects from Tepe Yahya, in south-eastern Iran, carried out by Thornton *et al.* (2002), have provided considerable support for Tylecote's acceptance of the Gezer pin, and therefore the 'longer' history of brass. The site was excavated between 1967 and 1975, but the material not analyzed until 2002. In addition to the expected alloys, three jewellery fragments were found with 17–20% zinc, dated to the 15th Century BC (*i.e.*, earlier than the Palestinian pin), becoming some of the earliest brass objects known. Furthermore, since the objects contained no iron or manganese, the authors suggest that they were made by cementation with sphalerite (ZnS) rather than smithsonite ($ZnCO_3$), following the work of Ponting and Segal (1998), who described this process, but assigned its origin to 1st Millennium BC Anatolia. As succinctly stated by Thornton *et al.* (2002; 1459): '*The prologue to the history of brass is only now being written, and future analyses will hopefully shed more light on this elusive subject.*'

The early history of metallic zinc itself has been reviewed in some detail by Craddock (1990). Although a number of artefacts from the classical world have been reported as being made from zinc, the one which has excited most attention and has survived close scrutiny is the small sheet (6.6 × 4 cm) of zinc found in 1939 in the Athenian Agora, stratified with pottery and coins of the 4th to 2nd Centuries BC. It was subsequently analysed by emission spectrography, and found to be of zinc, with 1.3% Pb, 0.06% Cd and traces of Cu, Fe, Mn and Mg, with very faint traces of Sn, Ag, Si and Cr (thought to be a residue from the cleaning process applied after excavation). Although the levels of lead, iron and cadmium are similar to those of modern zinc, the presence of the other traces were not as would be expected in a modern sample, but it was recognized that a relatively short period of burial could have been responsible for their presence. The clinching argument was felt to be the metallographic structure, which showed moderate hammering, rather then the casting or hot rolling which has been common in European zinc manufacture since about AD 1805.

Craddock (1990) accepts this piece as evidence that metallic zinc, although apparently extremely rare, was known in antiquity, and turns to the classical authors for support. The Greek author Strabo in his Geography (drawing on the now lost 'Philippica' of the 4th Century BC writer Theopompus) writes of a stone from Andeira (in Phrygia, modern-day Turkey) which when '*treated in a furnace with certain earth it yields drops of false silver. This, added to copper forms the mixture which some call* oreichalkos' (Craddock, 1990; 4, from a translation of Strabo by Caley, 1964). 'Oreichalkos' is generally taken as referring to brass, and therefore the 'false silver' is identified as zinc. This reference by Strabo is thought to refer to the occasional condensation of reduced zinc on the walls of furnace flues, which has been observed to happen in later periods, and was referred to in the German mining regions as '*conterfei*'. It would appear that a direct process of zinc manufacture was known to the ancient Greeks, albeit possibly as a great rarity.

Another reference, this time probably to the calamine process, appears in the works referred to as a pseudo-Aristotelian compilation called 'On Marvellous Things Heard', probably written in the 3rd Century BC. As translated by Caley (1967; 67), the relevant passage runs:

> '*the bronze of Mossynoeci is very shiny and light in colour, though tin is not mixed with the copper, but a kind of earth which occurs there is smelted with it. But they say that the discoverer of the mixing process did not instruct anyone else, so that the bronze objects formerly produced there are superior, whereas those made subsequently are not.*'

The Mossynoeci were a people who lived in Asia Minor, on the shores of the Black Sea, again in modern-day Turkey.

It seems clear now from analytical evidence that the manufacture of brass was discovered some time during the 2nd Millennium BC, somewhere around modern-day Turkey, Iran or even Central Asia. Under the influence of the Greeks in the 1st Millennium BC, this knowledge spread throughout the Mediterranean, finally culminating in its use for the reformed coinage issues of Augustus in 23 BC, which meant that the alloy was widely available throughout the Roman Empire (Craddock *et al.*, 1980). Its popularity is attested by its widespread use as a coinage metal from this date onwards, and ultimately as the alloy of choice for a wide range of items of jewellery (Bayley, 1990) – presumably, because of its golden appearance, and the relative ease of manufacture using the calamine process.

Following the decline of classical influence in Europe, knowledge of the manufacture of brass by the calamine process became enmeshed in the Alchemist's search for *aurifaction* – the production of gold from base metals (Pollard, 1988). Some of this lore from Ptolemaic and Roman Alexandria has been preserved in the writings of the likes of Zosimus (3rd Century AD) and Mary the Jewess (2nd Century AD) who, significantly perhaps, is traditionally credited with the invention of the retort for distillation (Holmyard, 1957). This knowledge was translated into Arabic after the Umayyad conquest of Egypt (AD 641) for the use of the Islamic proto-chemists, and was eventually returned to Europe by way of the Moors in Spain (conquered AD 711), as one of the many great contributions of Islamic science to European culture. It is very easy to see the relationship between brassmaking – the addition of a special 'earth' to a base metal producing a psuedo-gold – and the 'great work' of alchemy – the isolation of the 'philosopher's stone', which transmutes base metal into gold (Pollard, 1988).

Needham (1974) gives a very full and well-documented account of the early history of brass and zinc in the Far East. Briefly, he concludes that the production of brass from calamine and copper was practised in China by the 3rd or 4th Centuries BC. The relative agreement of the dates of the first production by this method in both East and West caused Needham to speculate that knowledge of the process diffused from some intermediate place – he suggests Persia, on the grounds that brass was being imported into China from there before AD 590, and Chinese sources often say that Persia is the source of 'real' brass. Brass was also known in India by AD 646, but the work outlined above so far points to Asia Minor or Central Asia as the most likely source of such knowledge. In their analytical study of copper-based Chinese coinage, Bowman *et al.* (1989) showed the essential conservatism of the Chinese mint, in that, although copper alloys formed the basis of the official issue, brass was not introduced until sometime between AD 1503 and 1505, but became the sole alloy in use after AD 1527. Chinese literary sources, particularly the *T'ien Kung K'ai Wu* (written around AD 1637), refer to an alloying procedure involving six or seven parts of copper to three or four parts of zinc, allowing for 25% loss of the zinc through vaporization, giving an approximate composition of 30% zinc in the final alloy. Zinc metal was therefore clearly available at this time, and was referred to as *wo chienn* ('poor lead').

The actual isolation of zinc metal on an appreciable scale seems to have occurred first in China in the 10th Century AD (Xu, 1990), using an upwards distillation procedure from secondary (oxidized) zinc minerals. Earlier finds of metallic zinc (such as that at the Agora, noted above) are possibly explained by the chance condensation of small quantities of zinc in the furnace during the production of lead and silver from mixed ores. Much attention has been focused in recent years on northern India, particularly the Zawar region, following the spectacular finds of *in situ* downward distillation zinc furnaces that used sulfide ores (Craddock *et al.*, 1990; Hedges, 1989). The dating evidence from this site is somewhat tantalizing: radiocarbon dates from the associated deep mines show that these were in use during the 1st Millennium BC, but the evidence for the actual use of the downward distillation process dates to some time after the 10th Century AD, with the principal industrial phases at Zawar dating to the 14th Century AD (Craddock *et al.*, 1990). Metallic zinc, or '*spelter*', was exported to Europe from the Far East after AD 1605, with England importing about 40 tons per year between 1760 and 1780. In 1745, a ship called the *Gotheberg* sank with her cargo of porcelain, tea, silk and zinc. On recovery and analysis of the metal, around 1870, it was found to be better than 98.99% Zn, with traces of iron and antimony.

If we accept a date of around AD 1000 for the commencement of the distillation of zinc on a large scale, then, following the work of Craddock (1978), all earlier brasses should contain less than 28% Zn, as this is the approximate upper limit for the calamine process at around 1000 °C. Above this temperature, the process is more efficient, but it is said that the brass produced melts and the active surface area for the process is thus reduced. By granulating the copper and therefore increasing the surface area, the maximum can be pushed to around 33% Zn, but it is unlikely that this was done in Europe until the 18th Century (see Section 6.4). This model is supported by the analytical data: Craddock's work on Roman brass indeed shows an upper limit of about 28% zinc.

6.4 THE MEDIEVAL AND LATER EUROPEAN BRASS INDUSTRY

In Medieval Europe, brass continued to be made by the cementation process, particularly in Germany, which was famous for its skill in metalwork. Many of the early alchemical writers mention brass production – Albertus Magnus (AD 1193–1280), Thomas Aquinas (1225–1274) and Roger Bacon (1214–1292) [see Grant (1974) and Partington (1961)]. More importantly, cementation is mentioned in the famous technical treatises of Theophilus (*On Divers Arts*; Hawthorne and Smith, 1979), Agricola (*De Re Metallica*; Hoover and Hoover, 1950) and Biringuccio (*Pirotechnia*; Smith and Gnudi, 1959). Theophilus, probably a German monk and metalworker, in his *De Diversis Artibus* written around AD 1100 (dating according to Hawthorne and Smith), gives a very clear and well-known account of a calamine-type process (Book III, Chapter 66; 143–144):

'*...when the crucibles are red-hot take some calamine,...that has been [calcined and] ground up very fine with charcoal, and put it into each of the crucibles until they are about one-sixth full, then fill them up completely with the above mentioned [crude] copper, and cover them with charcoal....Now, when the copper is completely melted, take a slender, long, bent iron rod with a wooden handle and stir carefully so that the calamine is alloyed with the copper....Put calamine in them all again as before and fill* *them with copper and cover them with charcoal. . . . When it is once more completely melted, stir again very carefully and remove one crucible with tongs and pour out everything into [little] furrows cut in the ground. Then put the crucible back in its place. Immediately take calamine as before and put it in, and on top as much of the copper that you have [just] cast as it can hold. When this is melted as before, stir it and add calamine again and fill it again with the copper you have [just] cast and allow it to melt. Do the same with each crucible. When it is all thoroughly melted and has been stirred for a very long time pour it out . . . This alloy is called coarse brass and out of it are cast cauldrons, kettles and basins.'*

The process described here does not involve closed crucibles – it simply appears to rely on the depth of the crucible [see Hawthorne and Smith (1979; Figure 16), for a reconstruction of the furnace and crucibles used] and the charcoal cover to produce the necessary conditions for the solid–vapour reaction to occur. The above description is clearly written by a man with first-hand experience of the process. Elsewhere, however (Book III, Chapter 48), he states that there are many sorts of gold, and proceeds to describe 'Spanish Gold', which is an alchemical transmutation of copper treated with 'powder of Basilisk and human blood'. This ambivalent approach to the nature of brass seems to have vanished by the 16th Century AD, which saw the publication of Biringuccio's *Pirotechnia* (1540) and Agricola's *De Re Metallica* (1556) which give (particularly the latter) a wealth of practical detail which dispels much of the mythology associated with metal production.

Paracelsus, the Swiss physician (1493–1541 AD) appears to be the first European to describe metallic zinc: *'zinc, a bastard of copper, a peculiar metal, but often adulterated by foreign materials. It is of itself fusible, but does not admit to hammering.'* Libavius (1540–1616) obtained some zinc from the East Indies via Holland in 1597. Löhneyss, in 1617, described zinc ('contrefey') as a metal that serves to imitate gold, and reports the distillation of zinc in a lead-smelting furnace in Rammelsburg. At an earlier date, Erasmus Ebener showed that zinc could be used to make brass instead of calamine (*ca.* 1550–1560). The documentary evidence from this fascinating period in the history of chemistry and metallurgy is discussed by Partington (1961).

By 1529 in England, Henry VIII was so worried about the supplies of 'strategic metals' for ordnance that he forbade the export of brass, copper, bell metal and latten. During the reign of Elizabeth I (1558–1603) brass was made in England, but the mining of the rich English copper deposits was almost totally neglected (Hamilton, 1967). In 1566 she had persuaded German miners and metallurgists to come to Britain, and a copper mine was opened at Keswick in the Lake District under the auspices of the *Mines Royal Company* (1568), which was, like all other mining concerns of the time, a Crown monopoly. Following the discovery of suitable calamine in the Mendips, the *Mineral and Battery Works* was established, and produced the first English brass at Tintern Abbey wireworks in 1568 (Day and Tylecote, 1991). The Mineral and Battery Works were conferred with the sole right to mine calamine, and also to produce wire and battery products (largely vessels made from hammered and shaped sheets) using the new method of water-powered machinery, also imported from Germany. However, Tintern brass could not be made malleable enough for the satisfactory production of battery – possibly due to sulfur being added from the coal used, or the poor quality of the copper supplied by the Mines Royal Company (Day, 1973) – and the works soon ceased production of brass and turned to manufacturing iron wire for use in the local wool-combing industry. The zinc content of this first English brass was recorded as 20% (28 lb in weight added to 1 cwt of copper), but this was soon improved to 24% (35 lb weight added; Day, 1973).

The rights of brass production were subsequently leased by mills at Isleworth and Rotherhithe, although illegal brassworks were also in operation, including one in the Bristol area, which was soon to become an important centre. Under the Crown monopoly, however, the industry declined, and by 1660 good-quality calamine was being exported abroad and finished brass being imported. During the English Civil War, brass was scarce, and copper was imported from Sweden because the Keswick mines had closed. By about 1670 foreign competition (including copper from Japan!) had virtually closed the English brassworks, and import controls were imposed to try to protect the flagging domestic industry. The monopoly of the Mines Royal Company was abolished in the Mines Royal Act of 1689 by William and Mary, and many new mines and brassworks came into being – *e.g.*, the English Copper Company in Cornwall (1691) and the Dockwra(y) Copper Company, Esher, Surrey (1692). The export of copper was legalized in 1694 (but not to France!). In 1702 the influential Bristol Brass and Wire Company was established, again using Dutch and German labour. The calamine process was still poorly understood, and the quality of the brass was said to be low, resulting in the continued importation of better quality Dutch brass. The Bristol and Esher companies combined in 1709, and in 1711 they petitioned the House of Commons to give protection (by imposing 'additional duties') from imported brass, since the English manufacture had been brought 'to great perfection' (Hamilton, 1967; 112). A counter-petition was lodged by the Company of Armourers and Braziers of London and others, who, complaining of the poor quality of English brass, feared that the standard of available brass would fall as a result. The House considered copious conflicting evidence about the quality of English brass, but eventually the petition was refused. It was resubmitted in 1721, but was again refused, partly as a result of the evidence on quality submitted by the clockmakers and watchmakers (Hamilton, 1967; 116).

It appears that the principal obstacle to the production of good quality brass was the purity of the copper available – Day (1973) gives a figure of 92% for the purity of English copper around 1700. In the early 18th Century several technical improvements were made to the process of copper refining, such as the replacement of coal by coke in the reverberatory furnaces used at Bristol in 1710. The resulting improvement to the quality of English brass meant that by 1740 the amount of imported material was drastically reduced without legislation. Hamilton (1967) states that '*high purity copper was achieved by 1778.*' English copper was used for coinage from 1714 (copper coinage commenced in England in 1613) – before that, copper for coinage had been imported from Sweden.

It was the 18th Century which saw the first changes in European brass-manufacturing technology since Roman times. In 1723 Nehemiah Champion of Bristol obtained a patent (no. 454) for an improved version of the calamine process using granulated copper (prepared in the same way as lead shot) instead of broken pieces of copper. This was said to improve the uptake of zinc from 28.6% to 33.3% (40 lb of copper yielding 60 lb of brass instead of 56 lb). As noted above, zinc metal had been known in the East for at least 700 years by this time, and some had found its way to Europe, where chemists had realized its role in the production of brass. A small amount was used in the production of gilding metals (called *pinchbeck*, *etc.* – see below), which only contain 10–15% Zn. It must have been too expensive, however, to use in the production of true brass, particularly since calamine was very cheap. At Goslar, in Germany, Caspar Neumann (1683–1737) saw zinc being scraped off the furnace walls after the smelting of lead. It was called 'furnace cadmia', and sold for use in brass manufacture (Dawkins, 1950). Watson (1786) credits Henckel as being the first European to deliberately produce zinc from calamine, quoting from his publications in 1721 and 1737. Dawkins (1950) observes that Marggraf published the process in 1743.

In England, William Champion of Bristol patented the process for the production of zinc in 1738 (no. 564). Watson (1786), however, says that it was Dr Isaac Lawson who first produced the metal whilst working for Champion, and suggests that Lawson had visited China. As pointed out by Day (1990), however, derivation from India is more likely since Champion employed a downward distillation, as used in India, as opposed to the upward distillation traditional in China. Champion went on to produce zinc commercially, but traders to the East Indies lowered their prices in an attempt to force him out of business. He reduced production, and moved to the Warmley works where, according to Watson, he was still producing brass by the calamine process in 1748. His patent was extended until 1750, although this was opposed by metalworkers, on the grounds that it would affect production of 'brass, copper, Prince's metal and Bath metal'. He tried to expand and monopolize the industry but was eventually forced to close as a result of rising competition from the more forward-looking Birmingham brassworks, and went bankrupt in 1769.

John Champion patented in 1758 the process of producing brass and zinc from the common ore of *zincblende* (ZnS) or *black jack* (patent number 726). This ore had previously been considered worthless, but even with this advance it is obvious that metallic zinc was still far too expensive to use in brass production by direct mixing, a process which was patented by James Emerson in 1781 (no. 1297). Watson (1786), however, describes the use of zinc in the production of high-quality gilding brasses such as *pinchbeck*, *tomback* and *Mannheim gold* (see below).

By the 1780s, English production was dominated by the Birmingham brasshouses, producing mainly cast and stamped objects. Before this time, Birmingham had been a centre of highly skilled craftsmen producing finished goods, but the advent of the canals and the steam engine allowed the production of raw materials to move to the area. In 1786, Watson noted that *'great quantities of good brass are made by most realms in Europe, as well as by the English, but the English brass is more adapted to the Birmingham manufactures than any other sort is. The manner of mixing different sorts of brass, so as to make the mixture fit for particular manufactures is not known to foreigners; though this is a circumstance of great importance.'* Elsewhere, he records that German calamine brass contains 29% Zn (64 lb copper yielding 90 lb brass), but English brass contains about 34% Zn (45 lb copper giving 60–70 lb brass, but typically 68 lb). In addition, Bohemian calamine was said to contain iron, whilst English calamine contained lead. In a further reference to English brass, Watson also states that 1.5 tons of brass is made from 1 ton of copper, and that it is better than that of foreign manufacture due to the use of pure calamine and granulated copper. We must assume that this figure of 50% Zn in calamine brass, even allowing for granulation of the copper, is something of an exaggeration, although our understanding of the process is far from complete.

Watson describes the brass produced by Emerson (*i.e.*, by the direct mixing of metals) as *'the purest and finest brass in the world'*, saying that it is free from iron and therefore good for use in compasses. Despite this praise, Emerson went bankrupt in 1803. By the turn of the century however, large-scale zinc production was underway in Europe, with factories at Carinthia (1799), Ruhberg (1799–1800), Liege (1809) and the Vielle Montagne Company in 1837 (Lones, 1919). In addition, an alteration to English import tariffs in 1839 made it cheaper to import zinc, so the declining price of the metal resulted in the slow phasing out of the calamine process. This happened around 1830 in Cheadle and about 1840 in Bristol. The last brasshouse in Birmingham to produce brass by the calamine method closed in 1866 (Aitken, 1866), whilst Hamilton (1967) records that calamine furnaces were still in operation in South Wales in 1858. According to Lones (1919), brass made from spelter by the direct process had completely replaced calamine brass by 1870, but *'the old process yielded the best qualities of brass.'*

In the 19th Century, according to Aitken (1866), the practical brassworker *'could distinguish the difference between calamine brass, and the brass now made by direct mixture, in the peculiar appearance of its polished surface.'* He also asserts that calamine brass was superior. Day (1973) records that, in Bristol, the direct method used refined copper and metallic zinc plus scrap brass of high quality (*'shruff'* – off-cuts from other processes plus material purchased from outside). It was carefully tested and sorted to enable the correct percentages of copper and zinc to be obtained in the finished brass, allowing for volatilization. This was the responsibility of the 'mixer', who kept records of the various qualities and quantities of scrap brass required to give the desired product when mixed with new metals. As regards the calamine brass, however, Aitken states that *'it is worthy of note, that though various calamines were used – dug from Derbyshire, Flintshire, Somersetshire and Yorkshire – from long experience and care, the percentage of zinc in the different qualities of brass produced was secured with as much certainty as it is by the method of direct mixtures now practised.'*

In the closing years of the production of calamine brass, the following grades of brass are recorded by Aitken (1866):

BB – made of best copper. '*Latten*' or sheet brass with 33% Zn 'as nearly as possible';
BC – inferior copper, slightly more zinc;
AM – some copper and calamine – tainted with lead (?);
YY – made of ash metal and other inferior materials (?).

More helpfully, the various qualities of gilding metals in use during the 18th and 19th Centuries are listed by Lones (1919):

Tomback – derived from a Malay word – 86% Cu, 14% Zn + a little tin;
Pinchbeck – named after an 18th Century London clockmaker – 88% Cu, 12% Zn;
Mannheim gold – 80% Cu, 20 % Zn;
Leaf gold or '*Dutch metal*' – 84% Cu, 16% Zn.

Two other metals are referred to by various authors, but there is some disagreement about the composition:

Prince's metal – said to have been invented by Prince Rupert in 1680 for casting guns – 73% Cu, 27% Zn (Lones, 1919), but 'equal quantities copper and zinc' (Aitken, 1866).
Bath metal – 83% Cu, 17% Zn (Hamilton, 1967): 55% Cu, 45% Zn (Lones, 1919).

By the 19th Century, several grades of brass seem to have been in common use, although there is sometimes confusion between the terms employed (from Day and Tylecote, 1991, with additions):

62% Cu/37% Zn/1% Sn	*Naval Brass*
63% Cu/37% Zn	*Common Brass*
67% Cu/33% Zn	*Stamping Brass*
67% Cu/32% Zn/1% Pb	*Clock* or *engraving metal*
70% Cu/30% Zn	*Cartridge Brass*
75% Cu/25% Zn	*Sheet brass* (Hamilton, 1967), *Rolling brass, drawing brass* ('as free from impurities as possible') (Lones, 1919).

In the later 19th Century, the demand for hot working and special brasses resulted in a wide range of named alloys:

Muntz metal (*yellow metal, patent brass*) – 60% Cu, 40% Zn (patented 1832)
Sterro metal – 60% Cu, 38% Zn, 2% Fe
Delta metal – 56–58% Cu, 40–42% Zn, 1–2% Fe + Pb, Mn (1883)
Duranna metal – 65% Cu, 30% Zn, 1.75% Sb
Collins Yellow Sheathing Metal – 56% Cu, 44% Zn

This brief review has concentrated on the development of the production methods of brass in Europe, but it must be remembered that a number of other dates are important in the history of the various manufacturing processes, such as *ca.* AD 1697 for the introduction of the rolling mill to replace the battery process, which, however, continued in use in some places to the end of the 18th Century.

6.5 THE CHEMICAL ANALYSIS OF METAL OBJECTS

The analysis of metal objects of the type discussed here invariably poses a number of questions to the chemical analyst. There are two major sources of difficulty:

(i) Most Medieval brass objects are inherently valuable – the majority of the objects are either museum samples, or in the collection of an individual, who, understandably, requires that any analysis should be as near 'non-destructive' as possible.
(ii) Copper alloy objects are inherently inhomogeneous, particularly if the lead content is more than a fraction of a percent. This can be exacerbated if the object is from an excavated context, since it is well-known that electrochemical processes in water can cause severe loss of zinc (*dezincification*) from the surface of brass objects (Finnegan *et al.*, 1981; Polunin *et al.*, 1982; Trethewey and Pinwill, 1987).

Many of the analytical techniques described in Chapter 2 can be regarded as largely fulfilling these requirements, but two [atomic absorption spectrometry and X-ray fluorescence (XRF)] have been used for the majority of the work published until the previous decade or so on European Medieval museum objects made of brass. The experimental technique for atomic absorption has been well-described in the literature (Hughes *et al.*, 1976) and only requires a fine diameter drilling (approximately 1 mm) to be made into the metal, yielding a sample of around 10–20 mg. This is sufficient to measure around ten elements with a coefficient of variation (one standard deviation) of between 1 and 4% for the major and minor elements. The main drawback with this method is the possibility of erroneous results due to sampling if the metal being analysed is inhomogeneous. This is particularly a problem if the object is cast and contains a large amount of lead (as exemplified by Hughes *et al.*, 1982, but not on

Medieval brass), although the problems of electrochemical surface modification due to corrosion are minimized if the first few turns of the drilling are discarded and only bright metal turnings are used.

The damage as a result of the drilling may be unacceptable to museum curators if complete artefacts (or small coins) are being analysed. The analytical method chosen to cause the absolute minimum of damage to museum objects has been energy-dispersive XRF, which, with specially designed instrumentation to allow large objects to be positioned in front of the detector, is a rapid and virtually non-destructive means of analysing whole objects without sampling. The main disadvantages are the relatively poor sensitivity to the trace elements, and the necessity to clean an area large enough for the X-ray beam (typically a couple of millimetres in diameter) in order to reduce the problems of surface enrichment (in the case of excavated objects) or surface lacquering on museum objects. In the case of thin metal objects (such as coins, or sheet metal plates) it is possible simply to clean the edge and to allow for the reduced area irradiated by the primary beam by normalizing the analytical results to 100%. Providing the methodology is (as should always be the case) validated using 'secondary' standards – standards with certified analyses not used in the primary calibration which give an independent check on the method (Pollard *et al.*, 2007; 306) – this gives an acceptable compromise between analytical rigour and object preservation. In the analytical work reported here on scientific instruments and brass tokens (carried out in Oxford), the following peak intensities were measured – Fe K_α, Ni K_α, Cu K_α, Zn K_β, Pb L_α + As K_α, As K_β, Pb L_β, Ag K_α, Sn K_α and Sb K_β. Severe peak overlap between Cu K_β and Zn K_α, Pb L_α and As K_α, and Sn K_β and Sb K_α meant that the Zn K_β,. As K_β and Sb K_β peaks were used to quantify Zn, As and Sb, which necessarily gives reduced sensitivity for these elements. In the calibration procedure, all the measured intensities were ratioed to the Cu K_α intensity to reduce the problems of long-term drift and to minimize geometry and surface condition problems. Each instrument to be analysed was dismantled as much as possible to allow the analysis of all the major components, but small easily replaceable items such as screws were generally ignored. In almost all cases the instruments were not corroded, and sample preparation was limited to the removal of surface tarnish and lacquer with a sharp scalpel or a glass fibre brush. Experiments on a series of brass sheets treated with various chemical and mechanical cleaning agents also showed that this preparation was sufficient to remove any traces of the likely conservation treatments, and also that replicate analyses reproduced the bulk composition on a range of modern brass standards (Pollard, 1983a). Table 6.1 shows the estimated coefficients of variation (1 standard deviation) and minimum detectable levels for all the elements measured.

6.6 THE CHEMICAL STUDY OF EUROPEAN BRASS TOKENS AND COINS

The Oxford University Research Laboratory for Archaeology and the History of Art has had a long-standing interest in the analysis of metals. An example of

Table 6.1 Details of the XRF analyses of brass instruments and jettons.

Minimum detectable levels (wt%):								
Fe	Ni	Cu	Zn	As	Pb	Ag	Sn	Sb
0.05	0.05	–	0.25	0.18	0.05	0.05	0.22	0.10

Coefficients of variation (1σ):	
Cu, Zn	1–2%
Sn, Pb	5–10%
Rest	10–20%

this was the work on the 'Drake Plate' (Hedges, 1979), which is a small sheet of brass ($20 \times 14 \times 0.3$ mm) found in 1936 in the San Francisco Bay area. It is inscribed with a title claim to the land in the name of Elizabeth I, signed by Francis Drake and dated June 17 1579 (Michel and Asaro, 1970). The plate was found on analysis to have a zinc content of 34.8 ($\pm$0.4) %, together with negligible levels of Pb, Sb and Sn (<0.05%). Comparative analyses were undertaken in Oxford of 18 brass scientific instruments (from the History of Science Museum, Oxford) and four memorial brasses (from St John's College Chapel, Oxford) covering the date range AD 1540–1720. These, together with the compilation by Cameron (1974) of data from English memorial brasses between the 12th and 16th Centuries AD, showed an average composition for Elizabethan brass of around 20% for Zn with 0.5–1 % of both tin and lead [consistent with the 20% Zn quoted by Day (1973) from contemporary records]. On this basis, combined with the very important fact that the sheet was found to be of a thickness consistent with No. 8 gauge brass of the American Wire Gage standards used in the 1930s, it was concluded that the sheet of metal was unlikely to be of Elizabethan manufacture.

Although this particular piece of forensic work was, for a number of reasons, conclusive in condemning the Plate of Brass, it did reveal some potential problems with regard to our knowledge of Medieval and later European brassmaking. It is apparent that the argument is somewhat circular, because very few of the objects used to establish the criteria for authenticity have a comprehensively known history, and can therefore be assumed to be completely genuine. This can be a particular problem with scientific instruments or clocks, which can often have parts replaced as they wear out or are lost. It is clear that the compositional trends really need confirmation from some other well-dated brass objects of the same period, to compare with both the instruments and other published analyses. The published analyses of European brass objects of the period are unfortunately of little help in this situation, because they are mainly of cast objects such as candlesticks and statuettes, which contain much more lead than items made from sheet brass (Werner, 1977; 1980; 1982; Brownsword and Pitt, 1983). The trace element data contained in these analyses are, however, of interest and can be compared with those from sheet brass. The most relevant published data are those of Cameron (1974) on English monumental brasses, discussed below. Potentially, suitable objects for study are the small brass tokens known as jettons, described in detail by Barnard (1916). They are coin-like tokens are of no monetary significance, although they are usually considered as numismatic items. They were used as reckoning or gaming counters, and, being of little value, were unlikely to have been extensively copied or forged. By the same argument, however, they were also unlikely to have been made of the best-quality brass, or minted under strict regulation, although Barnard (1916) claims that many of the French tokens were made in the Royal Mints, with a large degree of regulation. Despite this, they can be dated and provenanced, at least to a country of origin, and as such are useful indicators of the state of brass production at a particular place and time.

A large number of tokens (and some copper coinage) dating from *ca.* AD 1280–1900 was analysed in the Oxford Laboratory during the 1980s, from England (Mitchiner *et al.*, 1985; 1987a), Nuremburg (Mitchiner *et al.*, 1987b), France (Mitchiner and Pollard, 1988) and other European countries (Mitchiner *et al.*, 1988). The total number of jetton and coin analyses carried out under this programme exceeded 600 (including approximately 300 from Nuremburg, 160 from England and 100 from France). Complete details and analyses are given in the publications cited above, but a brief summary of the major findings which have relevance to the technology of brass manufacture is given here.

A preliminary examination of the entire data shows that the results fall naturally into two main groups – an earlier group (before approximately AD 1450) characterized by high tin and low nickel, and a later group with low tin and high nickel. The early group has an average composition of 85.7% Cu, 8.0% Zn, 3.7% Sn and 1.4% Pb, with 0.39% Fe, 0.23 % As, 0.14% Ag, 0.30% Sb and <0.05% Ni. The (larger) later group, which predominantly consists of Nuremburg jettons, but also contains some later French and Tournai examples, has an average composition of 75.7% Cu, 22.6% Zn and 0.66% Pb, with 0.25% Ni, 0.17% Fe, <0.2% Sn and As, and <0.1% Ag and Sb. These results are illustrated in Figures 6.3 and 6.4, where the tin and nickel contents are plotted against time. Prior to AD 1450, the majority of jettons have a tin content in excess of 2%, whereas after that date very few have more than 1%. Before AD 1400, almost all jettons have a nickel content of less than 0.1%, whereas later examples have up to 0.4%. The 'spike' in the nickel concentration between AD 1600 and AD 1650 is very characteristic, and is reflected to a lesser extent in similar plots of the arsenic and antimony results, and may reflect the temporary use of some inferior copper ore. Technologically, however, the most important diagram is that of zinc, shown in Figure 6.5, which plots all jettons with more than 20% zinc. This group is dominated by the Nuremburg samples, and so the ensuing remarks may only apply to that production centre. Following the documentary sources and the analytical work of Craddock and others, the two 'critical' concentration levels of 28 and 33% are marked (relating to the presumed maximum uptake by the 'classical' and the 'granulated' calamine processes), showing that the 28% level is reached as early as AD 1450, and exceeded by AD 1560 (160 years before the patent awarded to Nehemiah

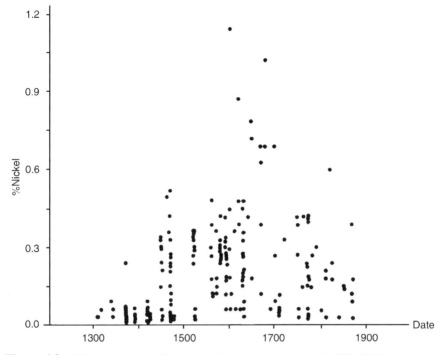

Figure 6.3 Nickel content of European brass jettons, *ca.* AD 1300–1850.

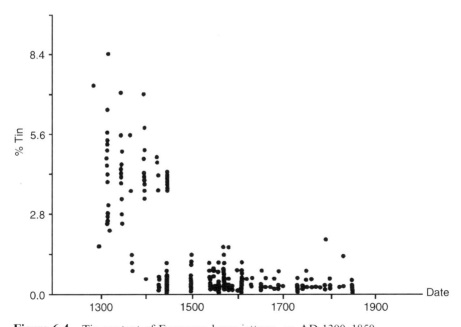

Figure 6.4 Tin content of European brass jettons, *ca.* AD 1300–1850.

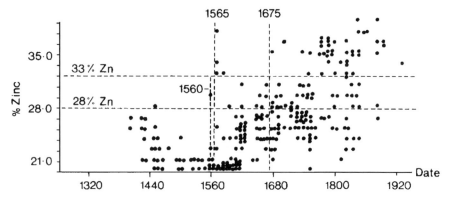

Figure 6.5 Zinc content of European brass jettons with more than 20% Zn, *ca*. AD 1400–1850.

Champion for the granulation process in England). Indeed, a couple of examples as early as AD 1565 appear to exceed 33%, but a clear step is seen at AD 1675, when 33% is routinely exceeded, suggesting manufacture by direct mixing. Even taking this later date, it again is 100 years before the patenting of the direct process in England. On the assumption that the technological interpretations are correct (see above), we must assume that the improvements to the calamine process patented in England during the 18th Century were known on the Continent some 100 years earlier. The study of these relatively insignificant little artefacts clearly may have an important influence on our interpretation of European brass manufacturing.

One further point can be made from this overview by considering Figure 6.6, the lead content of the later jettons in the series. From AD 1450 to AD 1750, the overwhelming majority of jettons contain less than 1% lead, but around AD 1760 a small group (eight) of high-lead examples emerge (with more than 1.5% lead). The average composition of these tokens is 63.1% Cu, 33.9% Zn and 2.2% Pb, where the high zinc concentration may suggest manufacture by some form of direct mixing. Possibly the increased lead is being added along with the zinc, or perhaps the copper being used for this process is less well refined.

The technological information derived from a study of the early jettons can be summarized quite simply – before *ca*. AD 1450, European jettons contained on average around 4% tin and 1–2% lead, with a low zinc content, typically less than 10%. Traces of antimony, silver and arsenic are often present at 0.1–02%. Nickel is very low, usually less than 0.05%. After AD 1450, the zinc content rises quickly to over 20%, with a corresponding fall in the lead (less than 1%) and tin (less than 0.2%). The trace elements are usually less than 0.1%, with the exception of nickel, present up to 0.5%. The combined copper plus zinc total is usually greater than 97%.

There is probably a twofold explanation for this change during the 15th Century. One is a change in the copper ore source that supplied much

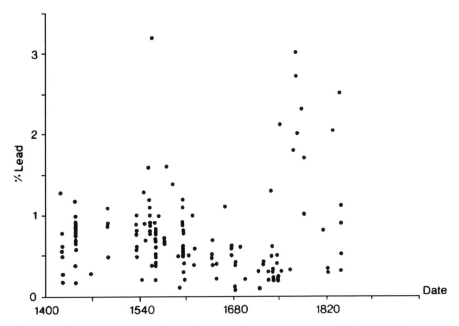

Figure 6.6 Lead content of later European brass jettons, *ca.* AD 1400–1850.

of north-west Europe, from an arsenic – antimony – silver rich ore to one containing more nickel, and the second is a change in manufacturing technique from one in which scrap bronze is normally included in the calamine process (which almost invariably used some scrap in the recipe), to one where only scrap brass or copper was allowed. It is interesting to speculate at this point as to whether this change can be related to the economic shifts occurring in north-west Europe. Before 1400, the Hanseatic League (founded in the northern German town of Lübeck around AD 1158) dominated trade in the Baltic and along the north coast of Europe, and supplied copper from the Falun mine in Sweden, which opened around AD 1200 (Tylecote, 1976). The League declined during the 15th Century, perhaps most severely in 1466 when the towns of western Prussia rose against the Teutonic Order with the help of Poland. Their position in the European copper trade was taken over by the Fuggers of Augsberg, supplying copper from Hungary and the Tyrol. By the early 16th Century Jakob Fugger virtually monopolized the copper supply, as well as that of lead and silver, and actually obtained a monopoly on mercury supplied from Spain. Political problems in Spain, however, brought about the decline of the Fugger family, resulting in the loss of the Hungarian mines by 1546. There is some additional evidence for the validity of this scenario, in the form of the analyses published by Werner (1982), which also draws on previous compilations by von Bibra and others. In his Table 11.4.1, Werner gives the analyses of 11 coppers extracted from Austro-Hungarian ores, which shows an average of 0.48% Ni, with only traces of As ($\leq 0.09\%$), and traces of Sb. His Table 11.10a gives the analyses of 10 Swedish copper artefacts of the 11th to 13th Centuries

AD, showing an average Ni content of 0.036%, with 0.16% As and 0.13% Sb. Although far from conclusive, these analyses suggest that the characteristics observed in the jettons do indeed reflect a change in the copper in circulation from sources in Sweden to somewhere in central Europe, some time around AD 1450.

In view of the documentary evidence for the introduction of new brassmaking processes into England in the 18th Century, and the analytical evidence presented above for the earlier invention of these processes in Europe, it is worth re-examining the jetton and coin evidence for the later period in England (Mitchiner *et al.*, 1985; 1987a). A plot of the zinc content of around 50 tokens minted in England between AD 1600 and 1850 shows that the 33% limit is 'breached' almost immediately (Figure 6.7). These tokens (two from Bridgwater, Somerset, dated 1654 and bearing the name William Sealy, with zinc contents of 34.2% and 34.6%, plus two with slightly less zinc – 32.4% in one from Taunton dated 1667 and 31.5% in one from Great Yarmouth, also dated 1667) suggest that high zinc brass, possibly made using some metallic zinc, was available in England by the mid-17th Century. As discussed above, zinc was first produced on a commercial scale in Europe in 1738 and the manufacture of brass by direct mixing (but still using some calamine) was patented in England in 1781. Zinc was, however, available on a small scale some time before the 18th Century – the Dutch East India Company (established AD 1602) had by the mid-17th Century replaced the Portuguese as the major trading company with the East, and were certainly importing zinc, if not brass. Smith and Gnudi (1959), in their annotated translation of Vannoccio

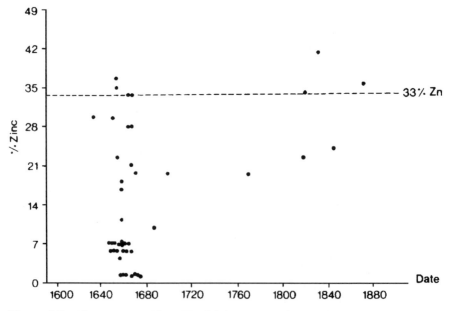

Figure 6.7   Zinc content of later English jettons, *ca.* AD 1600–1850.

Biringuccio's *Pirotechnia*, suggest that Glauber (in his *De Prosperitate Germanias*, Amsterdam, 1656) was the first European to realize that brass was an alloy of copper and zinc, and that calamine was a zinc ore. Prince Rupert, with whom Glauber was associated, is said to have made a harder metal than that obtainable from the calamine process by adding additional zinc to the alloy. This has the advantage of being cheaper than making a high-zinc brass entirely from pure copper and zinc. Watson (1786) confirms that cost was a major reason for the continuation of the calamine process, and says that the use of zinc was restricted to the manufacture of low-zinc gilding metals. On this basis, we feel reasonably confident in identifying Prince Rupert's metal as a brass with more than 34% zinc, available shortly after 1650, well in advance of the English patent for the full 'direct' process.

6.7 THE ANALYSIS OF EUROPEAN BRASS SCIENTIFIC INSTRUMENTS

Having accumulated a good deal of knowledge, both documentary and analytical, about the Medieval and later history of European brassmaking, it seems reasonable to turn cautiously from jettons to other more valuable brass objects, such as scientific instruments and clocks, the authenticity of which is a constant cause for concern. A number of problems may be anticipated, however, mostly related to the nature of the analytical evidence from the jettons – how representative was this metal of that available to the instrument or clock makers, and how geographically representative is it? It is clear that a large programme of analysis of 'authentic' scientific instruments is still required to corroborate the combined literary and analytical evidence for brassmaking practices in Europe from 1400 onwards. It is also appreciated that chemical information can only contribute to the overall study of these instruments, and is unlikely to be conclusive on its own. Chemical analysis of metals can, after all, only give evidence of compatibility with known objects of the same age. With the co-operation of the History of Science Museum, Oxford, therefore, the Research Laboratory for Archaeology and the History of Art began a programme of analysis of dated instruments, which resulted in the initial accumulation of 285 individual analyses from 65 instruments, ranging in date from *ca.* AD 1400 to AD 1770. By country, there were 15 instruments each from Italy, France and Germany, ten from Holland and five each from Spain and Flanders. A summary of these results was presented at a United Kingdom Institute of Conservation meeting in 1983 (Pollard, 1983b), which was only published in abstract form, and so the main findings are reproduced here. A further 69 English and German instruments were subsequently analysed, but again were only published in summary form (Mortimer, 1989).

The main technical interest as before is the maximum zinc content observed in the instruments, which is summarized in Figure 6.8, showing only the maximum zinc content as a function of date, classified by country of origin. For the purposes of this summary, all the analytical results have been accepted as valid, with the exception of one Dutch analysis with 42% zinc, which was rejected as modern, and the attributions have been assumed to be correct. Four countries apparently started using brass with more than 28% zinc at around the same time (*ca.* AD 1560) when the zinc content of English, French, Spanish and Flemish instruments jumps simultaneously from below 24% to about 33%. Politically, Spain and the Low Countries (the Spanish Netherlands) were united under Charles V and later Philip II, although rebellion resulted in the establishment of the Protestant Dutch Republic around 1600. The wars with France ended with a treaty in 1559, so we may surmise that after 1560 conditions for free trade in raw materials or finished metals were relatively stable. Nuremburg,

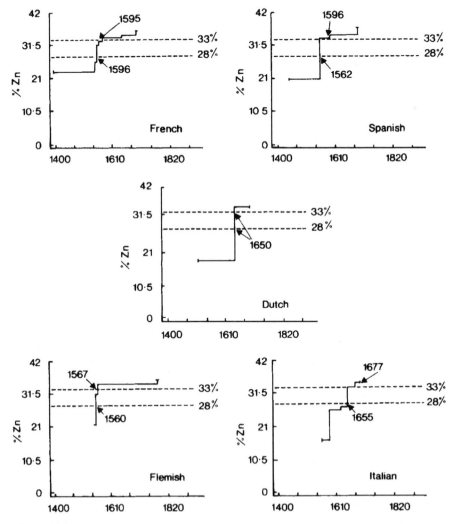

Figure 6.8   Summary of the maximum zinc content of European brass scientific instruments (*ca.* AD 1400–1800), classified by country of origin. (Redrawn from Pollard, 1983b, and Mortimer, 1989; Figure 1 by permission of Elsevier Science and the author.)

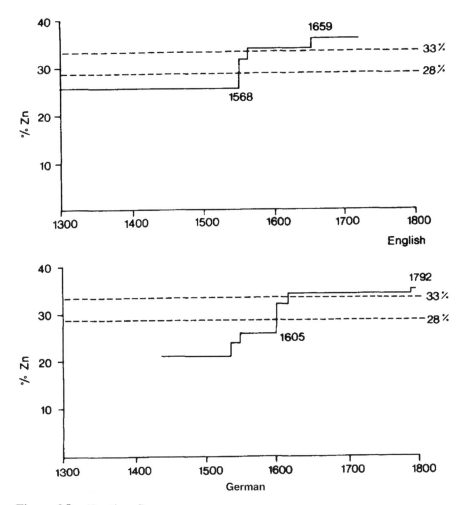

Figure 6.8 (Continued)

on the other hand, was by this time part of the Protestant states of Germany, and was certainly rising in importance as a brass producer, and therefore may not have imported any of this new alloy.

An early date of around 1560 is therefore claimed for the introduction of a process capable of producing brass with more than 28% zinc (presumably a modified calamine process using granulated copper) into the Catholic countries of Spain, France and Flanders, and also in the brass available in England. The data on the jettons offer good support for this, coming up with a similar date of 1560 for the introduction of this metal. Further support can be found for this model in the work of Cameron (1974) on English monumental brasses. Figure 6.9 shows a plot of zinc content against date taken from Cameron's results. Neglecting one point, which is a suspect 19th Century analysis, the first brasses with more than 28.5% zinc are found from 1570 onwards. The early

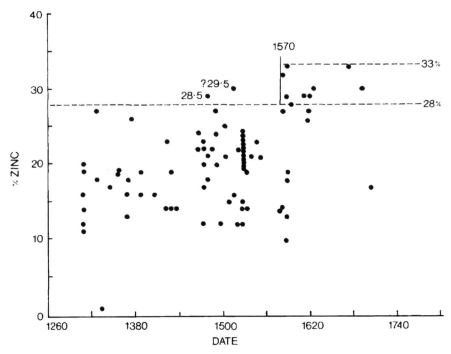

Figure 6.9   Zinc content of English monumental brasses, *ca.* AD 1250–1700. (Redrawn from Cameron, 1974; Figure 3, by permission of the Royal
Archaeological Institute).

trade between England and Flanders in finished monumental brasses is well-
known (Page-Philips, 1972; Chapter 4), thought to be due to the Hanseatic
League. After the glut of re-usable material supplied by the Reformation had
dried up, Page-Philips suggests that tomb makers were again turning to the
continent for supplies of sheet brass, some time after 1560. Table 6.2 lists the
average composition of post-1560 scientific instrument components with more
than 28% zinc from Spain, France and Flanders, plus the relevant results from
Cameron mentioned above. There is clearly a high degree of similarity between
all four sets of figures, and it is not therefore unreasonable to suggest a common
source for this brass – presumably Flanders, in view of the dominant position
of Flemish metalworkers at this time.

Table 6.3 (from Pollard, 1983b) lists the average composition of low
zinc brasses (<28% Zn) used for instruments, classified again by country of
manufacture. The majority are pre-1560, but some later instruments have been
included where they appear to be made from the same type of metal. Also
included is the average for Cameron's analyses of early monumental brasses.
There is some similarity between the French, Spanish and Dutch instruments,
particularly in the low nickel concentration, suggesting trade of raw materials
or finished products through the Hanseatic League, as discussed for the jettons.
The German instruments are quite different, having higher nickel and lower tin.

Table 6.2 Average analyses of post AD 1560 brass scientific instruments from France, Spain, and Flanders with more than 28% zinc.

		Fe	Ni	Cu	Zn	As	Pb	Ag	Sn	Sb
Flemish Instruments	$\bar{x}$	0.24	0.21	68.0	30.0	<0.2	0.85	<0.1	0.41	<0.1
1560–1770	s	0.06	0.16	3.0	2.9		0.15		0.10	
$n = 16$										
Spanish Instruments	$\bar{x}$	0.27	0.25	65.1	32.2	<0.2	1.43	<0.1	0.40	<0.1
1562–1596	s	0.27	0.07	2.3	2.3		0.87		0.24	
$n = 5$										
French Instruments	$\bar{x}$	0.22	0.20	66.8	31.2	<0.2	0.86	<0.1	0.30	<0.1
1560–1652	s	0.12	0.10	3.2	3.5		0.41		0.41	
$n = 33$										
English Monumental Brasses	$\bar{x}$	(0.4)	–	68.8	29.8	–	1.0	–	0.2	–
(Cameron) 1571–1678	s			2.4	2.1		0.5		0.2	
$n = 11$										

(– = not reported)
(n = number of analyses)

Table 6.3 Average analyses of low Zinc brasses (<28% Zn) classified by country. The majority date from before AD 1560.

		Fe	Ni	Cu	Zn	As	Pb	Ag	Sn	Sb
French Instruments	$\bar{x}$	0.25	<0.05	78.7	18.6	<0.2	0.66	0.12	1.37	<0.1
1400–1700	s	0.08		2.5	2.5		0.32	0.05	0.46	
$n = 46$										
German Instruments	$\bar{x}$	0.16	0.45	79.8	18.1	<0.2	0.72	<0.1	0.39	<0.1
1490–1677	s	0.08	0.33	3.1	3.0		0.68		0.19	
$n = 20$										
Spanish Instruments	$\bar{x}$	0.16	<0.05	76.5	20.3	<0.2	0.62	<0.1	1.81	<0.1
ca. 1450–1598	s	0.08		3.4	3.9		0.32		1.10	
$n = 10$										
Dutch Instruments	$\bar{x}$	0.15	<0.05	79.3	18.2	<0.2	0.23	<0.1	1.96	<0.1
1516	s									
$n = 2$										
Italian Instruments	$\bar{x}$	0.13	0.46	82.7	15.5	0.45	<0.1	0.08	0.54	<0.1
1580–1636	s	0.02	0.11	1.5	1.5	0.20		0.04	0.07	
$n = 9$										
Italian Instruments	$\bar{x}$	0.08	<0.05	71.3	25.6	<0.2	2.60	<0.1	<0.2	<0.1
1588–1694	s	0.04		1.8	1.6		0.50			
$n = 14$										
English Monumental	$\bar{x}$	–	–	75.9	18.9	–	3.2	–	1.4	–
Brasses (Cameron) 1300–1691	s			4.5	4.5		2.4		1.4	
$n = 70$										

(– = not reported)
(n = number of analyses)

The early Italian instruments are interesting, forming two distinct compositional groups, one with high nickel, but low zinc and lead, and the other with low nickel and tin, but high zinc and lead. It may be that these two represent northern and southern Italy – the Kingdom of Naples in the south being a

Spanish stronghold, and the north having economic ties with Germany, but a more detailed analysis is required to justify this suggestion.

The first appearance of high zinc brasses in scientific instruments – more than 33% zinc – is slightly more difficult to ascertain. Table 6.4 lists the 13 instruments found during the first programme of analyses which have components made from a metal containing more than 34% zinc. The figure of 34% was chosen to allow a small margin for error, either in the analytical results or the accuracy of the boundary figure, but where an instrument consisted of one part with more than 34% zinc, parts with more than 33% were included in the average. The fraction in brackets after the accession number represents the number of results included in the average compared with the total number of parts analysed on that particular instrument. The three earliest examples (dated 1567, 1595 and 1596) are all slightly different from the later examples, in that they all have a relatively high nickel and tin content (greater than 0.1 and 0.2% respectively). A summary of the difference between these pre-1650 analyses and those post-dating 1650 is given at the bottom of Table 6.4, together with an average of the compositions of the high-zinc jettons discussed above for comparison. A closer examination of these early instruments (together with those tokens in Figure 6.5 dating to before 1675 with more than 34% zinc) is called for, since these may represent either the very maximum achievable by the calamine process, or they may be the first evidence for the direct use of zinc in Europe, dating from around 1570 onwards. On present evidence, it is tempting to suggest that there is a close correlation between the date attributed by Smith and Gnudi (1959) to the work of Glauber and the first appearance of the low nickel, low tin, high zinc brass instruments, and that 1650 is therefore the most likely date for the first widespread production of high-zinc brasses (*i.e.*, made with the deliberate addition of some metallic zinc) in Europe. It is also likely that this alloy was known as Prince Rupert's metal.

In contrast to the situation described above, very few non-European scientific instruments have been analysed, and therefore there is little with which to compare these observations. In a recent paper, however, Newbury *et al.* (2006) have published the analyses of eight 17th Century AD astrolabes, mostly from Lahore in modern Pakistan, using three synchrotron X-ray techniques (diffraction, fluorescence and radiography): see Pollard *et al.* (2007; 290) for an introduction to synchrotron radiation. All eight showed some components which contained an $\alpha + \beta$ brass phase, implying the presence of Zn in excess of 33%, indicating use of the direct manufacturing process, and hence the availability of metallic zinc. The oldest of the these astrolabes from Lahore is dated to 1601, and is attributed to Isa ibn Allāhdād, the son of Allāhdād al Humāyūnī, who manufactured astrolobes in Delhi for the Moghul emperor Humāyūnī in the mid-16th Century. The majority of the parts of this astrolabe are of high zinc brass, clearly showing that metallic zinc was available and used for brass manufacture in the sub-continent by AD 1600 at the latest, and probably at least 50 years earlier, if the son inherited the father's knowledge.

Table 6.4 Analyses of scientific instruments with more than 34% zinc.

	Fe	Ni	Cu	Zn	As	Pb	Ag	Sn	Sb
Astronomical ring. Flemish, 1567 57–84/24 (2/3)	0.16	0.29	63.5	34.1	<0.2	1.37	<0.1	0.33	<0.1
Astrolabe. French, 1595 IC 211 (5/6)	0.25	0.24	64.4	33.7	<0.2	0.98	<0.1	0.22	<0.1
Equinoctial Dial. Spanish, 1596 S7 (2/2)	0.14	0.21	63.9	34.1	<0.2	0.81	<0.1	0.55	<0.1
Perpetual Calendar. Dutch, 1650 2497 (3/5)	0.13	<0.05	64.0	34.2	<0.2	1.25	<0.1	<0.2	<0.1
Circumferentor. French, 1652 57–84/253 (3/4)	0.23	0.12	63.9	34.6	<0.2	0.82	<0.1	<0.2	<0.1
Quadrant Italian, 1677 – (1/3)	0.27	<0.05	63.7	34.7	<0.2	0.94	0.10	<0.2	<0.1
Equinoctial Dial. Spanish, 1695 S10 (4/5)	0.11	<0.05	63.1	35.5	<0.2	0.96	<0.1	<0.2	0.11
Scaphe. Italian, 1697 I.98 (1/3)	0.16	<0.05	64.6	34.2	<0.2	0.45	<0.1	0.24	<0.1
Equinoctial Dial. French, 1700 F.208 (2/5)	0.16	0.21	63.1	35.7	<0.2	0.52	<0.1	<0.2	<0.1
Equinoctial Dial. Dutch, 1700 G.126 (2/6)	0.18	0.34	63.1	35.4	<0.2	0.59	<0.1	<0.2	<0.1
Dialling Instrument. Dutch, 1705 28 (2/3)	0.14	<0.05	63.4	35.0	<0.2	1.15	<0.1	<0.2	<0.1
Astronomical Compendium. Dutch, 1709 M3 & M29 (7/14)	0.15	<0.05	64.2	34.3	<0.2	1.03	<0.1	<0.2	0.10
Solar & Lunar Dial. Flemish, 1770 M18 (8/13)	0.14	<0.05	63.6	34.6	<0.2	1.29	<0.1	<0.2	<0.1
Pre-1650 average $\bar{x}$	0.18	0.25	63.9	34.0	<0.2	1.05	<0.1	0.37	<0.1
s	0.06	0.04	0.5	0.2		0.29		0.17	
Post-1650 average $\bar{x}$	0.17	<0.1	63.7	34.8	<0.2	0.90	<0.1	<0.2	<0.1
s	0.05		0.5	0.6		0.30			
High zinc jettons $\bar{x}$	<0.1	<0.1	62.0	36.3	<0.2	1.11	<0.1	<0.2	<0.1
s			3.3	3.2		0.82			

$n = 3$

$n = 10$

$n = 35$

6.8 THE ANALYTICAL AUTHENTICATION OF BRASS INSTRUMENTS

As an example of how this type of information can help in the study of scientific instruments, we present two previously unpublished case studies. One relates to some instruments from the Barberini Collection, now in the National Maritime Museum, London, and one relates to two 18th Century English clocks by famous London makers (Tompion and Graham) from a private collection.

The Barberini Collection comprises 24 various scientific instruments, supposedly put together by the Barberini family in Florence – particularly Maffeo Barberino (1568–1644), later Pope Urban VIII, and more particularly Cardinal Francesco Barberino (1597–1679). Unfortunately, only one piece of this magnificent collection can be definitely traced back to Francesco Barberino – a large concave burning mirror. The first complete list of the collection dates only to the First World War, and it was purchased by Michel and Landau of Paris in the 1930s, and some of it was exhibited at the Descartes Exhibition at the Bibliothèque Nationale in Paris in 1936. The complete collection was finally purchased by the National Maritime Museum in 1949 (Shaw, 1973).

Only part of the collection was analysed at the Research Laboratory for Archaeology in Oxford – five examples were chosen, as follows (descriptions from Shaw, 1973):

6. A.47/46–230C (860). Large Astrolabe, probably Italian. 41.5 cm diameter, in original case. Unsigned and undated, *ca.* 1630.
8. Cl/S.10/47–216C (883). Sector. Brass, 19 cm length. Signed *I Galli's fecit.* Undated *ca.* 1600.
14. D.359/47–223C (867). Bowl sundial, 32 cm diam., 50 cm high. Signed on back of bowl *I. Galli.* Undated, *ca.* 1600.
16. T.63/47–210C (870). Trigomètre. Barberini cypher on back. Brass, 37 cm length folded. Tooled morocco case embossed with Barberini cypher. Signed: *Philipus Danfrie: F: Anno Domini 1580.*
17. T.64/47–211C (871). Companion scale to trigomètre, with sights. Barberini cypher on back. Brass, 33 cm length folded. Tooled morocco case embossed with Barberini cypher similar to above.

A summary of the analyses of these five instruments is given in Table 6.5. The figure in brackets gives the total number of separate parts of the instrument analysed (the denominator), and the number of analyses included in the average figure (the numerator).

The large astrolabe (number 6) was dismantled into 16 parts, and was found to be made from broadly similar (but definitely not identical) metal. The average zinc content is around 13%, but the range on the 16 parts analysed is from 9.0 to 20.9%; hence the relatively large standard deviation. Similarly, the lead levels vary from 0.2 to 4.6%, and six of the components analysed had measurable amounts of As (0.2–0.3%). There is clearly some evidence that the instrument was not made from uniform material – whether this is significant in

Table 6.5 Average analyses of five instruments in the Barberini Collection.

Description		Fe	Ni	Cu	Zn	As	Pb	Ag	Sn	Sb
6. Astrolabe (16/16)	$\bar{x}$	0.34	0.17	82.0	13.2	<0.2	1.8	0.12	2.1	0.23
	s	0.13	0.06	2.7	3.4		1.2	0.06	0.8	0.20
8. Sector (3/3)	$\bar{x}$	0.23	0.10	75.1	21.9	<0.2	0.97	0.09	1.53	0.17
	s	0.02	0.02	2.1	2.3		0.12	0.02	0.12	0.08
14. Bowl sundial (7/8)	$\bar{x}$	0.35	0.27	77.7	17.8	<0.2	2.5	<0.1	1.0	0.21
	s	0.11	0.13	2.6	3.5		1.0		0.6	0.18
(bowl) (1/8)	$\bar{x}$	0.24	1.54	69.6	20.5	<0.2	2.9	0.06	4.4	0.58
16. Trigomètre (8/15)	$\bar{x}$	0.23	0.16	77.4	19.2	<0.2	0.97	<0.1	1.88	<0.1
	s	0.05	0.05	2.6	2.8		0.14		0.23	
(7/15)	$\bar{x}$	0.32	0.24	65.6	31.9	<0.2	1.43	<0.1	0.33	<0.1
	s	0.12	0.06	1.1	1.0		0.35		0.24	
(case hook)	$\bar{x}$	0.07	<0.05	55.4	43.3	<0.2	1.1	0.07	<0.2	<0.1
17. Scale (5/10)	$\bar{x}$	0.41	0.23	74.5	22.6	<0.2	0.98	<0.1	0.9	<0.1
	s	0.04	0.06	2.4	2.9		0.13		0.6	
(5/10)	$\bar{x}$	0.49	0.23	66.7	30.9	<0.2	1.04	<0.1	0.4	<0.1
	s	0.11	0.06	1.2	0.7		0.15		0.4	
(case hooks) (2/2)	$\bar{x}$	0.18	<0.05	68.7	29.6	<0.2	0.4	<0.1	1.08	<0.1
	s	0.01		0.5	0.1		0.2		0.16	

the context of the authenticity of an instrument is difficult to say. Simply taking the average analysis and comparing it with Italian instruments of a similar period (Table 6.3) shows that the analysis of the Barberini piece differs in detail from either of the two Italian groups identified above ('high' and 'low nickel'), but that the composition is not itself unreasonable for the proposed date. The sector signed by Galli (number 8) is relatively homogeneous, in that the three pieces analysed are compositionally similar. It does not match any of the other instrument groups exactly, but again its composition is not unlikely for the period. The same comments apply to the majority of pieces (seven out of eight) of the bowl sundial (number 14), except to note that the most important part of the sundial – the bowl itself – has a relatively unusual composition, with 1.54% nickel, which is as high as anything analysed during the whole programme.

The trigomètre and its companion scale (items 16 and 17) are both similar, in that they appear to be made from two qualities of brass – one approximately 75% Cu, 20% Zn with 1–2% each of Pb and Sn, and one about 65% Cu, 30% Zn, with 0.5–1.5% Pb and Sn. The lower zinc composition is similar to (but not identical with) contemporary instruments from parts of Europe, but the higher zinc composition, on the jetton evidence, would be unusual (but not impossible) for the period after 1560. It must be noted that the summary in Table 6.5 gives equal weight to all components of the instruments, ranging from major components such as the arms of the trigomètre, down to smaller items such as sites and mount plates. Only a detailed study of the analyses in relation to each component can be expected to give a true picture of the history of the instrument.

In summary, the analyses of these five instruments were essentially inconclusive, as far as authenticity is concerned. All of the compositional groups obtained were plausible for the period claimed (therefore making it impossible to declare any object a 'fake'), but the inconsistency of the analyses when compared with the larger data set of other contemporary scientific instruments leaves open the possibility that some, at least, of the instruments are not entirely genuine. This is particularly so for the trigomètre and its companion scale, which show two very distinct compositional groups, one of which is unusual for the late 16th Century. As a corollary to this study, two samples from the wooden cases from items 16 and 17 were subjected to radiocarbon dating by accelerator mass spectrometry in Oxford. The results were 275 ± 75 BP (OxA-536) and 310 ± 80 BP (OxA-537), respectively. Unfortunately, when these are calibrated, although the central dates are AD 1640 and AD 1630, the 95% confidence intervals on these dates extend right up to AD 1950 in each case, because of the flatness of the calibration curve from 1650 onwards. Although strictly inconclusive, it is possible to wring a little more information from the dates, by noting that there is an 84% probability that the true age lies before one standard deviation above the mean. On this basis, there is an 84% chance that the date of the first sample is before AD 1800, and the second before AD 1660 (Gowlett, pers. comm.). Albeit in a somewhat unsatisfactory manner, it is possible to suggest on this evidence that the cases at least are likely to be genuine.

A more convincing case study is provided by the chemical analyses of the brass components of two privately owned 18th Century English clocks by

Table 6.6 Summary of English 18th century clock analyses.

Description			Fe	Ni	Cu	Zn	As	Pb	Ag	Sn	Sb
Tompion, 1709	13/13	$\bar{x}$	0.26	<0.05	74.1	22.9	<0.2	1.5	<0.1	0.97	<0.1
		s	0.10		1.2	1.3		0.4		0.20	
Graham, 1722	9/15	$\bar{x}$	0.31	<0.05	72.4	25.1	<0.2	1.3	<0.1	0.62	<0.1
		s	0.06		1.0	1.3		0.3		0.15	
	3/15	$\bar{x}$	0.41	<0.05	74.2	21.4	0.2	2.73	~0.1	0.70	0.23
		s	0.06		0.3	0.3	0.2	0.09		0.07	0.03
	2/15	$\bar{x}$	0.16	<0.05	65.9	32.1	<0.2	1.6	<0.1	<0.2	<0.1
		s	0.04		0.4	1.1		0.5			
	1/15		0.29	<0.05	72.5	21.6	<0.2	2.30	<0.1	3.12	<0.1

famous makers – Thomas Tompion (1639–1713, working in Fleet Street, London) and George Graham (1673–1751, also working in London). As can be imagined, the analysis of the moving parts of clocks of this quality (and value) places the highest possible constraints on the analytical procedures in terms of minimizing damage – even the removal of very fine drillings would be unacceptable, since it would destroy the balance of the parts. At the time when these analyses were carried out (1983), surface analyses (with minimal cleaning using a solvent to remove lacquer) by XRF provided the only acceptable means of analysis, although now one might consider laser ablation inductively coupled plasma mass spectroscopy (ICP-MS) as a better alternative (see Chapter 2). The XRF analyses are summarized in Table 6.6. The Tompion (dated 1709) was dismantled into 13 pieces, and proved to have a remarkably uniform analysis, which is entirely consistent with English scientific instruments of a similar date (Mortimer, pers. comm.). The Graham, dated to 1722, was analysed in 15 separate components, and these were less homogeneous. The larger proportion of the components (nine out of 15) had an analysis very similar to that of the Tompion, and therefore also consistent with an early 18th Century date. A further three elements had a slightly higher lead content (2.73%) than the previous set, but were not too dissimilar from one component of the Tompion, which had a lead content of 2.24%. Two elements of the Graham (the escapement and the pendulum) had a much higher zinc content (32%) – not impossible for the early 18th Century, but more likely to be later. One part (a 49 mm diameter solid wheel) was very similar to the bulk of the clock, but had a significantly higher tin content (3.12%). The conclusion drawn from this work was that the Tompion was entirely consistent with having been made in 1709, and all the parts analysed were likely to be original. The Graham was less internally homogeneous, but was also consistent with a date of 1722, with the possibility that the escapement and pendulum had been replaced at a later date.

6.9   EUROPEAN COPPER AND BRASS IN NORTH AMERICA

At least two materials of European manufacture have been used as important indicators of trade and chronology around the time of European contact with

North America – glass beads and copper kettles. From 1497, when Cabot explored the Atlantic coast of North America, and particularly since Jacques Cartier's second voyage in 1535, when he explored for the first time the interior of the North American continent via the Saint-Lawrence River as far as Montréal, occasional meetings between native Americans and Europeans resulted in unsystematic trading. Once the fishing banks off Newfoundland and the whaling grounds in the Saint-Lawrence were being exploited by Europeans (especially the Basques) in the mid-16th Century, more organized trading took place, with furs being exchanged for European iron tools, copper vessels and glass beads (Moreau and Hancock, 1999). Glass beads from Holland, France and possibly England are found in large quantities on 17th Century archaeological sites in north-eastern North America, and are an important tool in documenting contact between native Americans and Europeans (Kidd and Kidd, 1970). For Iroquoian sites in the north-eastern United States and south-eastern Canada, for example, the chronology around the contact period is divided into three 'Glass Bead Periods' (Walker *et al.*, 1999). Given this importance, it is not surprising that glass beads have been subjected to chemical analysis to aid classification (*e.g.*, Sempowski *et al.*, 2001, and references therein).

Copper metallurgy was, of course, well-established in North America prior to European contact, but was based purely on native (*i.e.*, not smelted) copper exploitation. In North America, native copper objects, including ornaments and axes, are found from *ca.* 3000 BC onwards (Scott, 2002; 5). Although copper workers knew how to work and anneal copper, the smelting process for winning copper from ores was unknown. This is in stark contrast to South America (especially Peru), where a sophisticated indigenous smelting techno-logy emerged in the first few centuries BC. The discovery of smelted copper on a North American archaeological site is therefore taken to be an indicator of European contact, and the general pattern in north-eastern archaeology is that soon after the first contact in the late 16th Century European smelted copper rapidly displaces unsmelted native copper (Hancock *et al.*, 1991). It is therefore clearly critical to be able to distinguish between native and smelted copper. The most obvious difference is in the metallographic structure, with unworked native copper showing distinctive features such as long twin planes (Scott, 2002; 3). However, once native copper is worked and annealed, this distinctive metallography is lost, and other means of discrimination become necessary. One obvious approach is to determine trace element composition, and this has been done since the 1950s. Early studies (somewhat hampered by the poor analytical sensitivity of the techniques of the time, such as optical emission spectroscopy) concluded that native coppers were analytically 'clean' relative to European smelted coppers; elements such as Sb, Cr, Co, Fe, Ni, Se, Ag, Sn, Zn, Al and Mg are all very much lower in any samples of native copper analysed, although As was found to be variable and occasionally present in the percent region (Hancock *et al.*, 1991). Using the much more sensitive technique of neutron activation analysis (NAA), Hancock *et al.* (1991) identified three groups of native copper, characterized by very low total trace elements, Fe-rich or As-rich, respectively. They succeeded in distinguishing native coppers from European smelted copper, primarily using levels of Au, As, Sb, Ag and Ni, with loglog plots of ppm Au *vs.* ppm Ag being adequate to discriminate the majority of samples. Moreover, native coppers, despite being analytically 'clean' relative to smelted 16th to 17th Century copper, were found to contain higher levels of As, Ag and Hg than does modern commercial copper – an important observation when considering the authenticity of such artefacts.

The earliest European copper trade items appear to be kettles made from thick 'red copper', as a result of the activities of Basque whalers in the Gulf of Saint-Lawrence around AD 1540–1600 (Walker *et al.*, 1999). As the French, Dutch and English displaced the Basques in the North American fur trade around 1600, so the copper alloy changed from copper to brass. Although it is undoubtedly more complicated than this, the sequence of native copper, smelted Basque copper and finally European brass is an important chronological indicator in North American archaeology. In the early contact period, copper kettles appear to have been broken up for re-use as ornaments, possibly of a sacred nature, and were traded into the interior using established indigenous networks. As the items became more numerous, they were used intact as prestige goods for feasting and gift exchange.

Being relatively rare as finds in the early period, most analyses of brass objects from North American sites have been focussed either on simply determining the metal type for chronological purposes, or to try and determine from chemical groupings the minimum number of vessels represented (*e.g.*, Pavlish *et al.*, 2004). It has been acknowledged that the history of European copper production from the 16th Century is an important parameter in understanding the meaning of the distribution of copper alloys in America (*e.g.*, Walker *et al.*, 1999). To date, however, there has been little attempt to link the analytical work carried out on European copper alloys in North America with the data obtained in Europe, in order to identify the country of origin of the copper alloy. Although this should be possible, it is complicated by the fact that much of the recent data gathered in North America is obtained by NAA, whereas European data until recently tended to be XRF, so there is very little inter-comparability of the data. Clearly there is scope for a concerted analytical programme, perhaps using the same ICP technology, to address this interesting problem.

6.10 SUMMARY

The early history of brass and zinc production has engaged scholars for a number of years, and continues to provide surprises. The established view of the 1990s was that brass as a deliberate alloy originated somewhere in Asia Minor, sometime during the mid-1st Millennium BC. Recent analyses of the apparently insignificant material from Tepe Yahya has, however, suggested that this is not the case, and that the origin is at least a millennium earlier, and possibly further east – perhaps somewhere in Central Asia. Given the vast quantities of (mostly unanalysed) metalwork excavated during the 20th

Century from important sites in Iran, Mesopotamia, Anatolia and (to a lesser extent) Central Asia, it is tantalizing to think that important evidence is almost certainly sitting on some museum shelf or in a storeroom box somewhere. This is an instructive lesson on a number of levels. When using chemical analysis of archaeological materials to answer questions about human behaviour, it is important to remember that the answer is only ever as good as the data allow, and new information can easily overturn established thinking. It also points to the value of systematic analysis of excavated material, however uninspiring it might look at first sight. More specifically, the origin of brass production is particularly interesting because it marks an important stage in the history of human technological ingenuity – especially because the volatility of metallic zinc means that a completely new process had to be mastered to produce the alloy. Moreover, brass making is almost certainly the process which gave rise to the literature of alchemy, being the production of 'gold' from base metal and a magical 'earth'. It is, therefore, ultimately linked to the mysterious origins of chemistry itself. There is still a lot to learn.

By the Medieval period, brass had become much more commonplace, and has consequently been the subject of more intensive analysis. From both the European instrument and the jetton data, evidence has been obtained for two variations in the production of brass in later Medieval Europe via the calamine process – one, largely dating to before AD 1450, yielding a brass with appreciable levels of tin, which virtually disappears after this date. This may be the result of using scrap bronze in the melt (perhaps because of the high value placed on raw copper), which was discontinued in the later period (or perhaps only scrap copper and brass were allowed). The latter process results in a pure brass, with the combined copper and zinc total exceeding 97%. The data on the jettons, showing an increase in the nickel content of the later tokens, suggest that this might be roughly coincident with a change in ore source in addition to the use of an improved manufacturing technique. The early period probably represents the phase of domination of the copper trade, and also in finished metal, by the Hanseatic League, supplying copper from Sweden. The decline of the Hanseatic League resulted in the rise of the Fuggers of Augsberg, trading in central European copper, some of which has characteristically higher nickel. However, a change in trace element concentrations does not automatically mean a change in ore source – a change in foundry practice may well also affect the trace element levels (Pollard *et al.*, 1991). Nevertheless, these observations are in harmony with our knowledge of the economic history of Europe at the time.

Contemporary literary evidence suggests that two dates should mark improvements in the manufacturing process of English brass. One is the early 18th Century, when the use of granulated copper is said to have increased the zinc uptake of the calamine process from 28 to 33%, and the second is the introduction of direct mixing of metals as a way of making brass in the late 18th Century. Interpretation of the analytical data on these points is always difficult, because apparent anachronisms can be interpreted either as earlier introductions of these methods, or as evidence of the objects themselves being later copies. There is now, however, a reasonable amount of evidence to suggest that brass compatible with the granulated calamine process was being used in Europe by around 1560. The published analyses of Cameron (1974) also give figures in this category dated to *ca.* 1570. This supports the view that granulated copper was used in the calamine process well before the early 18th Century, if the technological interpretation of the reason for this increase in zinc content is correct.

The date of the earliest appearance of European brass made by direct mixing of metals (not including gilding metals, with low zinc contents) is rather more difficult to ascertain. From Figure 6.5, it could be as early as around 1565, but all the evidence points to its widespread use by 1675. 18th Century documentary evidence is quite plainly against the use of zinc for brasses other than gilding metals on account of cost, and hence the survival of the calamine process into the 19th Century. Craddock (1981; 16) suggests that the iron content should distinguish calamine brass from the later type, with calamine brass containing more iron (*i.e.*, 0.2–0.5%, possibly as high as 'several percent'). Watson (1786) also says that brass produced by the direct method is free from iron, and is therefore good for compasses. The data discussed here show relatively little iron for all the groups discussed above, with the majority of individual results lying between 0.1 and 0.4% Fe. These are within the range expected for calamine brass, and no significant decrease is observed in the high zinc brasses. In this case the iron therefore appears to give no clear indication of manufacturing technique.

The problem of using this information to detect later copies of scientific instruments and other brass artefacts is one which still requires considerable attention. It obviously cannot be claimed that chemical analysis can be used to date a piece of brass – all that can be done is to compare its composition with those of known genuine pieces. This makes it possible that anachronisms will be found, but the chemical evidence should be considered along with stylistic and other technical reports, simply as part of the overall assessment. Werner (1980) used his vast data bank of analyses of cast brass objects in an attempt to detect known modern castings – he did not succeed in every case. It is always possible, therefore, that some copies will meet the chemical requirements, either by accident or from some knowledge of the metallurgical background on the part of the faker. It is also possible that perfectly genuine instruments will have analyses outside the expected range for a particular country and date, but the maximum zinc content is believed to be an absolute thermodynamic (or kinetic) limitation in calamine brass – hence the emphasis on this aspect of the analyses in this work. Much more data need to be considered, especially from non-European material, before anything more than general statements about 'consistency with published analyses' can be made, but a good start has been made. If, as appears possible, the question of the provenance of European brass during the Medieval and later periods can be satisfactorily answered by chemical analysis (at least in broad terms to country of origin), then a whole range of interesting questions about contact, trade and colonialism could be addressed on a world-wide basis, as exemplified by the work already done in North America. This seems to be a challenging but worthwhile objective.

REFERENCES

Aitken, W.C. (1866). Brass and brass manufactures. In *The Resources, Products, and Industrial History of Birmingham and the Midland Hardware District*, ed. Timmins, S., Hardwicke, London, pp. 225–281.

Barnard, F.P. (1916). *The Casting-counter and the Counting Board*. Clarendon, Oxford.

Bayley, J. (1990). The production of brass in antiquity with particular reference to Roman Britain. In *2000 Years of Zinc and Brass*, ed. Craddock, P.T., British Museum Occasional Paper No. 50, British Museum, London, pp. 7–27.

Bowman, S.G.E., Cowell, M.R. and Cribb, J. (1989). Two thousand years of coinage in China: an analytical survey. *Historical Metallurgy* **23** 25–30.

Brownsword, R. and Pitt, E.E.H. (1983). Alloy composition of some cast 'latten' objects of the 15/16th centuries. *Historical Metallurgy* **17** 44–49.

Budd, P., Haggerty, R., Pollard, A.M., Scaife, B. and Thomas, R.G. (1996). Rethinking the quest for provenance. *Antiquity* **70** 168–174.

Caley, E.R. (1964). *Orichalcum and Related Ancient Alloys: Origins, Composition and Manufacture with Special Reference to the Coinage Of The Roman Empire*. Notes and Monographs No. 151, American Numismatic Society, New York.

Caley, E.R. (1967). Investigations on the origin and manufacture of orichalcum. In *Archaeological Chemistry*, ed. Levey, M., American Chemical Society, Division of the History of Chemistry, University of Pennsylvania Press, pp. 59– 74.

Cameron, H.K. (1974). Technical aspects of monumental brasses. *Archaeological Journal* **131** 215–237.

Craddock, P.T. (1978). The composition of the copper alloys used by the Greek, Etruscan and Roman civilizations. 3 The origins and early use of brass. *Journal of Archaeological Science* **5** 1–16.

Craddock, P.T. (1981). The copper alloys of Tibet and their background. In *Aspects of Tibetan Metallurgy*, ed. Oddy, W.A. and Zwalf, W., British Museum Occasional Paper No. 15, British Museum, London, pp. 1–32.

Craddock, P.T. (1990). Zinc in classical antiquity. In *2000 Years of Zinc and Brass*, ed. Craddock, P.T., British Museum Occasional Paper No. 50, British Museum, London, pp. 1–6.

Craddock, P.T. (1995). *Early Metal Mining and Production*. Edinburgh University Press, Edinburgh.

Craddock, P.T. (ed.) (1998). *2000 Years of Zinc and Brass*. British Museum Occasional Paper No. 50, British Museum, London (revised edition).

Craddock, P.T., Burnett, A.M. and Preston, K. (1980). Hellenistic copper base coinage and the origins of brass. In *Scientific Studies in Numismatics*, ed. Oddy, W.A., British Museum Occasional Paper No. 18, British Museum, London, pp. 53–64.

Craddock, P.T., Freestone, I.C., Gurjar, L.K., Middleton, A.P. and Willies, L. (1990). Zinc in India. In *2000 Years of Zinc and Brass*, ed. Craddock, P.T., British Museum Occasional Paper No. 50, British Museum, London, pp. 29–72.

Dawkins, J.M. (1950). *Zinc and Spelter*. Zinc Development Agency, Oxford.

Day, J. (1973). *Bristol Brass: A History of the Industry*. David and Charles, Newton Abbot.

Day, J. (1990). Brass and zinc in Europe from the Middle Ages until the 19th Century. In *2000 Years of Zinc and Brass*, ed. Craddock, P.T., British Museum Occasional Paper No. 50, British Museum, London, pp. 123–150.

Day, J. and Tylecote, R.F. (1991). *The Industrial Revolution in Metals*. The Institute of Metals, London.

Finnegan, J.E., Hummel, R.E. and Verink, E.D. (1981). Optical studies of dezincification in alpha-brass. *Corrosion* **37** 256–261.

Grant, E. (ed.) (1974). *A Sourcebook on Medieval Science*. Harvard University Press, Harvard.

Haedecke K. (1973). Gleichgewichtsverhältnisse bei der messingherstellung nach dem Galmeiverfahren. *Erzmetall* **26** 229–233.

Hamilton, H. (1967). *The English Brass and Copper Industries to 1800*. Cass, London, 2nd edn.

Hancock, R.G.V., Pavlish. L.A., Farquhar, R.M., Salloum, R. Fox, W.A. and Wilson, G.C. (1991). Distinguishing European trade copper and north-eastern North American native copper. *Archaeometry* **33** 69–86.

Hawthorne, J.G. and Smith, C.S. (trans.) (1979). *Theophilus, De Diversis Artibus*. Dover Publications, New York.

Hedges, R.E.M. (1979). Analysis of the Drake plate: comparison with the composition of Elizabethan brass. *Archaeometry* **21** 21–26.

Hegde, K.T.M. (1989). Zinc and brass production in Ancient India. *Interdisciplinary Science Reviews* **14** 86–96.

Holmyard, E.J. (1957). *Alchemy*. Penguin Books, Harmondsworth.

Hoover, H.C. and Hoover, H..L. (trans.) (1950). *Agricola, De Re Metallica*. Dover Publications, New York.

Hughes, M.J., Cowell, M.R. and Craddock, P.T. (1976). Atomic absorption techniques in archaeology. *Archaeometry* **18** 19–37.

Hughes, M.J., Northover, J.P. and Staniaszek, B.E.P. (1982). Problems in the analysis of leaded bronze alloys in ancient artefacts. *Oxford Journal of Archaeology* **1** 359–363.

Kidd, K.E. and Kidd, M.A. (1970). A classification system for glass beads for the use of field archaeologists. *Canadian Historic Sites: Occasional Papers in Archaeology and History* **1** 45–89.

Lones. T.E. (1919). *Zinc and its Alloys*. Pitmans, London.

Michel, H.V. and Asaro, F. (1970). Chemical study of the Plate of Brass. *Archaeometry* **21** 3–19.

Mitchiner, M.B. and Pollard, A.M. (1988). Reckoning counters: patterns of evolution in their chemical composition. In *Metallurgy in Numismatics 2*, ed. Oddy, W.A., Royal Numismatic Society, London, pp. 105–126.

Mitchiner, M.B., Mortimer, C. and Pollard, A.M. (1985). The chemical compositions of English seventeenth-century base metal coins and tokens. *British Numismatic Journal* **55** 144–163.

Mitchiner, M.B., Mortimer, C. and Pollard, A.M. (1987a). The chemical compositions of nineteenth-century copper-base English jettons. *British Numismatic Journal* **57** 77–88.

Mitchiner, M.B., Mortimer, C. and Pollard, A.M. (1987b). Nuremberg and its jettons, c. 1475 to 1888: chemical compositions of the alloys. *Numismatic Chronicle* **147** 114–155.

Mitchiner, M.B., Mortimer, C. and Pollard, A.M. (1988). The alloys of Continental copper-base jettons (Nuremburg and Medieval France excepted). *Numismatic Chronicle* **148** 117–128.

Moreau, J.-F. and Hancock, R.G.V. (1999). Faces of European copper alloy cauldrons from Québec and Ontario 'contact' sites. In *Metals in Antiquity,* ed. Young, S.M.M., Pollard, A.M., Budd, P. and Ixer, R.A., BAR International Series 792, Archaeopress, Oxford, pp. 326–340.

Mortimer, C. (1989). X-ray fluorescence analysis of early scientific instruments. In *Archaeometry: Proceedings of the 25th International Symposium*, ed. Maniatis, Y., Elsevier, Amsterdam, pp. 311–317.

Needham, J. (1974). *Science and Civilisation in China , Vol. 5 Chemistry and Chemical Technology. Part II: Spagyrical Discovery and Invention: Magisteries of Gold and Immortality.* Cambridge University Press, Cambridge.

Newbury, B.D., Notis, M.R., Stephenson, B., Cargill, G.S. III and Stephenson, G.B. (2006). The astrolabe craftsmen of Lahore and early brass metallurgy. *Annals of Science* **63** 201–213.

Page-Philips, J. (1972). *Macklin's Monumental Brasses.* Allen and Unwin, London.

Partington, J.R. (1961). *A History of Chemistry.* Macmillan, London.

Pavlish, L.A., Hancock, R.G.V. and Ross, B. (2004). Instrumental neutron activation analysis of copper-rich samples from the Bead Hill site, Ontario, Canada. *Historical Metallurgy* **38** 106–112.

Pollard, A.M. (1983a). Authenticity of brass objects by major element analysis? Paper presented at *Symposium on Archaeometry*, Castel Dell'Ovo, Naples, 18th–23rd April 1983.

Pollard, A.M. (1983b). An investigation of the brass used in Medieval and later European scientific instruments. Paper presented at *The Preservation of Historical Scientific Materials,* UKIC Meeting, Geological Museum, London, 14th November 1983.

Pollard, A.M. (1988). Alchemy – a history of early technology. *School Science Review*, June 1988, 701–712.

Pollard, A.M., Thomas, R.G. and Williams, P.A. (1990). Experimental smelting of arsenical copper ores: implications for Early Bronze Age copper production. In *Early Mining in the British Isles*, ed. Crew S. and Crew P., Occasional Paper No. 1, Plas Tan y Bwlch, Snowdonia National Park Study Centre, Gwynedd, pp. 72–74.

Pollard, A.M., Thomas, R.G. and Williams, P.A. (1991). Some experiments concerning the smelting of arsenical copper. In *Archaeological Sciences 1989*, ed. Budd, P., Chapman, B., Jackson, C., Janaway, R. and Ottaway, B., Oxbow Monograph 9, Oxbow Books, Oxford, pp. 169–174.

Pollard, A.M., Batt, C.M., Stern, B. and Young, S.M.M. (2007). *Analytical Chemistry in Archaeology*. Cambridge University Press, Cambridge.

Polunin, A.V., Pchelnikov, A.P., Losev, V.V. and Marshakov, I.K. (1982). Electrochemical studies of the kinetics and mechanism of brass dezincification. *Electrochimica Acta* **27** 465–475.

Ponting, M. and Segal, I. (1998). Inductively coupled plasma–atomic emission spectroscopy analyses of Roman military copper alloy artefacts from the excavations at Masada, Israel. *Archaeometry* **40** 109–122.

Preston, K. (1980). *Roman Brass Coinage: Its Origins and the Influence of its Introduction on First Century A.D. Provincial Mints*. Unpublished M.A. dissertation, Department of Archaeological Sciences, University of Bradford.

Rollason, E.C. (1973). *Metallurgy for Engineers*. Edward Arnold, London, 4th edn.

Scott, D.A. (1991). *Metallography and Microstructure of Ancient and Historic Metals*. Getty Conservation Institute, Los Angeles.

Scott, D.A. (2002), *Copper and Bronze in Art*. Getty Conservation Institute, Los Angeles.

Sempowski, M.L., Hohe, A.W., Hancock, R.G.V., Moreau, J.-F., Kwok, F., Aufreiter, S., Karklins, K., Baart, J., Garrad, C. and Kenyon, I. (2001). Chemical analysis of 17th-century red glass trade beads from north-eastern North America and Amsterdam. *Archaeometry* **43** 503–515.

Shaw, K. (1973). *Section 34 – Barberini Collection*. Department of Astronomy, National Maritime Museum, London.

Smith, C.S., and Gnudi, M.T. (trans.) (1959). *Biringuccio, Pirotechnia*. Basic Books, New York.

Thornton, C.P., Lamberg-Karlovsky, C.C., Liezers, M. and Young, S.M.M. (2002). On pins and needles: tracing the evolution of copper-base alloying at Tepe Yahya, Iran, via ICP-MS analysis of common-place items. *Journal of Archaeological Science* **29** 1451–1460.

Trethewey, K.R. and Pinwill, I. (1987). The dezincification of free-machining brasses in seawater. *Surface and Coatings Technology* **30** 289–307.

Tylecote, R. (1976). *A History of Metallurgy*. Metals Society, London.

Walker, C., Hancock, R.G.V., Aufreiter, S., Latta, M.A. and Garrad. C. (1999). Chronological markers? Chemical analysis of copper-based trade metal artefacts from Petun sites in southern Ontario, Canada. In *Metals in Antiquity*, ed. Young, S.M.M., Pollard, A.M., Budd, P. and Ixer, R.A., BAR International Series 792, Archaeopress, Oxford, pp. 317–325.

Watson, R. (1786). *Chemical Essays Vol. 4*. Evans, London.

Werner, O. (1970). Über das vorkommen von zink und messing im altertum und im mittelalter. *Erzmetall* **23** 259–269.

Werner, O. (1977). Analysen Mittelalterlicher bronzen und messinge I. *Archäologie und Naturwissenschaften* **1** 144–220.

Werner, O. (1980). Composition of recent reproduction castings and forgeries of Mediaeval brasses and bronzes. *Berliner Beitrage zur Archaeometrie* **5** 11–35.

Werner, O. (1982). Analysen Mittelalterlicher bronzen und messinge II und III. *Archäologie und Naturwissenschaften* **2** 106–170.

Xu Li (1990). Traditional zinc-smelting technology in the Guma district of Hezhang County. In *2000 Years of Zinc and Brass*, ed. Craddock, P.T., British Museum Occasional Paper No. 50, British Museum, London, pp. 103–121.

CHAPTER 7

The Chemistry and Use of Resinous Substances

7.1 INTRODUCTION

Naturally occurring organic molecules are as important in the contemporary world as they were in the past. Although the synthetic chemical industry plays a major role through the production of a vast range of molecules tailored to particular needs, the study of natural products in the living world continues unabated. For example, chemical exploration of the constituents of plant and animal tissues has major pharmacological implications. The pharmacological activities of a single albeit diverse group of molecules known as alkaloids has long been known; '*since early times selected plant products (many containing alkaloids) have been used as poisons for hunting, murder and euthanasia; as euphoriants, psychedelics, and stimulants (e.g., morphine and cocaine); or as medicines (e.g., ephedrine, for respiratory problems)*' (Mann *et al.*, 1994; 389). Other naturally occurring plant and animal tissues can be used as adhesives, disinfectants, sealants, dyestuffs, perfumes, incenses, waterproofing agents and so on. Direct archaeological evidence for such activities in the past is difficult to obtain and many researchers rely on textual sources, the behaviour of contemporary industrial and non-industrial peoples and whatever fragmentary evidence the archaeologist can place into the equation in terms of botanical remains, material culture (through the presence of specific artefacts, and so on) and occasional chemical investigations. In contrast to the limited chemical evidence for the use of alkaloids, another class of naturally occurring chemical compounds, the terpenoids, has been found to survive in a large number of archaeological contexts from around the world. Terpenoids are the major constituents of resins. The contrast in abundance may be due to the survivability and visibility of the latter compound class in a range of burial environments, although the fact that few explicit methodologies have been applied to explore the survival of alkaloids in archaeological contexts is significant.

Archaeological Chemistry, Second Edition
By A. Mark Pollard and Carl Heron

The aim of this chapter is to review generally the potential for the study of the organic chemistry of amorphous deposits associated with artefacts (such as pottery vessels, stone tool surfaces and other materials such as basketry). Artefacts such as pottery vessels were used frequently to prepare, store, transport, serve and consume foodstuffs and other natural products in the past. Consequently, traces of these substances may survive, preserved on the surface of the vessel. Similarly, organic molecules can occlude in the permeable ceramic matrix. These chemical remnants offer valuable clues to the use of pottery and other artefacts and may provide novel identifications of food items and other organic substances in the archaeological record (see Chapter 11). One category of natural product is reviewed in this chapter: higher plant resins and related substances. The aim is to give examples of the identification of aged samples and to consider the range of uses to which these substances were put. The case study focuses on the Neolithic of northern Europe. The summary then concludes with a broader statement on the potential for study of other organic substances.

7.2 RESINS: DEFINITION AND USES

A *resin* is one of a number of natural products defined as a *plant exudate*. Other exudates include latex, gum and kino. Resins are non-cellular, water-insoluble substances and serve to protect higher plants, if wounded, from excessive water loss and the invasion of micro-organisms. Resins often comprise both volatile and non-volatile fractions. Gianno (1990; 5) distinguishes exudates from extractives (the latter include tannins, alkaloids and essential oils and require, for example, a solvent to enable their isolation from plant fluids). The derivatives produced by heating resin, as well as resinous wood, are collectively referred to as *pyroligeneous* substances. These include tar, the initial pyrolysate, and pitch, a thicker substance derived from further heating of the tar to drive off remaining volatiles. Tar and pitch are technological terms not often used by botanists and chemists (Gianno, 1990; 8).

Resin-producing trees are found over vast areas encompassing much of the torrid and temperate parts of the world. Resin preserves reasonably well and possesses a wide array of functional attributes. The properties of '*adhesiveness, insolubility in water, inflammability, healing and poisoning properties, fragrance, plasticity, vitreosity, colorability, pigment mediability, and resistance to spoilage are qualities that apply, to a greater or lesser degree, to all resins*' (Gianno, 1990; 1). As such, these natural substances have played a role in most communities. In the modern world, synthetic chemicals, including those obtained from petroleum or coal, have replaced many of the uses of natural resins.

In the past, natural resins served as universal adhesives for fixing stone points and blades to hafts of wood or antler, to glue feathers to arrow shafts and for repairing materials such as broken pottery and bone combs. As sealants, resins were used to coat the surfaces of pottery and basketry in order to provide an impermeable lining. Resins were also used to waterproof hunting equipment and fishing nets, as well as canoes and ships from Noah's Ark (Genesis 6:14) to the Mary Rose (Robinson *et al.*, 1987). A wide range of less visible uses have been documented historically and/or ethnographically. The inflammability of resins was exploited as a source of light and even liquid fire. Burning resin could be used as incense and its disinfectant properties found use in wine (notably retzina) and embalming. Resins contributed to medicinal preparations. Finally, resins were undoubtedly the earliest chewing gums: mastic from *Pistacia lentiscus*, birch bark tar, pine resins and many other resinous substances were chewed to alleviate toothache and sore throats.

7.3 CHEMISTRY OF RESINS

A comprehensive survey of the chemistry of natural resins relevant to art historical and archaeological contexts is given in Mills and White (1994; 95–128) and is not repeated here. Rather, those aspects of structure and chemistry relating to molecular transformation and to the composition of aged resins encountered in archaeological contexts will be emphasized.

Terpenes or terpenoids are distributed widely in plants from marine and terrestrial sources. The term terpene derives from 'terpen' and is attributed to Kekulé who used it to describe $C_{10}H_{16}$ hydrocarbons in turpentine oil. Terpenes designate molecules made up of isoprene units and are collectively referred to as secondary metabolites, since it has long been considered that these molecules do not play a role in primary metabolism. However, this view is being eroded as an increasing number have been found to play active roles in ecological interactions and as defence and attack chemicals (Dev, 1989; 790–791). For example, the volatile or turpentine fraction of resin contains a range of insect and microbial toxins and other agents that act to discourage insect predation (Phillips and Croteau, 1999).

Isoprene (2-methylbuta-1,3-diene [Structures 7.1a and 7.1b]) is a C_5 unit. Structure 7.1a shows the full structural formula where each line between the atoms represents two shared electrons in a covalent bond. In the case of more complex molecules, skeletal structures are used, as in Structure 1b, where carbon atoms are normally represented by an intersection of bonds. Carbon–hydrogen bonds are not shown, although all other atoms (O, N, P and so on) are indicated.

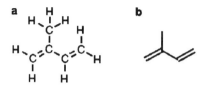

Structures 7.1a and 7.1b   2-Methylbutadiene (isoprene). a) The full structural formula, and b) the abbreviated skeletal structure

The term *terpenoid* is now the preferred generic name for this class of natural products. Many thousands of individual terpenoid molecules are known, belonging to a range of skeletal types. These can be classified in terms of the number of isoprene units. Isoprene, although a convenient model to help understand the structural chemistry of terpenoids, is a simplification of much more complex biosynthetic pathways (Phillips and Croteau, 1999). For the purposes of this chapter the following groups noted in this section are relevant [see Dev (1989) and Banthorpe (1994) for comprehensive reviews]. Trivial names for terpenoids and a great many other natural products remain much used since the systematic names are cumbersome and are seldom used unless the structures are sufficiently simple.

7.3.1 Monoterpenoids and Sesquiterpenoids

Mono- and sesquiterpenoids are the usual constituents of essential oils. Monoterpenoids are C_{10} compounds which are distributed widely in the plant kingdom. Many structures are known, including a number of oxygenated molecules with alcohol, ketone and ether groups. The actual chemical properties are determined by the nature of the functional group. The volatile fraction obtained by distilling pine resin is known as *oil of turpentine* and is composed largely of monoterpenes. Monoterpenoids and sesquiterpenoids (C_{15} compounds), although volatile, do survive over considerable timespans in favourable preservation environments. In fossil resins, such as amber, combined gas chromatography/mass spectrometry (GC-MS) has demonstrated the survival of monoterpenoids (such as camphor [Structure 7.2] and fenchone [Structure 7.3]) and sesquiterpenoids held within the natural polymer (Mills *et al.*, 1984/ 85). The survival of the sesquiterpenoids calamenene [Structure 7.4] and cadelene [Structure 7.5] has been demonstrated in the softwood (probably pine) tar samples that had poured out of transport vessels (amphoras) at the site of the wreck of a 6–7th Century BC Etruscan vessel located off the coast of the Italian island of Giglio (Robinson *et al.*, 1987). These molecules will survive if protected from oxidation within a solidified mass of resinous material.

Structure 7.2 Camphor   **Structure 7.3** Fenchone

Structure 7.4 Calamenene   **Structure 7.5** Cadelene

7.3.2 Diterpenoids

Diterpenoids (C_{20} compounds) comprise the bulk composition of resins from the families Coniferae (encompassing Pinaceae, Cupressaceae and Araucariaceae) and Leguminosae. The most abundant sources of resin in temperate regions are trees of the genus *Pinus*. Diterpenoid compounds possess mainly abietane, pimarane and labdane skeletons. In 'soft' resins (*i.e.*, those containing no polymerized structures), such as those derived from Pinaceae, abietane and pimarane compounds are predominant. The Pinaceae, and especially *Pinus*, generally have resins with a high content of abietic acid [Structure 7.6], a tricyclic molecule, and a small number of abietane isomers (Mills and White, 1994; 98–102), but resins from *Abies* and *Picea* species also contain large amounts of labdanes, as do Cupressaceae resins. Some *Pinus* resins are rich in pimarane-type acids (Fox *et al.*, 1995).

Structure 7.6   Abietic acid

The double bonds in abietane acids are conjugated and in fresh *Pinus* resins will undergo significant modification during treatment. Warming of the resin (*e.g.*, during distillation to remove oil of turpentine) induces isomerization reactions leading to a mixture enriched in abietic acid at the expense of other abietane molecules (Mills and White, 1977; 14: see Figure 7.1). The solid product remaining is referred to as *rosin* or *colophony*.

The fossil resin amber comprises a complex mixture of molecules based primarily on diterpenoid and monoterpenoid structures. Amber is a hard resin (between 2 and 2.5 on Moh's scale) and behaves largely as a polymer. This fraction is formed from the esterification of a polyvalent alcohol with a dibasic acid (Mills and White, 1994; 110). The polyvalent alcohol is the co-polymer of communol [Structure 7.7] and communic acid [Structure 7.8] (only the *cis* forms are shown), as found in kauri resin from *Agathis australis* (Araucariaceae), although additional modifications may have occurred as a result of geological conditions. The dibasic acid is succinic acid [Structure 7.9], the molecule used by Otto Helm to base his scheme for separating Baltic amber (succinite) from other fossil resins in the late 19th Century (see Chapter 1). The solvent (ether) soluble portion of amber contains hundreds of individual molecular species, many of which can be identified using GC-MS (Mills *et al.*, 1984/85). Amber occurs naturally over large areas of northern Europe, although the richest sources remain the east Baltic coast and the west Jutland peninsula.

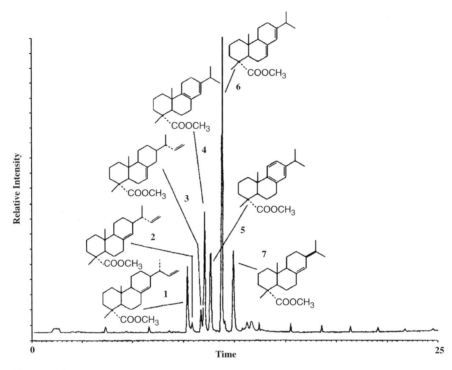

Figure 7.1 Total ion current (TIC) chromatogram obtained by GC-MS analysis of a resin (*Pinus sylvestris*). The diterpenoid resin acids were methylated (using diazomethane) to improve chromatographic performance. Peak identities: 1, Methyl pimarate; 2, Methyl sandaracopimarate; 3, Methyl isopimarate; 4, Methyl palustrate; 5, Methyl dehydroabietate; 6, Methyl abietate; 7, Methyl neoabietate. For GC-MS operating conditions, see Heron and Pollard (1988).

Structure 7.7 *cis*-Communol   **Structure 7.8** *cis*-Communic acid

Structure 7.9 Succinic acid

7.3.3 Triterpenoids

Triterpenoids (C_{30} compounds) are the most ubiquitous of the terpenoids and are found in both terrestrial and marine flora and fauna (Mahato *et al.*, 1992). Diterpenoids and triterpenoids rarely occur together in the same tissue. In higher plants, triterpenoid resins are found in '*numerous genera of broad-leaved trees, predominantly but not exclusively tropical*' (Mills and White, 1994; 105). They show considerable diversity in the carbon skeleton (both tetracyclic and pentacyclic structures are found) which occur in nature either in the free state or as glycosides, although many have either a keto or a hydroxyl group at C-3, with possible further functional groups and/or double bonds in the side-chains.

A number of important resins are composed of triterpenoids, including the dammar resins which derive from a sub-family of the family Dipterocarpaceae. Dammar resins are fluid, balsamic oleoresins highly suited for caulking and waterproofing. Frankincense (olibanum) is known as a gum-resin collected from various *Boswellia* spp. and contains amyrin epimers and triterpenoid acids. The gum component is polysaccharide in origin and is water soluble. The Anacardiaceae family contains the genus *Pistacia* (Mills and White, 1977; 21; Mills and White, 1989).

7.4 ANALYSIS OF RESINS IN ARCHAEOLOGICAL CONTEXTS

Preservation of resin is generally favoured in anaerobic or near-anaerobic locations (as in the case of waterlogged deposits, as well as in marine and lacustrine areas), under permafrost or in arid environments. These conditions provide protection against atmospheric oxidation and photoxidation and reduce the activity of micro-organisms. The extent of post-deposition chemical alteration will also be dictated by the physical state of the resin (whether it is present as a thin film or as a thick deposit of material). Chemical analysis of putative resin samples usually proceeds with the aim of identifying the botanical source, although precise species-specific identification of aged samples is problematic. To a lesser extent, geographical origin is important if it is suspected that resins may have been transported. Since resins can be observed on artefact (ceramic, stone, bone and wood) surfaces, analysis offers opportunities for assessing the ways in which artefacts were used. The analysis of ancient resins is also relevant to procurement and methods of preparation as well as in determining the range of uses that these substances were put to. Identification of ancient resin samples is not a straightforward task. Although ancient resins have been investigated for many decades, their chemical complexity has hindered confident assignments. Visual characteristics and examination of simple chemical or physical properties may offer little or no clue as to the identity of resin samples, whether ancient or modern. Consequently, chemical analysis must be performed in order to characterize which molecular species are present.

The development of sensitive and specific analytical techniques has improved significantly the opportunities for identifying organic molecules. In particular,

GC and GC-MS (see Chapter 2), are ideal for the separation and characterization of individual molecular species. Characterization generally relies on the principle of chemotaxonomy, where the presence of a specific compound or distribution of compounds in the ancient sample is matched with its presence in a contemporary authentic substance. The use of such '*molecular markers*' is not without its problems, since many compounds are widely distributed in a range of materials, and the composition of ancient samples may have been altered significantly during preparation, use and subsequent burial. Other spectroscopic techniques offer valuable complementary information. For example, infrared (IR) spectroscopy and ^{13}C nuclear magnetic resonance (NMR) spectroscopy have also been applied.

Terpenoids are susceptible to a number of alterations mediated by oxidation and reduction reactions. For example, the most abundant molecule in aged *Pinus* samples is dehydroabietic acid [Structure 7.10], a monoaromatic diterpenoid based on the abietane skeleton which occurs in fresh (bleed) resins only as a minor component. This molecule forms during the oxidative dehydrogenation of abietic acid, which predominates in rosins. Further atmospheric oxidation (autoxidation) leads to 7-oxodehydroabietic acid [Structure 7.11]. This molecule has been identified in many aged coniferous resins such as those used to line transport vessels in the Roman period (Heron and Pollard, 1988; Beck *et al.*, 1989), in thinly spread resins used in paint media (Mills and White, 1994; 172–174) and as a component of resin recovered from Egyptian mummy wrappings (Proefke and Rinehart, 1992).

Structure 7.10   Dehydroabietic acid   **Structure 7.11**   7-Oxodehydroabietic acid

Distinguishing between genera, for example, of larch, fir and pine resins should be possible even though they contain many of the same acids. In addition to abietane and pimarane acids, larch resins contain labdane alcohols. Fir resins have large amounts of a labdane alcohol (*cis*-abienol [Structure 7.12]) which, although susceptible to alteration, should give rise to recognizable oxidation products (White, 1992; 8). Given the changes in fresh resins during preparation, use and deposition, linking an aged sample to a species-specific origin may not be feasible. However, White (1992; 7) suggests that in certain cases the ratio of pimaradiene (pimarane, sandaracopimaric and isopimaric) acids might be instructive. Since *Pinus* spp. are the most abundant source of resin, the assumption is frequently made that a pine source is suspected.

When resin or resinous wood is heated strongly, significant changes in resin composition occur, leading to the formation of tar and pitch. Chemical changes include thermal dehydrogenation, decarboxylation and demethylation which give rise to a large number of potential alteration products of varying aromaticity. Stable end products of these reaction pathways include retene [Structure 7.13], a triaromatic defunctionalized diterpenoid with the formula $C_{18}H_{18}$. Intermediates include dehydroabietane, dehydroabietin, simonellite, the *nor*-abietatrienes and tetrahydroretene, although some of these molecules are present in low abundance in relatively fresh bleed resins.

Beck *et al.* (1997) used GC-MS to monitor the increase in the proportion of retene in pine tars with increasing temperature, and ^{13}C-NMR to monitor the increased aromatic signal resulting from dehydrogenation and decarboxylation reactions. These data allow approximate determinations of the production temperature of a tar from the molecular composition. Diterpenoid molecules of probable pine origin have been detected in many archaeological contexts and some detailed compositional studies have appeared (Robinson *et al.*, 1987; Heron and Pollard, 1988; Beck *et al.*, 1989; Reunanen *et al.*, 1989; Beck *et al.*, 1994).

Structure 7.12   *cis*-Abienol   **Structure 7.13**   Retene

The potential complexity of softwood tars is exemplified by the analysis of a black resinous substance which filled a ceramic container recovered from a late Roman shipwreck dating to the late 4th Century AD (Beck *et al.*, 1994). It is thought that the tar was used on board the ship as naval stores. The tar, separated into acid and neutral fractions, was analysed using GC-MS. The methylated acid fraction comprised 54 components with the unusual methyl abiet-13-en-18-oate [Structure 7.14] present in greater abundance than methyl dehydroabietate. In the unmethylated neutral fraction, comprising 61 recognizable peaks, sesquiterpenoids, diterpenoid hydrocarbons and methyl dehydroabietate were identified. Whereas compositional data may inform on the specific nature of the production process (precise source of the tar, temperature, presence of admixtures and so on), there '*remains much to be learned about the relationship between the present composition and the history, both pre- and post-depositional, of tars and pitches*' (Beck *et al.*, 1994; 119).

In another study carried out by Biers *et al.* (1994), the residual contents of intact Corinthian 'plastic' vases of the 7–6th Centuries BC were analysed non-destructively by pouring solvent into the vessels and decanting. A large number of mono-, sesqui- and diterpenoids were identified in the solvent washes. The diterpenoid, manoyl oxide [Structure 7.15] was identified in 16 vases. This molecule is found in the bark of *Pinus* and *Abies* spp. and in the essential oils of the Cupressaceae family including *Juniperus oxycedrus*.

Structure 7.14   Methyl abiet-13-en-18-oate   **Structure 7.15**   Manoyl oxide

In general, triterpenoids identified in the archaeological record have not been subjected to significant structural alteration from fresh authentic samples, although these have occurred in favourable preserving environments, such as marine contexts. The identity of resin preserved in around 100 Caananite amphoras from the near anaerobic conditions of the Late Bronze Age wreck at Ulu Burun in southern Turkey has been demonstrated using GC-MS (Mills and White, 1989). Triterpenoids congruent with the tree *P. atlantica* Desf. (which grows extensively in the eastern Mediterranean, the Near East and North Africa) were identified. A number of neutral and acidic triterpenoids were identified, including β-amyrin [Structure 7.16] and oleanonic acid (as its methyl derivative [Structure 7.17]). This resin, also known as *Chios turpentine*, was possibly used as incense. This is an important find since it represents a substantial cargo of resin (and myriad other materials) and indicates trade in resins during the Late Bronze Age in the eastern Mediterranean.

Structure 7.16   β-Amyrin   **Structure 7.17**   Oleanonic acid

Serpico (2000) has reviewed the growing chemical and other evidence of resin, as well as of amber and bitumen in ancient Egypt. Here, the results of several lines of investigation when combined indicate certain trends that can be subject to further testing and refinement in the future. One example is the relationship between the resin of *Pistacia* spp. and its use as an incense and varnish during the New Kingdom period (see also Serpico and White, 2000; Stern *et al.*, 2003). Another lies in the selection of softwood resins and tars, beeswax and other substances in Egyptian mummification practices (Buckley and Evershed, 2001).

> '*Such relationships are important to note because they can help us to formulate strategies for further research. The benefit of multiple, rather than one-off analyses is also evident. However, it is unlikely that the lines*

can be so rigidly drawn. What will probably emerge is a simple predilection for a particular choice, dependent on a range of factors such as availability, current fashion, cost, and the suitability of the product for that purpose. These connections will be much looser than a strict adherence of a specific substance for a specific usage, but will provide a better understanding of the products as commodities.' (Serpico, 2000; 468–469).

Although one-off analyses will continue to be reported, the results of these can be placed in context more easily in a wider framework. Koller *et al.* (2003) identified sesquiterpenoids and phenolics most likely obtained from cedar wood (*Cedrus* spp.) and preserved as an unused embalming material recovered adjacent to a mummy (Saankh-kare) from the 18th Dynasty (1500 BC) at Deir el-Bahari. Mathe *et al.* (2004) identified a sample of frankincense (genus *Boswellia*) from an ointment vase in the tomb of princess Sat-mer-Hout (12th Dynasty *ca.* 1897–1844 BC). The sample also contained a softwood resin and a putative vegetable oil offering insights into the kinds of organic preparations accompanying the funerary ritual. The occurrence of resins in these contexts reinforces their importance in ritualistic contexts, such as incense and anointing the dead. Finds of frankincense from Qasr Ibrîm, Egypt (AD 400–500), have also been reported (Evershed *et al.*, 1997a; van Bergen *et al.*, 1997).

The importance of certain resins led to their transport over long distances. Applications to ancient Egyptian contexts appear profitable since indigenous sources of natural resins would have been extremely limited. The most successful study to date is the import of *Pistacia* spp. resin in Late Bronze Age Canaanite jars into Egypt. This investigation has highlighted links between pistacia resin and certain jar fabrics (Serpico *et al.*, 2003). Petrological study of the fabrics has indicated the most likely sources along or near to the Levantine coast. Moreover, at Amarna, Egypt, it appears that the resin was used as an incense as it occurs in association with a large number of shallow bowls manufactured in local Nile silt clay; in some examples, clear evidence of heating is discernible (Serpico and White, 2000; 889; Stern *et al.*, 2003).

A study of 15–17th Century AD shipwrecks in south-east Asia provides another example of the trade in resins (Gianno *et al.*, 1990). Resin composed largely of triterpenoids, possibly from a tree of the family Dipterocarpaceae, was identified by Fourier transform infrared (FTIR) and GC-MS. Three other samples of resin found within jars that were still sealed have been interpreted as representing a valuable commodity which was being traded; an interpretation which is supported by their identification as being derived from a *Styrax* spp. tree, possibly *S. benzoin*. This is known from historical and ethnographic sources as a valued incense and medicinal resin, and is believed to have been traded over a wide area since the 13th Century AD. Resins of Dipterocarpaceae have been identified in much older contexts from south-east Asia. Recent GC-MS and radiocarbon dating of samples adhering to pottery sherds from Spirit Cave and Noen-U Loke in Thailand has been undertaken (Lampert *et al.*, 2002; 2003; 2004).

Terpenoids occur widely in the sedimentary record, such as deep-sea sediments, fossil resin, petroleum, coal and so on (Simoneit *et al.*, 1986). Diterpenoids serve as valuable marker compounds of terrigenous resinous plants while sesquiterpenoids can have either a higher plant or an algal origin. In sediments, organic molecules are subjected to increases in temperature and pressure and considerable structural modifications, rearrangements and polymerization reactions may occur. ten Haven *et al.* (1992) have suggested a scheme for the geological fate of terrigenous triterpenoids during the earliest stages of diagenesis. The extent of preservation of individual molecules depends on their discrete chemical structure (presence of double bonds, labile side-chains and so on). Indeed, the degree of functionality retained in terpenoids in the sedimentary record provides a means for considering the extent of diagenesis (Grimalt *et al.*, 1989). Generally, free terpenoids are converted into a similar suite of defunctionalized molecules found in wood tars. Polyterpenoids based on labdane structures are generally referred to as resinite. These 'fossilized' products of resins are widespread and abundant in the geosphere. Pyrolysis GC-MS techniques have been applied with a view to characterizing the composition of the source resin acid (Anderson and Winans, 1991). These results have permitted the classification of fossil resins into distinct classes based on molecular composition (Anderson *et al.*, 1992; Anderson and Botto, 1993).

7.5 NEOLITHIC TAR

On Thursday 19 September 1991, tourists Erika and Helmut Simon discovered a body protruding from glacial ice below the Hauslabjoch in the Tyrolean (Ötzaler) Alps between Austria and Italy. The realization that the body (later named *Ötzi*) had lain there for several thousand years took some time and a remarkable account has been written by Konrad Spindler (1993). Radiocarbon dates now available suggest the most likely date of the body to be around 3330 BC. The finds associated with the body are as astounding; in total 20 different items of equipment, including weapons, tools and containers, were recovered. One of these was an almost pure copper axe attached to a haft of yew glued and bound with leather or hide. The glue was spread over 4.3 cm, bonding the axe to the wood. Also, in a quiver, trimmed halves of feathers were found attached to arrows made from long shoots of the wayfaring tree (*Viburnum lantana*) with flint tips attached to the arrows using adhesive material (Spindler, 1993; 123–124). The analysis of the glue was carried out by a group of chemists from Vienna Technical University (Sauter *et al.*, 1992) concluding that the tar was produced by heating bark of the family Betulaceae, possibly *Betula pendula* (also referred to as *B. alba* or *B. verrucosa* in other parts of Europe) in a sealed vessel, thereby limiting the amount of oxygen.

Birch bark tar has been identified in numerous investigations of prehistoric samples, as residues of hafting on stone tools, as visible surface deposits on pottery vessels and as isolated finds, sometimes displaying clear evidence of chewing. A suite of triterpenoid biomarkers, including betulin, lupeol, lupenone and betulone, has been used to characterize birch bark tar. The preparation of bark and wood tars represents one of the earliest chemical–technological processes. Indeed, the archaeological literature devoted to prehistoric Europe contains many references to black, sticky substances and lumps of 'glue' recovered from excavations of sites from the Mesolithic onwards, but only recently has scientific analysis become common as a means of identification. These analyses have stimulated a wider evaluation of the uses of tars and of the artefacts with which these substances are associated.

The use of birch bark to produce a tar appears to have been known from the Palaeolithic. Two stone flakes partly covered in birch bark tar have been recovered from fluvial gravels and clay not far from Florence, Italy (Mazza *et al.*, 2006). The flakes were associated with bones of *Elephas antiquus* and a number of small mammals. The research team considered the deposits, on the basis of the faunal evidence and the stratigraphy, to date to before Oxygen Isotope Stage 6 (*ca.* 200 000 years ago). Solidified pieces of tar bearing imprints of the stone tools and wooden hafts have been recovered from the Palaeolithic of Germany (Mania and Toepfer, 1973; Koller *et al.*, 2001; Grünberg, 2002). One of the two finds from Königsaue (Saxony-Anhalt, Germany) is thought to have been attached to a bifacial tool – in this case, radiocarbon dates were obtained directly on the lumps [yielding dates of $43\,800 \pm 2100$ BP and $48\,400 \pm 3700$ BP – see Koller *et al.* (2001; 387–388) for further consideration of the chronological aspects of these finds]. According to Koller *et al.* (2001), one of the finds (Königsaue B) belongs to the Mousterian, whereas the other (Königsaue A) is associated with lithic artefacts of the Central European Micoquian. The key point of interest is the apparent preference for the deliberate manufacture of birch bark tar – such material cannot be collected or harvested directly from a tree (unlike, for example, pine resin). Birch bark must be collected and then subjected to heating in the absence of oxygen to temperatures of around 350°C – how this was actually achieved is not yet known.

The apparent deliberate selection and preparation of certain resources over others (birch bark tar over softwood products) is repeated throughout later European prehistory. Whilst we lack systematic comparative surveys of the physical properties of resin, heated wood and bark products, and bitumen, both choice and preference were being expressed. Yet, even if we had data of this nature, other factors are likely to have come into play. The ability to transform natural materials (such as wood or bark) into discrete organic substances then subject to myriad uses (hafting of tools and weapons just happens to be the most visibly persistent role) would have had a dramatic impact on those who made these substances.

This evidence is part of a growing record of the hafting of lithic tools into wooden or bone handles carried out by Neanderthals and early modern humans, although evaluating the importance of hafting in the two groups remains difficult. Surviving biomolecular traces of hafting combined with evidence of polish resulting from contact with a hafting adhesive has been suggested on European Mousterian tools as well as from sites in the Near East and South Africa (Churchill, 2001). Hardy *et al.* (2001) have used use-wear to identify hafted tools on Palaeolithic sites apparently made by Neanderthal populations in Crimea (Ukraine). Possible differences in the nature of hafting type and its frequency between Neanderthal and early modern human populations has been suggested by Churchill (2001).

Boëda *et al.* (1996) identified bitumen on a flint scraper and a Levallois flake, discovered in Mousterian levels (about 40 000 BP) at the site of Umm el Tlel in Syria. The occurrence of polyaromatic hydrocarbons such as fluoranthene, pyrene, phenanthrenes and chrysenes suggested that the raw bitumen had been subjected to high temperature. The distribution of the sterane and terpane biomarkers in the bitumen did not correspond to the well-known bitumen occurrences in these areas. In other studies of bitumen associated with a wide variety of artefacts of later date, especially from the 6th Millennium BC onwards, molecular and isotopic methods have proved successful in recognizing different sources of bitumen enabling trade routes to be determined through time (Connan *et al.*, 1992; Connan and Deschesne, 1996; Connan, 1999; Harrell and Lewan, 2002).

The preparation of bark and wood tars represents one of the earliest technological processes. The evidence suggests that the preparation and use of birch bark tar in particular is a pan-European phenomenon of considerable antiquity. In recent years, samples have been identified on artefacts recovered from sites in a number of countries including Finland, Sweden, France, Germany, Austria, Italy and Slovenia, and even finds of birch bark tar on Neolithic pottery from northern Greece (Urem-Kotsou *et al.*, 2002). Whilst Palaeolithic finds are rare, many more examples are known from the Mesolithic (e.g., Aveling and Heron, 1998; 1999). Birch bark tar has also been identified in later (Roman and Medieval) periods. In certain regions of Europe, birch bark tar survived as an important natural product until the onset of the synthetic chemical industry, and can still be found today in commodities such as birch bark tar soap. The specific origin of the tar from bark is, from chemical evidence alone, an assumption since detailed compositional studies of birch wood and its ability to produce tar with the same properties as that obtained from the bark have yet to appear in print. However, the assumption that bark itself is a rich source of 'tar-forming compounds' is backed up by the ethno-historical record and experimental simulations.

Birch is widespread in the cooler regions of Europe. Although the wood was not valued as timber and was rarely used for building, the bark was much exploited (Vogt, 1949). It could be detached readily in thin layers and was easy to work and sew. Its properties as a waterproofing and insulating material ensured widespread use in roofing and flooring, for making containers which were efficient for food storage and preparation, by both indirect and direct heating methods (Vogt, 1949; Dimbleby, 1978; 36) and even for making canoes and tents. Two birch bark containers were found in Ötzi's tool kit, although one was inadvertently destroyed soon after the body was found.

7.5.1 The Chemistry of Birch Bark and Birch Bark Tars

The bark of *B. pendula*, especially the outer part, contains a high level of triterpenoid compounds, in particular betulin or betulinol (lup-20(29)-ene-3ß,28-diol [Structure 7.18]; Hayek *et al.*, 1989; Ekman, 1983; Ukkonen and Erä, 1979; O'Connell *et al.*, 1988; Pokhilo *et al.*, 1990; Reunanen *et al.*, 1993; Binder *et al.*, 1990), which contributes more than 70% of the total triterpenoids. Betulin is a white crystalline compound which gives silver birch bark its characteristic coloration. High proportions ($>25\%$ dry weight of the solvent soluble fraction) of betulin are known to occur in the outer bark of *B. pendula*. The other triterpenoid components include lupeol (lup-20(29)-en-3ß-ol [Structure 7.19]), lupenone (lup-20(29)-en-3-one [Structure 7.20]) and occasionally betulinic acid (lup-20(29)-en-3ß-ol-28-oic acid [Structure 7.21]). The bark of young trees contains much larger quantities of terpenoids than that of older trees (Ukkonen and Erä, 1979; 217), which appears to confirm the statement by Rajewski (1970) that historically only the bark of young trees was selected for the preparation of tar. Triterpenoids in black birch (*B. lenta*) are much reduced in comparison. In one study, the total triterpenoids amounted to only 0.8% dry weight of the bark with lupeol as the most abundant of the terpenoids (Cole *et al.*, 1991).

Structure 7.18
Betulin (lup-20(29)-ene-3β,28-diol)

Structure 7.19
Lupeol (lup-20(29)-en-3β-ol)

Structure 7.20
Lupenone (lup-20(29)-en-3-one)

Structure 7.21
Betulinic acid (lup-20(29)-en-3β-ol-28-oic acid)

In addition to triterpenoids in the bark of Betula spp., suberin (a biopolyester comprising primarily hydroxy, epoxy and dicarboxylic acids) is also present, as demonstrated by GC-MS of modern birch bark extracts following alkaline hydrolysis (Ekman, 1983). Suberin monomers include 9,10-epoxy-18-hydroxyoctadecanoic acid [Structure 7.22] and ω-hydroxyacids (e.g., 22-hydroxydocosanoic acid [Structure 7.23]). Suberin is quite resistant to biodegradation by fungi, but, unless protected, extracellular enzymes of fungi will gradually degrade it.

Structure 7.22 9,10-Epoxy-18-hydroxyoctadecanoic acid

Structure 7.23 22-Hydroxydocosanoic acid

The connection of hafting adhesives with birch bark subjected to a process of 'distillation' has been suggested for a long time [see Weiner (1999) for a comprehensive review of the early chemical literature]. For example, Grewingk (1882; 25) and Lidén (1938; 25–28 and 1942; 99–100) considered birch bark to be the source of prehistoric tar all quoted in Clark, 1975; 210. Experimental studies have demonstrated that dry distillation of birch bark in a sealed vessel readily produces a viscous tar at temperatures between 250 and 350 °C (Charters *et al.*, 1993). According to these early investigations birch bark tar could be modified through the addition of beeswax and animal fat in order to lower its melting point. Grewingk also suggested possible admixtures of pine resin and iron oxide. The chemist, Berlin, proposed that archaeological samples were birch bark tar mixed with amber (Ruthenberg, 1997). Very little information is given regarding the criteria used to characterize these materials. Ruthenberg (1997) considers that smell during combustion ('*Juchtengeruch*') and certain distillation properties formed the basis for differentiation. Consequently, these results must be considered with caution and, at best, should serve only as possible guides towards identification.

Spectroscopic and chromatographic investigations of ancient resinous substances began during the 1960s. In 1965, Sandermann, using IR spectroscopy, concluded that birch bark derivatives were present on potsherds of Neolithic/Bronze Age date from Germany. Similarly, an adhesive used to mend Iron Age pottery was ascribed a birch origin by analysis of carbon, hydrogen and nitrogen elemental ratios and IR spectroscopy (Sauter, 1967). Using thin layer chromatography (TLC), a birch bark origin has been attributed, on the basis of betulin detection, to adhesives used to haft flint implements (Funke, 1969) and to lumps of tar with human tooth impressions (Rottländer, 1981). It should be emphasized that detections of betulin using TLC alone are questionable because very many triterpenoid molecules exhibit similar R_f values (Hayek *et al.*, 1989; 2229). For detailed compositional information GC and GC-MS are the required techniques. IR and NMR spectroscopy were used to identify the origin of resinous material on lithics and ceramics from northern Yugoslavia (Hadzi and Orel, 1978). The ancient samples resembled spectra from freshly prepared birch bark tar.

Analysis of tars of archaeological origin by combined GC-MS (Hayek *et al.*, 1990; 1991; Heron *et al.*, 1991; Reunanen *et al.*, 1993; Charters *et al.*, 1993) has demonstrated that betulin is retained as a major component in those ascribed a birch origin (Figure 7.2). This is to be expected, given its high thermal stability (Hayek *et al.*, 1989). Betulin seems to be highly resistant to fungal attack as is indicated by the exceptional preservation of materials fabricated from birch bark, although biodegradation products of betulin have been reported in the natural products literature (*e.g.*, Fuchino *et al.*, 1994).

Hayck *et al.* (1990) reported the identification of 14 samples of birch bark tar of Chalcolithic to Early Iron Age date from sites in Austria and Denmark. In this study, authentic tars from bark samples of a number of different species (*e.g.*, birch, oak, alder, hazel, elm and so on) were produced under laboratory conditions (using Kugelrohr distillation) and compared with the aged samples

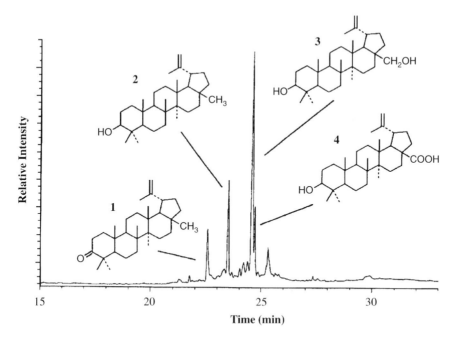

Figure 7.2 Partial capillary gas chromatogram displaying the principal lipid solvent soluble constituents of a black tarry deposit on the surface of a Neolithic potsherd from Ergolding Fischergasse, Bavaria, Germany. The sample was trimethylsilylated prior to GC (for further details see Heron *et al.*, 1997). Peak identities (confirmed by GC-MS analysis): 1, lupenone; 2, lupeol; 3, betulin; 4, betulinic acid.

using GC-MS. Principal components analysis was used to compare aged and modern samples. In each case, the compositional data supported a birch bark origin. However, it was also found that the bark and tar of many species (including beech) contain triterpenoids such as betulin, although relative abundances are not given. More recently, pyrolysis GC has been shown to be a rapid method for discriminating wood and bark tars from hardwood and softwood sources (Puchinger *et al.*, 2007).

The chemical composition of birch bark tar is dependent on the temperature at which tar is produced. In producing simulated tars in the laboratory for comparison with an adhesive used to repair a Roman jar from Stanwick, Charters *et al.* (1993) found that tars prepared at 350 °C displayed an increase in triterpenoid hydrocarbons as well as unresolved components presumably resulting from pyrolysis, although the precise nature of these molecules has not been elucidated. Binder *et al.* (1990) and Charters *et al.* (1993) also report the presence of allobetul-2-ene [Structure 7.24] in aged birch bark tars. Since this molecule has not been reported in extracts from fresh birch bark, it could be formed during heating to produce the tar (Regert *et al.*, 2003).

Structure 7.24   Allobetul-2-ene

Analysis of a large number of amorphous deposits surviving on artefact surfaces at the Neolithic waterlogged settlement at Ergolding Fischergasse, Germany (mid-4th Millennium BC; Ottaway, 1995) has confirmed the abundance of birch bark tar (Heron, 1995; Heron *et al.*, 1997). Although the majority of finds sampled are from potsherds, similar deposits have been found on other artefacts from the site, including stone implements (sickle blades and whetstones) and worked bone. Evaluation of the position of the deposits on the potsherds suggested the most common application of the tar was as a layer on the inner surface, but layers on the outside wall are also found, including an unusual example where a smoothed layer appears to have been applied deliberately to the outer surface of the base of a pot. Interior linings may have served to seal the permeable fabric, allowing storage of liquids, although the pottery was well fired and of apparent low porosity. Alternatively, Schiffer (1990) has shown that interior resin-lined pottery possesses excellent heating effectiveness, so these linings could represent use for direct boiling containers. Ethnographic evidence exists for the application of resin to the outer surfaces of pots (*e.g.*, Foster, 1956), which may have improved the transmission of heat. In truth, a great diversity of organic substances have been used to coat and seal pottery (*e.g.*, Rice, 1987; 163–164). For example, traditional potters of sub-Saharan Africa coat their pots with aqueous extracts of stem bark from a number of plants, including *Bridelia furruginea*; a phytochemical study has investigated the procyanidins thought to be important in the process of providing an impermeable layer (Diallo *et al.*, 1995). Tar may also have been applied in order to facilitate handling when the vessel was full and heavy, as is suggested for the characteristic rough exteriors to the vessels from Ergolding Fischergasse.

Deposits on three sherds of pottery from Ergolding Fischergasse were identified as softwood tars. Resin acids found in abundance in bleed resins are absent, although this is not unexpected in strongly heated samples. A high proportion of dehydroabietic acid and defunctionalized diterpenoids, such as retene, suggest that these samples are wood tars. Their composition is very similar to those reported for the putative pine tars by Robinson *et al.* (1987) used on the Mary Rose. In one study, Beck and Borromeo (1990) have suggested that the presence of methyl esters in a tar indicates that it is produced from the wood since methanol in the resinous wood reacts with the free resin acids during heating to form the methyl derivative. Tars formed by heating resin alone (obtained by tapping) comprise resin acids but no esters. Of course, if methylation is the chosen method for derivatization prior to GC separation then it is not possible to distinguish whether there are any methyl esters originally present in the tar. However, by separating the soluble fraction into acid and neutral constituents, the presence of methyl esters in the non-methylated neutral fraction can be used to assess the production method. Alternatively, if the samples are trimethylsilylated then no fractionation is required.

Research into samples of Mesolithic date, particularly those in Scandinavia, reveals the same preponderance of birch bark tar. Tar may not have only been used in hafting (Aveling and Heron, 1998) as some isolated lumps display clear evidence of human tooth impressions [see Aveling and Heron (1999), for consideration of the reasons]. Nine lumps of tar with human tooth impressions have been found at the Neolithic lake dwelling at Hornstaad-Hornle I, in southern Germany (Rottländer, 1981; Schlichtherle and Wahlster, 1986; 92) and others are known from Mesolithic bog sites in Scandinavia (Larsson, 1983; 75–76). It is plausible that birch bark tar served as a mild stimulant. A more prosaic interpretation is that chewing the tar rendered it more ductile for use.

Regert (2004) has shown that it is possible to piece together findings from a wide variety of sites based on relatively small numbers to construct a broader history of use of these natural resources. In this study, 90 samples from eight sites in France, covering material from the Neolithic, Bronze Age and Iron Age, were analysed. The samples included hafting adhesives, deposits associated with pottery vessels, a bronze sword and isolated lumps of organic matter. The Neolithic samples were dominated by birch bark tar, although in one case Pinaceae (possibly from *Pinus* spp.) diterpenoids were also identified in a mixture with birch bark tar. In addition, wood or bark from other trees yet to be fully characterized were also used to produce adhesives. A sample from the Late Bronze Age (La Fangade, Hérault, France) proved to be derived from *Pinus*.

At Grand Aunay (Sarthe, France), eight brown–black lumps and 16 visible surface deposits on pottery vessels dating to the 3rd to 1st Centuries BC were identified as birch bark tar (Regert *et al.*, 2003). Regert *et al.* (2003) categorized the distribution of MD (*marquers degradation*) with that of the original biomarkers. These molecules (including lupa-2,20(29)-diene and lupa-2,20(29)-dien-28-ol) are familiar products of defunctionalization associated with heating. This allowed further insight into the nature and preparation of the tars from the source bark. According to the authors, the production of birch bark tar was not standardized – if this were the case, one might expect a much more consistent molecular composition. One sample from a pottery vessel was a mixture of birch bark tar and beeswax interpreted by the authors as an adhesive – the absence of *n*-alkanes in the beeswax was considered a loss resulting from heating. The funnel shape and perforation in the base of one of the vessels led the authors to consider that at least some of the vessels were used to produce birch bark tar [see Kurzweil and Todtenhaupt (1990) for images of vessels used in producing birch bark tar]. The appearance of other surface coatings suggested the use of birch bark tar as an interior sealant or lining, either for functional or decorative purposes. The authors also consider the possibility that the beeswax or the tar–wax mixture may have been used in metalworking (in the lost wax production process) or in fixing composite artefacts. Chemical evidence of birch bark tar in later periods is sparse. Charters *et al.* (1993) and Dudd and Evershed (1999) have identified birch bark tar of Romano-British date used to repair a broken pottery vessel and as a tar–animal fat mixture in a small, enamelled vessel from Roman Catterick (Yorkshire, UK).

Whereas the geochemical sourcing of restricted numbers of bitumen deposits has been applied to archaeological samples for some time (*e.g.*, Connan *et al.*, 1992), it is understandable that evaluating the geographical origin of tars from birch trees is likely to be highly problematic. Stern *et al.* (2006) have evaluated bulk isotope measurements (δD, $\delta^{13}C$ and $\delta^{18}O$) undertaken on archaeological samples of birch bark tar over a wide geographical (*e.g.*, Greece to Norway) and chronological (9500–3000 BP) range. The $\delta^{13}C$ values in samples from Greece were less depleted than those from northern Europe, indicating that in extreme cases birch bark tar from different regions could be distinguished from those produced locally. Long-distance movement of birch bark tar is unlikely, although the possibility of birch bark tar applied to artefacts that were transported over long distances is, perhaps, the more likely scenario.

7.5.2 The Production and Uses of Neolithic Tars

The precise methods for the preparation of tar in prehistory are not well understood. Archaeological and documentary evidence exists for sophisticated and large-scale methods of manufacture in the Medieval period (Kurzweil and Todtenhaupt, 1991). A number of simpler methods involving heating in sealed pottery vessels have been proposed and tested (Kurzweil and Todtenhaupt, 1990). However, these processes would leave little trace in the archaeological record. Evidence for the use of tars pre-dates the introduction of pottery by many millennia, so questions remain as to the full range of production processes. Experiments simulating the preparation of tar without the use of pottery vessels have yet to meet with complete success (Czarnowski and Neubauer, 1990). As a footnote to the ubiquitous production and use of birch bark tar in prehistoric Europe the subject is not complete without reference to its familiar smell. In an investigation of the physiological response to various odours (birch tar, galbanum, heliotropine, jasmine, lavender, lemon and peppermint) as measured by electroencephalography, birch tar was amongst the most consistently unpleasant smells recorded (Klemm *et al.*, 1992).

In the Neolithic of northern Europe, birch bark and other tars served many of the functions noted above. As adhesives, tars could be used to mend broken combs or pottery vessels (Sauter, 1967; Abb 3), to reattach handles and to fix lithic implements to their hafts (Funke, 1969; Larsson, 1983). Some pottery vessels display possible adhesive around the rim, plausibly to fix a skin or lid over the top of the pot. In some areas of Switzerland and southern Germany, tar was used to attach cut-out decorations of birch bark to the outside of pots (Vogt, 1949). Sauter *et al.* (2002) identified birch bark tar in shallow incised grooves on the hips and stomach of the Venus from Brunn am Gebirge, an Early Neolithic ceramic figurine from Austria dating to the 6th Millennium BC. The finding of birch bark tar in the grooves suggests that an additional, perhaps perishable, organic material was glued to the figurine.

Lucquin *et al.* (2007) identified birch bark tar in two 'coupes-à-socle' or 'footed cups' from the Neolithic passage grave at La Hougue Bie (Jersey, UK). These vessels have been associated with ceremonial activities at Neolithic ritual monuments, and Andrew Sherratt (amongst others) has speculated about their possible use as braziers for burning psychoactive substances (Sherratt, 1991). Lucquin and co-workers, whilst acknowledging the pungent smell of tar mentioned above, note that the burning of tar at burial monuments could have helped to mask the smell of decomposition and to cleanse and purify. It also points to the wider symbolic role of products of this nature, in addition to the functional roles already discussed.

Further uses of tars are documented much later in the Medieval period. Tar was also painted over doors or beds, where it was thought to ward off evil spirits, perhaps due to its pungent smell. Rajewski (1970) observes that references to tar can be encountered in many proverbs of at least Medieval date in both rural and urban populations. This remarkable ethnohistoric record has its roots in much earlier populations of northern Europe.

7.5.3 Alternatives to Birch Bark and Softwood Tar

Whereas the range of uses to which resin exudates and tar–pitch extractives has been emphasized, some of these functions could have been plausibly served by a range of other natural products. For example, knowledge of other prehistoric adhesives could be biased by the poor survival of protein-based (such as fish, bone and horn) glues. However, it should be remembered that these substances are water-soluble and in many cases would not have been particularly useful. A single burnt residue on a potsherd from the Neolithic site at Ergolding Fischergasse, Germany, has been identified as beeswax (Heron *et al.*, 1994). GC analysis showed the presence of wax esters in the range of 40–50 carbon atoms. The partial high-temperature gas chromatograms displayed in Figure 7.3

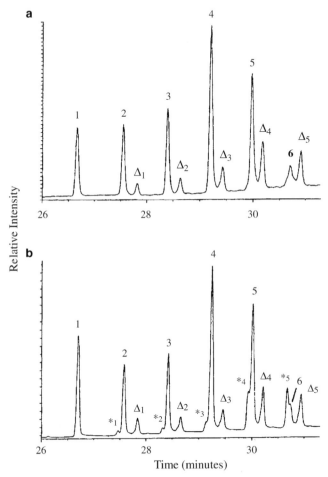

Figure 7.3   Partial gas chromatograms comparing the wax ester distribution in (a) Neolithic sample from Ergolding Fischergasse with (b) authentic beeswax (*Apis mellifera*). Peak identities: Peaks 1–6 are wax esters in the range C_{40} (peak 1) to C_{50}, (peak 6) comprising hexadecanoic (palmitic) acid esterified with alcohols of increasing chain length (C_{24} to C_{34}). Peaks Δ_1 to Δ_5 may represent co-elution of two molecular species one of which is a series of hydroxyesters. Peaks $*_1$ to $*_5$ are present only in the authentic sample and represent wax esters comprising an unsaturated (octadecenoyl) fatty acid moiety. Their absence in the ancient sample is not unexpected given the susceptibility of the double bond to oxidation or reduction reactions. For further details see Heron *et al.* (1994). (Reproduced by permission of Springer-Verlag.)

compare the wax ester distribution in authentic beeswax (*Apis mellifera*) with the lipid extract of the surface deposit coating the Neolithic sherd. Confirmation of the peak identities was provided by GC-MS. The principal wax esters comprise even-carbon number aliphatic chains of saturated alcohols and fatty carboxylic acids with total carbon numbers in the range from C_{40} to C_{52} with the C_{46} wax ester (triacontyl hexadecanoate [Structure 7.25]) the most abundant. Beeswax is produced by various *Apis* spp. The European honeybee (*A. mellifera*) produces honeycombs of almost pure wax. The principal constituents are 14% odd-carbon number *n*-alkanes with a range of 21–33 carbon atoms (with heptacosane $C_{27}H_{56}$ as the most abundant alkane), 35% C_{40}–C_{52} monoesters, 14% C_{56}–C_{66} diesters, 3% triesters, 4% C_{24}–C_{34} hydroxymonoesters, 8% hydroxypolyesters, 12% free acids, 1% C_{16}–C_{20} acid monoesters, 2% acid polyesters and 7% unidentified material (Schulten *et al.*, 1987). In the Neolithic sample, the *n*-alkane fraction was severely depleted, suggesting their combustion when the beeswax was burned. The sealing and waterproofing properties of beeswax would have made it a useful commodity in prehistoric Europe. The presence of wax on the inner surface of the potsherd suggests that a thin layer of hot wax may have sealed the permeable fabric of the vessel, enabling the storage of liquids. Alternatively, beeswax could have been stored in the vessel, awaiting use in a number of activities. The early occurrence of beeswax also implies the availability of honey to Neolithic communities in Europe. Other finds of beeswax of prehistoric and later date have been reported (Evershed *et al.*, 1997b; Regert *et al.*, 2001; Colombini *et al.*, 2003; Evershed *et al.*, 2003; Roumpou *et al.*, 2003). However, interpretation is not as straightforward as it might seem. For example, the identification of beeswax in association with a pottery sherd could suggest its application as a sealant, as a residue of honey storage or as a vessel used to prepare candles, cosmetics, adhesives and so on. Clearly, other evidence must come into play, such as the vessel form and volume, evidence of use alteration and the context of recovery (Charters *et al.*, 1995; Evershed *et al.*, 2003; Regert *et al.*, 2001). The identification of a bituminous substance as an adhesive to fix an Early Bronze Age knife handle to a double T-shaped handle (from Xanten-Wardt, Germany; Koller and Baumer, 1993) also suggests occasional use of other substances.

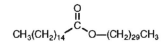

Structure 7.25   Triacontyl hexadecanoate

7.6   SUMMARY: EVIDENCE FOR OTHER ORGANIC SUBSTANCES

Aside from the characterization of birch bark tar, higher plant resins and their heated derivatives and beeswax, what potential remains for the identification of other organic substances used in prehistoric Europe? This short summary intends to whet the appetite by briefly reviewing recent investigations and considering possible future avenues for research. The association between pottery vessels and organic substances has been emphasized. Determining how pottery was actually used in the past is a key area for current and future research (Rice, 1987; 207–243). Archaeologists commonly use the form of reconstructed vessels to infer possible use. For example, in the Early Neolithic of southern England, Thomas (1991; 89–90) uses size and shape criteria for distinguishing vessels used primarily for consumption, perhaps associated with feasting activity, and those used for storage and other domestic activities, and notes the predominance of either one or the other in assemblages at certain site types. Such suggestions could benefit from chemical investigations aimed at defining pottery use [see Heron and Evershed (1993) for a review]. However, as Hodder (1990; 204) has lamented, few systematic studies have been carried out on Neolithic pottery and this point has been reiterated in Vencl's review (1994) of the possible identification of liquids, such as milk (see Chapter 11), mead, beer and wine in prehistoric Europe.

The situation is not irretrievable. Until recently, and unless substantial liquid–solid remains of wine are present, as is the case in one or two very rare examples of stoppered Roman amphoras and Medieval glass bottles, claims for the detection of wine associated with pottery vessels have been based on a combination of a tartaric acid 'spot test', FTIR and high performance or high pressure liquid chromatography (HPLC; *e.g.*, McGovern, 2003; Boulton and Heron, 2000). More recently, LC-MS-MS analysis of samples of pottery vessels, some inscribed with wine as the contents from well-dated tombs, including that of Tutankhamun (1332–1322 BC) in the Valley of Kings in Egypt, has been undertaken (Guasch-Jané *et al.*, 2004; 2006a, 2006b). Tartaric acid and syringic acid could be identified. The latter compound is released from the anthocyanin, malvidin-3-glucoside, which provides the colour in red wine; if absent, the authors believed the wine to be white. In another study Garnier *et al.* (2003) have explored thermally assisted hydrolysis and methylation of the products of polyphenols, which are then analysed by GC-MS. Analysis of ancient grape pits and wine residues demonstrated a complex series of phenolic components in the samples, enabling confirmation of the presence of wine in amphoras of Roman date.

Andrew Sherratt (1991) explored the use of specific pottery vessel types in later Neolithic Europe for burning or infusing psychoactive substances, in particular opium (*Papaver somniferum*) and cannabis (*Cannabis sativa*). The proposition is supported with botanical evidence from a number of sites. According to Sherratt (1991; 52), '*Any account of prehistoric Europe which omits a consideration of such substances is likely to be incomplete.*' Specific chemical evidence from the examples presented is lacking, although recognition of these substances in later contexts has been proposed. There is unambiguous archaeological evidence for burning cannabis seeds in the Iron Age of the Steppe region. The remarkable finds from the frozen tombs of Pazyryk (4th Century BC) in southern Siberia include paraphernalia for inhaling cannabis vapour – a fur bag filled with cannabis seed, a censer filled with stones and a hexapod frame of an inhalation tent. Possible medicinal use of cannabis has been reported from the Middle East (Zias *et al.*, 1993). The remains of a female aged about 14 years were found in a family burial tomb near Jerusalem. Coins recovered from the tomb suggest it was in use during the 4th Century AD. In the pelvic area of the girl lay the skeletal remains of a 40 week foetus. It was suggested that the girl may have died during pregnancy. In the abdominal area of the skeleton, nearly 7 g of a grey, carbonized material was recovered. GC-MS and NMR spectroscopy confirmed preliminary suggestions that the substance was cannabis. The principal constituent was identified as Δ^{6-}tetrahydrocannabinol (THC [Structure 7.26]). Although a minor constituent of cannabis, its presence indicated the conversion of the major constituents such as Δ^{9-}THC and cannabidiol into Δ^{6-}THC, a conversion apparently known to occur during the burning process. The burnt residue was assumed to be cannabis administered to the girl as an inhalant to facilitate the birth process by alleviating pain.

Structure 7.26 Δ^6-Tetrahydrocannabinol (THC)

Possible opium use has been associated with Bronze Age Cypriot Base-Ring I juglets, which resemble in shape an upturned opium-poppy capsule's head and may have been used to hold opium for export to Egypt where these vessels are also found (Merrillees, 1962; Knapp, 1991; 25–26, Merrillees with Evans, 1989). Although Bisset *et al.* (1994) could find no trace of morphine and hence opium in the tomb of the ancient Egyptian chief royal architect Kha, a later study by the same group (Bisset *et al.*, 1996) claimed opium in a Base-Ring 1 juglet from Cyprus.

In other areas of the world, a number of recent investigations underline the potential for providing new evidence for the early use of such substances. For example, analysis of snuff samples from the grave of a mummified body in northern Chile dating to around the end of the 8th Century AD identified the psychoactive alkaloids dimethyltryptamine [Structure 7.27], 5-methoxydime-thyltryptamine [Structure 7.28] and 5-hydroxy-*N,N*-dimethyltryptamine (bufotenine; [Structure 7.29]) suggesting the source of the snuff was a species of the genus *Anadenanthera* (Torres *et al.*, 1991). The snuff was part of kit which comprised wooden snuff trays, a snuffing tube, usually of wood or bone, a spoon or spatula and a mortar and pestle. Cocaine metabolites have been detected in samples of human hair from Chile using radioimmunoassay (Cartmell *et al.*, 1991), although a more recent GC-MS study, also under-taken on Chilean mummy samples, proved to be less successful (Báez *et al.*, 2000). Elsewhere, the alkaloids caffeine [Structure 7.30] and theobromine

[Structure 7.31] have been identified using liquid chromatographic techniques in Mayan pottery vessels used to consume cocoa (Hurst *et al.*, 1989; 2002) and lactones congruent with the intoxicating drink kava (prepared from the roots of *Piper methysticum*) have been identified in residues adhering to archaeological ceramics from Fiji (Hocart *et al.*, 1993). A wide range of molecules were characterized, including 7,8-dihydrokawain [Structure 7.32].

Structure 7.27
Dimethyltryptamine

Structure 7.28
5-Methoxydimethyltryptamine

Structure 7.29
5-Hydroxy-*N,N*-dimethyltryptamine

Structure 7.30 Caffeine

Structure 7.31 Theobromine

Structure 7.32 7,8-Dihydrokawain

Clearly there is considerable further potential for using archaeological chemistry to identify a wide range of naturally occurring organic molecules used in the past. This will undoubtedly contribute substantially to our understanding of the relationship between human populations and their use of plant and animal resources, as well as in the determination of the myriad ways in which artefacts were used.

REFERENCES

Anderson, K.B. and Botto, R.E. (1993). The nature and fate of natural resins in the geosphere III: re-evaluation of the structure and composition of highgate copalite and glessite. *Organic Geochemistry* **20** 1027–1038.

Anderson, K.B. and Winans, R.E. (1991). Nature and fate of natural resins in the geosphere I: evaluation of pyrolysis–gas chromatography/mass spectrometry for the analysis of natural resins and resinites. *Analytical Chemistry* **63** 2901–2908.

Anderson, K.B., Winans, R.E. and Botto, R.E. (1992). The nature and fate of natural resins in the geosphere II: identification, classification and nomenclature of resinites. *Organic Geochemistry* **18** 829–841.

Aveling, E.M. and Heron, C. (1998). The chemistry of birch bark tars at Mesolithic Star Carr. *Ancient Biomolecules* **2** 69–80.

Aveling, E.M. and Heron, C. (1999). Chewing tar in the early Holocene: an archaeological and ethnographic evaluation. *Antiquity* **73** 579–584.

Báez, H., Castro, M.M., Benaventeb, M.A., Kintze, P., Cirimelee, V., Camargo, C. and Thomas, C. (2000). Drugs in prehistory: chemical analysis of ancient human hair. *Forensic Science International* **108** 173–179.

Banthorpe, D.V. (1994). Terpenoids. In *Natural Products: Their Chemistry and Biological Significance*, ed. Mann, J., Davidson, R.S., Hobbs, J.B., Banthorpe, D.V. and Harborne, J.B., Longman, London, pp. 289–359.

Beck, C.W. and Borromeo, C. (1990). Ancient pine pitch: technological perspectives from a Hellenistic shipwreck. *MASCA Research Papers in Science and Archaeology* **7** 51–58.

Beck, C.W., Smart, C.J. and Ossenkop, D.J. (1989). Residues and linings in ancient Mediterranean transport amphoras. In *Archaeological Chemistry IV*, ed. Allen, R.O., Advances in Chemistry Series 220, American Chemical Society, Washington, pp. 369–380.

Beck, C.W., Stewart, D.R. and Stout, E.C. (1994). Analysis of naval stores from the Late Roman ship. In *Deep Water Archaeology: a Late Roman Ship from Carthage and an Ancient Trade Route near Skerki Bank off Northwest Sicily*, ed. McCann, A.M. and Freed, J., Journal of Roman Archaeology, Supplementary Series No. 13, Michigan, Ann Arbor, pp. 109–121.

Beck, C.W., Stout, E.C. and Janné, P.A. (1997). The pyrotechnology of pine tar and pitch inferred from qualitative analyses by GC-MS and carbon-13 NMR spectrometry. In *Proceedings of the First International Symposium on Wood Tar and Pitch, Biskupin, Poland*, ed. Brzezinski, W. and Piotrowski, W., State Archaeological Museum in Warsaw, Warsaw, pp. 181–192.

Biers, W.R., Gerhardt, K.O. and Braniff, R.A. (1994). *Lost Scents: Investigations of Corinthian 'Plastic' Vases by Gas Chromatography–Mass Spectrometry*. MASCA Research Papers in Science and Archaeology Volume 11,

University of Pennsylvania Museum of Archaeology and Anthropology, Philadelphia, Pennsylvania.

Binder, D., Bourgeois, G., Benoist, F. and Vitry, C. (1990). Identification de brai de bouleau (*Betula*) dans le Néolithique de Giribaldi (Nice, France) par la spectrométrie de masse. *Revue d'Archéométrie* **14** 37–42.

Bisset, N.G., Bruhn, J.G., Curtos, S., Holmstedt, B., Nyman, U. and Zenk, M.H. (1994). Was opium known in 18th dynasty? An examination of materials from the tomb of the chief royal architect Kha. *Journal of Ethnopharmacology* **41** 99–114.

Bisset, N.G., Bruhn, J.G. and Zenk, M.H. (1996). The presence of opium in a 3,500 year old Cypriote base ring juglet. *Egypt and the Levant* **6** 203.

Boëda, E., Connan, J., Dessort, D., Muhesan, S., Mercier, N., Valladas, H. and Tisnérat, N. (1996). Bitumen as a hafting material on Middle Palaeolithic artefacts. *Nature* **380** 336–338.

Boulton, N. and Heron, C. (2000). Chemical detection of ancient wine. In *Ancient Egyptian Materials and Technologies*, ed. Nicholson, P.T. and Shaw, I., Cambridge University Press, Cambridge, pp. 399–403.

Buckley, S.A. and Evershed, R.P. (2001). Organic chemistry of embalming agents in Pharaonic and Graeco-Roman mummies. *Nature* **413** 837–841.

Cartmell, L.W., Aufderheide, A.C., Springfield, A., Weems, C. and Arriaza, B. (1991). The frequency and antiquity of Prehistoric coca-leaf-chewing practices in Northern Chile: radioimmunoassay of cocaine metabolite in human mummy hair. *Latin American Antiquity* **2** 260–268.

Charters, S., Evershed, R.P., Goad, L.J., Heron, C. and Blinkhorn, P. (1993). Identification of an adhesive used to repair a Roman jar. *Archaeometry* **35** 91–101.

Charters, S., Evershed, R.P., Blinkhorn, P. and Denham, V. (1995). Evidence for the mixing of fats and waxes in archaeological ceramics. *Archaeometry* **37** 113–127.

Clark, J.G.D. (1975). *The Earlier Stone Age Settlement of Scandinavia*. Cambridge University Press, Cambridge.

Cole, B.J.W., Bentley, M.D. and Hua, Y. (1991). Triterpenoid extractives in the outer bark of *Betula lenta* (black birch). *Holzforschung* **45** 265–268.

Churchill, S.E. (2001). Hand morphology, manipulation, and tool use in Neandertals and early modern humans of the Near East. *Proceedings of the National Academy of Sciences USA* **98** 2953–2955.

Colombini, M.P., Giachi, G., Modugno, F., Pallecchi, P. and Ribechini, E. (2003). The characterization of paints and waterproofing materials from the shipwrecks found at the archaeological site of the Etruscan and Roman harbour of Pisa (Italy). *Archaeometry* **45** 659–674.

Connan, J. (1999). Use and trade of bitumen in antiquity and prehistory: molecular archaeology reveals secrets of past civilizations. *Philosophical Transactions of the Royal Society of London* **354** 33–50.

Connan, J, and Deschesne, O. (1996). *Le Bitume à Suse: Collection du Musée du Louvre*. Département des Antiquités Orientales, Musée du Louvre, Paris.

Connan, J., Nissenbaum, A. and Dessort, D. (1992). Molecular archaeology: export of Dead Sea asphalt to Canaan and Egypt in the Chalcolithic–Early Bronze Age (4th–3rd Millennium BC). *Geochimica et Cosmochimica Acta* **56** 2743–2759.

Czarnowski, E. and Neubauer, D. (1990). Aspekte zur produktion und verarbeitung von birkenpech. *Acta Praehistorica et Archaeologica* **23** 11–14.

Dev, S. (1989). Terpenoids. In *Natural Products of Woody Plants II: Chemicals Extraneous to the Lignocellulosic Cell Wall*, ed. Rowe, J.W., Springer-Verlag, Berlin, pp. 691–807.

Diallo, B., Vanhaelen, M. and Gosselain, O.P. (1995). Plant constituents involved in coating practices among traditional African potters. *Experientia* **51** 95–97.

Dimbleby, G.W. (1978). *Plants and Archaeology*. John Baker, London, 2nd edn.

Dudd, S.N. and Evershed, R.P. (1999). Unusual triterpenoid fatty acyl ester components of archaeological birch bark tars. *Tetrahedron Letters* **40** 359–362.

Ekman, R. (1983). The suberin monomers and triterpenoids from the outer bark of *Betula verrucosa* Ehrh. *Holzforschung* **37** 205–211.

Evershed, R.P., van Bergen, P.F., Peakman, T.M., Leigh-Firbank, E.C., Horton, M.C., Edwards, D., Biddle, M., Kjølbye-Biddle, B. and Rowley-Conwy, P.A. (1997a). Archaeological frankincense. *Nature* 390 667–668.

Evershed, R.P., Vaughan, S.J., Dudd, S.N. and Soles, J.S. (1997b). Fuel for thought? Beeswax in lamps and conical cups from Late Minoan Crete. *Antiquity* **71** 979–985.

Evershed, R.P., Dudd, S.N., Anderson-Stojanovic, V.R. and Gebhard, E.R. (2003). New chemical evidence for the use of Combed Ware pottery vessels as beehives in Ancient Greece. *Journal of Archaeological Science* **30** 1–12.

Foster, G.M. (1956). Resin-coated pottery in the Philippines. *American Anthropologist* **58** 732–733.

Fox, A.F, Heron, C. and Sutton, M.Q. (1995). Characterisation of natural products on Native American archaeological and ethnographic materials from the Great Basin region, USA. *Archaeometry* **37** 363–375.

Fuchino, H., Sou, K., Imai, H., Wada, H. and Tanaka, N. (1994). A biodegradation product of betulin. *Chemical and Pharmaceutical Bulletin* **42** 379–381.

Funke, H. (1969). *Chemische-analytische Untersuchungen verschiedener archäologische Funde*. Dissertation Hamburg.

Garnier, N., Richardin, P., Cheynier, V. and Regert, M. (2003). Characterization of thermally assisted hydrolysis and methylation products of polyphenols from modern and archaeological vine derivatives using gas chromatography–mass spectrometry. *Analytica Chimica Acta* **493** 137–157.

Gianno, R. (1990). *Semelai Culture and Resin Technology*. Memoirs of the Connecticut Academy of Arts and Sciences, Volume XXII, Newhaven, Connecticut.

Gianno, R., Erhardt, D., von Endt, D.W., Hopwood, W. and Baker, M.T. (1990). Archaeological resins from shipwrecks off the coasts of Saipan and

Thailand. In *Organic Contents of Ancient Vessels: Materials Analysis and Archaeological Investigation*, ed. Biers, W.R. and McGovern, P.E., MASCA Research Papers in Science and Archaeology, Volume 7, The University Museum of Archaeology and Anthropology, University of Pennsylvania, Philadelphia, pp. 59–67.

Grimalt, J.O., Simoneit, B.R.T. and Hatcher, P.G. (1989). Chemical affinities between the solvent extractable and bulk organic matter of fossil resin associated with an extinct podocarpaceae. *Phytochemistry* **28** 1167–1171.

Grünberg, J.M. (2002). Middle Palaeolithic birch-bark pitch. *Antiquity* **76** 15–16.

Guasch-Jané, M.R., Ibern-Gómez, M., Andrés-Lacueva, C., Jáuregui, O. and Lamuela-Raventós, R.M. (2004). Liquid chromatography with mass spectrometry in tandem mode applied for the identification of wine markers in residues from Ancient Egyptian vessels. *Analytical Chemistry* **76** 1672–1677.

Guasch-Jané, M.R., Andrés-Lacueva, C., Jáuregui, O. and Lamuela-Raventós, R.M. (2006a). The origin of the ancient Egyptian drink *Shedeh* revealed using LC/MS/MS. *Journal of Archaeological Science* **33** 98–101.

Guasch-Jané, M.R., Andrés-Lacueva, C., Jáuregui, O. and Lamuela-Raventós, R.M. (2006b). First evidence of white wine in ancient Egypt from Tutankhamun's tomb. *Journal of Archaeological Science* **33** 1075–1080.

Hadzi, D. and Orel, B. (1978). Spektrometricne raziskave izvora jantarja in smol iz prazgodovinskih najdisc na Slovenskem. *Vestnik Sloveneskega Kemitskega Drustva* **25** 51–62.

Hardy, B.L., Kay, M., Marks, A.E. and Monigal, K. (2001). Stone tool function at the paleolithic sites of Starosele and Buran Kaya III, Crimea: behavioral implications. *Proceedings of the National Academy of Sciences USA* **98** 10972–10977.

Harrell, J.A. and Lewan, M.D. (2002). Sources of mummy bitumen in ancient Egypt and Palestine. *Archaeometry* **44** 285–293.

Hayek, E.W.H., Jordis, U., Moche, W. and Sauter, F. (1989). A bicentennial of betulin. *Phytochemistry* **28** 2229–2242.

Hayek, E.W.H., Krenmayr, P. Lohninger, H., Jordis, U., Moche, W. and Sauter, F. (1990). Identification of archaeological and recent wood tar pitches using gas chromatography/mass spectrometry and pattern recognition. *Analytical Chemistry* **62** 2038–2043.

Hayek, E.W.H., Krenmayr, P. Lohninger, H., Jordis, U., Sauter, F. and Moche, W. (1991). GC/MS and chemometrics in archaeometry; investigation of glue on copper age arrowheads. *Fresenius Journal of Analytical Chemistry* **340** 153–156.

Heron, C. (1995). Analysen organischer rückstände an Altheimer Scherben. In *Ergolding Fischergasse: Eine Feuchtbodensiedlung der Altheimer Kultur in Niederbayern*, ed. Ottaway, B.S., Materialhefte zur bayerischen vorgeschichte A68, Verlag Michael Laßleben, Kallmünz, pp. 206–209.

Heron, C. and Evershed, R.P. (1993). The analysis of organic residues and the study of pottery use. In *Archaeological Method and Theory V*, ed. Schiffer, M.B., University of Arizona Press, Tucson, pp. 247–286.

Heron, C. and Pollard, A.M. (1988). The analysis of natural resinous materials from Roman amphoras. In *Science and Archaeology, Glasgow 1987*, ed. Slater, E.A. and Tate, J.O., British Series 196, British Archaeological Reports, Oxford, pp. 429–447.

Heron, C., Evershed, R.P., Chapman, B.C.G. and Pollard, A.M. (1991). Glue, disinfectant and 'chewing gum' in prehistory. In *Archaeological Sciences, Bradford 1989*, ed. Budd, P., Chapman, B.C.G., Jackson, C., Janaway, R.C. and Ottaway, B.S., Oxbow Publications, Oxford, pp. 325–331.

Heron, C., Nemcek, N., Bonfield, K.M., Dixon, D. and Ottaway, B.S. (1994). The chemistry of Neolithic beeswax. *Naturwissenschaften* **81** 266–269.

Heron, C., Bonfield, K.M. and Nemcek, N. (1997). Wood tar in the Neolithic of Europe. *Proceedings of the International Symposium on Wood Tar and Pitch, Biskupin, Poland*, ed. Brzezinski, W. and Piotrowski, W., State Archaeological Museum in Warsaw, Warsaw, pp. 203–211.

Hocart, C.H., Fankhauser, B. and Buckle, D.W. (1993). Chemical archaeology of kava, a potent brew. *Rapid Communications in Mass Spectrometry* **7** 219–224.

Hodder, I. (1990). *The Domestication of Europe*. Blackwell, Oxford.

Hurst, W.J., Martin, R.A., Tarka, S.M., and Hall, G.D. (1989). Authentication of cocoa in Maya vessels using high performance liquid chromatographic techniques. *Journal of Chromatography* **466** 279–289.

Hurst, W.J., Tarka, S.M., Powis, T.G., Valdez, F. and Hester, T.R. (2002). Cacao usage by the earliest Maya civilization. *Nature* **418** 289–290.

Klemm, W.R., Lutes, S.D., Hendrix, D.V. and Warrenburg, S. (1992). Topographical EEG maps of human response to odors. *Chemical Senses* **17** 347–361.

Knapp, A.B. (1991). Spice, drugs, grain and grog: organic goods in Bronze Age East Mediterranean trade. In *Bronze Age Trade in the Mediterranean*, ed. Gale, N.H., Paul Åström's Förlag, Jønsered, pp. 21–68.

Koller, J. and Baumer, U. (1993). Analyse einer Kittprobe aus dem Griff des Messers von Xanten-Wardt. In Koschik, H., Messer aus dem Kies: Zu zwei Messern der jüngeren Bronzezeit aus dem Rhein bei Xanten-Wardt und aus der Weser bei Petershagen-Hävern. *Acta Praehistorica et Archaeologica* **25** 117–131.

Koller, J., Baumer, U. and Mania, D. (2001). High-tech in the middle Palaeolithic: Neandertal-manufactured pitch identified. *European Journal of Archaeology* **4** 385–397.

Koller, J., Baumer, U., Kaup, Y., Schmid, M. and Weser, U. (2003). Analysis of a pharonic embalming tar. *Nature* **425** 784.

Kurzweil, A. and Todtenhaupt, D. (1990). *Teer aus Holz*. Museumsdorf Düppel, Berlin.

Kurzweil, A. and Todtenhaupt, D. (1991). Technologie der Holzteergewinnung. *Acta Praehistorica et Archaeologica* **23** 63–79.

Lampert, C.D., Glover, I.C., Heron, C.P., Stern, B., Shoocongdej, R. and Thompson, J.B. (2002). The characterization and radiocarbon dating of archaeological resins from Southeast Asia. In *Archaeological*

Chemistry – Materials, Methods, and Meaning, ed. Jakes, K.A., Symposium series No. 831, American Chemical Society, Washington D.C., pp. 84–109.

Lampert, C.D., Glover, I.C., Hedges, R.E.M., Heron, C.P., Higham, T.F.G., Stern, B., Shoocongdej, R. and Thompson, G.B. (2003). Dating resin coating on pottery: the Spirit Cave early ceramic dates revised. *Antiquity* **77** 126–133.

Lampert, C.D., Glover, I.C., Hedges, R.E.M., Heron, C.P., Higham, T.F.G., Stern, B., Shoocongdej, R. and Thompson, G.B. (2004). AMS radiocarbon dating of resins on ceramics: results from Spirit Cave, Thailand. In *Radiocarbon and Archaeology: Fourth International Symposium*, ed. Higham, T., Bronk Ramsey, C. and Owen, C., Oxford University School of Archaeology Monograph, Oxford, pp. 161–169.

Larsson, L. (1983). Ageröd V: an Atlantic bog site in central Scania. *Acta Archaeologica Lundensia 12*, Lund, Sweden.

Lucquin, A., March, R.J. and Cassen, S. (2007). Analysis of adhering organic residues of two 'coupes-à-socles' from the Neolithic funerary site 'La Hougue Bie' in Jersey: evidences of birch bark tar utilisation. *Journal of Archaeological Science* **34** 704–710.

McGovern, P.E. (2003). *Ancient Wine: The Scientific Search for the Origins of Viniculture*. Princeton University Press, Princeton.

Mahato, S.B., Ashoke, K.N. and Roy, G. (1992). Triterpenoids. *Phytochemistry* **31** 2199–2249.

Mania, D. and Toepfer, V. (1973). *Königsaue: Gliederung, Oekologie und Mittelpaläolithische Funde der Letzten Eiszeit*. VEB Deutscher, Berlin.

Mann, J., Davidson, R.S., Hobbs, J.B., Banthorpe, D.V. and Harborne, J.B. (ed.) (1994). *Natural Products: Their Chemistry and Biological Significance*. Longman, London.

Mathe, C., Culioli, G., Archier, P. and Vieillescazes, C. (2004). Characterization of archaeological frankincense by gas chromatography–mass spectrometry. *Journal of Chromatography A* **1023** 277–285.

Mazza, P.P.A., Martini, F., Sala, B., Magi, M., Colombini, M.P., Giachi, G., Landucci, F., Lemorini, C., Modugno, F. and Ribechini, E. (2006). A new Palaeolithic discovery: tar-hafted stone tools in a European Mid-Pleistocene bone-bearing bed. *Journal of Archaeological Science* **33** 1310–1318.

Merrillees, R.S. (1962). Opium trade in the Bronze Age Levant. *Antiquity* **36** 287–292.

Merrillees, R.S. with Evans, J. (1989). Highs and lows in the Holy Land: opium in Biblical times. In *Yigael Yadin Memorial Volume*, ed. Ben-Tor, A., Greenfield, J. and Malamat, A., Eretz Israel 20, Israel Exploration Society, Hebrew University, Jerusalem, pp. 148–154.

Mills, J.S. and White, R. (1977). Natural resins of art and archaeology. *Studies in Conservation* **22** 12–31.

Mills, J.S. and White, R. (1989). The identity of resins from the Late Bronze Age shipwreck at Ulu Burun (Kaş). *Archaeometry* **31** 37–44.

Mills, J.S. and White, R. (1994). *The Organic Chemistry of Museum Objects*. Butterworths, London, 2nd edn.

Mills, J.S., White, R. and Gough, L.J. (1984/85). The chemical composition of Baltic amber. *Chemistry and Geology* **47** 15–39.

O'Connell, M.M., Bentley, M.D., Campbell, C.S. and Cole, B.J.W. (1988). Betulin and lupeol in bark from four white-barked birches. *Phytochemistry* **27** 2175–2176.

Ottaway, B.S. (1995). *Ergolding Fischergasse: Eine Feuchtbodensiedlung der Altheimer Kultur in Niederbayern.* Verlag Michael LaBleben, Kallmünz.

Phillips, M.A. and Croteau, R.E. (1999). Resin-based defenses in conifers. *Trends in Plant Science Reviews* **4** 184–190.

Pokhilo, N.D., Makhenev, A.K., Demenkova, L.I. and Uvarova, N.I. (1990). Composition of the triterpene fraction of outer bark extracts of *Betula pendula* and *Betula pubescens* (in Russian). *Khimiya Drevesiny* **6** 74–77.

Proefke, M.L. and Rinehart, K.L. (1992). Analysis of an Egyptian mummy resin by mass spectrometry. *Journal of the American Mass Spectrometry Society* **3** 582–589.

Puchinger, L., Sauter, F., Leder, S. and Varmuza, K. (2007). Studies in organic archaeometry VII: differentiation of wood and bark pitches by pyrolysis capillary gas chromatography (Py-CGC). *Annali di Chimica* **97** 513–525.

Rajewski, Z. (1970). Pech und teer bei den Slawen. *Zeitschrift für Archäologie* **4** 46–53.

Regert, M. (2004). Investigating the history of prehistoric glues by gas chromatography–mass spectrometry. *Journal of Separation Science* **27** 244–254.

Regert, M., Colinart, S., Degrand, L. and Decavallas, O. (2001). Chemical alteration and use of beeswax through time: accelerated ageing tests and analysis of archaeological samples from various environmental contexts. *Archaeometry* **43** 549–569.

Regert, M., Vacher, S., Moulherat, C. and Decavallas, O. (2003). Adhesive production and pottery function during the Iron Age at the site of Grand Aunay (Sarthe, France). *Archaeometry* **45** 101–120.

Reunanen, M., Ekman, R. and Heinonen, M. (1989). Analysis of Finnish pine tar and tar from the wreck of frigate St. Nikolai. *Holzforschung* **43** 33–39.

Reunanen, M., Holmbom, B. and Edgren, T. (1993). Analysis of archaeological birch bark pitches. *Holzforschung* **47** 175–177.

Rice, P.M. (1987). *Pottery Analysis: A Sourcebook.* University of Chicago Press, Chicago.

Robinson, N., Evershed, R.P., Higgs, W.J., Jerman, K. and Eglinton, G. (1987). Proof of a pine wood origin for pitch from Tudor (Mary Rose) and Etruscan shipwrecks: application of analytical organic chemistry in archaeology. *Analyst* **112** 637–644.

Roumpou, M., Heron, C., Andreou, S. and Kotsakis, K. (2003). Organic residues in storage vessels from the Toumba Thessalonikis. In *Prehistoric Pottery: People, Pattern and Purpose*, ed. Gibson, A.M., International Series 1156, British Archaeological Reports, Oxford, pp. 189–199.

Rottländer, R.C.A. (1981). A Neolithic 'chewing gum'. Poster presentation, 21st Archaeometry Symposium, Brookhaven, USA (abstract only).

Ruthenberg, K. (1997). Historical development and comparison of analytical methods for the identification of tars and pitches. In *Proceedings of the First International Symposium on Wood Tar and Pitch, Biskupin, Poland*, ed. Brzezinski, W. and Piotrowski, W., State Archaeological Museum in Warsaw, Warsaw, pp. 173–179.

Sandermann, W. (1965). Untersuchung vorgeschichtlicher, Gräberharze' und Kitte. *Technische Beiträge zur Archäologie* **2** 58–73.

Sauter, F. (1967). Chemische untersuchung von, Harzüberzügen' auf hallstattzeitlicher keramik. *Archaeologia Austriaca* **41** 25–36.

Sauter, F., Jordis, U. and Hayek, E. (1992). Chemsiche untersunchungen der Kittschäftungs-materialien. In *Der Mann im Eis, Band 1, Bericht über das Internationale Symposium 1992 in Innsbruck*, ed. Höpfel, F., Platzer, W. and Spindler, K., Eigenverlag der Universität Innsbruck, Innsbruck, pp. 435–441.

Sauter, F., Varmuza, K., Werther, W. and Stadler, P. (2002). Studies in organic archaeometry V: chemical analysis of organic material found in traces on a Neolithic terracotta idol statuette excavated in lower Austria. ARKIVOC 2002 http://www.arkat-usa.org/ark/journal/2002/General/1–343E/343E.pdf. Accessed 20-7-2007.

Schiffer, M.B. (1990). The influence of surface treatment on heating effectiveness of ceramic vessels. *Journal of Archaeological Science* **17** 373–381.

Schlichtherle, H. and Wahlster, B. (1986). *Archäologie in Seen und Mooren – Den Pfahlbauten auf der Spur*. Konrad Theiss Verlag, Stuttgart.

Schulten, H.-R., Murray, K.E. and Simmleit, N. (1987). Natural waxes investigated by soft ionization mass spectrometry. *Zeitschrift für Naturforschung C: Biosciences* **42** 178–190.

Serpico, M. (2000). Resins, amber and bitumen. In *Ancient Egyptian Materials and Technology*, ed. Nicholson, P.T. and Shaw, I., Cambridge University Press, Cambridge, pp. 430–474.

Serpico, M. and White, R. (2000). The botanical identity and transport of incense during the Egyptian New Kingdom. *Antiquity* 74 884–897.

Serpico, M., Bourriau, J., Smith, L., Goren, Y, Stern, B. and Heron, C. (2003). Commodities and containers: a project to study Canaanite amphorae imported into Egypt during the New Kingdom. In *The Synchronisation of Civilisations in the Eastern Mediterranean in the Second Millennium BC II*, ed. Bietak, M., Proceedings of the SCIEM2000 Euro-Conference, Haindorf, pp. 365–375.

Sherratt, A. (1991). Sacred and profane substances: the ritual use of narcotics in Later Neolithic Europe. In *Sacred and Profane*, ed. Garwood, P., Jennings, D., Skeates, R. and Toms, J., Oxford Committee for Archaeology, Oxford, pp. 50–64.

Simoneit, B.R.T., Grimalt, J.O., Wang, T.G., Cox, R.E., Hatcher, P.G. and Nissenbaum, A. (1986). Cyclic terpenoids of contemporary resinous plant detritus and of fossil woods, ambers and coals. *Organic Geochemistry* **10** 877–889.

Spindler, K. (1993). *The Man in the Ice*. Weidenfeld and Nicholson, London.

Stern, B., Heron, C., Corr, L., Serpico, M. and Bourriau, J. (2003). Compositional variation in aged and heated pistacia resin found in Late Bronze Age Canaanite amphorae and bowls from Amarna, Egypt. *Archaeometry* **45** 457–469.

Stern, B., Clelland, S.J., Nordby, C.C. and Urem-Kotsou, D. (2006). Bulk stable light isotopic ratios in archaeological birch bark tars. *Applied Geochemistry* **21** 1668–1673.

ten Haven, T.L., Peakman, T.M. and Rullkötter, J. (1992). 2-Triterpenes: early intermediates in the diagenesis of terrigenous triterpenoids. *Geochimica et Cosmochimica Acta* **56** 1993–2000.

Thomas, J. (1991). *Rethinking the Neolithic*. Cambridge University Press, Cambridge.

Torres, C.M., Repke, D.B., Chan, K., McKenna, A.L. and Schultes, R.E. (1991). Snuff powders from Pre-Hispanic San Pedro de Atacama: chemical and contextual analysis. *Current Anthropology* **32** 640–649.

Ukkonen, K and Erä, V. (1979). Birch-bark extractives. *Kemia-Kemi* **6** 217–220.

Urem-Kotsou, D., Stern, B., Heron, C. and Kotsakis, K. (2002). Birch-bark tar at Neolithic Makriyalos, Greece. *Antiquity* **76** 962–967.

van Bergen, P.F., Peakman, T.M., Leigh-Firbank, E.C. and Evershed, R.P. (1997). Chemical evidence for archaeological frankincense: boswellic acids and their derivatives in solvent soluble and insoluble fractions of resin-like materials. *Tetrahedron Letters* **38** 8409–8412.

Vencl, S. (1994). The archaeology of thirst. *Journal of European Archaeology* **2** 299–326.

Vogt, E. (1949). The birch as a source of raw material during the Stone Age. *Proceedings of the Prehistoric Society* **5** 50–51.

Weiner, J. (1999). European Pre- and Protohistoric tar and pitch: a contribution to the history of research 1720–1999. *Acta Archaeometrica* **1** 1–109.

White, R. (1992). A brief introduction to the chemistry of natural products in archaeology. In *Organic Residues in Archaeology: Their Analysis and Identification*, ed. White, R. and Page, H., UK Institute for Conservation Archaeology Section, London, pp. 5–10.

Zias, J., Stark, H., Seligman, J., Levy, R., Werker, E., Breuer, A. and Mechoulam, R. (1993). Early medical use of cannabis. *Nature* **363** 215.

CHAPTER 8

Amino Acid Stereochemistry and the First Americans

8.1 INTRODUCTION

In 1989, D.J. Meltzer published an article in the journal *American Antiquity* entitled '*Why don't we know when the first people came to North America?*' (Meltzer, 1989). He noted that, despite over 100 years of intense debate, the question had failed to be resolved. There are two polarized positions which are so far apart that the only point of agreement is that humans first entered North America from Siberia. One school maintains that this took place at the end of the most recent glaciation [sometime between 14 000 and 12 000 uncalibrated radiocarbon (^{14}C) years before present – signified as 'BP' or 'radiocarbon years BP', as opposed to 'cal. BP' for calibrated dates], and that the vast continent was empty (in human terms) at this time. The second school, citing a number of sites with radiocarbon dates which appear to signify human habitation before this time, insist that the colonization took place some time before 15 000 BP, possibly as early as 35 000 BP. The 'late' (or 'short-chronology') school reject the various sites cited as early on grounds such as poor dating or poor evidence of human occupation. They point to the apparent explosion of human activity starting some time between 11 500 and 11 000 radiocarbon years BP, character-ized by stone tools known as Clovis fluted projectile points, and interpret this as evidence for the spread of a vigorous population of efficient large mammal hunters – the 'Clovis folk' – who spread from North to South America in a few hundred years, and gave rise to 'the Clovis horizon' in American prehistory (Fagan, 1991; Dillehay and Meltzer, 1991). This short-chronology model is also viewed favourably by those palaeontologists seeking a simple explanation for the sudden demise of the American megafauna (mammoths, camelids, sloths, *etc.*) at approximately the same time, on the grounds that human 'overkill' is supported by historical observations of similar extinction events as humans have colonized unpopulated islands [*e.g.*, the extinction of the New Zealand

Archaeological Chemistry, Second Edition
By A. Mark Pollard and Carl Heron
© The Royal Society of Chemistry 2008

moa after the arrival of the Maori about 1000 years ago (Martin, 1984)]. Supporters of the 'long' chronology dispute this 'blitzkrieg' model of human expansion and megafaunal extinction, and point to a handful of sites scattered throughout the continent which they say yield substantial evidence for pre-Clovis occupation – sites referred to by J.M. Adovasio (one of the major supporters of an early date for the migration) as '*the ones that will not go away*' (Adovasio, 1993).

Much of the debate centres around the exact dating of these archaeological sites, and the quality of the association between the evidence for human occupation and the material dated (Haynes, 1992). The backbone for the chronology of all late Quaternary geology and archaeology (back to around 40–50 000 BP) is radiocarbon dating (Bowman, 1990). In the early days (up to the 1970s), radiocarbon required relatively large samples (either of charcoal, wood, or bone – in the latter case, up to 100 g was required) from which to purify the carbon and measure the residual activity of the ^{14}C atoms. Developments in nuclear accelerator physics during the 1970s resulted in the availability of an alternative mass spectrometric technique for radiocarbon dating, known as *accelerator mass spectrometry* (AMS; Aitken, 1990). One of the major advantages of AMS ^{14}C dating is that the size of the sample is reduced from hundreds of grams to a few milligrams, making it possible to date unique objects (such as the Shroud of Turin), as well as human skeletal remains, without their total destruction. This has the twofold advantage that rare human remains can be dated directly (using the older methods, such material was often dated by contextual association with something less valuable, such as charcoal), and that the chemical purification process can be carried out much more rigorously, to reduce the chances of error due to contamination.

Before AMS ^{14}C dating in the 1980s, there was always some room to doubt the quality of radiocarbon dates produced on human bone, either on the grounds of poor association between the bone and the material actually dated (if this was not bone), or, if the bone itself was used, because of the actual procedure employed. In the early stages of the debate (before AMS dating in the 1980s), many of the dates were produced from the 'whole bone' (mineral plus organic phase), which may include secondary carbonates from the mineral fraction of the bone, which are therefore liable to be heavily contaminated by circulating groundwaters during burial. The announcement, therefore, in the early 1970s of another technique, apparently independent of radiocarbon, which dated bone directly using minute samples, was greeted with great enthusiasm by archaeologists, especially in North America, since it seemed that the long-running colonization debate could be solved once and for all by dating directly a number of the putatively early Paleoindian bones. This technique is known as *amino acid racemization* (AAR) dating. It relies on the fact that the amino acids which constitute bone protein are formed *in vivo* as the left-handed (L) enantiomer (see below), but, after death, racemize slowly to produce measurable quantities of the right handed (D) form (Masters, 1986a). One particular molecule – aspartic acid – was singled out for extensive study, and apparently yielded conclusive evidence of the antiquity of humans in North America when it was finally announced that some of the Californian Paleoindian bones were as old as 70 000 years BP (Bada *et al.*, 1974). For a moment, it seemed as if the debate had been settled decisively in favour of the longer chronology.

The sceptics were, however, still unconvinced – partly because the skeletons that were given such old dates didn't *look* that old. Morphologically, they were deemed to be identical to modern Native American skeletons, and, by analogy with Europe, skeletons which were that old should show some archaic traits. In Europe, skulls dating before about 40 000 BP show features such as brow ridges, which are associated with Neanderthal populations. This debate coincided with the first archaeological ^{14}C measurements using the new accelerator technique, and it was subsequently confirmed that an error had been made, and that none of the bones were as old as had been suggested by AAR. In fact, none appeared to be older than mid-Holocene (*i.e.*, around 5–6000 BP; Taylor *et al.*, 1985). As a result of this, doubt was inevitably cast on the validity of the AAR method, but subsequent investigations showed that this was not the prime cause of the problem. Because racemization is a chemical reaction, the rate of which depends on the ambient temperature, it had been decided to determine the racemization reaction constant using the measured D/L aspartic acid ratio in a bone of known date – known, that is, from radiocarbon dating. It turns out that this calibration date was in serious error, and, if the racemization measurements are re-calculated using the corrected radiocarbon date, the ages predicted by AAR are reasonably in line with those produced by other techniques (Bada, 1985). Nevertheless, this cameo within the wider debate on the origins of North American Paleoindians has had a lasting effect on the reputation of the applications of amino acid stereochemistry in archaeology. As a technique to date bones from the past 45 000 years at least, it has largely been superseded by AMS ^{14}C dating, which offers a more established technique on samples of similar size. The measurement of enantiomeric ratios of various amino acids has, however, been considered as an indicator of age at death in forensic and archaeological investigations, since some tissues, exceptionally, racemize *in vivo* (Gillard *et al.*, 1990). Measurement of AAR in the protein fraction of mollusc shells is also widely used to give stratigraphic relationships and dating evidence in quaternary geochronology (Bada, 1991). There is still considerable scientific interest in the phenomenon of racemization of amino acids in proteins contained in calcified tissue, and how this process is affected by interaction with a burial environment over geological time periods. In this chapter we give a brief introduction to the structure of mammalian bone protein, and to the process of AAR. The factors surrounding the Californian Paleoindian controversy are reviewed and brought up to date, followed by a summary of some of the current uses of amino acid stereochemistry in archaeological and forensic science.

8.2 THE STRUCTURE OF BONE

Bone is a natural composite material, with an average composition for dry compact human bone of 70% (by weight) insoluble inorganic matter, 20%

organic matter and 10% water (*i.e.*, water lost below 105 °C). It has a structure consisting of a fibrous organic matrix (collagen) which contains within it the finely crystalline inorganic mineral phase. The principal components of the inorganic phase are calcium and phosphate ions. It used to be thought of as poorly mineralized calcium hydroxyapatite $[Ca_{10}(PO_4)_6(OH)_2]$, but is best described as *dahllite*, a carbonate hydroxyapatite mineral with an approximate stoichiometry $(Ca,X)_{10}(PO_4,CO_2)_6(O,OH)_{26}$, with $CO_2 > 1\%$ and $F < 1\%$ by weight (McConnell, 1973; 88), where $X = $ Na, Mg, K, Sr, *etc.* (McConnell, 1960; 214). Work using transmission electron microscopy has revealed that the mineral takes the form of thin plates or tablets, with a length and breadth of a few hundred Ångstrom units but a thickness of only 20–30 Å (Weiner and Traub, 1992). (The Ångstrom is a non-SI unit corresponding to 10^{-10} m, or $10\text{ Å} = 1$ nm, but it is still widely used in mineral chemistry. The 'correct' units are nm (10^{-9} m), giving a thickness of 2–3 nm.) These mineral plates are embedded in the collagen fibres (see below) in a highly ordered way. As a living tissue, human bone is constantly being destroyed and reformed and contains a number of specialized cells responsible for the production of bone tissue (osteoblasts), the maintenance of tissue (osteocytes) and the removal of bone (osteoclasts). Two types of bone structure are distinguished – *cancellous* or spongy bone and compact or *cortical* bone (see Figure 10.1 in Chapter 10). Cancellous bone is characterized by a porous structure consisting of a network of trabeculae. The distribution of cortical and cancellous bone throughout the skeleton is governed largely by biomechanical considerations, which require the unique combination of lightness and strength that is a characteristic of bone (McLean and Urist, 1986).

Approximately 90% (by weight) of the organic fraction in bone is made up of a fibrous structural protein of the *collagen* family. The remaining 10% are either other proteins (collectively termed 'non-collagenous proteins' – ncp) or various lipids (fats). Collagens of several types are widely distributed throughout the connective tissue of the body, and are largely responsible for the strength and elasticity of such tissues. Proteins are biopolymers consisting of one or more chains of amino acids, linked together by peptide bonds. Amino acids are a family of organic compounds with the general formula:

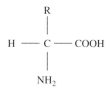

where R is one of a number of organic groups, termed *radicals* – the simplest being the hydrogen atom (H), making the amino acid glycine (gly) which is the smallest member of the amino acid family. The most important aspect of the chemistry of the amino acids as a family is that they can link together to form long chains via the peptide bond. A peptide bond is formed when the amine

(NH$_2$) radical of one amino acid links to the acid (COOH) radical of the next, with the elimination of a water molecule:

Peptide Bond

$$\underset{\text{Amino Acid 1}}{HOOC-\underset{R_1}{\overset{H}{\underset{|}{\overset{|}{C}}}}-NH_2} + \underset{\text{Amino Acid 2}}{HOOC-\underset{R_2}{\overset{H}{\underset{|}{\overset{|}{C}}}}-NH_2} = \underset{\text{Dipeptide}}{HOOC-\underset{R_1}{\overset{H}{\underset{|}{\overset{|}{C}}}}-\underset{H}{\overset{|}{\underset{|}{N}}}-\overset{O}{\overset{||}{C}}-\underset{R_2}{\overset{H}{\underset{|}{\overset{|}{C}}}}-NH_2} + \underset{\text{Water}}{H_2O}$$

These elimination reactions can continue until a protein chain has been formed containing many hundreds of amino acids all linked together – each amino acid in the chain is termed an amino acid *residue*. The protein that is formed is characterized by the specific sequence of amino acids, labelling from the nitrogen-containing radical end of the chain. There are 23 'natural' amino acids found in animal proteins, but collagens typically only contain about 17 of these – a full list of the structure of these amino acids can be found in most text books on organic chemistry or biochemistry, such as Morrison and Boyd (1983; Table 30.1). The dominant sequence found in collagens is the repeated unit:

$$-gly - X - Y - gly - X - Y-$$

where gly signifies the amino acid glycine, and X and Y are any of the other amino acids found in collagen, but most often proline (pro) and hydroxyproline (hypro: see Figure 8.1a). This results in an average composition in most kinds of human collagen (by number of residues) of 33–35% glycine, 7–13 % proline and 9–13 % hydroxyproline, with the other amino acids making up less than half of the total residues. Hydroxyproline is an unusual constituent in that it only appears to be a significant component of collagen and does not generally occur in other proteins. This fact has relevance for the radiocarbon dating of potentially contaminated bone, since if hydroxyproline can be extracted from the collagen then it is reasonably certain to be endogenous. Aspartic acid (where R $=$ CH$_2$COOH) is the amino acid most commonly used in archaeological racemization studies. Normal bone collagen contains about 4–5 % of aspartic acid residues by number (Miller and Gay, 1982; Table 8.3).

Collagen is made up of a rope-like structure consisting of three of these polypeptide chains, twisted together in a right-hand helix (Figures 8.1b and 8.1c). Each individual amino acid chain is twisted to the left, one turn per three residues (thus aligning the glycine molecules above each other at every third residue), with ten turns of each chain per turn of the triple helix. There are seven or eight common sequences of amino acid chains in collagens, and collagens in different tissues are made up of different combinations. The most common collagen (Type I) makes up approximately 90% of body collagen, and occurs in bone, tendon, dentine, cornea, soft tissue and scar tissue. It is made up of two chains of the same type [labelled α1(I)], plus one different chain [α2(I)], and is therefore described as α1(I)$_2$α2(I). The majority of the rest of the collagens

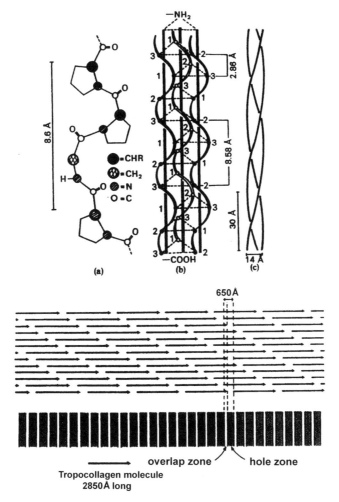

Figure 8.1 Structure of collagen in bone at different scales: (a) part of the single peptide chain (two pro or hypro and one glycine residues in sequence). (b) three of these peptide chains link together via hydrogen bonds and twist together to form a triple helix, (c) simplified view of the triple helix forming a single collagen molecule, (d) alignment of collagen molecules in a collagen fibril, showing how this alignment gives rise to the characteristic 650 Å banding. (Adapted from Williams and Elliott, 1989; Figures 2.5 and 2.6.)

found in human tissue are made up of three identical chains, such as Type III, occurring in blood vessel walls, which consists of $\alpha 1(III)_3$ (Miller and Gay, 1982).

The individual collagen 'ropes' are called fibrils, and have an average molecular weight of around 300 000 Daltons (atomic mass units), with a length of 260 nm and a diameter of 1.4–2.0 nm. In the collagen fibres which can be seen in bone, the individual fibrils are aligned head to tail, with a gap of 40 nm between fibrils, and a stagger of 65 nm (one-quarter of the length of the molecule) between adjacent rows (Figure 8.1d). This gives rise to the characteristic 65 nm banding that is visible in electron microscope photographs of collagen. Mature collagen is insoluble in water because of the covalent cross-links between adjacent polypeptides in the 'rope' helix. Solubility increases as the protein is denatured – *i.e.*, as the overall molecular weight is reduced.

As noted above, the other organic components of bone are often grouped together as *non-collagenous proteins* (ncp's). Common in tooth dentine (but not in bone) are phosphoproteins, with an unusual amino acid composition of 50% serine and 40% aspartic acid, and a total phosphorus content of 26 weight percent (wt %). Other proteins include osteocalcin in which the glutamic acid side-chain has been carboxylated. These so-called *gla proteins* are the major component of the ncp fraction in bone. The lipid component in dentine and bone makes up only about 0.1 wt % of the tissue: in bone, three-quarters of this is triglyceride (triacylglycerol), with the rest being predominantly cholesterol (Williams and Elliott, 1989; 366).

8.3   STEREOCHEMISTRY OF AMINO ACIDS

One of the interesting properties of amino acids (except glycine, the simplest member of the family) is known as *chirality*, which gives rise to the subject of *stereochemistry*. In the general formula for the amino acids given above, the molecule is drawn as planar, with a central carbon atom having four different chemical groups (radicals) attached to the bonds, each of which projects at right angles to its neighbour. In reality, the molecule is not flat, and the carbon atom is at the centre of a three-dimensional molecule in which each of the attached radicals projects outwards in a tetrahedral structure, as shown in Figure 8.2. Because each radical is different, there are two chemically identical but structurally different isomers of this molecule, related to each other in the same way as an object and its mirror image (Figure 8.2). Just as our own left and right hands are (in principle) identical but different (in that they cannot be superimposed on each other by rotation), these two molecules can be regarded as the left- and the right-handed forms of the amino acid (termed

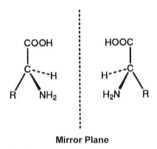

Figure 8.2   Stereo pair of amino acids.

enantiomorphs). They cannot be distinguished from each other using simple tests such as melting point or refractive index, but they do differ in their power (when in aqueous solution) to rotate plane polarized light – a property known as *optical activity*, and measured by *polarimetry*. In the older chemical notation for these molecules, the two enantiomorphs are termed D- and L-optical isomers, because a solution of one isomer in water will rotate plane polarized light to the right (the D-, or dextro-rotatory form), and the other to the left (the L-, or laevo-rotatory form). The nomenclature D- and L- are termed *relative configurations*. Strictly speaking, we should refer to (+) and (−) instead of D- and L-, since plus and minus refer directly to the direction of rotation of plane polarized light. It was Fischer in the 19th Century who arbitrarily gave the notation D- to that isomer of glyceraldehyde which gave rise to positive (right-handed) rotation. All compounds which could be converted to this particular enantiomorph by degradation or synthesis were termed 'D' compounds (Dawber and Moore, 1980; 51).

Unfortunately, it is not possible to predict from the molecular structure which way plane polarized light will be rotated. Worse, it is sometimes possible to alter the direction of rotation of plane polarized light by changing the structure of the molecule away from the chiral centre (in the case of amino acids, that is the central carbon atom), without actually altering the bonding at this centre. To deal with this, the modern (*absolute*) notation is systematically based on the molecular configuration, and is known as the *Cahn–Ingold–Prelog* convention. To use this, the four radicals attached to the chiral carbon atom are labelled a, b, c and d in decreasing order of atomic weight (*i.e.*, the heaviest group is labelled a, in sequence to the lightest which is d). The molecule is then imagined as if the lightest group (d) is pointing towards the observer, and the orientation of the remaining groups is noted. If it has the sequence a–b–c in a clockwise fashion, the absolute configuration of the molecule is *R*. If it is anti-clockwise, it is *S*. For most naturally occurring amino acids in proteins, the absolute configuration of L-amino acids is usually *S* (Morrison and Boyd, 1983; 343). Although absolute configurations are becoming the norm in organic chemistry, the majority of the literature on amino acid stereochemistry in archaeology, geology and forensic science still uses the relative terminology, and we have therefore retained that convention for the remaining discussion.

Further complications may arise with the larger amino acids such as isoleucine, where the R side-chain itself contains a chiral carbon atom [R = $CH_3CH_2C^*H(CH)_3$, where the asterisk denotes the second chiral centre]. This molecule is an example of a *diastereomer* – a molecule with more than one chiral centre. Diastereomers have different physical and chemical properties, and their interconversion is more complicated, and is termed *epimerization*.

8.4 RACEMIZATION OF AMINO ACIDS

Racemization is the process by which one enantiomer converts to the other. In normal circumstances, organic compounds which may exhibit optical activity (such as simple sugars) exist in solution as equal numbers of D- and L- forms.

The overall optical activity of such a solution is zero, since equal but opposite rotations cancel out, and the solution is termed *racemic*. It is, in fact, in a state of dynamic equilibrium, with the rate of conversion (racemization) of the D- to D- form the same as the conversion of the L- to D- form – thus there is no change overall. The situation in biochemical reactions is quite different, particularly in relation to the amino acids which make up proteins such as collagen. Biochemical syntheses (*e.g.*, the production of biomolecules by the human body) are often asymmetric, in that they favour the production of one optical isomer over the other. In fact, the manufacture of amino acids for bone collagen is so asymmetric that *in vivo* virtually all of the amino acids making up the molecules are only in the L- form: no D- amino acids should be present at all. The physiology of living bone also ensures that this asymmetry is maintained throughout life, since bone is constantly being resorbed and remodelled by cell activity, which results in the destruction of any proteins containing D-amino acids which may be formed.

At death, physiological control ceases, and the bone can be thought of as making the transition from a biochemically to a chemically (or geochemically) controlled system. (In fact, death itself might be viewed as the process of transition from the biosphere to the geosphere.) Under these conditions, the equilibrium condition of the amino acids is the racemic mixture, and in theory the process of racemization begins to increase the number of D- amino acids present in the collagen. Thus the percentage of D- amino acids will increase gradually until the ratio of D/L is equal to one, at which point the system is again in dynamic chemical equilibrium.

The true situation is more complex than this, in that amino acids within the protein chain might be expected to racemize at a different rate to terminal amino acids, which in turn will be different to free amino acids (*i.e.*, amino acids existing as discrete molecules). The actual mechanism of racemization of free amino acids is thought to be via the formation of an unstable planar intermediate – the proton (hydrogen ion) attached to the chiral carbon atom at the centre of the molecule can be removed in aqueous solution, forming a carbanion intermediate in which the other three radicals temporarily re-orientate themselves into a triangular configuration which is planar (see Figure 8.3). This structure is unstable in water, where there are plenty of protons available to reconstitute the original molecule. If the new proton

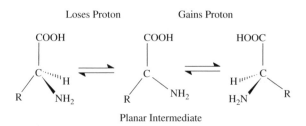

Figure 8.3   Schematic postulated mechanism of racemization.

approaches from the same side as that from which the original one was lost, then the amino acid will reconfigure in the same enantiomorphic form, with the net result being that nothing has changed. If, however, a proton happens to approach from the other side, then the other enantiomer will result – effectively the molecule will have racemized (Smith and Evans, 1980). Most of the studies of racemization of amino acids have been carried out on free amino acids dissolved in water, which is not the situation which pertains in bone protein. It is well-known that one of the effects of joining amino acids together via peptide bonds is to slow down the rate that a particular amino acid will racemize compared to its free form. Moreover, research has shown position within the chain can be important: terminal amino acids (*i.e.*, those situated at the end of the peptide chain) racemize at a different rate to the same amino acid in an 'interior' position, although the evidence is equivocal about which is faster (Smith and Evans, 1980; 265). The factor which affects the rate is largely geometrical, and is called *stearic hindrance* – molecules in some chemical bonding environments are effectively more free to move than the same molecule in a different environment. This can be most easily demonstrated using molecular models. As discussed further below, the process of racemization in archaeological and geological bone is therefore complicated, and until recently has been relatively poorly understood.

Because of these factors, it is very difficult to predict the rates of racemization of the various amino acids in a particular chemical environment. Furthermore, because racemization is a chemical reaction (unlike radioactive decay, which is a nuclear process, unaffected by chemical environment), the rate is also affected by environmental conditions, such as temperature, water availability and pH, chemical complexation and the presence of other organic and inorganic species in solution. As a general rule of thumb, it is expected that the rate of a chemical reaction will double for every 10 °C increase in temperature, showing just how sensitive these processes are to environmental factors. Certain generalizations are possible which will cover most circumstances encountered when using racemization as a dating or ageing technique: for example, it has been found that aspartic acid derived from collagen racemizes roughly twice as fast as alanine, which in turn is twice as fast as leucine (Smith and Evans, 1980; 274). Bada and Shou (1980; 252) have reported that the rate of racemization of aspartic acid in 'typical uncontaminated fossil bone' is nearly twice that of free aspartic acid at 142 °C, and that in fossil bone the sequence of rate constants is:

aspartic acid $\gg$ glutamic acid $\approx$ phenylanaline $>$ alanine $>$ leucine
$\approx$ isoleucine $\approx$ valine $>$ proline

There is now considerable interest in the relationship between the preservation of collagen and the rate of racemization of its constituent amino acids, and further data may be expected to clarify these problems.

Because at ambient temperatures the racemization rates of all amino acids are slow, it is usually found that aspartic acid is most useful archaeologically, but over much longer geological timescales aspartic acid may become racemic (and therefore useless as a chronological indicator), and the slower rate of leucine becomes more useful.

8.5   AMINO ACID RACEMIZATION DATING OF THE CALIFORNIAN PALEOINDIANS

The phenomenon of AAR came to the attention of the archaeological world, especially in North America, with the publication of the first AAR dates on some of the Paleoindian bones from California in 1974 (Bada *et al.*, 1974), although this was not the first published work using AAR of relevance to archaeology. In the previous year, Schroeder and Bada (1973) had used the measured aspartic acid enantiomeric ratio in radiocarbon-dated bones from Spain, Kenya and Olduvai Gorge, Tanzania, to estimate that the average temperature rise between the glacial and the post-glacial period in the Mediterranean had been 4 °C, and between 5–6 °C for East Africa. In the same year, Bada and Protsch (1973) published age estimates for fossil bones from Olduvai Gorge, giving ages of between 5000 and 60–70 000 BP for a range of bones. Their method used 10 g of bone, as compared to the 412 g required for the radiocarbon date obtained on the same bone, and, of course, yielded dates which were older than could have been measured using radiocarbon (which then had a time depth limit of about 30–40 000 years, because of the decreasing amount of ^{14}C present with time). They used an automatic amino acid analyser (*i.e.*, a liquid chromatographic method), after first hydrolyzing the demineralized protein to give a solution of amino acids, and separating out the aspartic acid component by ion-exchange chromatography. The enantiomers were separated on the amino acid analyser after derivatizing with L-leucine-*N*-carboxyanhydride. Bones from the middle section of the beds at Olduvai were used to calibrate the rate of racemization for the site, which gave a radiocarbon date of 17 550 ± 1000 BP (UCLA-1695 – the radiocarbon laboratory and identification number). Subsequent AAR measurements for aspartic acid were done using gas chromatography after converting the purified amino acid to the volatile *N*-trifluoroacetyl-(+)-2-butyl ester (Bada *et al.*, 1973).

   This work demonstrated that AAR could give reasonable dates from smaller samples of bone than were necessary for radiocarbon, and had a time depth of at least 70 000 years, and possibly more if one of the more slowly racemizing amino acids such as alanine was used. The key paper came in 1974 (Bada *et al.*, 1974), which published dates of between ∼6000 and 48 000 BP for various samples of human bone from the Californian coast (Table 8.1). The SDM (San Diego Museum) samples from site W-2 were from a shell midden near La Jolla excavated in 1926. Subsequently, it appears that 19 individual burials were recovered in a rescue operation from this site, known as La Jolla Shores: SDM-16755 is thought to refer to more than one individual (La Jolla Shores I and II), with a third (La Jolla Shores III) identified as SDM-16740 (Taylor *et al.*, 1985; Table 8.1). Site W-34 was located between Del Mar and Solano Beach, from a shell midden which had been largely destroyed by coastal

Table 8.1   Aspartic acid racemization dates from Californian Paleoindian Skeletons. (Reprinted with Permission from Bada *et al.*, 1974; Tables 1 and 2. Copyright 1974 American Association for the Advancement of Science.)

Sample	Description and location	D/L *aspartic acid*	Predicted age *(yr)*
Laguna Skull	Skull and long bones found at Laguna Beach in 1933	0.25	(used as calibration)
Los Angeles Man	Skull fragment from north of Baldwin Hills, found 1936	0.35	26 000
SDM-18402	Long bones from base of shell midden site W-2	0.16	~6000
SDM-16755	Rib and fragments in fill of site W-2, found 1926	0.36	28 000
SDM-16742	Human frontal found in white sand at W-2, 1926	0.50	44 000
SDM-16704 (femur)	Skull and mandible, long bones and scapula fragments from lower midden W-34–A, 1929	0.53	48 000

erosion. Details of the other samples are given in an earlier publication (Berger *et al.*, 1971), where the relevant radiocarbon dates had already been published. The so-called Los Angeles Man was found during excavation work in 1936 in a location north of the Baldwin Hills: subsequently, the remains of a mammoth were found at the same depth, but 370 m away from the skull. This gave rise to speculation that the skull might be of a late Pleistocene date. Amino acids from the collagen extracted from 100 g of skull were radiocarbon dated at University of California, Los Angeles (UCLA), giving a date of >23 600 BP (UCLA-1430). Despite the relatively large sample, there was insufficient carbon extracted to give a finite date. The Laguna Skull was the first Paleoindian skull to be found in California, and was discovered in 1933 during road building operations at 255 St Ann's Drive, Laguna Beach. It was reported to be lying alongside some long bone fragments. A purified organic sample from 78.5 g of the skull was dated by radiocarbon to give a date of 17 150 ± 1470 years BP (UCLA-1233A). A smaller sample (23 g) of the associated long bone fragments gave a minimum age of >14 800 years (UCLA-1233B), supporting the Pleistocene age of the Laguna human. The association between the skull and the long bones was confirmed by measurements of their fluorine, nitrogen and uranium contents (the measurement of which in Oxford had finally revealed the Piltdown hoax in Britain), which were found to be closely similar. Given that these were radiocarbon dates carried out to the best standards of the day, there appeared to be no doubt on the radiocarbon evidence alone that humans were present in California during the late Pleistocene.

Given this evidence, Bada and co-workers used the date on the Laguna Skull to calibrate the rate of racemization of aspartic acid in human bone in a Californian coastal environment, giving a rate constant (k_{asp}) of $1.08 \times 10^{-5}\,yr^{-1}$. They then applied this rate constant to the other samples listed in Table 8.1, yielding even older dates, and pushing back the arrival of humans to around 50 000 years ago. Corroboration was provided by comparing an AAR date of 33 000 years obtained on a dwarf mammoth bone from Santa Rosa Island (after correcting the racemization constant for the present-day temperature difference between coastal California and the southern Californian Channel Islands) with a radiocarbon date obtained on the same charred bone of $30\,400 \pm 2500$ years (UCLA-1898). These radiocarbon dates are uncalibrated, and are therefore likely to be too young (see below). In principle AAR dates are absolute and do not require calibration. It is possible, however, that AAR dates should be viewed as equivalent to uncalibrated radiocarbon dates if the racemization constant has been determined by reference to the radiocarbon age of a 'calibration' bone.

Further work followed on both Californian and other human material, summarized by Bada and Helfman (1975). Relevant to the Californian Paleoindian debate are the additional data listed in Table 8.2. The result on the SDM-16704 is a duplicate measurement (by NASA) of that listed in the first set of data (Table 8.1), and is in close agreement. Sample SDM-19241 was given a date of around 6000 years using a 'southern California post-glacial' k_{asp} value of $1.5 \times 10^{-5}\,yr^{-1}$. This agreed with a radiocarbon date on associated shells of 6700 ± 150 yr (LJ-79, La Jolla laboratory). Using this 'post-glacial average value' for k_{asp} also gave Stanford Man a date of ~ 7000 years. The Sunnyvale skeleton was also measured by NASA, using a slightly different value for the racemization constant ($\sim 7 \times 10^{-6}\,yr^{-1}$). Different values of k_{asp} were argued to be necessary for very old or very young samples to account for the different average deposition temperatures experienced by these samples from ice age to

Table 8.2 AAR dates on other Californian Paleoindian skeletons. (Bada and Helfman, 1975; Table 7, by permission of Routledge Publishers.)

Sample	Description and location	D/L *aspartic acid*	Predicted age (yr)
SDM-16704 (skull)	Del Mar W-34A	0.520	47 000
SDM-16704 (femur)		0.470	41 000
SDM-16740	W-12A Cliffs north of	0.458	39 000
SDM-16724	Scripps	0.347	27 000
SDM-16706	Batiquitos Lagoon	0.505	45 000
SDM-19241	W-9 Cliff N of Scripps	0.154	~ 6000
Sunnyvale skeleton		0.522 (ulna) 0.498 (skull)	70 000?
Stanford Man I		0.14	~ 7000

postglacial times. All samples measured by Bada were also assayed for nitrogen, as an approximate check on their antiquity (Oakley, 1963), and the paper reasonably concludes that these data constitute definitive evidence for the presence of humans in North America before 40 000 BP.

As noted above, the sceptics were still distinctly unconvinced, dismissing this evidence as coming from an unproven dating technique, and noting the lack of archaic features on these supposedly very old human skulls. In a scholarly review of the nature of the evidence for and against the presence of pre-Clovis humans in North America, Dincauze (1984) investigated the AAR evidence closely. She noted that the radiocarbon age on the Laguna Skull used to calibrate the AAR method was not secure, on the basis that earlier studies of the alluvial fan in which it was found produced bracketing radiocarbon dates in the 9th Millennium BP from marine molluscs, with a reverse relationship between stratigraphy and date. It therefore did not have the required geological or archaeological integrity for such a key calibration. She also noted that '*the great ages claimed for the skeletons were seriously incompatible with their wholly modern physical types*' (Dincauze, 1984; 289). Already by 1984 evidence from other dating techniques was beginning to suggest that the AAR dates were insupportable. The Sunnyvale skeleton, in particular, had been re-dated using first of all uranium series dating (Bischoff and Rosenbauer, 1981) and then by radiocarbon dating (Taylor *et al.*, 1983). The uranium series work (for a description of the method, see Aitken, 1990) used two independent decay schemes ($^{238}U \rightarrow {}^{230}Th$ and $^{235}U \rightarrow {}^{231}Pa$) as an internal check on the consistency of the measurements, and concluded from measurements on postcranial fragments that the Sunnyvale skeleton dated to 8300 ($+230$, -100) years BP (absolute) using the ^{230}Th method, or 9000 (±600) using ^{231}Pa. The Del Mar tibia (SDM-16704) was similarly re-dated to either 11 000 ($+500$, -100) or 11 300 ($+1300$, -1200) years BP. These are internally consistent, but compare very badly with 70 000 BP for Sunnyvale by AAR, and 48 000 for Del Mar. Furthermore, two radiocarbon dates on molluscs thought to stratigraphically bracket the Sunnyvale burial yielded dates of 10 100 to 10 400 BP (Bischoff and Rosenbauer, 1981; Table 8.1). Taylor *et al.* (1983) reported four ^{14}C measurements on organic fractions taken from the postcranial bones of the Sunnyvale skeleton, including three by the (then) novel method of AMS, all of which gave relatively recent dates: 4390 $\pm$ 150 BP (UCR-1437A) using the 'conventional' method, and 3600 $\pm$ 600 (UCR-1437A/AA-50), 4850 $\pm$ 400 BP (UCR-1437B/AA52) and 4650 $\pm$ 400 BP (UCR-1437D/AA-51) by AMS ('direct') dating. They concluded that this evidence was consistent with other geological, archaeological and anthropometric evidence – by implication, therefore, there was a serious problem with the AAR dates.

As a result of these inconsistencies, the very same amino acid extracts that had been used to produce the contentious AAR dates were independently dated by the AMS method at the Oxford Radiocarbon Accelerator Unit of the Research Laboratory for Archaeology and the History of Art, University of Oxford (OxA numbers: Bada *et al.*, 1984) and the NSF Accelerator Facility for Radioisotope Analysis, University of Arizona, Tucson (AA numbers).

The resulting combined publication (Taylor *et al.*, 1985) was given the conclusive sub-title 'none older than 11 000 C-14 years BP'. It is a good example of how advances in another scientific discipline (in this case, the development of accelerator physics) can come along at exactly the right time to apparently resolve the questions thrown up in another area entirely. In the early days of radiocarbon, as noted above, sample requirements were so large that it was normal to combust the entire bone – mineral, collagen, plus whatever else might be present – to give what was known as a 'whole bone' date. In the 1970s, it was realized that the mineral fraction was potentially subject to post-depositional contamination by dissolved carbonates in circulating groundwaters, and that purified collagen was a much better material to date (the 'collagen' dates discussed above). With the advent of AMS, sample preparation could go even further – down to the individual amino acid fraction, especially hydroxyproline which is virtually unique to mammalian collagen – thus enabling contaminating compounds to be removed, hopefully giving even more reliable dates (Stafford *et al.*, 1991). Without doubt, AMS dating revolutionized our approaches to prehistory during the 1980s, just as conventional radiocarbon dating had done so two decades earlier (Gowlett and Hedges, 1986).

The revisions published by Taylor *et al.* (1985), also summarized by Gowlett (1986), as well as assigning all the Californian Paleoindian remains to the Holocene, also pointed to the apparent cause of the problem – not the AAR methodology itself, but in the use of a seemingly erroneous radiocarbon date of 17 150 BP for the Laguna Skull as the calibration. Unfortunately, AAR was to receive all of the blame for the furore, and has hardly yet recovered in the eyes of most archaeologists! Nevertheless, the revised dates themselves were taken as clear support – or at least, as not offering contradictory evidence – for the 'late school' of thought about the colonization of the Americas. Table 8.3 is compiled from a number of published sources giving relevant data (Stafford *et al.*, 1984; Bada *et al.*, 1984; Taylor *et al.*, 1985, Gowlett, 1986) with an attempt made to cross-correlate the dates (and remove some apparent discrepancies!) as far as possible. A more complete (and first-hand) review of the story regarding the dating of the Sunnyvale and Yuha skeletons has subsequently been published by Taylor (1991).

It was immediately obvious that the AAR dates were serious overestimates, if the AMS dates were correct. Bada (1985) lost no time in pointing out that, if the correct calibration date had been used, the AAR dates were nowhere near as outrageous as they seemed, and specifically observed that the calibration date of 17 150 BP had already provided evidence for the presence of humans in America in the late Pleistocene, and that the AAR dates were consistent at the time with other available radiocarbon dates. He re-calculated the racemization constant for aspartic acid, this time using the AMS dates on both the Laguna and Los Angeles (Baldwin Hills) material, to arrive at a value of $k_{asp} = 6.0 \pm 2 \times 10^{-5} \, yr^{-1}$, which he felt was applicable to bones with poor amino acid preservation. Applying this to the other data gave dates which were uniformly within the Holocene, and are 'compatible with their AMS radiocarbon ages (Bada, 1985; 645). Table 8.4 shows the results of this recalculation, adapted

Table 8.3 Comparison of radiocarbon AMS and AAR dates on Paleoindian skeletons, compiled from various sources listed in the text.

Skeleton	AAR Age (yr)	Revised AMS C - 14 Age (yr)		Other Dates (yr)
Laguna	calibration	OxA-189	5100 ± 500	> 14 800–17 150(C-14)
Sunnyvale	70 000?	AA-60 UCR-1437A	3600 ± 600	8300–9000(U-series)
		AA-51 UCR-1437D	4650 ± 400	UCR-1437A 4390 ± 150
		AA-52 UCR-1437B	4850 ± 400	
		OxA-187	6350 ± 400	
La Jolla SDM-16755	28 000	AA-186 LJS II	5600 ± 400	LJS I 1770 ± 790
		AA-610A LJS II	4820 ± 270	LJS I 1850 ± 200
		AA-610B LJS II	5370 ± 250	LJS I 1930 ± 200
		AA-611 LJS II	6330 ± 250	
Del Mar SDM-16704	41–48 000	OxA-188	5400 ± 120	11 000–11 300 (U-series)
		OxA-774	5270 ± 100	
Stanford Man I (UCLA-1425)	~7000	OxA-152	4850 ± 150	
		OxA-153	4950 ± 130	
			5130 ± 70	
Yuha	23 600		1650 ± 250	5800 (U-series)
			2820 ± 200	
			3850 ± 250	
Taber, Alberta		OxA-773	3390 ± 90	
		Chalk River	3550 ± 500	
Los Angeles Man, Baldwin Hills	26 000		3560	[22 000–60 000 on geological grounds]

Table 8.4 Revised aspartic acid racemization dates for Californian Paleoindians. (After Bada, 1985; Table 1, reproduced by permission of the
Society for American Archaeology from *American Antiquity*, **50** no.
3 1985.)

Sample	Conventional C-14 age (yr)	AMS age (yr)	Previous AAR age (yr)	Revised AAR age (yr)
Laguna	$17\,150 \pm 1470$	5100 ± 500	calibration	calibration
	$> 14\,800$			
Los Angeles	$> 23\,600$	3560 ± 220	26 000	calibration
Del Mar	–	5400 ± 120	46 000	7500 ± 3000
La Jolla				
SDM-16724			27 000	5000 ± 2000
SDM-16740			39 000	7100 ± 3000
SDM-16742			44 000	8000 ± 3000
SDM-16755	1770 ± 800	5330 ± 245	28 000	5100 ± 2000
	1850 ± 200	5600 ± 400		
	1930 ± 200	6326 ± 250		
Batiquitos			45 000	8000 ± 3000
Lagoon				
San Jacinto	3020 ± 140		37 000	5100 ± 2000
Sunnyvale	4390 ± 600	3600 ± 600	70 000?	8200 ± 3000
		4650 ± 400		
		4850 ± 400		
		6300 ± 400		

from Bada (1985). One feature of the revised AAR dates in Table 8.4 is that the
error estimates have increased considerably, in some cases by an order of
magnitude (probably as a result of increased uncertainty in k_{asp}), and are now
significantly larger than the error estimates attached to the AMS dates.

There are a number of observations to be made at this stage. The first is to
note the poor concordance between the older (in the sense of those measurements carried out earlier) radiocarbon dates and the AMS dates. This is largely
attributable to either the use (or incorporation) of the mineral fraction in the
bone, or to the fact that the bones are poorly preserved – the total organic
content is much lower than one would expect with fresh bone. This is a salutary
reminder to archaeologists working in regions of the world where bone is
generally poorly preserved, or to those tempted to use older (sense as above)
radiocarbon dates from the literature. The second is that, if we accept Bada's
assertion that the revised AAR dates are consistent with the AMS dates, then
this is only because the AAR dates have an unacceptably large associated error.
Nor is there particularly good agreement between the other dating evidence
(principally uranium series dates) for the same material (Table 8.3) with either
the AMS or the AAR dates, beyond the fact that they are all Holocene.
Furthermore, as Bada *et al.* (1984) themselves pointed out, there appears to be
no straightforward relationship between the AMS radiocarbon age of the
skeletons and the measured D/D ratio of the aspartic acid extracts, in contrast
to the general correlations found in other parts of the world. They note that in

the Paleoindian material, the best correlation appears to be between the degree of organic preservation in the bone (as measured by the total amino acid content of the bones) and the degree of racemization. They attribute this to faster rates of racemization of aspartic acid in *N*-terminal (end-chain) positions in the protein, which could occur if hydrolysis was cleaving the chain to leave the aspartic residues exposed. This point has been further studied by Taylor *et al.* (1989), who also correlate this non-age-related variability to the state of preservation of the bone – specifically, the total nitrogen content and the ratio of glycine to glutamic acid. Clearly, the state of preservation is an extremely important factor in any dating technique when applied to bone.

In fact, developing this hypothesis, the story has yet one more twist. Further work by Stafford *et al.* (1990; 1991) on the AMS radiocarbon ages of mammoth and human bone in various states of preservation has shown that, depending on the level of collagen surviving in the bone, significantly different ages can be obtained on different amino acid fractions from the same bone. The problem is so severe in what they term 'non-collagenous' bone (bone with less than about 10% of the original amino acids remaining, or less than 0.3% total nitrogen) that age estimates varying between 2500 and 8000 years BP can be obtained on the same bone. Because the Californian Paleoindian skeletons are generally 'non-collagenous', they conclude that the revised Holocene dates themselves are questionable, and may be a result of poor preservation. Even if the Paleoindian bones were actually late Pleistocene, they might be expected to give Holocene AMS dates because of their poor preservation. The circle appears to be complete, and the long-standing question of 'why don't we know when the first humans entered North America' may still be unanswered!

8.6   THE END OF A PARADIGM: EARLY HUMANS IN THE AMERICAS

But what of this original question – when did humans first enter the Americas? The removal of early AAR dates was a blow to the 'long chronologists', although the addendum with regard to the reliability of even AMS dates from poorly preserved bone (above) certainly counsels caution in the interpretation of the apparently Holocene dates. In fact, the question of interpretation is even more complex than this, since it has long been known that all radiocarbon dates must be calibrated to give correct calendar dates, usually against dendrochronological measurements (Bowman, 1990). Uncalibrated dates are uniformly several hundred years (or even thousands of years) too young by the late Pleistocene. This calibration process opens up the possibility of multiple probabilistic calibrated dates for a single radiocarbon date (because of 'wiggles' in the calibration curve), and usually means that apparently precise uncalibrated dates may get 'smeared' out by the calibration process. Up until recently this was of little concern to late Pleistocene geologists or archaeologists, since data for calibrating dates earlier than around 9000 calendar years BP were not available. It is now possible, with some certainty, to calibrate radiocarbon dates back to 26 000 cal. BP (cal. BP is the accepted terminology for calibrated radiocarbon dates, equivalent to calendrical years) using IntCal04 (Reimer *et al.*, 2004). This is a composite calibration curve, consisting of a high-resolution tree-ring curve back to 12 400 cal. BP, and data from marine corals corrected for reservoir effects giving a terrestrial calibration curve for the earlier period. {This curve can be accessed using free calibration software (OxCal) available from [http://www.rlaha.ox.ac.uk]}. Using an earlier version of this software, Batt and Pollard (1996) carried out a calibration of the radiocarbon dates associated with the 'Clovis horizon' in North American prehistory. This suggested that, instead of Clovis being confined to a very tight timescale of 11 200 to 10 900 radiocarbon years BP (as suggested by many authors prior to the late 1990s, *e.g.*, Haynes, 1991), the true time period represented by these Clovis sites might be closer to a range of around 1850 calendar years, between about 14 000 and 12 150 cal. BP. This has very significant implications for the interpretation of the 'Clovis horizon' in American archaeology. From the uncalibrated dates it appears that the spread of Clovis was an extremely rapid phenomenon – covering most of continental USA in less than a few hundred (radiocarbon) years, making it a rate of dispersion without known parallels in the archaeological record. Some authors went to great lengths to explain this by the vision of 'Clovis the mammoth hunter', rapidly spreading through an unpopulated landscape in pursuit of these huge mammals. When calibrated, however, the Clovis phenomenon appears to extend over millennia rather than centuries, which, although still rapid, is perhaps more explicable. The full impact of calibrated dates in the Late Pleistocene is still being felt in the analysis of North American prehistory (as elsewhere), although the debate has moved on considerably in recent years. It could also mean that some of the chrono-logical problems which have been debated for decades are not capable of resolution using radiocarbon alone – particularly, for example, the question of the synchronicity between the arrival of the Clovis people and the demise of the American megafauna. When calibrated dates (and their associated larger error terms) are taken into consideration, the old certainties which underpinned the blitzkrieg model are gradually unravelling [but see Guthrie (2006), where this debate is re-considered using data on mammoth, horse, bison wapiti, moose and humans from Alaska and the Yukon].

The 'Clovis first' model – the view that the first occupants of the Americas crossed the Bering Straits via the land bridge and entered continental USA by the 'ice free corridor' which opened up between the Wisconsin and Cordilleran ice sheets at the end of the most recent glaciation – has, however, finally been overthrown, at least in the minds of the majority of americanists. Recent reviews (*e.g.*, Dillehay, 2000; Meltzer, 2004, Pedler and Adovasio, 2006) list a minimum of five sites in the Americas which are clearly pre-Clovis in date (most significantly, Monte Verde in Chile and Meadowcroft Rockshelter in Pennsylvania, but also Cactus Hill and Saltville in Virginia, Topper in South Carolina, plus the Nenana Complex sites in Alaska), and suggest that there may be as many as 20. The view that the Clovis culture was the first human occupation in the Americas is no longer tenable. Monte Verde, a site in southern Chile with remarkable preservation of artefacts of indisputable human origin, such as wooden house-foundation timbers, stone, bone, tusk and wooden artefacts, plus food items such as pieces of meat and hide, and a wide range of plant remains (Dillehay, 1989; 1997), is key to this. It has no cultural associations with Clovis in the USA, and the MVII layer (which contains the above objects) is dated to an average radiocarbon age of 12 500 C^{14} years BP, which calibrates to around 15 500–14 200 cal. BP. Such a well-established cultural horizon at this date and this far south cannot possibly be explained as the southern extent of migration of the Clovis people (even on the uncalibrated timescale, it pre-dates Clovis by 1000 radiocarbon years). In fact, there are no Clovis-type cultures in South America, and a number of other sites now attest to pre-Clovis occupation. These sites must represent the remains of an earlier migration into the Americas, certainly some time before 15 000 cal. BP, and possibly much earlier.

Indeed, there is tantalizing evidence at Monte Verde itself of earlier occupation. Layer MVI is dated to *ca.* 33 000 ^{14}C years BP, and yielded stone tools of probable human origin. Even one of the excavators of Monte Verde, Tom Dillehay, as late as 2000, was unwilling to accept this as evidence of early human occupation: '*Although the stratigraphy is intact, the radiocarbon dates are valid, and the human artifacts are genuine, I hesitate to accept this older level without more evidence and without sites of comparable age elsewhere in the Americas*' (Dillehay, 2000; 167). Recent evidence from Mexico (González *et al.*, 2006) has now offered some support for this. Several hundred supposedly human and animal footprints have been reported from Valsequillo Basin, south of Puebla, Mexico, embedded in Xalnene volcanic ash dated by a number of methods to around 40 000 BP. However, Renne *et al.* (2005) have dated the same deposit, using argon–argon dating, to 1.30 ± 0.03 million years old, and therefore assert that the footprints are very unlikely to be human. If the complex chronological issues surrounding this site can be resolved, and it is shown to push back the human entry into the Americas to at least 40 000 BP, then it will have substantial implications for our understanding of the rate of global dispersal of modern humans out of Africa, and hence the evolution of our species.

8.7 'SOME KIND OF JOKE'? CURRENT USES OF AMINO ACID RACEMIZATION

For a brief time, it had looked as if the racemization of amino acids in bone protein might provide the clinching factor in this important archaeological debate, but it was not to be. The episode so damaged the reputation of AAR as a means of dating that Milford H. Wolpoff, an eminent anthropologist, was reported as saying in 1990 that it must be '*some kind of joke*' (Marshall, 1990). Fortunately, however, this has not been uniformly the opinion. A small number of dedicated scientists (including Ed Hare in the USA, D.Q. Bowen in Wales and Matthew Collins in England) continued to study the racemization of amino acids, and have restored it to scientific respectability (but not as a means of dating bone). In the wider field of Quaternary geochronology, amino acid chronostratigraphy has continued to be applied. For example, the work of Bowen *et al.* (1989) on British Pleistocene deposits, where the epimerization of L-isoleucine to D-alloisoleucine is used to create a relative stratigraphic sequence which has been related to the oxygen isotope timescale using independent dating techniques [uranium series, thermoluminescence and electron spin resonance techniques; for details of these, see Aitken (1990)].

One important key to this resurgence has been a better understanding of the process of racemization itself. The model of interconversion between D- and L-amino acids presented above has proved to be inadequate when considering racemization in proteins. It appears that amino acids bound into collagen chains are probably completely incapable of racemization, because of steric hindrance (Van Duin and Collins, 1998). Racemization is most likely only at the terminal amino acids in the chain – therefore the degree of racemization reflects the amount of damage (probably by hydrolysis) suffered by the collagen over time. This will, of course, itself be time-dependent, but not necessarily related to any measured racemization constant. In fact, it had already been observed, during the attempts made to understand the failure of AAR dating of early American humans (see above), that racemization was more related to the degree of damage suffered by the collagen than to time directly. This in itself is a very useful observation, and has resulted in AAR measurements being proposed as a proxy for the biomolecular integrity of the protein, and, by extension, for other biomolecules in the same context. Thus AAR has been proposed as a screening method for predicting which bones might be most suitable for ancient DNA amplification (Poinar *et al.*, 1996). These authors showed that ancient DNA sequences could not be extracted from animal bone in which the D/L ratio of aspartic acid was greater than 0.08. Moreover, since these processes are temperature-dependent, the possibility exists that the degree of racemization can be taken as a proxy for the integrated thermal history of the protein, which is important not only for predicting DNA survival (Smith *et al.*, 2003), but also for predicting which bones might contain enough surviving endogenous protein to yield reliable radiocarbon dates.

Furthermore, work on shell has revealed another interesting observation. It appears that some protein is 'protected' from environmental degradation because it is contained within the crystalline phase of the shell, rather than between crystals (Sykes *et al.*, 1995). This occluded protein can be extracted by treatment with a strong chemical oxidant (NaOCl) after initial acid treatment, and appears to give much more reliable AAR data than the more exposed (and therefore more degraded) protein. Biomineralized tissue such as eggshell and mollusc shell appears to contain suitable occluded material for dating. Bone does not, because the structure is too open to contain any protected protein. Dental enamel, the most resistant of all mammalian biominerals, unfortunately contains too little protein to yield measurable racemization using current techniques. One recent significant advance has been to extract protein from the opercula (the little mineralized flap used by snails to close the shell when they withdraw inside) of *Bithynia* (a freshwater gastropod). This mineralizes purely as calcite, unlike the shell itself, which may contain aragonite phases which are likely to convert over geological time to calcite, thus potentially exposing the occluded protein to degradation. Preliminary measurements on seven operculae from Pakefield, Suffolk (Parfitt *et al.*, 2005) were consistent with other evidence for the date of the site at *ca.* 700 000 BP, therefore confirming the flint artefacts found there as the oldest recorded human activity in northern Europe. Unpublished racemization measurements by Collins and co-workers on operculae from geological deposits covering the past million years appear to show remarkable linearity with time, suggesting that AAR (on the right material) might indeed be a useful and practical dating tool – not a joke at all!.

Of considerable interest in recent years to archaeologists and forensic scientists has been the observation that not all mammalian tissues maintain their protein amino acids in the L- form. It has been known for some time that D-aspartyl residues accumulate *in vivo* with time in the metabolically stable proteins found in tooth enamel and dentine, and in eye-lens tissue, due to the racemization of L-aspartyl residues (Helfman and Bada, 1976; Masters *et al.*, 1977). Pioneering work in Cardiff and Bradford demonstrated that the ratio of D/L aspartic acid in dental collagen from human first premolars gives a very good prediction of the biological age of the individual, when using modern (extracted or fresh *post-mortem*) teeth (Gillard *et al.*, 1990; 1991; Child *et al.*, 1993). Figure 8.4 (from Gillard *et al.*, 1991) shows the calibration curve obtained on modern teeth, and Table 8.5 shows the result of using this

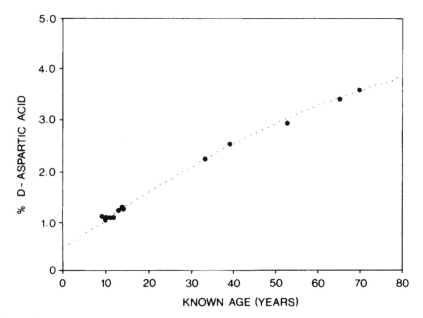

Figure 8.4 Calibration curve for racemization of aspartic acid in modern dental collagen. (From Gillard *et al.*, 1991; Figure 1, by permission of Birkhäuser Verlag AG.)

Table 8.5   Blind test of aspartic acid calibration data on teeth extracted from living individuals.

Tooth	%D-asp	Predicted age (yr)	Known age (yr)
Lower first premolar	3.30	61.7	60.5
Lower lateral incisor	2.27	33.5	31.0
Upper canine	2.67	43.5	42.0

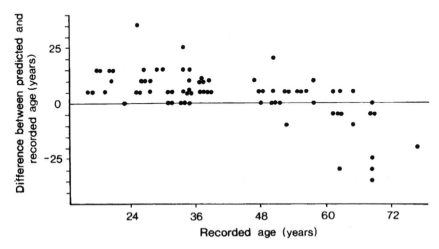

Figure 8.5   Difference between 'real' age and age estimated from racemization measurements on first premolars plotted as a function of 'real' age in the Spitalfields population. (Redrawn from Gillard *et al.*, 1991; Figure 3, by permission of Birkhäuser Verlag AG.)

calibration curve on modern teeth in a blind test, with very satisfactory results. However, when the same technique was applied to the known-age dental remains from identified 18th and 19th Century inhumations in the crypt of Christ Church Spitalfields, London (Reeve and Adams, 1993; Molleson and Cox, 1993), the results began to show some discrepancies (Figure 8.5, redrawn from table II in Gillard *et al.*, 1991). This shows a general trend of the predicted age being an overestimate for individuals younger than about 50 years, and an underestimate for the age of individuals older than 60 years. Assuming that the 'known ages' are correct, the reason for this is not entirely clear. In fact, exactly the same pattern of predicted versus real ages is observed in the data obtained from combined skeletal age determinations in the Spitalfields collection (Figure 8.6, adapted from Molleson and Cox, 1993; 171). These estimates were obtained by combining four well-established criteria for estimating age (based on observations of the pelvis, the femur, cranial sutures and humerus). The Spitalfields work is almost unique in archaeology, in that it allows the established archaeological methods to be compared against the 'true' answer, and the observed discrepancies are therefore disturbing. From the amino acid

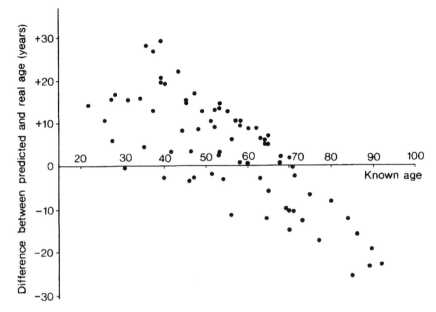

Figure 8.6 Difference between 'real' age and age estimated from skeletal measurements plotted as a function of 'real' age in the Spitalfields population. (Adapted from Molleson and Cox, 1993; Figure 12.4, by permission of the CBA and the author.)

viewpoint, it is at least encouraging to observe that the pattern and magnitude of the deviations observed are very similar to those using other methods, suggesting that the problem might have a common origin. It has since been observed that this common cause might be the use of regression-based methods for converting age indicators into estimated ages (Aykroyd *et al.*, 1999), which has been shown to introduce such bias. An alternative method has been proposed for such calibrations using kernel density procedures followed by the application of Bayes's theorem (Lucy *et al.*, 2002) which, when applied to the data used to derive the estimates discussed above, gives age estimates showing less bias than those from regression.

One of the more surprising results of the Spitalfields study is the data shown in Table 8.6, obtained when three of the four first premolars from the same individual were analysed independently. Obviously, it is expected that three teeth from the same dentition should give the same estimated age: it is also clear that in some cases they do not! Individual number 899, for example, gives an average age of 31.2 years (range = 3.5 years), comparing satisfactorily with the known age of 31; number 545 gave an average of 33.6 year (range only 3.2 years), but the true age was 21 (*i.e.*, a precise but not accurate age estimate); number 643, with a true age of 63 years, gave a range from 33.0 to 58.0 years (both inaccurate and imprecise). Similar variability has also been observed in the use of aspartic acid age estimates in modern forensic cases. In the case of an

Table 8.6 Reproducibility of Spitalfields aspartic acid age estimates from multiple teeth in the same dentition.

Catalogue no.	Recorded age (yr)	Age estimates for different teeth (yr)	Range (yr)
070	35	37.0, 39.0, 46.2	9.2
301	35	37.8, 51.0, 41.0	13.2
545	21	34.8, 34.5, 31.6	3.2
643	63	58.0, 57.0, 33.0	3.5
670	68	33.0, 37.2, 41.6	8.6
708	37	46.0, 45.6, 40.2	5.8
714	34	33.0, 34.6, 37.8	4.8
899	31	33.5, 30.2, 30.0	3.5

individual burned to death in a motorway crash, the age obtained from all four first premolars was found to be 24.2 years with a standard deviation of 4.5 years, compared with the subsequently known true age of 25 years. In a second case, of an individual exposed for three months after death on the Brecon Beacons (South Wales), two estimates (from two first premolars) were obtained of 40.4 and 38.9 years; subsequently the true age was found to be 25 years (Child *et al.*, 1993). It is clear from these data, and from forensic work carried out by other investigators (*e.g.*, Masters, 1986b), that the technique sometimes becomes unreliable when applied to potentially partially degraded dental collagen. In their review of the application of aspartic acid racemization to the determination of age at death in forensic cases, Waite *et al.* (1999) concluded that an accuracy of $\pm$ 3 years should be achievable if standardized laboratory procedures are used, making AAR as good if not better than conventional forensic methods of determining age. They conclude, however, that problems remain if the individual has been burned or exposed for a long period *post-mortem*.

An obvious limitation of the use of racemization-derived age at death estimates of older archaeological material is the magnitude of the contribution of *post-mortem* racemization to the measured ratio. On the (possibly erroneous) assumption that racemization can still be represented by a rate constant, we have calculated that the post-mortem contribution to the measured aspartic acid racemization in an individual aged 50 years at death in AD 1740 and buried in the Spitalfields vaults should be around 0.1% D-aspartic acid (Gillard *et al.*, 1990), compared to the 3% expected *in vivo* (this latter figure also includes an element of acid-induced racemization during the extraction procedure, estimated to be about 0.5%). Because of the lower temperature during burial (estimated to be 10–15°C on average, compared to 37°C *in vivo*), the *post-mortem* contribution can therefore be neglected over a period of 250 years, but clearly this would not be the case for human remains of greater antiquity, nor if the average burial temperature were significantly higher. For example, a Roman person aged 50 years should have accumulated approximately 2.8% D-aspartic acid *in vivo*, but could have a *post-mortem* contribution of about 1% (2000 years at a burial temperature of 10 °C). It is not difficult to envisage a case where the *post-mortem* contribution outweighs the *in vivo* racemization. In 1986, in an ambitious grant application, Pollard, Whittaker and Gillard proposed that this problem could be overcome by measuring the racemization of two different amino acids and solving the following simultaneous equation:

$$(D/L)_1 = K_1^{37} \times t_a + K_1^{10} \times t_{pm} + const$$

$$(D/L)_2 = K_2^{37} \times t_a + K_2^{10} \times t_{pm} + const$$

where 1 and 2 refer to two different amino acids, K^{37} and K^{10} are the rate constants for racemization at 37 and 10 °C, respectively (assuming an average burial temperature of 10°C), t_a is the age at death and t_{pm} is the post-mortem interval (time since death). Thus, from dental measurements, we suggested that both age at death and time since death could be determined, providing the rate constants were sufficiently well-known. Unsurprisingly, at the time this proved to be too ambitious for the funding body! Now, with better insights into the racemization process from the work of Collins and others, and improved analytical protocols, it is possible that it might be achievable.

8.8   SUMMARY

The racemization of amino acids – the interconversion of one form to its mirror image – sprang to archaeological prominence as a potential dating technique at a time when radiocarbon dating of bone required relatively large samples, and when the time depth limit of radiocarbon (then around 35 000 years BP) was seen as a serious drawback. A technique was devised which allowed a date to be obtained on very small samples of amino acids extracted from bone collagen, but which, unfortunately, did not yield 'absolute' dates. The influence of changing temperature on the rate constant meant that 'calibration' against a 'known age' (*i.e.*, radiocarbon dated) sample was deemed to be the best option. With hindsight, the sample chosen to calibrate the amino acid measurements to address one of the most important archaeological problems – the first date of human occupation in the Americas – was unfortunate, in that it turned out to be in error, and the resulting controversial dates have subsequently been withdrawn. The main impetus for using AAR measurements as a dating technique for bone has now largely gone. The advent of accelerator-based methods of radiocarbon dating has meant that excellent dates can be obtained on very small samples, although the age limit of ^{14}C dating still remains at around 50 000 years BP. There has been a lasting impression created in some quarters that AAR as a dating technique is 'some kind of joke'. Thanks, however, to the perseverance of a handful of scientists, we now know more about the circumstances under which racemization measurements might provide reliable dates. It now looks to be an extremely promising technique for dating the intra-crystalline protein in snail shells prior to the time period accessible by radiocarbon. If this can be verified, then AAR still has an extremely important contribution to make to Quaternary science. It has also been shown that measurements of the D-aspartic acid accumulated *in vivo* in certain tissues, especially dentine, are systematically related to the biological age of the individual, and this has profound implications, both archaeologically and, increasingly, forensically. Again, with improved protocols and under-standing such approaches are now becoming more accepted.

The episode of AAR dating relating to the arrival of the First Americans is interesting as a case study for a number of reasons. Essentially, it is a good illustration of the scientific process at work in archaeology – a problem is identified, a new technique applied to solve it, the result (in this case, archaeo-logically unpopular) is tested by other methods and the solution is evaluated (and, in this case, rejected). Some 20 years later, however, deeper understanding and better analytical equipment shows that the technique is not 'a joke', but that it was applied to the wrong material (bone) in undue haste to solve an important problem. As far as the archaeological question is concerned, further field work at important sites such as Monte Verde in Chile has resolved the issue for the majority of archaeologists – Clovis was not the first culture to reach the Americas. We still do not know when this was, and nor do we know what impact the earlier arrival of people into the New World might have on our models of human dispersal out of Africa, but at least some progress has been made after more than a century of impasse.

In his masterly review of the state of archaeological science, Renfrew (1992; 292) noted that:

> '*For is it not, these days, a defining character of real science that it is testable?*' . . . '*That archaeological science should sometimes give the wrong answers, and that these can later be shown to be indeed erroneous, must be counted one of the subjects greatest strengths.*' . . . '*Archaeologi-cal science has certainly now come of age, and can take such differences of opinion as these as a characteristic feature of scientific progress.*'

The AAR dating of early human bone in North America certainly illustrates this statement!

REFERENCES

Adovasio, J.M. (1993). The ones that will not go away. In *From Kostenki to Clovis*, ed. Soffer O. and Praslov N.D., Plenum Press, New York, pp. 199–218.

Aitken, M.J. (1990). *Science-based Dating in Archaeology*. Longmans, London.

Aykroyd, R.G., Lucy, D., Pollard, A.M. and Roberts, C.A. (1999). Nasty, brutish, but not necessarily short. *American Antiquity* **64** 55–70.

Bada, J.L. (1985). Aspartic acid racemization ages of California Paleoindian skeletons. *American Antiquity* **50** 645–647.

Bada, J.L. (1991). Amino acid cosmogeochemistry. *Philosophical Transactions of the Royal Society of London B* **333** 349–358.

Bada, J.L. and Helfman, P.M. (1975). Amino acid racemization dating of fossil bones. *World Archaeology* **7** 160–173.

Bada, J.L. and Protsch, R. (1973). Racemization reaction of aspartic acid and its use in dating fossil bones. *Proceedings of the National Academy of Sciences of the USA* **70** 1331–1334.

Bada, J.L. and Shou, M-Y. (1980). Kinetics and mechanism of amino acid racemisation in aqueous solution and in bones. In *Biogeochemistry of Amino Acids*, ed. Hare P.E., Hoering T.C. and King K. Jr, John Wiley, New York, pp. 235–255.

Bada, J.L., Kvenvolden, K.A. and Peterson, E. (1973). Racemization of amino acids in bones. *Nature* **245** 308–310.

Bada, J.L., Schroeder, R.A. and Carter, G.F. (1974). New evidence for the antiquity of man in North America deduced from aspartic acid racemization. *Science* **184** 791–793.

Bada, J.L., Gillespie, R., Gowlett, J.A.J. and Hedges, R.E.M. (1984). Accelerator mass spectrometry radiocarbon ages of amino acid extracts from Californian Palaeoindian skeletons. *Nature* **312** 442–444.

Batt, C.M. and Pollard, A.M. (1996). Radiocarbon calibration and the peopling of North America. In *Archaeological Chemistry V*, ed. Orna, M.V., Symposium Series 625, American Chemical Society, Washington D.C., pp. 415–433.

Berger, R., Protsch, R., Reynolds, R., Rozaire, C. and Sackett J.R. (1971). New radiocarbon dates based on bone collagen of California Paleoindians. *Contributions of the University of California Archaeological Research Facility* **12** 43–49.

Bischoff, J.L. and Rosenbauer, R.J. (1981). Uranium series dating of human skeletal remains from the Del Mar and Sunnyvale sites, California. *Science* **213** 1003–1005.

Bowen, D.Q., Hughes, S., Sykes, G.A. and Miller, G.H. (1989). Land–sea correlations in the Pleistocene based on isoleucine epimerization in non-marine molluscs. *Nature* **340** 49–51.

Bowman, S.G.E. (1990). *Radiocarbon Dating*. British Museum Press, London.

Child, A.M., Gillard, R.D., Hardman, S.M., Pollard, A.M., Sutton, P.A. and Whittaker, D.K. (1993). Preliminary microbiological investigations of some problems relating to age at death determinations in archaeological teeth. In *Archaeometry: Current Australasian Research*, ed. Fankhauser B.L. and Bird J.R., Occasional Papers in Prehistory, Australian National University, Canberra, pp. 85–90.

Dawber, J.G. and Moore, A.T. (1980). *Chemistry for the Life Sciences*. Macmillan, London, 2nd edn.

Dillehay, T.D. (1989). *Monte Verde: A Late Pleistocene settlement in Chile. Vol. 1 The palaeoenvironmental context*. Smithsonian Institution Press, Washington D.C.

Dillehay, T.D. (1997). *Monte Verde: A Late Pleistocene settlement in Chile. Vol. 2 The archaeological context and interpretation*. Smithsonian Institution Press, Washington D.C.

Dillehay, T.D. (2000). *The Settlement of the Americas: A New Prehistory*. Basic Books, New York.

Dillehay, T.D. and Meltzer, D.J. (ed.) (1991). *The First Americans*. CRC Press, Boca Raton.

Dincauze, D.F. (1984). An archaeo-logical evaluation of the case for pre-Clovis occupations. *Advances in World Archaeology* **3** 275–323.

Fagan, B.M. (1991). *Ancient North America*. Thames and Hudson, London.

Gillard, R.D., Pollard, A.M., Sutton, P.A. and Whittaker, D.K. (1990). An improved method for age at death determination from the measurement of D-aspartic acid in dental collagen. *Archaeometry* **32** 61–70.

Gillard, R.D., Hardman, S.M., Pollard, A.M., Sutton, P.A. and Whittaker, D.K. (1991). Determination of age at death in archaeological populations using the D/L ratio of aspartic acid in dental collagen. In *Archaeometry '90*, ed. Pernicka, E. and Wagner, G.A., Birkhäuser, Basel, pp. 637–644.

González, S., Huddart, D., Bennett, M.R. and González-Huesca, A. (2006). Human footprints in Central Mexico older than 40,000 years. *Quaternary Science Reviews* **25** 201–222.

Gowlett, J.A.J. (1986). Problems in dating the early human settlement of the Americas. In *Archaeological Results from Accelerator Dating*, ed. Gowlett, J.A.J. and Hedges, R.E.M., Oxford University Committee for Archaeology Monograph 11, Oxford, pp. 51–59.

Gowlett, J.A.J. and Hedges, R.E.M. (ed.) (1986). *Archaeological Results from Accelerator Dating*. Oxford University Committee for Archaeology Monograph 11, Oxford.

Guthrie, R.D. (2006). New carbon dates link climatic change with human colonization and Pleistocene extinctions. *Nature* **441** 207–209.

Haynes, C.V. Jr (1991). Geoarchaeological and paleohydrological evidence for a Clovis-age drought in North America and its bearing on extinction. *Quaternary Research* **35** 438–450.

Haynes, C.V. Jr (1992). Contributions of radiocarbon dating to the geochronology of the peopling of the new world. In *Radiocarbon after Four Decades*, ed. Taylor, R.E., Long, A. and Kra, R.S., Springer-Verlag, New York, pp. 355–374.

Helfman, P.M. and Bada, J.L. (1976). Aspartic acid racemization in dentine as a measure of ageing. *Nature* **262** 279–281.

Lucy, D., Aykroyd, R.G. and Pollard, A.M. (2002). Nonparametric calibration for age estimation. *Journal of the Royal Statistical Society C, Applied Statistics* **51** 183–196.

Marshall, E. (1990). Racemization dating: great expectations. *Science* **247** 799.

Martin, P.S. (1984). Prehistoric overkill: the global model. In *Quaternary Extinctions: A Prehistoric Revolution*, ed. Martin, P.S. and Klein, R.G., University of Arizona Press, Tucson, pp. 354–403.

Masters, P.M. (1986a). Amino acid racemisation dating – a review. In *Dating and Age Determination of Biological Materials*, ed. Zimmerman M.R. and Angel J.L., Croom Helm, London, pp. 39–58.

Masters, P.M. (1986b). Age at death determinations for the autopsied remains based on aspartic acid racemization in tooth dentin: importance of postmortem conditions. *Forensic Science International* **32** 179–184.

Masters, P.M., Bada, J.L. and Zigler, J.S. (1977). Racemization in dentine as measure of ageing and in cataract formation. *Nature* **268** 71–73.

McConnell, D. (1960). The crystal chemistry of dahllite. *American Mineralogist* **45** 209–216.

McConnell, D. (1973). *Apatite. Its Crystal Chemistry, Mineralogy, Utilization, and Geologic and Biologic Occurences.* Springer-Verlag, Vienna.

McLean, F.C. and Urist, M.R. (1968). *Bone. Fundamentals of the Physiology of Skeletal Tissue.* University of Chicago Press, Chicago, 3rd edn.

Meltzer, D.J. (1989). Why don't we know when the first people came to North America? *American Antiquity* **54** 471–490.

Meltzer, D.J. (2004). Peopling of North America. In *The Quaternary Period in the United States,* ed. Gillespie, A.R., Porter, S.C. and Atwater, B.F., Developments in Quaternary Science 1, Elsevier, Amsterdam, pp. 539–563.

Miller, E.J. and Gay, S. (1982). Collagen: an overview. *Methods in Enzymology* **82** 3–32.

Molleson, T. and Cox, M. (with Waldron A.H. and Whittaker, D.K.) (1993). *The Middling Sort. The Spitalfields Project Volume 2 – the Anthropology.* Research Report 86, Council for British Archaeology, York.

Morrison, R.T. and Boyd, R.N. (1983). *Organic Chemistry.* Allyn and Bacon, Boston, 4th edn.

Oakley, K.P. (1963). Fluorine, uranium and nitrogen dating of bones. In *The Scientist and Archaeology,* ed. Pyddoke, E., Roy Publishers, New York, pp. 111–119.

Parfitt, S.A., Barendregt, R.W., Breda, M., Candy, I., Collins, M.J., Coope, G.R., Durbidge, P., Field, M.H., Lee, J.R., Lister, A.M., Mutch, R., Penkman, K.E.H., Preece, R.C., Rose, J., Stringer, C.B., Symmons, R., Whittaker, J.E., Wymer, J.J. and Stuart, A.J. (2005). The earliest record of human activity in northern Europe. *Nature* **438** 1008–1012.

Pedler, D.R. and Adovasio, J.M. (2006). After Clovis: some thoughts on the slow death of a paradigm. Paper presented at 36th International Symposium on Archaeometry, 2–6 May 2006, Quebec City, Canada.

Poinar, H.N., Höss, M., Bada, J.L. and Pääbo, S. (1996). Amino acid racemization and the preservation of ancient DNA. *Science* **272** 864–866.

Reeve, J. and Adams, M. (1993). *Across the Styx. The Spitalfields Project Volume 1 – the Archaeology.* Research Report 85, Council for British Archaeology, York.

Reimer, P.J., Baillie, M.G.L., Bard, E., Bayliss, A., Beck, J.W., Bertrand, C.J.H., Blackwell, P.G., Buck, C.E., Burr, G.S., Cutler, K.B., Damon, P.E., Edwards, R.L., Fairbanks, R.G., Friedrich, M., Guilderson, T.P., Hogg, A.G., Hughen, K.A., Kromer, B., McCormac, G., Manning, S., Ramsey, C.B., Reimer, R.W., Remmele, S., Southon, J.R., Stuiver, M., Talamo, S., Taylor, F.W., van der Plicht, J. and Weyhenmeyer, C.E. (2004). IntCal04 terrestrial radiocarbon age calibration, 0–26 cal kyr BP. *Radiocarbon* **46** 1029–1058.

Renfrew, A.C. (1992). The identity and future of archaeological science. In *New Developments in Archaeological Science,* ed. Pollard, A.M., Proceedings of the British Academy 77, Oxford University Press, Oxford, pp. 285–293.

Renne, P.R., Feinberg, J.M., Waters, M.R., Arroyo-Cabrales, J., Ochoa-Castillo, P., Perez-Campa, M and Knight, K.B. (2005). Age of Mexican ash with alleged 'footprints'. *Nature* **438** (7068) E7–E8.

Schroeder, R.A. and Bada, J.L. (1973). Glacial–postglacial temperature difference deduced from aspartic acid racemization in fossil bones. *Nature* **182** 479–482.

Smith, G.G. and Evans, R.C. (1980). The effect of structure and conditions on the rate of racemisation of free and bound amino acids. In *Biogeochemistry of Amino Acids*, ed. Hare P.E., Hoering T.C., and King K. Jr, John Wiley, New York, pp. 257–282.

Smith, C.I., Chamberlain, A.T., Riley, M.S., Stringer, C. and Collins, M.J. (2003). The thermal history of human fossils and the likelihood of successful DNA amplification. *Journal of Human Evolution* **45** 203–217.

Stafford, T.W. Jr, Jull, A.J.T., Zabel, T.H., Donahue, D.J., Duhamel, R.C., Brendel, K., Haynes, C.V. Jr, Bischoff, J.L., Payen, L.A. and Taylor, R.E. (1984). Holocene age of the Yuha burial: direct radiocarbon determinations by accelerator mass spectrometry. *Nature* **308** 446–447.

Stafford, T.W. Jr, Hare, P.E., Currie, L., Jull, A.J.T. and Donahue, D. (1990). Accuracy of North American human skeletal ages. *Quaternary Research* **34** 111–120.

Stafford, T.W. Jr, Hare, P.E., Currie, L., Jull, A.J.T. and Donahue, D.J. (1991). Accelerator radiocarbon dating at the molecular level. *Journal of Archaeological Science* **18** 35–72.

Sykes, G.A., Collins, M.J. and Walton, D.I. (1995). The significance of a geochemically isolated intracrystalline organic fraction within biominerals. *Organic Geochemistry* **23** 1059–1065.

Taylor, R.E. (1991). Frameworks for dating the late Pleistocene peopling of the Americas. In *The First Americans*, ed. Dillehay, T.D. and Meltzer, D.J., CRC Press, Boca Raton, pp. 77–111.

Taylor, R.E., Payen, L.A., Gerow, B., Donahue, D.J., Zabel, T.H., Jull, A.J.T. and Damon, P.E. (1983). Middle Holocene age of the Sunnyvale human skeleton. *Science* **220** 1271–1273.

Taylor, R.E., Payen, L.A., Prior, C.A., Slota, P.J. Jr, Gillespie, R., Gowlett, J.A.J., Hedges, R.E.M., Jull, A.J.T., Zabel, T.H., Donahue, D.J., Stafford, T.W. and Berger, R. (1985). Major revisions in the Pleistocene age assignments for North American human skeletons: none older than 11 000[14]C years BP. *American Antiquity* **50** 136–140.

Taylor, R.E., Ennis, P.J., Slota, P.J. Jr and Payen, L.A. (1989). Non-age-related variations in aspartic acid racemization in bone from a radiocarbon-dated late Holocene archaeological site. *Radiocarbon* **31** 1048–1056.

Van Duin, A.C.T. and Collins, M.J. (1998). The effects of conformational constraints on aspartic acid racemization. *Organic Geochemistry* **29** 1227–1232.

Waite, E.R., Collins, M.J., Ritz-Timme, S., Schutz, H.W., Cattaneo, C. and Borrman, H.I.M. (1999). A review of the methodological aspects of aspartic acid racemization analysis for use in forensic science. *Forensic Science International* **103** 113–124.

Weiner, S. and Traub, W. (1992). Bone structure: from Ångstroms to microns. *FASEB Journal* **6** 879–885.

Williams, R.A.D. and Elliott, J.C. (1989). *Basic and Applied Dental Bio-chemistry*. Churchill Livingstone, Edinburgh, 2nd edn.

CHAPTER 9

Lead Isotope Geochemistry and the Trade in Metals

9.1 INTRODUCTION

The possibility of using some form of chemical fingerprinting to trace metal objects back to their ore source, and hence reconstruct prehistoric economic contacts, has long been one of the great goals of archaeological chemistry. Ever since the advent of instrumental methods for the rapid chemical analysis of metals in the 1930s, large programmes of analysis of prehistoric metal objects have been undertaken. This approach is fraught with problems – far more than have been encountered with the study of other archaeological materials, with the possible exception of glass. The relationship between the trace element composition of a metalliferous ore and that of a metal object derived from it is an extremely complicated one, which is influenced by a number of factors, considered below.

Most metallic elements exist naturally as different isotopes – atoms of the same element which have the same chemical characteristics, but vary in weight (see Appendix 2). Normally, however, the relative abundance of the different isotopes of the same metal varies very little across the surface of the Earth. Lead is unusual in that it has a large range of natural isotopic compositions, due to the fact that three of its four stable isotopes (^{206}Pb, ^{207}Pb and ^{208}Pb, with the fourth being ^{204}Pb) lie at the end of major radioactive decay chains. Thus, the ratios of these four isotopes vary with time, which allows them to be used as a chronometer. Geologists have been quick to exploit this fact – firstly, to obtain an estimate of the age of the Earth, and subsequently to estimate the geological age of the various metalliferous deposits. The discovery, therefore, that the ratios of the stable isotopes of lead vary measurably from metal deposit to metal deposit, and are apparently unaffected by anthropogenic processing, was naturally hailed as a major breakthrough in the scientific study of archaeological metals. Since then, it has been widely applied, particularly to

Archaeological Chemistry, Second Edition
By A. Mark Pollard and Carl Heron
© The Royal Society of Chemistry 2008

Late Bronze Age metal production in Anatolia and the Eastern Mediterranean. Although the results of lead isotope analysis have been widely quoted in the archaeological literature, during the 1990s the technique entered a phase of fierce debate regarding the archaeological interpretation of the data, which involves a great deal of technological discussion about the genesis of lead isotope deposits and, ultimately, the uniqueness of a particular isotopic signature for a particular 'ore field'. This chapter gives the geochemical background to the technique of lead isotope analysis, in order that this debate can be more easily interpreted. The chapter ends with a brief résumé of the debate, and suggests a new framework for the future applications of lead isotope data in archaeology.

9.2   THE TRACE ELEMENT APPROACH TO METAL PROVENANCE

As noted above and in Chapter 6, a number of theoretical difficulties are associated with the archaeological interpretation of the chemical analyses of metal objects, putting aside for current purposes the practical difficulties relating to sampling precious objects, and producing representative chemical analyses of potentially inhomogeneous artefacts. The first source of confusion is the use of the term 'ore'. In modern parlance, an ore is defined as '*a natural aggregation of minerals from which a metal or metallic compound can be recovered with profit on a large scale*' (Richards, 1909; 1). The application of this concept to prehistoric mineral processing is problematical, in that we have virtually no way of telling what value was placed upon the metal produced. It is quite possible that, at least in the early part of the Bronze Age, the symbolic value of possessing a metal object far outweighed any modern perception of value, which suggests that the current classification of 'economic mineral deposits' may not be useful in archaeometallurgy. Again, the modern term 'ore' is usually defined in terms of 'grams of metal extracted per tonne of ore processed', which is a measure of the abundance of metalliferous minerals in a particular vein, and therefore takes account of the presence of a large amount of *gangue* (vein material or parent rock, which has no significant metalliferous content). Modern commercially exploitable ores for, say, gold, may only contain a few parts per million of the metal. In addition, most metal-bearing deposits contain mixed mineralization – either mixtures of different species of minerals of the metal required, or, more commonly, a mixture of different metalliferous minerals. In antiquity, it is likely that miners were able to exploit extremely rich mineral deposits, in which case the 'ore' may well have contained very little gangue, and may even have been a relatively pure mineral species. The importance of this is that when one discusses the trace element composition of an ancient metal deposit, it is important to keep clearly in mind the possible differences between the minerals which make up the deposit and the mineralogical constitution of any 'ore' which may have been extracted from the deposit. This makes the use of modern-day quantitative data on ore deposits a hazardous business when attempting to interpret the archaeological record.

Further interpretative complications arise if the ore is reduced to metal by means of some sort of furnace technology, which is necessary for all but a few of the more inert metals, which can occur *native* (primarily gold, silver and sometimes copper and meteoric iron). Almost certainly prior to reduction the ore extracted from the mine (or removed from an open cut vein, or simply scraped from the surface) would have been enriched via some sort of *beneficia- tion* process (involving crushing or washing, or perhaps just hand picking), which may influence the mineralogical composition of the ore. Depending on the nature of the ore, it may also have been necessary to employ some form of *roasting* prior to reduction – partly to mechanically break up the ore, and partly to convert sulfides into oxides. This step would almost certainly result in the loss of some of the more volatile components of the ore, especially elements such as arsenic. The furnace reduction process itself may have caused further volatilization, and the addition of any other material to liquefy the slag produced (if any) may have added additional trace elements to the final metal. Thermodynamic considerations of the partition of trace elements between different liquid phases in the melt may also need to be considered. In all but the earliest stages of metallurgy, the properties of the metal were modified via the addition of other metals to produce alloys, *e.g.*, tin was added to copper to produce bronze shortly after the beginning of the Bronze Age, with lead being added at a later stage to produce an alloy which was sufficiently fluid at moderate temperatures to produce complex castings. Any such additions will certainly complicate the interpretation of trace element data.

There has been considerable debate about the nature of these earliest alloying processes. Adherents to the 'minimalist school' of early technology believe that many of the early alloys – certainly the copper–arsenic alloys which usually predate the use of tin bronzes, and possibly even the earliest tin bronzes themselves – were the result of the co-smelting of particular mixed mineral deposits (Pollard *et al.*, 1991; Lechtman and Klein, 1999). That is not to say that the production of these alloys may not itself have been deliberate, but that it may have been deliberate by geographical choice of deposit rather than by the careful mixing of smelted metals. Others believe that control over the composition of the alloys was exerted by subsequent additions of the appro- priate quantities of alloying metal to a relatively pure smelted metal. Half way between these two views is the possibility that control took place by careful blending of ores from different geographical locations before smelting, to give the appropriate balance of composition in the final metal. Whatever the true situation (and any of these may have been true in different places and times), the important point from our perspective is that the metal which emerged from the primary manufacturing process almost certainly represents the smelting product of several minerals, possibly from several geographical locations. The manufacturing processes of metal goods from these primary metals can then only further complicate the picture, since these would almost certainly have involved further high-temperature fabrication processes (with associated volatilization), and may have required additional mixing of primary metals.

Combine these possibilities with the almost inevitable practice of recycling metals in most metal-using communities, and the likelihood of post-depositional corrosion resulting from the electrochemical modification of composition of buried metalwork, and it is not surprising that many archaeo-metallurgists have come to the view that it is extremely unlikely that anything positive can be said about the relationship between the trace element composition of a metal object and its precise ore source. On one level, clearly, some statements can be made, such as if a copper alloy object contains nickel, then the most likely explanation is that the copper comes from a nickeliferous copper deposit, as discussed in Chapter 6. The reverse, however, is much less certain. The reason why a copper object does not contain nickel may indeed be the result of a lack of nickel in the ore source, but it may also be influenced by the manufacturing processes. Laboratory simulation work has shown, for example, that nickel-bearing copper ores do not contribute nickel to the smelted copper unless the temperature is in excess of 950 °C (Thomas, 1990; Pollard *et al.*, 1990). We must conclude, therefore, that the measured chemical composition of a metal artefact is a complex function of the chemistry of the ore source(s) from which it is derived, the thermodynamics and kinetics of the high-temperature processes employed, anthropogenic factors such as alloying and recycling of metals, and the electrochemistry of the corrosion processes acting during burial (and possibly after burial, either as a result of conservation processes, or as the object equilibrates with a new environment). Little wonder that few archaeological scientists place much faith in the general application of trace element provenancing to metals! Although these comments are primarily aimed at smelted non-ferrous metals (in practice, archaeologically this means copper alloy objects), the same considerations apply to ferrous metals and, to a lesser degree, to those metals which occur '*native*' (principally gold and platinum, although silver and copper are also known in the native state, and meteoritic iron can be considered a native metal). Few people have seriously attempted to provenance iron objects on the trace element composition of the metal. Some work has been done on the composition of the slag inclusions (*e.g.*, Hedges and Salter, 1979; Coustures *et al.*, 2003) which is more promising, but it is still likely to pose interpretative problems in terms of the partitioning of elements between metal and slag in the liquid state, and the trace element contribution of the fluxes added to the furnace charge. More recent work on iron has turned to the use of strontium and lead isotopes (Degryse *et al.*, 2007), with the observation that Sr isotopes are much less ambiguous than those of Pb in providing coherent signatures for iron objects. A great deal of work has also been done on the trace element and inclusion patterning (especially the platinum group elements) in gold, but with only limited success in archaeology (*e.g.*, Guerra *et al.*, 1999). The main problem, given the obvious need for non-destructive approaches to archaeological gold objects, is the extremely low abundance of many of the trace metals. The high sensitivity of inductively coupled plasma mass spectroscopy (ICP-MS), either in liquid or laser ablation mode, is beginning to make a contribution here, and substantial provenancing success has been claimed, at least in a forensic context (Watling *et al.*, 1994).

When it was demonstrated in the late 1960s that the lead isotopic composition of metal objects might give directly an indication of the ore source, it was therefore eagerly applied by archaeological scientists with access to the necessary high-precision heavy element mass spectrometers [thermal ionisation mass spectrometry (TIMS); see Chapter 2]. This was even more the case when it was realized that the method could be applied not only to lead artefacts (relatively rare in the archaeological record), but also to the traces of lead left in silver extracted from argentiferous lead ores by the cupellation process, and also to the lead impurities left in copper objects smelted from impure copper ores. It certainly appeared as if a prayer had been answered, and that some of the key questions asked of archaeometallurgists (such as the origin of the silver used in classical Athenian coinage, and the sources of the primary copper used in the Aegean Bronze Age) could now be answered with considerable confidence. Nearly 50 years later, there is a growing feeling that some of this confidence might have been misplaced, and that the time is ripe for a re-assessment.

9.3 NATURAL RADIOACTIVITY AND THE STABLE ISOTOPES OF LEAD

The basic concepts of nuclear structure and isotopes are explained Appendix 2. This section derives the mathematical equation for the rate of radioactive decay of any unstable nucleus, in terms of its *half life*.

The rate of decay of an unstable parent nucleus at any time t is proportional to the number (N) of atoms left (Faure, 1986; 38). In other words, the rate at which the number of radioactive nuclei decline is proportional to the number left at that time. Expressed mathematically, this becomes:

$$-\frac{dN}{dt} \propto N \tag{9.1}$$

The symbol λ is introduced as the constant of proportionality, which is termed the *decay constant* of the parent nucleus, and is characteristic of that nucleus (with units of inverse time):

$$-\frac{dN}{dt} = \lambda N \tag{9.2}$$

Rearranging terms:

$$-\frac{dN}{N} = \lambda \, dt \tag{9.3}$$

which integrates to give:

$$-\ln N = \lambda \, t + C \tag{9.4}$$

The constant of integration C can be defined by setting the starting condition so that $N = N_o$ at time $t = 0$, giving $C = -\ln N_o$:

$$-\ln N = \lambda t - \ln N_o \tag{9.5}$$

Rearranging this gives:

$$\ln N - \ln No = -\lambda t$$

$$\ln \frac{N}{No} = -\lambda t$$

$$\frac{N}{No} = e^{-\lambda t}$$

$$N = N_o e^{-\lambda t} \tag{9.6}$$

Equation (9.6) is the basic equation describing the decay of all radioactive particles, and, when plotted out, gives the familiar exponential decay curve. The parameter λ is characteristic of the parent nucleus, but is not the most readily visualized measure of the rate of radioactive decay. This is normally expressed as the *half life* ($T_{1/2}$), which is defined as the time taken for half the original amount of the radioactive parent to decay. Substituting $N = N_o/2$ into the Equation (9.6) gives:

$$\ln\left(\frac{1}{2}\right) = -\lambda T_{\frac{1}{2}}$$

$$\ln 2 = \lambda T_{\frac{1}{2}}$$

$$T_{\frac{1}{2}} = \frac{\ln 2}{\lambda} = \frac{0.693}{\lambda} \tag{9.7}$$

The half lives of natural radioisotopes vary widely, between fractions of a second up to many billions (10^9) of years, and are widely tabulated (*e.g.*, Littlefield and Thorley, 1979; appendix C: see also WebElements [http://www.webelements.com]).

Nuclei which are radioactively unstable usually decay by the emission of one of three particles from the nucleus, traditionally labelled α, β and γ particles. The largest, slowest and least penetrating of these are the α particles, which turn out to be the nucleus of the helium atom – *i.e.*, two protons and two neutrons, with an overall charge of $+2$. Decay by α emission is restricted to the heavier elements, and can be summarized in the following general equation:

$$_Z^A X \rightarrow \, _{Z-2}^{A-4} Y + {}_2^4 He$$

The ejection of the α particle (labelled as a helium nucleus in the above equation) from the nucleus of element X results in the transmutation of X into Y, which has an atomic number two less than that of X (*i.e.*, two positions below it in the periodic table). The particular isotope of element Y which is formed is that with an atomic mass of four less than that of the original isotope of X.

The next heaviest of the particles which can be emitted during radioactive decay is the β particle, which has been identified as being the same as an electron – a much lighter particle, with a mass of approximately 1/1840 of that of either the proton or the neutron, but carrying a single negative charge. It is important to realize that this is still a particle which has been ejected from an unstable nucleus, and not to confuse it with the orbital electrons, which are (initially at least) unaffected by these nuclear transformations. The general equation for β decay is:

$$\prescript{A}{Z}{X} \rightarrow \prescript{A}{Z+1}{Y} + \prescript{0}{-1}{\beta}$$

Here the effect of the β emission is to increase the atomic number by 1 (*i.e.*, to transmute X into the next heaviest element in the periodic table, Y), to leave the atomic weight unchanged (a so-called *isobaric* transmutation), and to emit the β particle, which is conventionally given a mass of 0 and a charge of -1. Although it does not actually happen like this, it is often useful to think of the β process as being the conversion of a neutron into two equal but oppositely charged particles, the proton and the electron, as follows:

$$\prescript{1}{0}{n} \rightarrow \prescript{1}{1}{p} + \prescript{0}{-1}{e}$$

In this notation, the upper line of superscripts refers to the mass of the particles in Daltons, and the subscripts refer to the charge. The resulting electron is then ejected from the nucleus as the β particle.

For the current discussion, the third type of particle, the γ particle, is of less interest, even though it is the most energetic and penetrating of the radioactive particles. It is in fact not a particle in the same sense that α and β are particles – it is a quantum of high-energy electromagnetic radiation, which can nevertheless be thought of as a particle as a result of particle–wave duality (see Pollard *et al.*, 2007; 279). It effectively has zero mass, and no electrical charge, and therefore the emission of a γ particle leaves the nucleus unchanged in terms of A and Z. It is best thought of as being a mechanism for removing energy from a nucleus which is in an energetically excited state as a result of other processes.

Although many nuclei are naturally radioactive, there are three main radioactive series in nature, all of which are relevant to a discussion of the isotopic composition of natural lead. These start with the elements uranium and thorium (^{238}U, ^{235}U and ^{232}Th) and all end in one of the three stable isotopes of lead (^{206}Pb, ^{207}Pb and ^{208}Pb, respectively). Although each chain goes through a large number of intermediate unstable nuclei (Figure 9.1), the three chains can be summarized as follows:

$$^{235}_{92}\text{U} \rightarrow {}^{207}_{82}\text{Pb} + 7{}^{4}_{2}\text{He} + 4\beta^- + energy$$

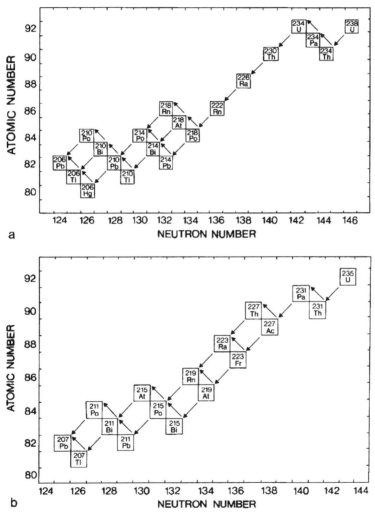

Figure 9.1 Radioactive decay chains of ^{238}U, ^{235}U and ^{232}Th. (a) decay of ^{238}U to stable ^{206}Pb; (b) decay of ^{235}U to stable ^{207}Pb; (c) decay of ^{232}Th to stable ^{208}Pb. On these plots, a downward-pointing arrow (to the left) generally signifies decay by α emission, and an upward-pointing arrow (also to the left) decay by β emission. Note that all three chains 'branch' at several points, and that all involve isotopes of the radioactive gas radon (Rn) at some stage. (After Faure, 1986; Figures 18.1–18.3. Copyright 1986 John Wiley & Sons, Inc. Reprinted by permission of the publisher.)

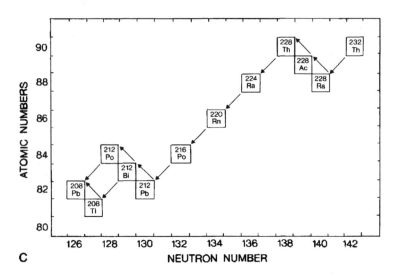

ATOMIC NUMBERS

NEUTRON NUMBER

C

Figure 9.1 (Continued)

$$\ce{^{238}_{92}U} \rightarrow \ce{^{206}_{82}Pb} + 8\,\ce{^{4}_{2}He} + 6\beta^- + energy$$

$$\ce{^{232}_{90}Th} \rightarrow \ce{^{206}_{82}Pb} + 6\,\ce{^{4}_{2}He} + 4\beta^- + energy$$

Thus the chain that starts with ^{238}U goes through eight radioactive decay processes each of which results in the emission of an α particle, and six involving β particles, and the stable end member is ^{206}Pb, at which point the series ends. The ^{206}Pb produced as a result of these radioactive processes is termed *radiogenic*, to distinguish it from any other ^{206}Pb which may exist. As can be seen in Figure 9.1, in detail each of these decay chains involve a number of radioactive intermediates (termed *daughters*), all of which have a particular half life. As an example, the full list of the half lives of the elements involved in the decay chain of ^{238}U is given in Table 9.1. Inspection of this shows that the first decay step in the chain (^{238}U to ^{234}Th) has by far the longest half life (approximately 4.5×10^9 years, compared to the next longest, which is 2.5×10^5 years). This is also true for the other two decay chains (listed in full in Russell and Farquhar, 1960; 4–5). Consideration of the behaviour of these chains under conditions of secular equilibrium (*i.e.*, when the rate of decay of the daughter isotope becomes equal to the rate of decay of the parent, and assuming a closed system) shows that it is possible to consider the decay chain simply in terms of the parent decaying directly to the stable lead end point, with a half life essentially the same as the longest half life in the system, which happens to be that of the parent isotope (Faure, 1986; 285). Thus each of the three decay chains can be simplified to the following, which is particularly

Table 9.1 Decay series of ^{238}U. (From Russell and Farquhar, 1960, Table 1.2, with half-lives modified by data in Littlefield and Thorley, 1979; Appendix C.)

Isotope	Particle emitted	Particle energy (MeV)	Half-life
^{238}U	α	4.18	4.5×10^9 yr
^{234}Th	β	0.205, 0.111	24.1 d
^{234}Pa	β	2.32, 1.50, 0.60	1.14 min
^{234}U	α	4.763	2.5×10^5 yr
^{230}Th	α	4.68, 4.61	8.0×10^4 yr
^{226}Ra	α	4.77	1620 yr
^{222}Rn	α	5.486	3.825 d
^{218}Po	α	5.998	3.05 min
^{214}Pb	β	0.65	26.8 min
^{214}Bi	α 0.04%	5.46	19.7 min
	β 99.96%	1.65, 3.17	
^{214}Po	α	7.680	160 μs
^{210}Tl	β	1.8	1.32 min
^{210}Pb	β	0.018	25 yr
^{210}Bi	β	1.17	4.8 d
^{210}Po	α	5.298	140 d
^{206}Pb			Stable

useful when considering the evolution of the isotopic compositions of terrestrial lead deposits:

$$^{238}U \rightarrow {}^{206}Pb \quad T_{1/2} = 4.468 \times 10^9 \text{ years}$$
$$^{235}U \rightarrow {}^{207}Pb \quad T_{1/2} = 0.7038 \times 10^9 \text{ years}$$
$$^{232}U \rightarrow {}^{208}Pb \quad T_{1/2} = 14.01 \times 10^9 \text{ years}$$

9.4 THE LEAD ISOTOPIC COMPOSITION OF METALLIFEROUS DEPOSITS

Having now established that three of the four stable isotopes of lead lie at the end of very long-lived radioactive decay chains, it is now appropriate to consider models for the development of the lead isotope composition of metallic ores. The fourth stable isotope (^{204}Pb) is not produced radiogenically, and is therefore termed *primeval* – its existence is the result of being present at the beginnings of the solar system, and therefore being incorporated into the earth as it solidified. The abundance of the three radiogenic isotopes also has a primeval component, to which has been added a radiogenic component. It is conventional to use isotopic ratios when discussing lead isotope geochemistry – geologists use the ratios ^{206}Pb/^{204}Pb, ^{207}Pb/^{204}Pb and ^{208}Pb/^{204}Pb, since ^{204}Pb is non-radiogenic, and these ratios occur in the equations for the isotopic evolution of ore bodies (see below), but there is also a practical reason. Until recently, the main method of making these very high precision isotope measurements was to use a machine called a *thermal ioniza-tion mass spectrometer*, and the best way of achieving the precision necessary with this method is to measure all the four isotope abundances simultaneously as three ratios (using a multicollector instrument), since this minimizes varia-tions due to small fluctuations in the ion beam (see Chapter 2). Even using ICP technology, the most precise measurements are still made using multicollector instruments. The raw data therefore consists of three isotopic ratios, and it is convenient to retain these in subsequent considerations.

It is now conventional to classify lead-bearing deposits into two types – *ordinary* or *common* lead deposits, and *anomalous* deposits. Since these names were first applied in the 1960s, it has become apparent that anomalous deposits are more common than ordinary deposits, but the names are still in use. *Ordinary lead* is found in the '*conformable*' mineral deposits of volcanic island arcs, where the mineralization is hosted in stratigraphic sequences of marine volcanic and sedimentary rocks. These deposits have simple lead evolution histories, with lead being derived by volcanic activity from the lower crust and mantle without radiogenic lead contamination from the upper crust. *Anomalous lead* occurs in deposits that have had more complex evolu-tionary histories experiencing radiogenic lead contamination from the upper crust. In general the isotopic composition of these deposits cannot be explained by the simple models described below. The earliest model for the isotopic evolution of lead minerals is called the *Holmes–Houtermans* model. The impetus for developing such a model came from a desire to be able to calculate the age of the Earth from the isotopic composition of common lead ores (Holmes, 1946).

The Holmes–Houtermans model makes a number of assumptions which are important to enumerate (Faure, 1986; 310):

   (i) originally the Earth was fluid and homogeneous, at which time U, Th and Pb were evenly distributed;

  (ii) the isotopic composition of this *primeval lead* was the same everywhere;

 (iii) on cooling, the Earth became rigid, and local variations arose in the U/Pb and Th/Pb ratios;

 (iv) in any region, the U/Pb and Th/Pb ratios subsequently change only as a result of radiogenesis;

  (v) at the time of formation of a *common (ordinary) lead* mineral, the Pb was separated from the U and Th, and there was no further change in its isotopic composition.

It is important to distinguish clearly in this scenario between the general solidification of the Earth's crust, which had the effect of 'freezing in' variations in the U/Pb and Th/Pb ratios, and the specific mineralization event which created the galena (lead sulfide, PbS) deposits, which removed the lead from the uranium and thorium, and effectively therefore 'froze' the isotopic composi-tion of the lead in the galena at the values representative of the time of mineralization.

The equation for the growth of a stable daughter from a radioactive parent can be easily derived from Equation (9.6) above, which is the familiar radioactive decay curve. We can write that:

$$D = N_o - N$$

where D and N are the abundances of the daughter and parent after time t, with N_o being the quantity of parent present at $t = 0$. Combining these two gives:

$$D = N_o - N_o(e^{-\lambda t})$$

or:

$$D = N_o(1 - e^{-\lambda t}) \tag{9.8}$$

This equation expresses the growth of the daughter from the radioactive parent in terms of the amount of the parent originally present (N_o). It is more useful if we use Equation (9.6) to replace N_o with N (the number of parent nuclei remaining after time t) in Equation (9.8), since N is the quantity which is actually measurable. Thus:

$$N_o = Ne^{\lambda t}$$

and therefore:

$$D = N(e^{\lambda t} - 1) \tag{9.9}$$

This is the equation for the growth of a stable radioactive daughter, assuming that no primeval D was present at time $t = 0$. If this was the case (and the amount is termed D_o), then the equation becomes:

$$D = D_o + N(e^{\lambda t} - 1) \tag{9.10}$$

Using the simplification outlined in the previous section (*i.e.*, that the radiogenic production of ^{206}Pb from ^{238}U can be regarded as a single step with a half life equal to that of the decay of ^{238}U), the equation for the growth of radiogenic ^{206}Pb can therefore be written as:

$$^{206}\text{Pb} = (^{206}\text{Pb})_i + {}^{238}\text{U}(e^{\lambda_1 t} - 1)$$

where the subscript i denotes the initial (primeval) amount of ^{206}Pb present, and λ_1 is the effective decay constant of ^{238}U to ^{206}Pb. Conventionally, as noted above, this is expressed as a ratio to the abundance of ^{204}Pb, as follows:

$$\frac{^{206}\text{Pb}}{^{204}\text{Pb}} = \left(\frac{^{206}\text{Pb}}{^{204}\text{Pb}}\right)_i + \frac{^{238}\text{U}}{^{204}\text{Pb}}(e^{\lambda_1 t} - 1)$$

Similar equations can be written for ^{207}Pb and ^{208}Pb using their appropriate radioactive parents and decay constants. If $t = 0$ is taken to represent the time of the formation of the Earth's crust, then these three equations describe the trajectory of the isotopic composition of terrestrial lead from that time. If T is the time elapsed since the formation of the Earth, (*i.e.*, the age of the Earth), and t_m is the time before present at which the lead minerals were formed, then, using the assumptions of the Holmes–Houtermans model given above, the isotopic composition of a common lead deposit formed t_m years ago is given as follows:

$$\frac{^{206}\text{Pb}}{^{204}\text{Pb}} = \left(\frac{^{206}\text{Pb}}{^{204}\text{Pb}}\right)_i + \frac{^{238}\text{U}}{^{204}\text{Pb}}(e^{\lambda_1 T} - e^{\lambda_1 t_m})$$

According to the model, this isotopic composition was fixed t_m years ago (on the separation of the lead minerals from the uranium- and thorium-bearing environment), is unchanged to the present day and is therefore measurable. There are two similar equations for the other two radiogenic stable isotopes of lead. To simplify the manipulation, we can use the following notation:

$$\frac{^{206}\text{Pb}}{^{204}\text{Pb}} = a, \quad \frac{^{207}\text{Pb}}{^{204}\text{Pb}} = b, \quad \frac{^{208}\text{Pb}}{^{204}\text{Pb}} = c$$

and:

$$\frac{^{238}\text{U}}{^{204}\text{Pb}} = \mu, \quad \frac{^{232}\text{Th}}{^{204}\text{Pb}} = \omega$$

The third ratio required (^{235}U/^{204}Pb) can be calculated from the expression for μ, since the ratio of ^{238}U/^{235}U is accepted to be 137.88. Using these symbols, the three equations can be reduced to:

$$a = a_o + \mu(e^{\lambda_1 T} - e^{\lambda_1 t_m})$$

$$b = b_o + \frac{\mu}{137.88}(e^{\lambda_2 T} - e^{\lambda_2 t_m}) \tag{9.11}$$

$$c = c_o + \omega(e^{\lambda_3 T} - e^{\lambda_3 t_m})$$

which form the basis for the Holmes–Houtermans model for the isotopic composition of common lead deposits. Providing we have a value for T (the age of the earth) we can use these equations to predict the so-called *model age* (t_m) for such deposits. The age of the Earth has been relatively precisely estimated using lead isotopic measurements on meteorites. Certain meteorites contain an iron sulfide phase called *troilite* which contains appreciable lead but virtually no uranium or thorium, and which is believed to be the most primeval lead available to us on Earth. This allowed equations derived from the above model to be solved for *T*, and the currently accepted value is now 4.55×10^9 years (Faure, 1986; 312).

Using this value, we can solve the set of equations (9.11), enabling us to determine the age of common lead deposits with single-stage histories (*i.e.*, formed by a single metallogenic event, with no subsequent alteration). This results in a pair of bivariate plots (conventionally $^{207}Pb/^{204}Pb$ *vs.* $^{206}Pb/^{204}Pb$, and $^{208}Pb/^{204}Pb$ *vs.* $^{206}Pb/^{204}Pb$), the first of which results from solving the pair of equations involving *a* and *b*, and the second from *b* and *c*. A graphical representation of the diagram for $^{207}Pb/^{204}Pb$ *vs* $^{206}Pb/^{204}Pb$ is shown in Figure 9.2. The solution of these equations is a set of growth curves, depending on the initial value of the parameter μ (the ratio $^{238}U/^{204}Pb$) at the time of the formation of the deposit (which is usually assumed to be the same as the present day overall value in the geographical region of interest,

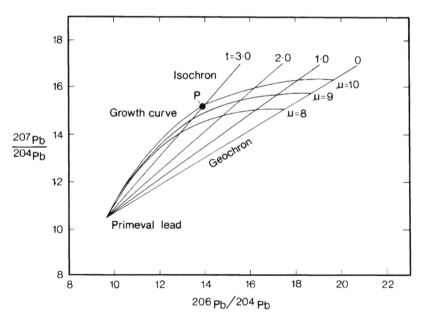

Figure 9.2 Holmes–Houtermans model for the evolution of common lead deposits. Growth curves for values of μ (ratio of $^{238}U/^{204}Pb$) equal to 8, 9, and 10 are shown. Isochrons (see text) corresponding to model ages of 1, 2, and 3 billion (10^9) years are plotted. The point P corresponds to the isotope ratios of a deposit formed 3 billion years ago with a μ value of 10. The isochron corresponding to a model age of zero gives the range of compositions expected from deposits forming now, depending on the μ value of the deposit, and is termed the geochron. (From Faure, 1986; Figure 19.2. Copyright 1986 John Wiley & Sons, Inc. Reprinted by permission of the publisher.)

since ^{238}U decays only very slowly, and ^{204}Pb is not radiogenic). All of these trajectories start at a point which represents the isotopic composition of primeval lead (corresponding to $t_m = T$), and end at a point equivalent to the isotopic composition of common lead minerals formed today (when $t_m = 0$). The exact location of any deposit along any of these trajectories depends on the time of formation of the deposit, but all deposits formed at the same time lie along straight lines starting at the primeval point, termed *isochrons*. The equation of these isochrons can be calculated by eliminating μ from the first two of equations (9.11), and the isochron gradient (m) is then given by:

$$m = \frac{1}{137.88} \left(\frac{e^{\lambda_2 T} - e^{\lambda_2 t_m}}{e^{\lambda_1 T} - e^{\lambda_1 t_m}} \right)$$

which shows that the gradient of the isochron is purely a function of t_m. It follows, therefore, that all common lead deposits formed at the same time should lie on the same isochron, cutting through the primeval composition. The isochron which defines a model age of zero is termed the *geochron*, and corresponds to the predicted isotopic composition of lead minerals forming now. Distance from the primeval isotopic value along the isochron is therefore simply dependent on the value of μ in the ore-forming region. A value of $\mu = 0$ (an unlikely occurrence on Earth) would give a mineral whose isotopic composition was the same as that of primeval lead. In theory therefore, no isotopic value of lead should plot in the region to the right and below the geochron, since this, according to the model, corresponds to a deposit which will form some time in the future! In fact, such plots do occur, but the explanation is somewhat more prosaic. In principle, however, plotting the isotopic ratios of a single-stage common lead deposit onto this diagram should allow the age of formation and the value of μ to be predicted.

Unfortunately, it turns out that very few terrestrial deposits actually conform to this model. Faure suggests (1986; 316) that only about ten deposits (mostly in Australia) were so classified in 1968, and these all fitted the $\mu = 8$ growth line. He noted that these deposits all occurred in volcanic and sedimentary stratigraphic sequences of marine origin in volcanic island arcs. The lead in these deposits is thought to represent metal which was extruded through the crust and emplaced in an environment which virtually eliminated any mixing with crustal rocks (Russell and Farquhar, 1960; 53). These so-called *conformable* deposits therefore represent the isotopic composition of lead in the upper mantle at the time of formation, which was uniform on a large scale, and are uncontaminated with radiogenic leads from the crust. This explains why they are common leads, and also why they have been observed to have a very narrow range of isotopic compositions. The majority of terrestrial deposits were not formed in this way. All deposits that are younger than the rocks hosting them, for example in mineral veins, cannot be explained using the single-stage model so far described. These deposits have complex lead evolution histories which involve mixing with upper crustal radiogenic lead.

The Holmes–Houtermans model does, however, form a basis for our further understanding of the majority of these deposits. Stacey and Kramers (1975) suggested a two-stage model for the isotopic evolution of common leads, which in general gives dates of mineralization which are closer to the correct age, as judged from the geological age of the surrounding rocks. The evolution of lead deposits started as primeval lead some 4.5×10^9 years ago, as envisaged in the Holmes–Houtermans model, but at a more recent time (calculated to be 3.7×10^9 years ago) the U/Pb and Th/Pb ratios of the reservoir changed, and subsequent isotopic evolution can be defined in terms of isochrons focusing not on the primeval 'origin' but on this secondary event (see Figure 9.3). The date of the formation of a common ore deposit can therefore be calculated from a knowledge of the U/Pb and Th/Pb ratios in the appropriate reservoir, as in the Holmes–Houtermans model. The cause of the change in the U/Pb and Th/Pb ratios of the reservoir is not certain, but it is thought to represent a switch in the source of ore fluid from the mantle to a mixed mantle–subducted continental crust from which some lead had already been extracted.

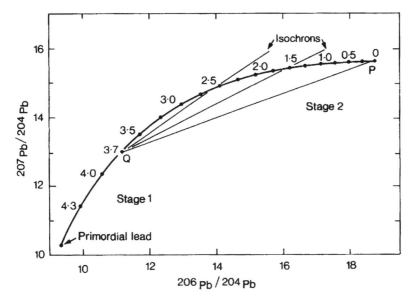

Figure 9.3 Stacey and Kramers' two-stage evolutionary model for lead isotopes. Point Q (at 3.7 billion years ago) indicates geochemical differentiation in the reservoir, and development subsequently occurs with a μ value of 9.735, compared to μ = 7.192 between P and Q. Isochrons grow from point Q – values for 0, 1.5 and 2.5 billion years are shown. (From Faure, 1986; Figure 19.6. Copyright 1986 John Wiley & Sons, Inc. Reprinted by permission of the publisher.)

More general models for the isotopic evolution of lead ores have been devised by Cumming and Richards (1975), amongst others [see Gulson (1986; Chapter 8) for a more complete summary]. These are termed *continuous evolution models*, in which it is recognized that the U/Pb and Th/Pb ratios have been constantly increasing in the source reservoir with time (*i.e.*, the value of μ changes systematically, unlike the previous models which assume a fixed value). The most sophisticated of these models is that due to Amov (1983), which can be simplified to give the other models in certain circumstances. It is based on the assumption that the ratios $^{238}U/^{204}Pb$ and $^{232}Th/^{204}Pb$ have changed continually during the evolution of the Earth, and that temperature is the controlling factor determining chemical fractionation between the crust and mantle. During the early stage of the Earth's history, the crust was at a relatively high temperature and the mobility of uranium and thorium was higher than that of lead under these conditions, and so the ratios listed above increased. On cooling, the replenishment of these elements from the mantle diminished, and the ratios decreased. The equations for this model are complex, but can be reduced to either the single-stage (Holmes–Houtermans) model if it is assumed that crustal formation and the beginnings of lead ore formation are contemporary, or the two-stage model if the date provided by Stacey and Kramers (1975) for the secondary differentiation of the ore reservoir is accepted. The power of the Amov model is that other scenarios can also be considered.

These improved models still principally apply only to common leads, in which mixing with crustal rocks during emplacement was negligible. Most lead and base metal deposits are of *hydrothermal* origin, implying that the metals were transported to the site of ore deposition by warm aqueous solutions (typically 50–400 °C) containing high quantities of dissolved solids [up to 30 weight percent (wt %)]. During the process of ore genesis hydrothermal fluids circulate in the surrounding country rocks leaching metals which are subsequently re-deposited to form the mineralization. The leaching of metals from a variety of sources, with various U/Pb and Th/Pb ratios, gives the lead an anomalous nature. Mineral deposits with complex lead evolution histories may be characterized by highly variable isotopic compositions and/or provide model ages that lie in the future due to a high radiogenic lead content, such as the *Mississippi Valley-type* deposits (Gulson, 1986; 154). These can be studied using multi-stage models which are discussed in detail by Gale and Mussett (1973), and reviewed by Gulson (1986; section 8.4). Several processes can be responsible for these deposits, but the simplest to visualize is the mixing of lead from different sources. Russell and Farquhar (1960) noticed that the isotope ratios of anomalous deposits from a particular ore body or mining region tended to lie along straight lines when plotted on an isotope ratio diagram similar to Figure 9.2. They attributed these 'anomalous lead lines' to the mixing of ordinary (single-stage) lead with radiogenic lead derived from crustal minerals with different levels of uranium and thorium. Although this model can allow the age of formation of the single-stage deposit to be predicted, together with

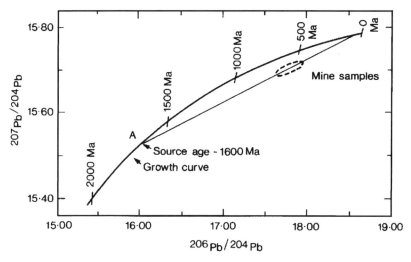

Figure 9.4 Mixing model for the galena deposit at Bingham, Utah. The observed isotope ratios suggest that the deposit was formed from the mixing of a source of age 1.6 billion years with material of a much more recent age. (Gulson, 1986; Figure 8.4, by permission of Elsevier Science and the author.)

the origin and age of formation of the contaminating radiogenic lead (Faure, 1986; 319), it is still limited to lead deposits which are primarily single stage in origin, and is ultimately therefore of restricted value. This approach is illustrated in Figure 9.4 (from Gulson, 1986). This shows that the isotopic composition of lead from Bingham Mine, Utah, falls on a *secondary isochron* attributed to the mixing of leads of two ages, one approximately 1.6 billion years old and the other modern. It must be stated that not all anomalous deposits can be interpreted as simply as this, and not all linear distributions of isotopic data have such chronological significance.

Figure 9.5 shows a simplified schematic interpretation for some of these models. It is now generally agreed that no model is particularly appropriate at the detailed level, but, whether it can be modelled or not, it is still true to say that the measured ratios are related strongly to the age of the ore body and the geological processes such as mixing which have taken place. Therefore the isotopic 'signature' of a particular ore body may or may not be unique – ore bodies formed at the same geological time in the same manner and in the same country rock would be expected to have very similar isotopic ratios. Furthermore, the values of the isotopic ratios are constrained, in that they should conform to theoretically predictable growth curves when plotted appropriately. These constraints may well mean that isotopic data cannot be treated as if they were randomly distributed variables. There is also good reason to suppose that the isotopic homogeneity of a deposit will be strongly influenced by its geological history. All of these factors are particularly important when the

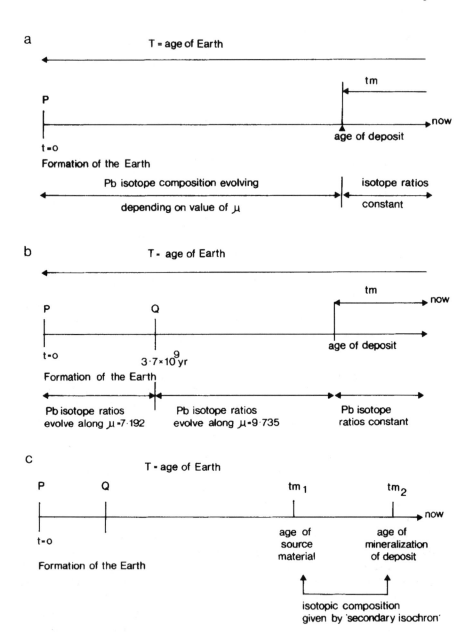

Figure 9.5   Schematic diagrams to illustrate some of the models for the evolution of the lead isotope ratios in lead deposits: (a) Holmes–Houtermans, (b) Stacey–Kramers, (c) mixing model.

measurement of lead isotope ratios is applied to archaeology, as discussed in the next section.

9.5 LEAD ISOTOPES IN ARCHAEOLOGY

The use of lead isotopes in geology has become extremely well-established since the 1930s – originally to provide an estimate of the age of the Earth and subsequently for dating the emplacement of metalliferous deposits, and ultimately as a geochemical prospection technique for locating suitable deposits for commercial exploitation (Gulson, 1986). Given some of the problems regarding metal provenance identified above, it is not surprising that the pioneering lead isotope work of Brill and Wampler (1967) was eagerly seized upon by archaeologists as a potential way out of the growing impasse. For reasons not explained, in this work the isotope ratios were plotted as $^{206}Pb/^{204}Pb$ *vs.* $^{206}Pb/^{207}Pb$ and $^{208}Pb/^{207}Pb$ *vs.* $^{206}Pb/^{207}Pb$ instead of the usual parameters used by isotope geologists. Nevertheless, it was clear that even with the analytical resolution available at the time, it was possible to differentiate between lead from various sources – on the basis of ore analyses, they separated samples from Laurion (Greece), England and Spain into distinct groups, although they noted, for example, that an ore sample from north-eastern Turkey fell into the same space as that occupied by three ores from England, presaging difficulties to come.

The two great advances in archaeological terms were the realization that lead isotope studies were applicable far beyond the study of metallic lead artefacts, which are relatively rare in the archaeological record, and also the demonstration that anthropogenic high-temperature processing did not, apparently, affect the isotopic ratio of the lead. The first metal to be studied in detail was silver, since the sources of silver are of great interest archaeologically (especially for coinage), and also because silver is normally extracted by cupellation from argentiferous lead sulfide deposits [galena, PbS; *e.g.*, Tylecote (1976; 38)]. Argentiferous galena may only contain around 0.1% Ag, but the ancient cupellation process, in which the molten lead is selectively oxidized to litharge and removed, was quite capable of producing relatively high purity silver [$\sim$95% Ag (Tylecote, 1976; 50)] from such ores. As might be expected, lead is one of the major impurities in this silver (often of the order of a few percent), and this lead is sufficient to produce a lead isotope 'fingerprint' characteristic of the source of the argentiferous galena.

Silver items, however, are also relatively rare in the archaeological record. The most common metal found is either copper, usually alloyed with either tin (bronze) or, in the later periods, zinc (brass), or iron. The latter contains very little lead and, because of severe corrosion problems, its survival rate is often low (but see Degryse *et al.*, 2007). Fortunately, copper can also be characterized from its lead isotope signature, since the primary ore of copper is chalcopyrite ($CuFeS_2$), which often co-occurs with galena (PbS) and sphalerite (ZnS). Even if the ore used is a secondary mineral formed by the oxidation of the primary deposit, the copper smelted from such a deposit would normally be expected to contain trace amounts of lead, which can therefore be compared with the isotope signature of the ore. Apart from the potential complications discussed below, this assumes two things:

(i) that the only source of lead in the copper alloy is that which is derived from the primary copper ore (*i.e.*, deliberately leaded copper alloys, such as those common from the Late Bronze Age onwards, cannot be expected to give information about the source of the copper);

(ii) that mixing of copper from different sources (or the addition of re-cycled metal during processing) is negligible.

The second of these assumptions has been the subject of some debate (Budd *et al.*, 1995a), and is discussed further below. Despite these possible complications, the method of lead isotope provenancing was applied enthusiastically to copper alloy artefacts, especially those from the Late Bronze Age of the Aegean (*e.g.*, Gale and Stos-Gale, 1992, and references therein) up until the late 1990s, when this activity virtually ceased, in part because of the contradictory interpretations which were being proposed.

Although the technique of lead isotope analysis for archaeological provenancing has been in use for more than 40 years, it is only in the past 15 years or so that some of the fundamental assumptions have been seriously reconsidered. The major areas that have been questioned can be classified under three headings:

(i) the assumption that anthropogenic processing such as roasting the ore, extraction of the metal or cupellation produce no isotopic fractionation, and therefore that ore and artefact can always be compared directly;

(ii) the geological and statistical definition of the extent of a 'lead isotope field';

(iii) the archaeological interpretation of lead isotope data, bearing in mind point (ii) and the possibility of significant pooling and recycling of metals. This is discussed below in the context of the debate relating to the reconstruction of the Bronze Age trade in metals in the eastern Mediterranean.

9.5.1 Isotopic Fractionation by Non-Equilibrium Evaporation

The first of these points is the simplest to address. The reference most often cited in defence of the assumption that anthropogenic fractionation is unimportant is Barnes *et al.* (1978). In this pioneering work they compared the lead isotope ratios of galena with that of the lead smelted from it, with litharge (PbO) prepared from the smelted lead, and of a $K_2O–PbO–SiO_2$ glass and a pigment ($Pb_2Sb_2O_7$) prepared from the lead. As reported in that paper, all the yields were high for each stage – 98.6% for lead recovery from galena, 95% for the production of litharge and presumably virtually 100% for the other stages.

The authors reported no measurable difference in any of the measured ratios, at least within the precision obtainable by the measurement techniques of the day. This latter point is an important one, since Gale and Stos-Gale (1992) have observed that all of Brill's analyses published prior to 1974 '*are of an accuracy too low to be of any use*' (Gale and Stos-Gale, 1992; 70). (This is a comment purely on the precision of the measurements and also applies to all of the earlier compilations of lead isotope data.) In this case, the publication by Barnes *et al.* post-dates that cut-off and, although the reference in their text to the method employed is dated 1973 (Barnes *et al.*, 1973), it has to be assumed that this work conforms to modern standards of measurement acceptability. Given this, with hindsight it is possible to show that the experiments they carried out were not those which might be expected to show significant fractionation, and their evidence cannot therefore necessarily be taken as convincing proof of the lack of anthropogenic fractionation in all processes.

In a similarly pioneering piece of work, Scaife (1993) demonstrated that the theoretical considerations of the thermodynamics of *non-equilibrium evaporation* from a liquid, first published by Mulliken and Harkins (1922), could have a significant effect on the isotopic ratio of lead as it is processed, providing the non-equilibrium losses themselves are sufficiently large. Kinetic theory predicts that the lightest isotope will preferentially enter the vapour phase on evaporation, leaving the liquid enriched in the heavier isotope and the vapour phase in the lighter. If evaporation goes to completion (*i.e.*, all the liquid is evaporated) or the system is left to equilibrate, then there is no change in the isotopic ratio of the liquid or condensate. Non-equilibrium evaporation is the situation in which the vapour phase is removed [by condensation, as in the original experiments of Mulliken and Harkins (1922), or as a result of air flow, as is more likely in lead processing]. Under these conditions, the lighter vapour phase will be removed and the liquid will be gradually enriched in the heavier isotope. The calculation can be extended to predict the change in isotopic ratio as a function of percentage non-equilibrium loss if the liquid contains more than two isotopes. The result of this theoretical calculation for lead is best summarized diagrammatically, as shown in Figure 9.6. Here it has been assumed that a sample from close to the centroid of the published lead isotope field for Laurion (as defined by Gale *et al.*, 1980 and Stos-Gale *et al.*, 1986) has been subjected to increasing weight loss under non-equilibrium evaporation conditions, and the straight line shows the trajectory of the isotopic ratio of that sample as a function of the percentage of lead removed. It is clear in this case that 40% non-equilibrium losses would be sufficient to bring the sample to the edge of the field, and above around 60% would effectively remove it completely from the field as normally defined.

Although this is theoretically convincing, the key question to be answered is whether any of the ancient processes for the production of lead and silver might be expected to give such a large non-equilibrium loss. The most likely contender for a process with such losses seems to be cupellation, in which a melt of around 99% Pb with less than 1% silver is converted via preferential oxidation into an alloy with greater than 95% silver – a total loss in excess of 99% of lead (Budd

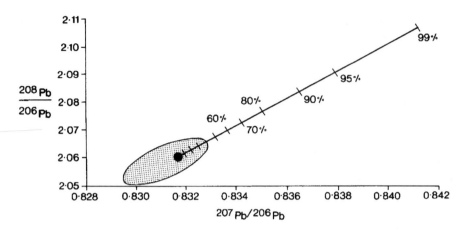

Figure 9.6   The theoretical effect of non-equilibrium evaporation of lead on the isotopic ratio of a typical sample from the Laurion field. The shaded area shows the extent of the Laurion field as commonly defined, and the diagonal line shows the effect of increasing non-equilibrium losses on the isotopic ratio of a sample from the centre of the field. (From Scaife, 1993, with permission of the author and the University of Bradford.)

et al., 1995b). However, experiments conducted by Pernicka and Bachmann in 1983 on the cupellation of silver from Laurion galenas concluded that the process did not induce fractionation. The Ancient Metallurgy Research Group in Bradford conducted a series of simulated laboratory metallurgical processes on galenas and lead, and, despite occasionally achieving significant lead losses, isotopic measurements showed no significant change in the isotopic ratios (Budd *et al.*, 1995c; McGill *et al.*, 1999). On this evidence, therefore, we must conclude that the circumstances under which non-equilibrium evaporation of lead may have taken place during most ancient processing techniques are unlikely to be significant. It remains possible, however, that given recent improvements in the sensitivity of lead isotope measurements using high-resolution ICP-MS (see Chapter 2 and below), small isotopic shifts as a result of realistic lead losses during processing might now be measurable – this is yet to be tested.

   This excursion into the theory of non-equilibrium thermodynamics was not, however, entirely wasted. The idea was subsequently applied to the possible anthropogenic fractionation of tin in bronze (Budd *et al.*, 1995c; Gale, 1997; Yi *et al.*, 1999; Clayton *et al.*, 2002) and zinc in brass (Budd *et al.*, 1999). The possible non-equilibrium evaporation of tin in ancient bronzes gave rise to speculation that it might be a fruitful method for quantifying the degree of recycling of copper alloys (Budd *et al.*, 1995d; Yi *et al.*, 1999). Although not focussing on this aspect of possible anthropogenic fractionation, Clayton *et al.* (2002) subsequently published precise determinations of the tin isotopic ratio ($^{122}Sn/^{116}Sn$) of metallic tin and a Malaysian cassiterite sample using high-resolution multiple collector ICP-MS and showed that natural isotopic fractionation of tin during ore formation can now be measured. If different deposits of cassiterite have different tin isotope signatures, then this opens up the prospect of scientifically addressing one of the most intractable questions in European archaeology – '*Where did the tin come from in the Bronze Age?*' (Budd *et al.*, 1994). Neither of these tantalizing prospects has yet been adequately followed up.

Perhaps the most interesting alloy to which process-induced fractionation might apply is brass (an alloy of copper and zinc), given the volatility of zinc and the variation in historical brass production methods (see Chapter 6). Brass melting experiments at 1100 °C followed by quadrupole ICP-MS zinc isotope ratio measurements ($^{68}Zn/^{64}Zn$) on the resulting alloys suggested that the non-equilibrium evaporation model is particularly appropriate to this system (Budd *et al.*, 1999). These data indicated that the change in zinc isotope ratio of the residual alloy should be measurable by quadrupole ICP-MS for evaporative zinc losses of more than about 30%. It is unlikely, however, that measurements at the precision available from quadrupole ICP-MS would be sufficient to distinguish between the two principal historical brassmaking processes (the cementation process, and the more recent direct process – see Chapter 6), although again more precise measurements using a high-resolution multi-collector ICP-MS may well be possible. Thus this may also be a profitable area to explore with newer instrumentation.

9.5.2 Defining an Ore Field

The second area of debate – the definition of the isotopic extent of a particular ore field – is complex, involving a discussion of the isotope geology ore deposits, the sampling strategy used in obtaining the samples to define the isotopic field and the nature of the statistical procedures used to define the field boundaries. The first question to be addressed is the extent of natural variation of the isotopic ratios in an ore deposit. TIMS is capable of making measurements accurate to an absolute 95% error of $\pm 0.05\%$ in the $^{207}Pb/^{206}Pb$ ratio, $\pm 0.1\%$ for $^{208}Pb/^{206}Pb$ and $^{206}Pb/^{204}Pb$ (Gale and Stos-Gale, 1993). The isotopic homogeneity of a conformable massive sulfide deposit is often quoted to be about ± 0.1 to $\pm 0.3\%$ in the $^{206}Pb/^{204}Pb$ ratio (Gulson, 1986; 30), although it is well-known that most deposits show much greater variation, such as the Pb–Zn sulfides in the Upper Mississippi Valley, which show a large gradation across the deposit [nearly 9%, with $^{206}Pb/^{204}Pb$ varying from 21.88 to 23.96 (Gulson, 1986; Figure 1.4)]. Moreover, work on large single crystals of galena (*e.g.*, Hart *et al.*, 1981) has shown variations across a 13 cm crystal from Buick Mine in south-east Missouri which are as large as the variations within the ore body itself (approximately 5% in $^{208}Pb/^{206}Pb$ and 4% in $^{207}Pb/^{206}Pb$). Although these are likely to be exceptional, it does highlight the necessity for good-quality geological sampling of a deposit.

It is clearly important to have a good knowledge of the ore geology of the area before isotopic provenancing is undertaken. Unfortunately, ancient mining is often in regions no longer considered economically viable, and the data required by isotope archaeologists are not often available in the published geological literature. This requires that the archaeological programme should include a fieldwork strategy for sampling the relevant deposits as well as measuring the isotopic ratios in archaeological artefacts. This involves a number of considerations, ranging from the necessity for the mineralogical characterization of the deposit to the collection of an adequate number of samples to isotopically characterize the deposit. It cannot be assumed that if the isotopic signatures of lead minerals from a deposit have been measured, then the traces of lead in, say, an associated copper mineral will be the same. Pernicka (1993) has noted that an increasing number of copper minerals can be shown to have a different lead isotopic signature from the co-existing lead minerals because of incorporation of uranium and/or thorium at the time of deposit formation. It is therefore important to compare metal artefacts only with the kind of ore from which they may have come, such as silver objects with argentiferous galenas, copper alloys with copper minerals, *etc.* This principle has been repeatedly stated in the archaeological literature (*e.g.*, Gale and Stos-Gale, 1992; 73), but has occasionally been sacrificed in the interest of increasing the number of measurements which can be used to define a 'field'. A second, and related, principle which has been repeatedly stated is that only those ores available to ancient miners need be considered when defining a field. From one point of view this is a perfectly valid position – clearly modern isotope data from deposits so deep that no ancient mining could possibly have exploited them, or from ores which are so finely dispersed that only modern extractive metallurgy can win metal, should not be included in any ore field constructed purely for archaeological purposes. In fact, one might wish to go further – only those ore deposits for which there is archaeological evidence of exploitation at the period in question should be considered as potential sources. (This raises a number of very difficult side issues, such as the lack of any direct means of dating ancient mining activity, and the validity of the assumption that we actually know – or will ever know – *all* of the sources of metal exploited in antiquity, which is very unlikely.) On the other hand, this dictum can be seen as being in direct contradiction with the statement made above – that a comprehensive geological understanding of an ore deposit is required in order to characterize it fully. Clearly any archaeological research strategy must balance out these requirements.

One of the most controversial issues is how the source isotopic fields are constructed from collected geological samples. There is wide agreement that a statistically valid number of samples must be measured. Gale and Stos-Gale (1992; 76) stress the need for '*at least 20 geologically well-selected ore samples*' per deposit, which seems to be an agreeable minimum level. There is also wide agreement that the data consists of precise measurements of three isotopic ratios, and that all three need to be taken into account when attempting to discriminate between deposits. Suggestions that the data should be converted into abundance measurements of the four isotopes (*e.g.*, Reedy and Reedy, 1992), although perhaps soundly based statistically, are unlikely to be helpful in the light of the method used to produce the data (*i.e.*, direct measurement of three isotope ratios). The debate therefore centres around how these three ratios are manipulated, and, as an important side issue, the nature of 'outliers' in such data (Budd *et al.*, 1993; Scaife *et al.*, 1996). Key questions are how can the isotopic extent of ore fields be predicted from a relatively small number of samples, what shape should they be, do they form discrete or overlapping groups and how can 'unknowns' (isotopic measurements from archaeological samples) be assigned to the appropriate source group?

Some have approached this problem as if they were dealing with analytical data from, for example, archaeological ceramics (*e.g.*, Sayre *et al.*, 1992). This leads to the definition of isotopic field boundaries as 95% 'confidence limits' around groups of points, calculated on the assumption of multivariate normality. Others have favoured the use of linear discriminant function analysis (DFA) to distinguish between pre-defined groups (*e.g.*, Gale and Stos-Gale, 1992; 73). Scaife *et al.* (1996) have challenged both these approaches, not on the grounds of statistical propriety *per se*, but as a result of considering the nature of the data themselves. The multivariate normality of lead isotope data has to be questionable, on the grounds that the data are not a random sample drawn from a multivariately normal parent population. The parent population (the isotopic data characteristic of the ore deposit) is a constrained set – constrained by the equations of isotopic evolution discussed above. Scaife *et al.* (1996) have argued that one procedure for constructing ore fields might involve modelling the isotopic data, rather than treating it statistically as if it were unconstrained trace element data. Baxter and Gale (1998) independently concluded that lead isotope ratios cannot safely be assumed to be normally distributed. Clearly this will have a strong influence over the separability of the different isotopic fields, and therefore the degree of certainty with which archaeological samples can be identified to source.

Lead isotope data are completely represented using a pair of bivariate diagrams – one plotting $^{208}Pb/^{206}Pb$ against $^{207}Pb/^{206}Pb$ and another, less often used, showing $^{204}Pb/^{206}Pb$ against $^{207}Pb/^{206}Pb$. Improvements in desk-top computing power over the past ten years mean that lead isotope data can now be fully visualized in a rotatable three-dimensional plot defined by the three independent isotopic ratios. If all the available data points from an ore body are plotted (with their error bars) onto such a diagram, this gives the most valid description of that ore body's isotope field that is possible without geochemical modelling (Scaife *et al.*, 1996). If an unknown sample falls within the limits of this isotope field then it cannot, on isotopic evidence alone, be excluded from that field. If, on the other hand, it lies outside these limits, then it cannot securely be attributed to that field. The problem then reduces to one of defining the limits of the isotope field, taking into account geochemical ore evolution models, together with sampling and measurement uncertainties.

It follows from this that the concept of what constitutes an 'outlier' is vitally important, since it is common practice for 'outliers' to be omitted from the definition of isotopic fields. Scaife *et al.* (1996) argue that if an accurate and precise measurement is made on a sample which is geologically securely tied to a specific ore deposit, then it cannot be classified as an outlier, no matter how isotopically different it is from other samples in that deposit. Given the isotopic inhomogeneity of many lead sulfide deposits (discussed above), it is not inconceivable that large variation might be found within a small number of samples, which should not lead to the extreme values being discarded as 'outliers'. Rather, it should be taken as an indication that the relevant 'isotopic field' has previously been too tightly defined. This might be too idealistic in some cases (*e.g.*, if the sample is very low in lead, in which case the isotope measurement is more difficult) but it must provide the starting point for the construction of isotopic fields. In general, all samples collected from a particular deposit must be used in the definition of its isotope field, unless an error has occurred, or the measurement is demonstrably dubious for some reason (*e.g.*, by reference to standards). Some concern has been expressed about the understandable but doubtful practice of re-analysing those samples (and only those samples) which fall outside some pre-conceived field boundaries (Begemann *et al.*, 1995). In such cases, a more scientific re-sampling procedure is necessary, otherwise there is an obvious danger of constructing the field to fit a pre-determined shape and size.

In the absence of an assumed underlying normal distribution, simple bivariate plotting does not lead to an estimate of the true extent of the parent isotope field. This is particularly a problem if only relatively few samples are available, as is usually the case. Kernel density estimation (KDE; Baxter *et al.*, 1997) offers the prospect of building up an estimate of the true shape and size of an isotope field whilst making few extra assumptions about the data. Scaife *et al.* (1999) showed that lead isotope data can be fully described using KDE without resort to 'confidence ellipses' which assume normality, and which are much less susceptible to the influence of outliers. The results of this approach are discussed in Section 9.6, after the conventional approach to interpreting lead isotope data in the eastern Mediterranean has been discussed.

9.6 LEAD ISOTOPES AND THE BRONZE AGE MEDITERRANEAN

Whilst these matters of detail have formed a lively debate in the literature and elsewhere, the important question to consider is how reliably can lead isotope data be interpreted archaeologically, and how does this impact on our knowledge of the ancient trade in metals. A good deal of the archaeological effort in lead isotope analysis historically has been devoted to studying the trade in metals in the Mediterranean in the Bronze Age and later. This has been the subject of several reviews – for example, Gale and Stos-Gale (1992) have summarized their lead isotope work carried out in Oxford as part of a larger British Academy project, which was an ambitious attempt to synthesize a range of scientific data with more traditional archaeological approaches to achieve an understanding of trade in the Bronze Age Mediterranean. A final report summarising the achievements of this laudable effort in relation to Cyprus has been published (Knapp and Cherry, 1994). Sayre *et al.* (1992) have summarized the work carried out largely under the auspices of the Smithsonian Institution, Washington, D.C., and the Brookhaven National Laboratory relating to the sources of metalliferous ores in ancient Anatolia, which clearly has an important relationship to the work in the Eastern Aegean. A group based in Heidelberg and Mainz has also been working for many years in the eastern Mediterranean and Anatolia (*e.g.*, Wagner *et al.*, 1986). Figure 9.7 shows a map of some of the more important prehistoric copper sources in the eastern Mediterranean.

Unfortunately, although substantial efforts have been made in recent years to publish the backlog of accumulated lead isotope data, particularly for the ore sources in and around the Aegean (*e.g.*, Rohl, 1996; Stos-Gale *et al.*, 1995, 1996, 1998; Gale *et al.*, 1997; Sayre *et al.*, 2001), much of the basic data still remain unpublished, and it is sometimes therefore difficult to evaluate the competing interpretations put forward by these three groups. Nevertheless, it is possible to summarize essentially two competing interpretations from the lead isotope data. On the one hand, Pernicka and co-workers contend that: '*...There is, in fact, hardly any need for multivariate statistical methods at all. Searching for subtle differences in various projections of the data is likely to lead to overinterpretation*' (Pernicka, 1993; 259). They claim that '*no convincing case has yet been presented where multivariate statistical methods revealed more information than the two diagrams* $^{208}Pb/^{206}Pb$ vs $^{207}Pb/^{206}Pb$ *and* $^{204}Pb/^{206}Pb$ vs $^{207}Pb/^{206}Pb$'', and that '*it is simply unrealistic to pretend that a unique characterization of ore deposits will eventually be possible, if one would only use more sophisticated methods of data analysis*' (Pernicka, 1993; 259). The outcome of this approach is the view that it is only possible (or indeed desirable) to define ore fields which cover a large geographical area, and which therefore provide little in the way of highly detailed provenance information. As an example, Figure 9.8 shows the 'Aegean field' defined by Pernicka *et al.* [1984; as published by Gale and Stos-Gale (1992; Figure 4)], which includes samples from the Troas, in western Anatolia, and many sites within the Aegean. This parsimonious approach is in direct contrast to the more ambitious attempts of the other groups to separate individual ore deposits within this area. An example of the level of discrimination seen by the Oxford group is shown in Figure 9.9 (also from Gale and Stos-Gale, 1992; Figure 13), where some of the copper ores from Lavrion near Athens, the Aegean and Anatolia are identified on one of the pair of lead isotope ratio plots. Apparent overlaps are resolved by discriminant analysis on selected groups, such as the resolution of the two Troad groups from the Kythnos samples, shown in Figure 9.10 (Gale and Stos-Gale, 1992; Figure 14). A similar level of separation for other sources is claimed by the Brookhaven–Washington group, as exemplified by Figure 9.11 (from Sayre *et al.*, 1992; Figure 2).

These figures have been selected for no other reason than to illustrate the varying levels of division in the data which the original authors themselves feel justifiable. It is noteworthy, however, that there is little agreement in terms of field definition between either of the two teams who wish to see the 'Aegean field' of Pernicka and co-workers broken down into smaller units. For example, Figure 9.9 (from the Oxford group) shows a single field for the lead isotope ratios in Cypriot copper ores, whereas Figure 9.11 from the Smithsonian group

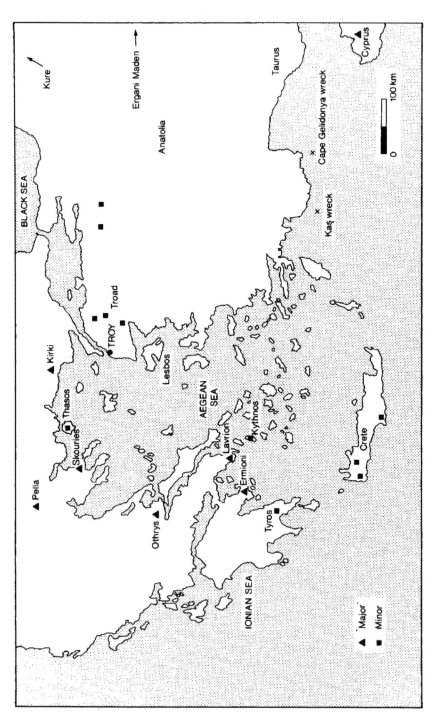

Figure 9.7 Map of some of the more important prehistoric copper sources in the eastern Mediterranean. (Adapted from Stos-Gale and Gale, 1990; Figure 1, in *Thera and the Aegean World III*, published with permission of the Thera Foundation, London, and the authors.)

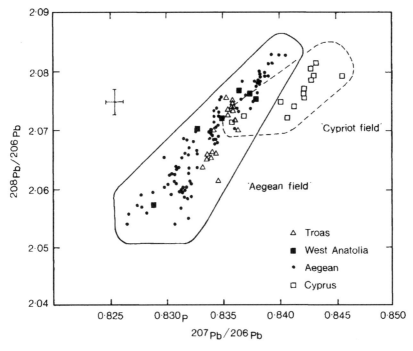

Figure 9.8 The generalized 'Aegean field' defined by Pernicka *et al.* (1984), as given by Gale and Stos-Gale, 1992, Figure 4. (Note: the horizontal axis was labelled as ^{208}Pb/^{206}Pb by Gale and Stos-Gale, and has been corrected here.) The cross in the upper left corner indicates the estimated precision of a single measurement. (Reproduced with permission from *Proceedings of the British Academy*, vol. 77, *New Developments in Archaeological Science*. © The British Academy 1992.)

shows a similar field, but divided into two separate groups. It has been known for many years that the general application of some multivariate methods of analysis to archaeological data may have a tendency to over-divide the data into apparently (statistically) valid sub-groups, but which may have no archaeological significance (Pollard, 1982), and this may be a further example of that problem.

As might be anticipated, with so little agreement about the definitions of source ore fields between the major practitioners, there has been a great deal of debate about the archaeological interpretations to be placed on such data. That debate is typified by the problems over the source attributions of the famous *oxhide ingots* recovered in large numbers from the Bronze Age Cape Gelidonya (13th Century BC) and Ulu Burun (Kaş; late 14th Century BC) shipwrecks (Bass, 1967; 1986; Bass *et al.*, 1989). These are large ingots (typically between 20 and 30 kg) of relatively pure copper, whose name refers to the similarity between their shape and that of the stretched out hide of an ox. It has been assumed that they represent trade in copper as a raw material, with an assumed

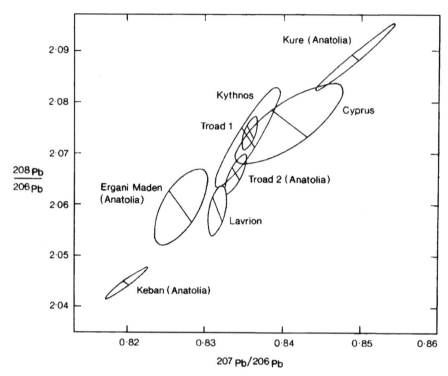

Figure 9.9 Bivariate lead isotope ratio diagram for copper ores from some Aegean and Anatolian deposits, as defined by the Oxford group (Gale and Stos-Gale, 1992; Figure 13). (Reproduced with permission from *Proceedings of the British Academy*, vol. 77, *New Developments in Archaeological Science*. © The British Academy 1992.)

origin of Cyprus, largely because of the known richness of this island in terms of copper ore, and also because of the extensive evidence for prehistoric working of these deposits (Gale and Stos-Gale, 1992; 87). The earliest oxhide ingots so far known (principally from Hagia Triadha on Crete, and dated to the 16th Century BC) were analysed isotopically for lead, and appeared to have different isotope signatures from the copper sources on Cyprus. They appeared to represent two sources with model ages calculated via the Cumming and Richards method to approximately 375 and 640 million years respectively (compared to the model age of around 100 million years for Cyprus). From a knowledge of the ore geology of the Mediterranean and Near East, Gale and Stos-Gale (1992; 91) concluded that the most likely sources were to be found in Iran or Afghanistan – a possibility not totally at odds with other archaeological evidence, in the light of the known presence of lapis lazuli (presumed to be from Afghanistan) in Minoan and Mycenean contexts. Further work demonstrated that the 13th Century BC oxhide ingots found on Cyprus were isotopically consistent with the Cyprus copper ores, suggesting that by this time the vast reserves of copper on Cyprus were indeed being exploited (Stos-Gale *et al.*, 1997).

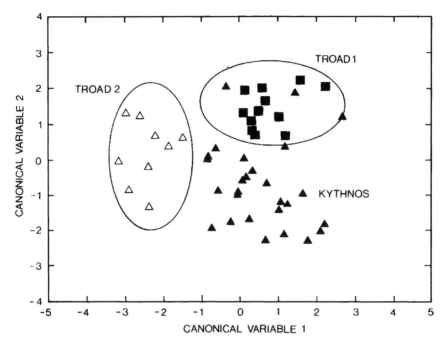

Figure 9.10   Discriminant analysis of three overlapping groups shown in Figure 9.9, showing apparent separation of Troad 1 and 2 (in Anatolia) and Kythnian copper and slags (Gale and Stos-Gale, 1992; Figure 14). (Reproduced with permission from *Proceedings of the British Academy,* vol. 77, *New Developments in Archaeological Science.* © The British Academy 1992.)

More significant was the data on the ingots from the two shipwrecks found off the south coast of Turkey, mentioned above. Fifteen of the Cape Gelidonya ingots were analysed in Oxford, and found to be isotopically consistent with the Cypriot field, as were the four samples taken from the 84 oxhide ingots on the Ulu Burun wreck. The latter ship was also carrying some much smaller copper ingots of a different shape, termed 'bun' ingots, some of which were also analysed, and not all were found to be consistent with Cyprus. This evidence was felt to be sufficiently strong to persuade Gale and Stos-Gale (1992; 94) to state that '*We can at least be sure that some Cypriot copper was being carried through the Mediterranean in the Late Bronze Age.*' Somewhat more controversial was the conclusion that the large number of oxhide ingot fragments found on Sardinia (some 2000 km distant in the western Mediterranean) were also isotopically identifiable as Cypriot products. Sardinia, whilst not being as rich in copper resources as Cyprus, still has extensive copper deposits. The proposal that copper was being shipped halfway across the Mediterranean to an island with its own abundant supply of copper was greeted with some scepticism in certain quarters (*e.g.*, Knapp *et al.*, 1988; Muhly, 1991).

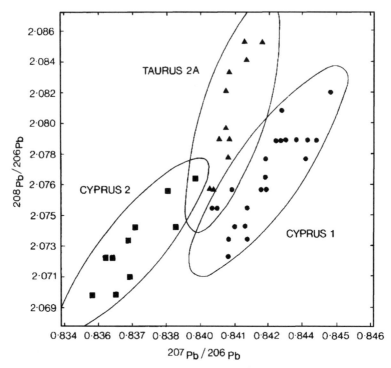

Figure 9.11   Bivariate isotopic ratio plot of Cypriot and Anatolian ore deposits.
(From Sayre *et al.,* 1992; Figure 2, with permission from the University
of Oxford.)

Recalling that one of the three major groups involved in this work in the
Mediterranean does not accept the divisions of the isotopic ore fields of
the kind proposed in these interpretations, it seemed appropriate to re-evaluate
the available data – remembering that much of the basic data remains
unpublished. The results of this reconsideration (Budd *et al.,* 1995a) sided very
heavily with the long-held beliefs of Pernicka *et al.* – *i.e.,* that it is essentially
impossible to subdivide the various ore deposits in the eastern Mediterranean
into individual separate 'fields'. There is some evidence, for example, that there
is a substantial overlap between the Sardinian and Cypriot isotopic signatures,
although this matter is still hotly disputed. An alternative model was subsequently put forward for the apparent isotopic homogeneity of the lead isotopes
in some of the oxhide ingots (Budd *et al.,* 1996), based on the possibility of
extensive mixing of copper sources, which remains to be tested on real data.
Simple modelling of the recycling and mixing of metals from different sources
suggests that the observed homogeneity of the isotopic signatures of objects
such as the oxhide ingots may actually be a natural consequence of recycling,
contrary to the popular view that it is a reliable indicator of metal from a single
source [*e.g.,* '*we do not observe the smeared out lead isotope compositions for
artefacts which would characterize mixing*' (Gale and Stos-Gale, 1992; 68)].

The possibility (Budd *et al.*, 1995c) that non-equilibrium fractionation of tin isotopes in recycled bronzes may allow the degree of recycling to be established (as discussed above) begins to assume great significance.

As discussed above, Scaife *et al.* (1999) proposed that KDEs applied to bivariate lead isotope plots offered the potential to avoid many of the assumptions inherent in the above approaches. They constructed KDE plots for the Troodos Mountain ore field, Cyprus, using published data from Gale *et al.* (1997). Firstly the data were rotated onto new axes parallel and orthogonal to the 121 Ma isochron. This represents a reasonable approximation to the main axis of variation of the data. A KDE was then carried out using kernels parallel to these new axes and a density estimation calculated. The density estimation was then rotated back to the original isotope ratio axes, resulting in Figure 9.12. Shown for comparison on this figure is the original 'Oxford ellipse' previously derived from the same data. The KDE model appears to be a much better representation of the original data than the imposed ellipse, and suggests that use of this ellipse might underestimate the true extent of the ore field, and hence lead to misclassification of the source of ancient artefacts. If this is true, then it implies that objects made from Cypriot copper may have been misclassified in the past as non-Cypriot. This shows that KDE techniques offer a positive and more reliable way forward for the presentation and interpretation of lead isotope data in the future.

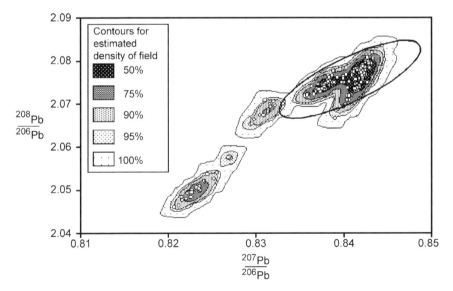

Figure 9.12 Kernel density estimate of the lead isotope data for part of the Troodos ore field, Cyprus (data from Gale *et al.*, 1997). The superimposed 'Oxford' ellipse has been used to represent the Cyprus ore field in several publications. (From Scaife *et al.*, 1999; Figure 6, with permission from the first author.)

9.7   EPILOGUE – 'WHAT A LONG STRANGE TRIP ITS BEEN'!

Provenance studies of inorganic artefacts, particularly using lead isotopes, are not currently the most fashionable branch of archaeological chemistry, and yet the capacity to carry out large-scale studies of this type is now more widely available than ever before. Advances in analytical instrumentation, and specifically the high throughput, multi-element sensitivity and isotopic resolution of the new generation of high-resolution inductively coupled plasma mass spectrometers means that capacity is no longer limited primarily by analytical restrictions. In particular, with ICP-MS the sensitivities for isotopic ratios such as lead are comparable to (if not better) than that of conventional TIMS instrumentation (Halliday *et al.*, 1998). More widespread use of high-resolution ICP-MS machines could herald a new age of rapid and relatively cheap isotopic and chemical studies of archaeological material. It is also likely that many of the questions raised above about the possibility of anthropogenically induced fractionation, not detectable on low-resolution machines, could be profitably revisited using these instruments, finally opening up the prospect of solving some of the more intractable questions about ancient metal processing.

The principle that the measurement of the isotopic ratios of lead in archaeological objects (initially glass, then lead itself, but more usefully silver and copper) can indicate the geological source of the lead has never been seriously disputed. What has been disputed is the degree of geographical specificity that can or should be achieved, and how archaeologically useful this information might be. Since the vigorous debate in the mid-1990s, very little has been published in the way of new archaeological interpretation for the Aegean using lead isotopes. Recent lead isotope studies of metals have focussed elsewhere, including South Asian metal icons (Srinivasan, 1999), Islamic copper objects (Al-Saad, 2000), bronze Punic coins from Sardinia (Attanasio *et al.*, 2001) and Roman silver coins (Ponting *et al.*, 2003). In particular, Ponting *et al.* (2003) use high-resolution ICP-MS with sampling by laser ablation, a clear example of the way forward using this new technology.

As the subject of study using chemical provenancing techniques, ancient metals suffer from all of the difficulties listed elsewhere (*e.g.*, Wilson and Pollard, 2001). The use of lead isotopes in theory circumvents many of these – the only fundamental drawback is if significant quantities of lead have been introduced from a different source, either by the recycling of scrap or by the deliberate addition of lead to improve fluidity during casting. Nevertheless, it does appear that the lead isotope method of provenancing does have some inherent and significant limitations. In their extensive study of the British sources of copper in the Bronze Age, Rohl and Needham (1998) show that, despite the fact that non-ferrous mineralization in the British Isles occurs in four different geological environments, the isotope data show no systematic differences between them. In fact, these authors present the data from all British and Welsh ore sources as 'EWLIO' – the *'English and Welsh Lead Isotope Outline'* – a single all-encompassing envelope. They also show how EWLIO relates to lead from some Scottish, Irish, French and German sources – in all

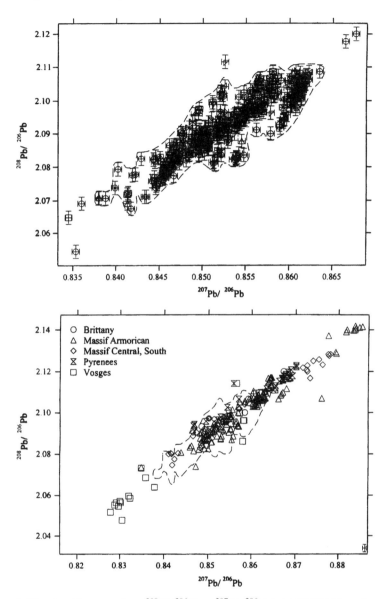

Figure 9.13 Lead isotope plots (^{208}Pb/^{206}Pb *vs* ^{207}Pb/^{206}Pb): (a) 'EWLIO' – galenas from all English and Welsh lead sources (Rohl and Needham, 1998; 37); (b) isotopes from French lead minerals superimposed against EWLIO, showing substantial overlap (Rohl and Needham, 1998; 71). (From Rohl and Needham, 1998, with permission. © The Trustees of the British Museum.)

cases, these data overlie EWLIO, although they often extend beyond it (Figure 9.13). Occasionally some particular attributions can be made (*e.g.*, certain parts of Cornwall have elevated uranium, which gives a highly distinctive lead isotope signature), but it seems clear that lead isotope data are insufficient to uniquely provenance copper alloys of British and nearby origin. This is an important but disappointing finding, confirming as it does the problem first identified in the pioneering work of Brill and Wampler (1967), who observed a single sample of Turkish ore overlapping the field defined by three English ores. The lesson is simple. Not all geographically discrete ore fields are isotopically distinct, and no amount of statistical manipulation can separate groups which fundamentally overlap in the parameters measured.

As a result of this ambiguity, attempts have been made by Rohl and Needham (1998) (amongst others) to combine isotope data with trace element information, specifically by introducing the concept of IMP-LIs – '*impurity composition and lead isotope composition*'. Each IMP-LI is said to represent '*a stock of metal in circulation which might either be similar to the character of a specific source, or alternatively be a complex but relatively coherent amalgam of different metals*' (Rohl and Needham, 1998; 84). Since isotopic and chemical characteristics are independent of each other, neither data set is given primacy in the analysis, and therefore these IMP-LIs are said to be more robust than either data set on their own. Objects from 12 major typological groupings of British Bronze Age metalwork assemblages are classified into 23 IMP-LIs. The authors state '*Previous classifications ... have certainly not been invalidated ... on the contrary, the chemical differences observed find much support in lead isotope ratios. However, the latter do show that at times the chemical patterns are not adequate discriminators between different inputs to the metal supply*' (Rohl and Needham, 1998; 176). The process occasionally also works in reverse – groups with similar isotopic ratios have different chemical compositions.

This work perhaps suggests that (with hindsight) we have been asking the wrong archaeological questions all along. It appears that, despite now having good archaeological evidence for early mining in several parts of the world [including the UK, where a number of Early Bronze Age (*ca.* 2100–1600 BC) mining sites have been investigated by the Early Mines Research Group], we still cannot easily determine the actual ore source (*i.e.*, the mining site) for well-established groups of metalwork. Does this matter? Perhaps we should be focussing more on the changes seen in the chemical and isotopic record in the metal objects themselves, and relating these to stratigraphical or typological variation (a view which might be termed '*processual provenance*'). These directly reflect a change in some aspect of human behaviour, which might indeed be the pattern of ore source exploitation, but could equally be a change in extractive technology, or the practice of metal circulation and recycling. Whatever the cause, the result is something tangible, and perhaps of more direct archaeological interest, ultimately, than knowing that the metal for a particular object came from a particular mine site. Although moving away from the original concept of provenance, it might be argued that the simple detection of change in the material record is a valuable observation when interpreting that record. This is particularly so if such a change is contemporaneous with other changes in the record – stylistic influences, mortuary patterns, stratigraphic sequence or whatever. It might then be reasonable to observe that a variation has occurred in the metal circulation as a result of some social change (*e.g.*, Begemann *et al.*, 1995), rather than necessarily assuming a simplistic shift in raw material extraction patterns or technological change.

9.8 SUMMARY

Although lead isotope techniques have been used in archaeology for more than 40 years, the time has clearly come to look forward, not least because advances in measurement techniques and the power of data manipulation have increased immeasurably in the past ten years, and new insights are undoubtedly possible. There remains, however, a fundamental issue, which goes to the heart of the long-running debate about the role of scientific methods in archaeology (*e.g.*, Pollard, 2004) – how can the social and economic questions posed by archaeology be translated into a geochemical research programme? And, of course, the reverse – how can the results of geochemical research be incorporated into a socio-economic model of past human behaviour? The key question is what kind and level of information is useful in the context of the current social models of trade and exchange for the region being studied. Much effort, particularly in the field of metal studies, has been expended in trying to identify the exact source of the raw materials, even down to a particular mine shaft. It might be argued that the interpretative archaeological framework is not sufficiently well-developed to accommodate such precise information. Perhaps, as suggested above, the knowledge that something has changed, rather than the specifics of what has changed, may be of more immediate archaeological relevance.

It is stated in the first sentence of this chapter that one of the major goals of chemical analysis of metals in general, and lead isotopic analysis in particular, is to '*trace metal objects back to their ore source, and hence reconstruct prehistoric economic contacts.*' Here 'hence' is a big word! With hindsight, it can be seen that the two parts of this aim are not always as well articulated as they need to be. Subject to the considerations outlined above, it is not disputed that the first part of this goal can be achieved to some degree. Whether the second part is achievable at all depends on how well expressed the archaeological problem has been in scientific terms, and how well the analytical programme has been constructed to address this question. We may look forward to a future in which the utility of heavy isotope studies in archaeology is no longer limited by analytical capability, but more by the quality of the articulation of the archaeological problem.

REFERENCES

Al-Saad, Z. (2000). Technology and provenance of a collection of Islamic copper-based objects as found by chemical and lead isotope analysis. *Archaeometry* **42** 385–397.

Amov, G.A. (1983). Evolution of uranogenic and thorogenic lead, 1. A dynamic model of continuous isotopic evolution. *Earth and Planetary Science Letters* **65** 61–74.

Attanasio, D., Bultrini, G. and Ingo, G.M. (2001). The possibility of prove-nancing a series of Bronze Punic coins found at Tharros (western Sardinia), using the literature lead isotope database. *Archaeometry* **43** 529–547.

Barnes, I.L., Murphy, T.J., Gramlich, J.W. and Shields, W.R. (1973). Lead separation by anodic deposition and isotopic ratio mass spectrometry of microgram and smaller quantities. *Analytical Chemistry* **45** 1881–1884.

Barnes, I.L., Gramlich, J.W., Diaz, M.G. and Brill, R.H. (1978). The possible change of lead isotope ratios in the manufacture of pigments: a fractionation experiment. In *Archaeological Chemistry II*, ed. Carter, G.F., American Chemical Society Advances in Chemistry Series 171, Washington D.C., pp. 273–277.

Bass, G.F. (1967). Cape Gelidonya: a Bronze Age shipwreck. *Transactions of the American Philosophical Society* **78** 1–177.

Bass, G.F. (1986). A Bronze Age shipwreck at Ulu Burun (Kaş): 1984 campaign. *American Journal of Archaeology* **90** 269–296.

Bass, G.F., Pulak, C., Collon, D. and Weinstein, J. (1989). The Bronze Age shipwreck at Ulu Burun: 1986 campaign. *American Journal of Archaeology* **93** 1–29.

Baxter, M.J. and Gale, N.H. (1998). Testing for multivariate normality via univariate tests: a case study using lead isotope ratio data. *Journal of Applied Statistics* **25** 671–683.

Baxter, M.J., Beardah, C.C. and Wright, R.V.S. (1997). Some archaeological applications of kernel density estimates. *Journal of Archaeological Science* **24** 347–354.

Begemann, F., Pernicka, E. and Schmitt-Strecker, S. (1995). Thermi on Lesbos: a case study of changing trade patterns. *Oxford Journal of Archaeology* **14** 123–136.

Brill, R.H. and Wampler, J.M. (1967). Isotope studies of ancient lead. *American Journal of Archaeology* **71** 63–77.

Budd, P., Gale, D., Pollard, A.M., Thomas, R.G. and Williams, P.A. (1993). Evaluating lead isotope data: further observations. *Archaeometry* **35** 241–263.

Budd. P., Gale, D., Ixer, R.A.F. and Thomas, R.G. (1994). Tin sources for prehistoric bronze production in Ireland. *Antiquity* **68** 518–524.

Budd, P., Pollard, A.M., Scaife, B. and Thomas, R.G. (1995a). Oxhide ingots, recycling and the Mediterranean metals trade. *Journal of Mediterranean Archaeology* **8**:1–32.

Budd, P., Pollard, A.M., Scaife, B. and Thomas, R.G. (1995b). The possible fractionation of lead isotopes in ancient metallurgical processes. *Archaeometry* **37** 143–150.

Budd, P., Haggerty, R., Pollard, A.M., Scaife, B. and Thomas, R.G. (1995c). New heavy isotope studies in archaeology. *Israel Journal of Chemistry* **35** 125–130.

Budd, P., Pollard, A.M., Scaife, B. and Thomas, R.G. (1995d). Lead isotope analysis and oxhide ingots: a final comment. *Journal of Mediterranean Archaeology* **8**:70–75.

Budd, P., Haggerty, R., Pollard, A.M., Scaife, B. and Thomas, R.G. (1996). Rethinking the quest for provenance. *Antiquity* **70** 168–174.

Budd, P.D., Lythgoe, P., McGill, R.A.R., Pollard, A.M. and Scaife, B. (1999). Zinc fractionation in liquid brass (Cu/Zn) alloy: potential environmental and archaeological applications. In *Geoarchaeology: Exploration, Environments, Resources*, ed. Pollard, A.M., Special Publication 165, Geological Society, London, pp. 147–153.

Clayton, R., Andersson, P., Gale, N.H., Gillis, C. and Whitehouse, M.J. (2002). Precise determination of the isotopic composition of Sn using MC-ICP-MS. *Journal of Analytical Atomic Spectrometry* **17** 1248–1256.

Coustures, M.P., Béziat, D., Tollon, F., Domergue, C., Long, L. and Rebiscoul, A. (2003). The use of trace element analysis of entrapped slag inclusions to establish ore-bar iron links: examples from two Gallo-Roman iron-making sites in France (Les Martys, Montagne Noire, and Les Ferrys, Loiret). *Archaeometry* **45** 599–613.

Cumming, G.L. and Richards, J.R. (1975). Ore lead isotope ratios in a continuously changing Earth. *Earth and Planetary Science Letters* **28** 155–171.

Degryse, P., Schneider, J., Kellens, N., Waelkens, M. and Muchez, Ph. (2007). Tracing the resources of iron working at ancient sagalassos (south-west Turkey): a combined lead and strontium isotope study on iron artefacts and ores. *Archaeometry* **49** 75–86.

Faure, G. (1986). *Principles of Isotope Geology*. John Wiley, New York, 2nd edn.

Gale, N.H. (1997). The isotopic composition of tin in some ancient metals and the recycling problem in metal provenancing. *Archaeometry* **39** 71–82.

Gale, N.H. and Mussett, A.E. (1973). Episodic uranium–lead models and the interpretation of variations in the isotopic composition of lead in rocks. *Reviews of Geophysics and Space Physics* **11** 37–86.

Gale, N.H. and Stos-Gale, Z.A. (1992). Lead isotope studies in the Aegean (The British Academy Project). In *New Developments in Archaeological Science*, ed. Pollard, A.M., Proceedings of the British Academy 77, Oxford University Press, Oxford, pp. 63–108.

Gale, N.H. and Stos-Gale, Z.A. (1993). Comments on P. Budd, D. Gale, A.M. Pollard, R.G. Thomas and P.A. Williams 'Evaluating lead isotope data: further observations', *Archaeometry*, **35** (2) (1993), and reply. Comments. II. *Archaeometry* **35** 252–259.

Gale, N.H., Gentner, W. and Wagner, G.A. (1980). Mineralogical and geographical sources of Archaic Greek coinage. In *Metallurgy in Numismatics Vol. 1*, ed. Metcalf, D.M., Royal Numismatic Society Special Publication 13, London, pp. 3–49.

Gale, N.H., Stos-Gale, Z.A., Maliotis, G. and Annetts, N. (1997). Lead isotope data from the Isotrace Laboratory, Oxford: Archaeometry data base 4, ores from Cyprus. *Archaeometry* **39** 237–246.

Guerra, M.F., Sarthre, C.O., Gondonneau, A. and Barrandon, J.N. (1999). Precious metals and provenance enquiries using LA-ICP-MS. *Journal of Archaeological Science* **26** 1101–1110.

Gulson, B.L. (1986). *Lead Isotopes in Mineral Exploration*. Elsevier, Amsterdam.

Halliday, A.N., Lee, D.C., Christensen, J.N., Rehkamper, M., Yi, W., Luo, X.Z., Hall, C.M., Ballentine, C.J., Pettke. T. and Stirling, C. (1998). Applications of multiple collector ICPMS to cosmochemistry, geochemistry and paleoceanography. *Geochimica et Cosmochimica Acta* **62** 919–940.

Hart, S.R., Shimizu, N. and Sverjensky, D.A. (1981). Lead isotope zoning in galena: an ion microprobe study of a galena crystal from the Buick Mine, Southeast Missouri. *Economic Geology* **76** 1873–1878.

Hedges, R.E.M. and Salter, C.J. (1979). Source determination of iron currency bars through analysis of the slag inclusions. *Archaeometry* **21** 161–175.

Holmes, A. (1946). An estimate of the age of the Earth. *Nature* **157** 680–684.

Knapp, A.B. and Cherry, J.F. (1994). *Provenance Studies and Bronze Age Cyprus: Production, Exchange and Politico-Economic Change*. Monographs in World Archaeology 21, Prehistory Press, Madison.

Knapp, A.B., Muhly, J.D. and Muhly, P.M. (1988). To hoard is human: the metal deposits of LC IIC – LC III. *Report of the Department of Antiquities of Cyprus*, pp. 233–262.

Lechtman, H. and Klein, S. (1999). The production of copper–arsenic alloys (arsenic bronze) by cosmelting: modern experiment, ancient practice. *Journal of Archaeological Science* **26** 497–526.

Littlefield, T.A. and Thorley, N. (1979). *Atomic and Nuclear Physics*. Van Nostrand Rheinhold, New York, 3rd edn.

McGill, R.A.R., Budd, P., Scaife, B., Lythgoe, P., Pollard, A.M., Haggerty, R. and Young, S.M.M. (1999). The investigation and archaeological implications of anthropogenic heavy metal isotope fractionation. In *Metals in Antiquity*, ed. Young, S.M.M., Pollard, A.M., Budd, P. and Ixer, R.A.F., BAR International Series 792, Archaeopress, Oxford, pp. 258–261.

Muhly, J.D. (1991). The development of copper metallurgy in Late Bronze Age Cyprus. In *Bronze Age Trade in the Mediterranean*, ed. Gale, N.H., Studies in Mediterranean Archaeology 90, Paul Åström's Förlag, Jønsered, pp. 180–196.

Mulliken, R.S. and Harkins, W.D. (1922). The separation of isotopes. Theory of resolution of isotopic mixtures by diffusion and similar processes. Experimental separation of mercury by evaporation in a vacuum. *Journal of the American Chemical Society* **44** 37–65.

Pernicka, E. (1993). Comments on P. Budd, D. Gale, A.M. Pollard, R.G. Thomas and P.A. Williams 'Evaluating lead isotope data: further observations', *Archaeometry*, **35** (2) (1993), and reply. Comments. III. *Archaeometry* **35** 259–262.

Pernicka, E and Bachmann, H.G. (1983). Archäometallurgische untersuchungen zur antiken silbergewinnung in Laurion III. Das verhalten einigerv spurenelemente beim abtreiben des bleis. *Erzmetall* **36** 592–597.

Pernicka, E., Seeliger, T.C., Wagner, G.A., Begemann, F., Schmitt-Strecker, S., Eibner, C., Öztunali, Ö and Baranyi, I. (1984). Archäometallurgische untersuchungen in Nordwestanatolien. *Jahrbuch des Römisch-Germanisches Zentralmuseums Mainz* **31** 533–599.

Pollard, A.M. (1982). A critical study of multivariate methods as applied to provenance data. In *Proceedings of the 22nd Symposium on Archaeometry*, ed. Aspinall, A. and Warren, S.E., University of Bradford Press, Bradford, pp. 56–66.

Pollard, A.M. (2004). Putting infinity up on trial: a consideration of the role of scientific thinking in future archaeologies. In *A Companion to Archaeology*, ed. Bintliff, J., Blackwell, Oxford, pp. 380–396.

Pollard, A.M., Thomas, R.G. and Williams, P.A. (1990). Experimental smelting of arsenical copper ores: implications for Early Bronze Age copper production. In *Early Mining in the British Isles*, ed. Crew S. and Crew P., Occasional Paper No. 1, Plas Tan y Bwlch, Snowdonia National Park Study Centre, Gwynedd, pp. 72–74.

Pollard, A.M., Thomas, R.G. and Williams, P.A. (1991). Some experiments concerning the smelting of arsenical copper. In *Archaeological Sciences 1989*, ed. Budd, P., Chapman, B., Jackson, C., Janaway, R. and Ottaway, B., Monograph 9, Oxbow, Oxford, pp. 169–174.

Pollard, A.M., Batt, C.M., Stern, B. and Young, S.M.M. (2007). *Analytical Chemistry in Archaeology*. Cambridge University Press, Cambridge.

Ponting, M., Evans, J.A and Pashley, V. (2003). Fingerprinting of Roman mints using laser-ablation MC-ICP-MS lead isotope analysis. *Archaeometry* **45** 591–597.

Reedy, T.J. and Reedy, C.L. (1992). Evaluating lead isotope data: comments on E.V. Sayre, K.A. Yener, E.C. Joel and I.L. Barnes, 'Statistical evaluation of the presently accumulated lead isotope data from Anatolia and surrounding regions'. *Archaeometry*, **34** (1) (1992), 73–105, and reply. Comments. IV. *Archaeometry* **34** 327–329.

Richards, R.H. (1909). *A Textbook of Ore Dressing*. McGraw-Hill, New York.

Rohl, B. (1996). Lead isotope data from the Isotrace Laboratory, Oxford: Archaeometry database 2, galena from Britain and Ireland. *Archaeometry* **38** 151–180.

Rohl, B. and Needham, S. (1998). *The Circulation of Metal in the British Bronze Age: The Application of Lead Isotope Analysis*. Occasional Paper 102, British Museum, London.

Russell, R.D. and Farquhar, R.M. (1960). *Lead Isotopes in Geology*. Interscience Publishers, New York.

Sayre, E.V., Yener, K.A., Joel, E.C. and Barnes, I.L. (1992). Statistical evaluation of the presently accumulated lead isotope data from Anatolia and surrounding regions. *Archaeometry* **34** 73–105.

Sayre, E.V., Joel, E.C., Blackman, M.J., Yener, K.A. and Ozbal, H. (2001). Stable lead isotope studies of Black Sea Anatolian ore sources and related Bronze Age and Phrygian artefacts from nearby archaeological sites. Appendix: new central Taurus ore data. *Archaeometry* **43** 77–115.

Scaife, B. (1993). *Lead Isotope Analysis and Archaeological Provenancing*. Unpublished B.Sc. dissertation, Department of Archaeological Sciences, University of Bradford.

Scaife, B., Budd, P., McDonnell, J.G., Pollard, A.M. and Thomas, R.G. (1996). A new statistical technique for interpreting lead isotope analysis data. In *Archaeometry 94. Proceedings of the 29th International Symposium on Archaeometry*, ed. Demirci, Ş., Özer, A.M. and Summers, G.D., Tübitak, Ankara, pp. 301–307.

Scaife, B., Budd, P., McDonnell, J.G. and Pollard, A.M. (1999). Lead isotope analysis, oxhide ingots and the presentation of scientific data in archaeology. In *Metals in Antiquity*, ed. Young, S.M.M., Pollard, A.M., Budd, P. and Ixer, R.A., BAR International Series 792, Archaeopress, Oxford, pp. 122–133.

Srinivasan, S. (1999). Lead isotope and trace element analysis in the study of over a hundred South Indian metal icons. *Archaeometry* **41** 91–116.

Stacey, J.S. and Kramers, J.D. (1975). Approximation of terrestrial lead isotope evolution by a two-stage model. *Earth and Planetary Science Letters* **26** 207–221.

Stos-Gale, Z.A. and Gale, N.H. (1990). The role of Thera in the Bronze Age trade in metals. In *Thera and the Aegean World III. Volume 1 Archaeology*, ed. Hardy, D.A., Thera Foundation, London, pp. 72–92.

Stos-Gale, Z.A., Gale, N.H. and Zwicker, U. (1986). The copper trade in the South-east Mediterranean region. Preliminary scientific evidence. *Report to the Department of Antiquities of Cyprus 1986*, pp 122–144.

Stos-Gale, Z., Gale, N.H., Houghton, J. and Speakman, R. (1995). Lead isotope data from the Isotrace Laboratory, Oxford: *Archaeometry* database 1, ores from the western Mediterranean. *Archaeometry* **37** 401–415.

Stos-Gale, Z.A., Gale, N.H. and Annetts, N. (1996). Lead isotope data from the Isotrace Laboratory, Oxford: *Archaeometry* database 3, ores from the Aegean, Part 1. *Archaeometry* **38** 381–390.

Stos-Gale, Z.A., Maliotis, G., Gale, N.H. and Annetts, N. (1997). Lead isotope characteristics of the Cyprus copper ore deposits applied to provenance studies of copper oxhide ingots. *Archaeometry* **39** 83–123.

Stos-Gale, Z.A., Gale, N.H., Annetts, N., Todorov, T., Lilov, P., Raduncheva, A. and Panayotov, I. (1998). Lead isotope data from the Isotrace Laboratory, Oxford: *Archaeometry* database 5, ores from Bulgaria. *Archaeometry* **40** 217–226.

Thomas, R.G. (1990). *Studies of Archaeological Copper Corrosion Phenomena*. Unpublished Ph.D. thesis, School of Chemistry and Applied Chemistry, University of Wales College of Cardiff, Cardiff.

Tylecote, R.F. (1976). *A History of Metallurgy*. Metals Society, London.

Wagner, G.A., Pernicka, E., Seeliger, T.C., Lorenz, I.B., Begemann, F., Schmitt-Strecker, S., Eibner, C. and Öztunali, Ö. (1986). Geochemische und isotopisch charakteristika fruher rohstoffquellen fur kupfer, blei, silber und gold in der Turkei. *Jahrbuch des Römisch-Germanischen Zentralmuseums Mainz* **33** 723–752.

Watling, R.J., Herbert, H.K., Delev, D. and Abell, I.D. (1994). Gold fingerprinting by laser-ablation inductively-coupled plasma-mass spectrometry. *Spectrochimica Acta* **B49** 205–219.

Wilson, L. and Pollard, A.M. (2001). The provenance hypothesis. In *Handbook of Archaeological Sciences*, ed. Brothwell, D.R. and Pollard, A.M., John Wiley and Sons, Chichester, pp. 507–517.

Yi, W., Budd, P., McGill, R.A.R., Young, S.M.M., Halliday, A.N., Haggerty, R., Scaife, B. and Pollard, A.M. (1999). Tin isotope studies of experimental and prehistoric bronzes. In *The Beginnings of Metallurgy*, ed. Hauptmann, A., Pernicka, E., Rehren, T. and Yalcin, U., Der Anschnitt Beiheft 9, Deutschen berbau-Museum, Bochum, pp. 285–290.

CHAPTER 10

The Chemistry of Human Bone: Diet, Nutrition, Status and Mobility

10.1 INTRODUCTION

The closest we can ever come to the physical presence of our ancestors is to excavate and study human bone, and it is therefore natural that human remains have acquired a special status in modern scientific archaeological research (*e.g.*, Larsen, 1997). Increasingly, however, the ethical and moral issues surrounding the excavation, study and, particularly, curation of human bones are constraining what can be done. In some parts of the world, it is regarded as preferable not to disturb human remains at all, or if they are accidentally disturbed then they must be immediately reinterred. In others it is possible to study excavated human material, but the presumption of reburial rather than long-term curation is dominant. Although understandable from a moral point of view, since there is clearly an ethical difference between a human skeleton and an archaeological ceramic, scientifically speaking, this is unfortunate. The methodologies for studying such material are continuously improving, and new questions can be answered, but only if there is reasonably free access to well-described material. The whole issue is obviously a tangled web of ethics, kinship, religious sensitivities and the politics of colonialism and land rights (*e.g.*, Gulliford, 1996). It is, however, ironic that, at the very time when scientific techniques have developed to allow questions of central importance to archaeology to be addressed, access to the appropriate material is increasingly being restricted. It is to be hoped, therefore, that the scientific study of human remains can continue within a sensitive consensual framework, because the value of such work in the context of understanding the lives of our ancestors, individually and collectively, is immense and irreplaceable.

The well-known saying 'you are what you eat' has been taken almost literally in archaeology for the past 40 years, and dietary reconstruction has been attempted using trace element levels in bone mineral, and stable isotope studies

Archaeological Chemistry, Second Edition
By A. Mark Pollard and Carl Heron
© The Royal Society of Chemistry 2008

on collagen and mineral in both bone and teeth [reviewed, for example, by Pate (1994), Katzenberg and Harrison (1997) and Hedges *et al.* (2006)]. It soon became apparent, however, that inorganic trace element studies in bone mineral were potentially bedevilled by *post-mortem* diagenetic effects (uptake or loss of trace elements to the surrounding medium), the magnitude and significance of which have been vigorously debated (*e.g.*, Hancock *et al.*, 1989; Price, 1989; Sandford, 1993; Radosevich, 1993; Katzenberg and Harrison, 1997). In most cases, the debate has centred primarily not on *whether* diagenetic effects occur, but whether such affects *can be removed* prior to chemical analysis – in other words, whether a reliable biogenic signal can be recovered if a potential geochemical overprint is present. Isotopic studies of the light elements (mainly carbon and nitrogen) in bone collagen have been far less controversial in this respect, and, for Holocene material at least, appear to avoid most of the diagenetic problems encountered with trace elements. Providing that more than about 10% of the original collagen survives [estimated from data presented in Ambrose (1990)], then the isotopic signal measured appears to be unchanged from that which would have been measured *in vivo*. Reviews of dietary reconstruction using isotopic techniques include those of DeNiro (1987), Keegan (1989), Schwarcz and Schoeninger (1991), Schoeninger and Moore (1992), van der Merwe (1992), Katzenberg (1992), Ambrose (1993), Pollard (1998) and Pollard *et al.* (2007; 180), as well as those listed above. The chemical study of the protein and mineral fraction of archaeological bone and teeth can reveal information on diet, health, status and mobility, providing that our knowledge of living bone metabolism is adequate, and that we can account for all of the changes which may occur during burial. Both of these factors can provide significant challenges to the archaeological chemist.

The isotopic method is based on the observation that carbon and nitrogen isotope ratios ($^{13}C/^{12}C$ and $^{15}N/^{14}N$) in bone collagen (and also $^{13}C/^{12}C$ in bone mineral carbonate, see Section 10.7.2) can reflect the corresponding isotopic ratios in the diet (see Appendix 2 for a definition of isotopes). On a broad scale, such as marine-based versus terrestrial-based, or carnivore versus herbivore and/or vegetarian, different diets might be expected to have different isotopic signals, since they reflect very different ecosystems or trophic levels within an ecosystem. In principle therefore these isotope ratios in bone can be used to distinguish between, for example, a heavy reliance on terrestrial or marine food resources in the diet of individual prehistoric humans. Carbon isotope ratios may further be used to differentiate between the consumption of terrestrial plants which photosynthesize using the C_3 and C_4 pathways. Virtually all land plants photosynthesize using the C_3 pathway, but a few plants, mainly tropical grasses, have evolved the C_4 pathway as an adaptation to hot dry environments. Archaeologically the most important C_4 plant is maize, which has its origins in Central America, and carbon isotope measurements have been used to trace with great success the spread of maize agriculture through prehistoric American cultures (van der Merwe, 1992). The nitrogen isotope ratios in bone can also reflect differential access to marine resources, since nitrogen is more heavily fractionated in ocean ecosystems with many more trophic levels than in terrestrial systems. These models of isotopic ecology have been validated using modern terrestrial and marine ecosystems, and have also been shown to have great value as ecological markers (Lajtha and Michener, 1994; Griffiths, 1998; Leng, 2006), as well as having significant archaeological potential.

10.2 DIETARY RECONSTRUCTION FROM TRACE ELEMENTS IN BONE MINERAL

Bone is a highly specialized composite structural tissue. In life, it performs a number of functions, such as providing support and protection for bodily soft tissue and organs, and allowing movement by providing points of attachment for muscles. Of more interest chemically, it acts as a reservoir for the storage of essential elements, and, to some extent, a repository for unwanted elements. The gross structure of bone is shown schematically in Figure 10.1, and the

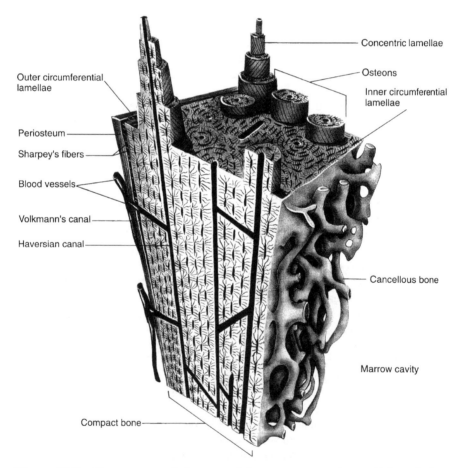

Figure 10.1 Gross structural features of human compact bone. (Adapted from Gartner and Hiatt, 1994; Graphic 4.1.)

structure of the principal organic component of bone, collagen, is shown in Figure 8.1 in Chapter 8. The mineral fraction of the bone is generally poorly mineralized, but is normally referred to as *biological calcium hydroxyapatite* [$Ca_{10}(PO_4)_6(OH)_2$], or, preferably, the carbonate hydroxyapatite mineral *dahllite* [approximate stoichiometry $(Ca,X)_{10}(PO_4,CO_2)_6(O,OH)_{26}$ (McConnell, 1973; 88)]. To reflect the many substitutions which can occur in the biological mineral, it is also referred to as carbonated hydroxyl apatite [$(Ca,Na,Mg)_5(HPO_4,PO_4,CO_3)_3(OH,CO_3)$ – see, for example, Berna *et al.* (2004)]. In life bone acts as a reservoir for calcium and phosphate ions – storing them until needed in bone remodelling and repair or as a result of breastfeeding, when they are remobilized via cellular activity. Other ingested elements, especially strontium and lead, are stored in the skeleton, and it is these elements which have been studied in archaeological bone for their presumed dietary and health significance (Sandford, 1993). The first element to be so studied was strontium (Sr), which is still the best-studied trace element in palaeodietary reconstruction. Other proposed indicators have included zinc (Zn), magnesium (Mg) and barium (Ba), although the physiological reasons for the dietary behaviour of these elements is not well understood (Ezzo, 1994). There are other elements deposited in bone which are physiologically harmful above certain levels, such as lead (Pb) and copper (Cu). The concentration of these elements might contain information about exposure to chemical insult such as might result from long-term work as a metal smelter or, perhaps more generally, the geographical location of the individual in life, but the suspicion must remain that in many cases elevated levels of such trace elements might be the result of *post-mortem* diagenesis.

In animals, strontium is discriminated against by the kidneys in favour of calcium for the synthesis of bone tissue. The ratio of strontium to calcium in the bone of a terrestrial herbivore is approximately five times lower than the ratio in the plants which form the diet. Carnivores feeding on these herbivores discriminate further against strontium, so that their bones exhibit a lower ratio still. Omnivores should have strontium levels intermediate between those of herbivores and carnivores, in proportion to the relative importance of plants and meat in their diet – this therefore should apply to human bone (Sillen and Kavenagh, 1982). The situation is different in the marine environment, where higher strontium levels are found because of the more concentrated mineral levels in oceanic waters, and we would therefore expect higher strontium levels in humans subsisting on a primarily marine diet. Controlled feeding studies have been carried out on animals to test the foundations of some of these assumptions [*e.g.*, Klepinger (1990) on magnesium, Burton and Wright (1995) on strontium and Lambert and Weydert-Homer (1993a; 1993b) on ten elements]. The study of Burton and Wright (1995) in particular shows the complexity of trace element incorporation into bone. They observed that the Sr/Ca ratio (calculated as ppm Sr/ppm Ca, sometimes multiplied by 1000) in bone reflects the available Sr/Ca ratio in food, and that food with high available Ca shows up disproportionately in the bone ratio, meaning that in mixed diets significant dietary shifts may not be reflected in the Sr/Ca ratio. Such non-linear relationships make dietary interpretation of trace elements difficult, but we should also remember that similar non-linearity can occur in protein metabolism, as discussed in Section 10.3 (Ambrose, 1993).

A further and potentially more serious complication with the use of trace elements is the behaviour of buried bone in contact with groundwater – so serious that the dietary significance of trace element analysis of excavated bone is completely discounted in some quarters. Several approaches have been used in order to estimate the degree of influence of this *post-mortem* diagenesis. Some authors have compared concentrations of a wide range of elements in bones and related soils (Lambert *et al.*, 1979) or, more correctly, soil solutions (Pate and Hutton, 1988), or in different skeletal elements of the same body (Lambert *et al.*, 1982). Other studies have concentrated on the variation in trace elements in bone cross-sections (Lambert *et al.*, 1983). These studies, although differing in detail, have all concluded that certain elements found in bone (*e.g.*, Fe, Al, K and Mn) are liable to post-depositional contamination, but that others (*e.g.*, Zn, Sr, Ba) are said to be sufficiently stable for use as dietary indicators. Other workers, however, have continued to challenge this view (*e.g.*, Hancock *et al.*, 1989), on the grounds that the levels of all these elements are always higher in buried than in fresh bone, and therefore that *post-mortem* contamination is ubiquitous. According to Hancock *et al.* (1989; 178):

> '*much of the published data . . . reflects major or minor diagenetic effects . . . Hence, many published dietary claims require critical re-evaluation. . . . unless one can establish unequivocally that the elements one has employed in a study were not subjected to diagenesis, dietary and other physiological implications will continue to be erroneous.*'

In terms of trace element studies and diet, we are inclined to agree with this view.

With the wide interest in the influence of the diagenetic alteration of bone on archaeological data (for example, in radiocarbon dating), many methods have been used to study the degree of alteration, reviewed by Hedges (2002). These include histological methods [the Oxford Histological Index (Hedges and Millard, 1995) is widely used as an overall indicator of diagenetic alteration], measuring the total amount of collagen left in the bone and the C/N ratio of that collagen, measurements of bone porosity and density or other physical properties of bone (tensile strength, *etc.*). One very common approach is to measure the 'Crystallinity Index', or infrared splitting factor (IRSF), using infrared spectroscopy, and quantifying the relative intensities of the absorption bands in the phosphate group (Weiner and Bar-Yosef, 1990; Pollard *et al.* 2007, 88). In living bone, cartilage, dentine and cementum, the typical apatite crystal has a platelet form of dimensions $15–30 \times 100–150 \times ca.$ $400 \,\text{Å}$ ($1 \,\text{Å} = 0.1 \,\text{nm}$). This small crystal size gives bone mineral a very large surface area/volume ratio – a surface area of $85–170 \,\text{m}^2 \,\text{g}^{-1}$ has been estimated (Lowenstam and Weiner, 1989), and makes the bone mineral surface chemically very reactive *in vivo*. Given this huge surface area, however, it is not at all surprising that mineralogical changes take place *post-mortem*, driven in part by Ostwald ripening – the tendency in solution of larger crystals to grow at the expense of small crystals. The 'Crystallinity Index' gives a direct indication of average crystal size (although it will also be affected by other factors such as the degree of substitution or crystal imperfection), and therefore allows the extent of diagenetic alteration to be quantified. A useful addition to the range of indicators for the study of bone diagenesis appears to be the rare earth element (REE) profile (Trueman, 1999). The REE profile in bone from terrestrial deposits is controlled by the early diagenetic environment and is affected by both sedimentological and taphonomic processes. It therefore provides a new tool for looking at the composition of mixed assemblages, and for studying the degree of alteration (*e.g.*, Trueman *et al.*, 2006).

Depending on how this *post mortem* recrystallization takes place, it might well be expected to affect the trace element composition of the bone mineral. If trace elements from the groundwater are incorporated into the growing crystals, or trace elements are lost to groundwater as the small crystals dissolve, then it is clear that these processes will affect the supposed dietary indicator elements in some way. It is well known that recrystallization is the standard method for purifying soluble products in synthetic chemistry, and one might therefore expect the recrystallized material to be less rich in trace elements than the *in vivo* material. This is not in accord with the analytical evidence, which shows that *post-mortem* bone is almost always richer in trace elements than is fresh bone (Hancock *et al.*, 1989). This suggests that the *post-mortem* contamination might be due more to exogenous mineral incorporation rather than to trace element incorporation into the recrystallized material. For example, Sillen (1989) has suggested that an increase in the Ca/P stoichiometric ratio to more than two may be due to recrystallization of the hydroxyapatite to other calcium phosphates, and/or the deposition of calcite crystals into the bone. This view gives rise to the possibility that diagenetic alteration might be physically removed from bone in the laboratory in order to reveal the *in vivo* signal.

Physical approaches to removing diagenetically altered bone have focussed initially on differential solubility (Sillen and Sealy, 1995), and subsequently on separation by bone density (Bell *et al.*, 2001). The former has been used particularly for the measurement of Sr isotopes in bone (see Section 10.8), and assumes that the altered mineral is more soluble than biomineralized apatite. The consensus of opinion amongst practitioners is that repeated acid washing removes any diagenetic mineral. Although this seems a rather simplistic assumption, there is a growing body of evidence in support of it (*e.g.*, Trickett *et al.*, 2003). Bone density separation involves grinding and deproteinating the bone, and separating the various density fractions using heavy liquid flotation techniques. It assumes that altered mineral is more dense than biomineral, and has been used to separate out unaltered bone carbonate for carbon isotope measurements on the carbonate fraction (see Section 10.7.1). Obviously *post-mortem* diagenesis can have a significant effect on bone mineral composition, and the geochemistry of the bone–groundwater interaction needs to be understood. Computer modelling techniques can be used to solve this complicated problem, with the ultimate aim of being able to predict the chemistry and mineralogy of bone buried in a particular environment. Recent work (Wilson, 2004) has shown that simple predictive models can be built, but there is still a considerable way to go. The question of the validity of dietary reconstruction from trace element data in bone continues to be scrutinized, with a strong tendency towards scepticism in most quarters. It is certainly the case that the onus is on the trace element researcher to demonstrate that the data are not diagenetically imprinted, rather than the other way round.

10.3   LIGHT ISOTOPE SYSTEMATICS AND TROPHIC LEVELS

Many light elements (*e.g.*, hydrogen and oxygen) occur naturally in more than one isotopic form. For example, the most abundant stable isotope of oxygen is ^{16}O, with eight protons and eight neutrons in the nucleus. However, two other heavier stable isotopes exist – ^{17}O, with nine neutrons, and ^{18}O, with ten neutrons. The natural abundance of ^{17}O is very low (0.04%) compared with 0.2% for ^{18}O, but the lightest isotope is by far the most abundant, at 99.76%. This is a common pattern for the isotopes of the lighter elements. Isotope systematics of the lighter elements – the relative behaviour of these isotopes as the element goes around the various biogeochemical cycles – have been extensively studied because of their importance in understanding a wide range of environmental, Earth science and bioscience processes (*e.g.*, Taylor *et al.*, 1991; Griffiths, 1998; Pollard and Wilson, 2001). For example, as water is cycled around the Earth's system (from atmosphere to terrestrial rivers to ocean and back to the atmosphere), the ratios of both $^2H/^1H$ and $^{18}O/^{16}O$ change. This is because the increased mass of the heavier isotope makes it relatively less likely to take part in evaporation, enriching slightly the liquid phase in the heavier isotope whilst depleting the resulting vapour. Such processes are termed *fractionation*, and the magnitude of fractionation for any given process and temperature depends on the mass difference between the two isotopes. Hydrogen, therefore, shows the largest fractionation, since the heavier isotope is twice the mass of the lighter, whereas in oxygen, it is only 12.5% heavier. Even for hydrogen, however, fractionation effects are small. Globally the average abundance ratio of ^2H to ^1H is 0.000156, and therefore any changes due to fractionation are only detected in the last digit. In order to magnify these small effects, isotope geochemists have adopted the δ (delta) notation, which for hydrogen is defined as:

$$\delta^2H = \delta D = \left(\frac{\left(^2H/^1H\right)_{sample} - \left(^2H/^1H\right)_{standard}}{\left(^2H/^1H\right)_{standard}} \right) \times 1000$$

In this notation, the ratio of ^2H to ^1H in the sample is compared to the same ratio in an internationally agreed standard material. If the ratio in the sample is identical to that in the standard, then the δ value (δ^2H, usually referred to as δD since ^{2}H is commonly called deuterium) is zero. If the sample is isotopically heavier (*i.e.*, has relatively more ^{2}H) than the standard, the ^{2}H/^{1}H ratio in the sample is larger than the standard, and δ becomes positive. If the sample is isotopically lighter, then the top line becomes negative, and δ becomes negative. The units (because of the multiplication by 1000) are known as 'per mil' (or 'parts per thousand'), symbolized as ‰. The advantage of the δ notation is that small changes due to fractionation are magnified and are therefore more easily appreciated, and the direction of change is made clear by comparing the values before and after fractionation, irrespective of the particular isotope system being studied. If δ becomes more positive as it is fractionated, this means that the sample is becoming isotopically heavier, and *vice versa*. The standard reference material used for hydrogen isotopic measurements in water is VSMOW (Vienna standard mean ocean water) or SLAP (standard light arctic precipitation). The equations for δ^{13}C, δ^{15}N, δ^{18}O and δ^{34}S are identical to that given above, with the appropriate isotope ratio (*i.e.*, ^{13}C/^{12}C, ^{15}N/^{14}N, ^{18}O/^{16}O and ^{34}S/^{32}S) replacing that of hydrogen. The same standards are used for the measurement of oxygen isotope ratios in water, but different standards are used for other isotopes in rocks and biominerals (Ehleringer and Rundel, 1988; IAEA, 1995).

Carbon has two stable isotopes – ^{12}C, with six protons and six neutrons, and ^{13}C, with seven neutrons. Isotopic fractionation of carbon as it is cycled around the biosphere has been studied in great detail for many years. The internationally agreed standard for carbon isotope measurements was originally the CO_2 produced from a Cretaceous belemnite rock in South Carolina, called the Peedee Formation (PDB). All the original stocks of this are now exhausted, and so the current standard adopted by the International Atomic Energy Agency (IAEA) is called VPDB, standing for Vienna PDB (IAEA, 1995; 65). Biological carbon fractionation begins when green plants photosynthesize, and combine CO_2 taken in from the atmosphere through the leaf stomata with H_2O taken up by the root system to produce sugars and eventually cellulose. A *very* simplified expression for this reaction is as follows:

$$6H_2O + 6CO_2 \rightarrow C_6H_{12}O_6 + 6O_2$$

For the majority of land plants, which photosynthesize using the C_3 metabolic pathway [the Calvin–Benson cycle (Schlesinger, 1997; 129)], this results in a measured range of δ^{13}C values for the plant tissue of between about –19‰ and –29‰, with an average value of –26.5‰ (van der Merwe and Medina, 1991). These numbers are negative simply because the plant tissue is isotopically depleted (has less ^{13}C) compared to the VPDB standard. The C_3 metabolic pathway is used by all trees and woody shrubs and temperate grasses, as well as by algae, autotrophic bacteria and aquatic plants (Dawber and Moore, 1980; 405). The first step in the photosynthetic process within the plant is the carboxylation of a five-carbon sugar molecule (ribulose 1,5-biphosphate) to form an unstable six-carbon molecule, which spontaneously hydrolyses into two molecules of 3-phosphoglyceric acid (3-PGA). The name (C_3) for this pathway comes from the fact that the first stable compound formed during the process is a three-carbon molecule. A second photosynthetic pathway was discovered as a result of isotopic studies of plants, termed C_4 (or the Hatch–Slack cycle), in which there is a primary carboxylation in the mesophyll cells of the leaf. Carbon dioxide is trapped in the leaf during the day and carboxylates the three-carbon organic acid phosphoenol pyruvic acid (PEP), giving malic acid, a four-carbon compound (hence C_4 photosynthetic pathway). This is subsequently decarboxylated to release CO_2, which is then assimilated by the Calvin–Benson cycle described above. This C_4 cycle is found in tropical grasses as an adaptation to deal with higher light levels, higher temperatures and limited water availability. This results in plants with $\delta^{13}C$ values of –12‰ and –16‰, with an average of –12.5‰ (Ehleringer and Monson, 1993). A third metabolic pathway, only of interest archaeologically in very arid environments such as the US south-west, is the crassulacean acid metabolism (CAM). This involves the synthesis of malic acid by carboxylation at night, and the subsequent daytime breakdown of the malic acid, releasing CO_2 for photosynthesis by the normal Calvin–Benson cycle. The advantages of CAM are improved photosynthetic performance in water (and/or CO_2) limited environments. CAM is present in a range of plants including desert cacti, many orchids and bromeliads in the tropical rainforest and in some aquatic angiosperms. Succulent plants suspected of having CAM photosynthesis show a wide range of carbon isotope ratios, with $\delta^{13}C$ values ranging from –14 to –33‰, but are typically similar to those of C_4 plants, *i.e.*, between about –12‰ and –16‰ (Szarek and Troughton, 1976).

Farquhar *et al.* (1982) showed that the carbon isotopic composition of plant material ($\delta^{13}C_p$) as a result of photosynthesis can be expressed empirically as:

$$\delta^{13}C_p = \delta^{13}C_{atm} - a - (b - a)^{c_i/c_a}$$

where a is the discrimination against $^{13}CO_2$ compared to $^{12}CO_2$ during the diffusion of CO_2 through air, and b is the discrimination in a particular plant species against ^{13}C during the carboxylation reaction. $\delta^{13}C_{atm}$ is the isotopic value of atmospheric CO_2, and c_i and c_a are the partial pressures of CO_2 within the intercellular spaces of the leaf and in the atmosphere, respectively. Typically $\delta^{13}C_{atm}$ has a value of around –7‰ (but this changes with latitude and over time), a is 4.4‰ and b varies from species to species, but is around 30‰. This gives rise to the range of $\delta^{13}C_p$ values for terrestrial C_3 plants as noted above (between –19‰ and –29‰, average –26.5‰). The C_4 photosynthetic pathway has a different value of b, and results in plants with $\delta^{13}C_p$ values of between – 12‰ and –16‰, with an average of –12.5‰.

Plants are the base of the terrestrial food chain. The body tissue of herbivores feeding on these plants, and subsequently of omnivores and carnivores, is dependent on the isotopic composition of the plant material ingested, and therefore reflects the $\delta^{13}C$ values of either a C_3 or C_4 biome. The exact values are modified by metabolically induced fractionation as the food is digested and incorporated into different body tissues. At its simplest, it has been assumed that the bone collagen $\delta^{13}C$ of a herbivore is $+5\permil$ enriched relative to the plant tissue consumed, and the bone collagen $\delta^{13}C$ of a carnivore is further enriched by $+7\permil$ relative to the meat it consumes. The detail is, however, much more complicated than this, and the debate has been hampered by a lack of detailed knowledge about the routing of various components of the diet into body tissue (Sealy, 2001). Controlled studies of rats (Ambrose and Norr, 1993) and mice (Tieszen and Fagre, 1993) fed on diets containing known ratios of protein, cellulose, starch and lipid of known carbon isotopic composition have shown, in general, that collagen appears to reflect the carbon isotope ratio of the dietary protein intake, whereas the carbon isotope ratio of biological carbonate reflects that of the total diet (*i.e.*, protein, carbohydrate and fats). This model for the routing of dietary carbon is illustrated schematically in Figure 10.2. Deviations from this model, when levels of protein intake fall, are shown in this figure.

Cycling of nitrogen can also be followed isotopically by measuring changes in $\delta^{15}N$, where the international standard is atmospheric nitrogen (AIR). Nitrogen isotope systematics are less well understood than those of carbon, but are frequently used in parallel to carbon to reconstruct diet from the isotopic analysis of skeletal collagen. The nitrogen isotope ratios in bone can reflect the differential utilization of nitrogen-fixing plants (*e.g.*, legumes) and non-nitrogen fixing plants. The principal variation, however, is between terrestrial and marine ecosystems. Terrestrial mammals and birds have a mean bone collagen $\delta^{15}N$ value of $+5.9\permil$, whereas marine mammals have an average value of $+15.6\permil$ (Schoeninger *et al.*, 1983). Subsequent studies (Wada *et al.*, 1991) have shown distinct trophic level discrimination against nitrogen in

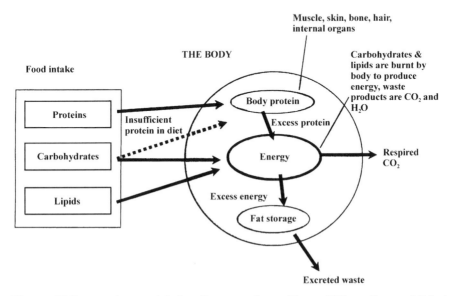

Figure 10.2   Routing model for dietary carbon. (From O'Connell, unpublished Figure, with permission.)

marine ecosystems. This results in a clear distinction in $\delta^{15}N$ between humans who live predominantly on marine resources (*e.g.*, the traditional diet of the North American Inuit), who have $\delta^{15}N$ between $+17$ to $+20‰$, and terrestrial agriculturists who have values in the range $+6$ to $+12‰$.

10.4   ISOTOPIC DIETARY RECONSTRUCTION FROM HUMAN BONE

It is clear from the above that the study of the isotopes of carbon and nitrogen in collagen derived from human bone has the potential to give information about the diet of an individual. This is important for at least three reasons. One is that large components of foodstuff are often invisible in the archaeological record – butchered animal bone may survive, and mollusc shells may survive, but the vast majority of what makes up the human diet, in general, will not. Reconstructing diet from direct archaeological evidence of what was eaten, therefore, often relies on inferences made from animal bone assemblages, or shell middens, or pollen spectra or the occasional find of charred cereal grain. Of necessity, this can only lead to a global and partial estimate of what was eaten – it shows some of what was *available* to be eaten, but not *what* was eaten by *whom*. This is the second advantage of isotopic dietary studies – it is specific to an individual. It allows the possibility of dietary discrimination between individuals – males compared to females, juveniles to adults, *etc.* A third advantage is that, within limits, the relative importance of dietary sources of protein can be quantified. If two sources, such as terrestrial C_3 and terrestrial C_4 biomes, give rise to different isotopic values, then the relative importance of each in a mixed diet can be estimated by interpolation between the two end points. This is the method adopted when attempting to calculate the proportion of marine carbon present in a dated sample in order to correct a radiocarbon date for the effects of the marine reservoir, and is discussed in more detail on page 363. A generalized plot of $\delta^{15}N$ against $\delta^{13}C$ for the major trophic levels in terrestrial and marine foodwebs is shown in Figure 10.3. Diagrams such as this form the starting point for all dietary interpretations from bone collagen stable isotope measurements.

But even the best ideas need to be demonstrated in order to be convincing, and the choice of case study can be crucial. In a now classic study of the spread of maize agriculture (maize being a C_4 plant) from Central America into the otherwise predominantly C_3 biomes of pre-Columbian eastern and central North America, van der Merwe and Vogel (1977) found exactly the right medium to demonstrate the power of this new technique. Archaeological opinion, prior to this work, was that maize had been introduced into the lower Illinois Valley area by AD 400 or perhaps earlier (van der Merwe, 1992), but isotopically it was shown not to arrive until the end of the Late Woodland period (*ca.* AD 1000). This work single-handedly established isotopic human palaeodietary studies as a legitimate and fruitful area of research within archaeology.

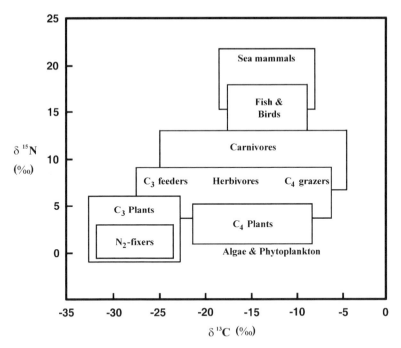

Figure 10.3   Generalized isotopic trophic level diagram for marine and terrestrial foodwebs. (Adapted from O'Connell, 1996, with permission.)

As a result of the relative ease with which the 'maize signal' can be detected in human collagen, and because of the cultural importance of the spread of maize agriculture, many isotopic studies of bone collagen have been carried out in the New World since the original work of van der Merwe and his colleagues. These have been reviewed by Tykot (2006). Katzenberg *et al.* (1995) compiled published $\delta^{13}C$ data from human bone collagen from central North America (the Mississippi, Illinois and Ohio valleys, and the Great Lakes region into southern Ontario, Canada) between AD 500 and AD 1300 (about 130 measurements). This showed a general increase in collagen $\delta^{13}C$ from around –20‰ in AD 600 to –10‰ by AD 1200, consistent with the adoption of maize subsistence, but with regional variations, which do not support a simple linear model for the spread of maize agriculture from the south. Southern states such as Missouri, Arkansas and Tennessee show little increase in maize utilization until near the end of the period, whereas Illinois appears to have early and continued exploitation of maize. As the picture begins to grow, we can see regional complexity developing, including perhaps a reversion to more traditional agricultural resources during times of social or environmental stress in some areas (Larsen *et al.*, 1992). Whatever emerges from such studies in the future, it is quite clear that the application of isotopic palaeodietary techniques is absolutely crucial to our understanding of these important cultural changes.

Compared to the intensity of such research in North America, it is surprising to see how relatively little work has been done in Central and South America.

White and Schwarcz (1989) measured stable isotopes in human bone collagen from the Lowland Maya site at Lamanai, Belize, from the pre-Classic (1250 BC to AD 250) to Historic (AD 1520 to 1670) periods. The $\delta^{13}C$ values suggested a maize input to the diet of 50% at the beginning of the pre-Classic period, falling to 37% at the end, and rising again to 70% in post-Classic times. The decline of maize in the diet at the end of the pre-Classic was taken to signify a reduction in maize production, perhaps as a result of social or climatic change. Tykot (2002) has reviewed stable isotope palaeodietary studies of the Maya. In South America, van der Merwe *et al.* (1981) established that maize was a significant dietary component along the Orinoco River in Venezuela between 800 BC and AD 400. They concluded that by AD 400 the diet consisted of 80% or more of maize, and that this allowed a 15-fold increase in population density. It is, however, a little hard to imagine what a diet in which 80% of the protein comes from maize would have been like! It is not necessarily the case, of course, that this implies an endless diet of corn on the cob. Animals fed on maize would acquire a C_4 signal in their flesh, which would be passed on (slightly fractionated) when they were eaten. Likewise, consumption of *chicha* (maize beer) would also induce a C_4 signal. It might not have been quite such a dreary diet as first appears!

From a European perspective, however, there is an apparent drawback to this technique – the native flora of Europe is overwhelmingly C_3 (as is that of the USA and Canada), and the major C_4 domesticates (maize, sorghum, millet and sugar cane) are a relatively recent introduction. In European archaeology, therefore, we should not expect to see such large changes in the collagen carbon isotope signals. Other sources of variation, however, might be detectable – particularly the difference between a reliance on terrestrial C_3 and marine protein sources. Studying these more subtle isotopic differences brings $\delta^{15}N$ measurements into play, because nitrogen is heavily fractionated through the many trophic levels in the marine ecosystem (see Figure 10.3). Isotopic dietary work in prehistoric Europe has therefore largely focused on determining the balance between terrestrial and marine subsistence strategies. The first such study was that of Tauber (1981), who measured $\delta^{13}C$ in collagen from human bones dating from the Danish Mesolithic (*ca.* 5200–4000 BC) through to the end of the Iron Age (*ca.* AD 1000), and also some Historic period Inuit samples from Greenland. This showed a clear dietary transition from the hunter-gatherer lifestyle of the Mesolithic ($\delta^{13}C$ approximately –11 to –15‰) to the agricultural lifestyle of the Neolithic and later populations ($\delta^{13}C$ approximately –13 to –27‰). In the absence of any known C_4 terrestrial dietary component in the Mesolithic, the higher (*i.e.*, less negative) $\delta^{13}C$ in the Mesolithic was attributed to a greater dependence on marine resources. Subsequent isotopic dietary studies of the Mesolithic–Neolithic transition in Atlantic Europe, such as Lubell *et al.* (1994) for Portugal, Richards and Mellars (1998) for western Scotland and Schulting and Richards (2001) for Brittany, have all demonstrated a shift away from marine exploitation to more reliance on terrestrial food resources at this boundary. This conclusion is supported by the archaeological evidence for the dating to the Mesolithic of large shell midden deposits around Denmark (and, indeed, many other coastal areas of Atlantic Europe) suggesting heavy marine resource exploitation (Price, 1991; Arias, 1999). It is also supported by the data published by Tauber (1981) on the isotopic composition of the collagen from three pre-contact Inuit from Greenland, who had a diet which consisted of around 75% marine resources, and who had isotopic values similar to those of the Mesolithic populations. It is hard to avoid the conclusion, derived from both isotopic and archaeological data, that the Mesolithic was characterized by a diet dominated by marine food, and that this declined rapidly as agriculture spread at the beginning of the Neolithic.

This view was intensified recently by the publication of Richards *et al.* (2003), which claimed that, in Britain at least, on the evidence of bone collagen δ^{13}C the transition from the Mesolithic to Neolithic lifestyles (at *ca.* 4000 cal BC) was a dramatic one. Figure 10.4 shows δ^{13}C in bone collagen plotted against radiocarbon ages for 183 British Mesolithic and Neolithic humans from coastal (squares) and inland sites (crosses). The sharp change in isotope ratio is interpreted as a shift from a marine diet to one dominated by terrestrial protein: '*marine foods, for whatever reason, seem to have been comprehensively abandoned from the beginning of the Neolithic in Britain*' (Richards *et al.*, 2003). Perhaps even more surprisingly, isotopic reconstructions of diet from Neolithic coastal sites and even from islands in Greece (Papathanasiou, 2003) also suggest that diet was based primarily on terrestrial agriculture. Why did Neolithic peoples across Europe apparently '*turn their back on the sea*' (Schulting and Richards, 2002)? Or are dietary interpretations based on the isotopic method giving misleading results – how sensitive is the method? In

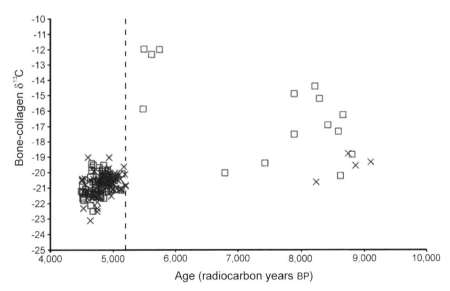

Figure 10.4   Carbon isotope ratios in bone collagen plotted against radiocarbon ages for British Mesolithic and Neolithic humans. (Reprinted by permission from Macmillan Publishers Ltd: Richards *et al.*, 2003. Copyright 2003.)

other words, how much marine protein can be consumed without affecting the 'terrestriality' of the signal, remembering that it is a time-average for each individual? If they did abandon marine resources, was it because terrestrial agriculture was easier and sufficiently bountiful? Was it because the new agriculturists were not adept at harvesting marine resources? Was it because marine resources were thought of as 'old fashioned'? These questions can only be answered from a knowledge of how agriculture came to be adopted – was it population replacement by a new breed of farmers, or was the *idea* of agriculture adopted by hunter-gatherer communities, who 'settled down'? Again, the isotopic dietary evidence, as articulated by Richards *et al.* (2003), appears to show a 'rapid and complete change' – contrary to the conventional archaeological view of a gradual assimilation of ideas, and perhaps more the sort of change that might be associated with population replacement rather than acculturation.

Not surprisingly, this radical suggestion has provoked a sharp debate on both the isotopic side, in terms of the interpretation of evidence, and the archaeological side, based on contrary evidence from other parts of Europe (*e.g.*, Milner *et al.*, 2004; Lidén *et al.*, 2004). Lidén and her colleagues, reporting isotopic data from Mesolithic and Neolithic southern Sweden, detect '*no clear cut dietary transition from the Mesolithic to Neolithic*' (Lidén *et al.*, 2004; 30). Indeed, they see just the reverse – equal dietary diversity on both sides of the boundary, characterized by $\delta^{13}C$ values ranging from –14 to –21‰ and $\delta^{15}N$ from 10 to 17‰. There is also some evidence from an early isotopic study in Britain to suggest that this proposed abandonment of marine resources was not uniform. In a pioneering isotopic analysis of human bone from Neolithic and Bronze Age Wessex and Orkney carried out in Cardiff between 1986 and 1990 [Table 10.1: published in summary form in Antoine *et al.* (1988a, 1998b), and Pollard *et al.* (1991)], it was shown that there were small differences between the average $\delta^{13}C$ values of sites on Orkney (Isbister, Holm of Papa Westray North, Quanterness and Point of Cott) and southern England (Wor Barrow, Shrewton, Hambledon Hill, Fussell's Lodge and Irthlingborough). Although there is some overlap between the two groups, Figure 10.5a shows that the Orkney samples have slightly higher (less negative) $\delta^{13}C$ values, consistent with a slightly higher marine component in the diet than in the southern English samples – as might be expected given the island nature of Orkney, but somewhat contrary to the suggestion of Richards *et al.* (2003). Although these data were obtained before the advent of modern continuous-flow isotope ratio mass spectrometers, there is sufficient evidence here to suggest that Neolithic and later peoples may not uniformly have turned their back on the sea. It would also appear (Fig. 10.5b) that the Sr/Ca ratio [here calculated as (ppm Sr/ppm Ca) × 1000] is also a very effective discriminator between Orkney and Wessex diets, and also suggestive of much higher marine input into the Orkney diet. It is equally possible, however, that this is a result of different diagenetic effects between Orkney and Wessex, as discussed above.

Milner *et al.* (2004) cite evidence for continuity of occupation from the Mesolithic to the Neolithic at coastal sites in Denmark, pointing out the

Table 10.1 Isotopic and trace element data for Neolithic and Bronze Age human bone from Orkney and Southern England.

	$\delta^{13}C_{pdb}$	$\delta^{15}N_{air}$	Ca (%)	P (%)	Sr (ppm)	Mg (ppm)	Zn (ppm)	Sr/Ca
South of England:								
Shrewton (n=19)	−21.9 ± 0.8	4.6 ± 2.8	16.8 ± 1.2	—	209 ± 35	708 ± 222	712 ± 737	1.25 ± 0.23
Wor Barrow (n=6)	−22.8 ± 0.8	7.0 ± 0.4	25.1 ± 4.5	—	180 ± 60	690 ± 132	254 ± 86	0.72 ± 0.20
Hambledon Hill (n=25)	−21.3 ± 2.6	6.7 ± 4.1	22.5 ± 4.4	—	270 ± 48	907 ± 212	201 ± 76	1.22 ± 0.23
Fussell's Lodge (n=95)	−23.2 ± 1.3	12.3 ± 3.4	36.5 ± 1.3	17.1 ± 0.6	456 ± 91	518 ± 256	139 ± 27	1.25 ± 0.25
Irthlingborough (n=70)	−22.9 ± 1.1	11.7 ± 3.4	34.4 ± 2.4	17.4 ± 0.5	676 ± 127	893 ± 887	148 ± 49	1.98 ± 0.44
Orkney:								
Isbister Cairn (n=30)	−20.7 ± 0.6	10.6 ± 2.1	27.1 ± 2.2	—	1828 ± 395	7603 ± 1593	417 ± 155	6.72 ± 1.26
Holm of Papa Westray (n=22)	−20.4 ± 1.2	9.9 ± 3.9	28.9 ± 5.8	—	1931 ± 817	4453 ± 895	515 ± 234	6.82 ± 2.53
Quanterness (n=97) (*n=45 δ^{13}C, **n=55 δ^{15}N)	*−20.4 ± 0.6	**8.3 ± 2.9	33.4 ± 3.3	—	1201 ± 242	3233 ± 1042	611 ± 209	3.62 ± 0.75
Point of Cott, Westray (n=10)	−20.1 ± 0.3	9.0 ± 4.1	34.4 ± 2.5	—	827 ± 275	2745 ± 580	377 ± 87	2.38 ± 0.66
Modern bone (n=5)	−19.5 ± 0.5	9.4 ± 2.6	40.7 ± 1.4	19.5 ± 0.7	286 ± 29	4750 ± 540	143 ± 11	0.70 ± 0.06

— = not determined.

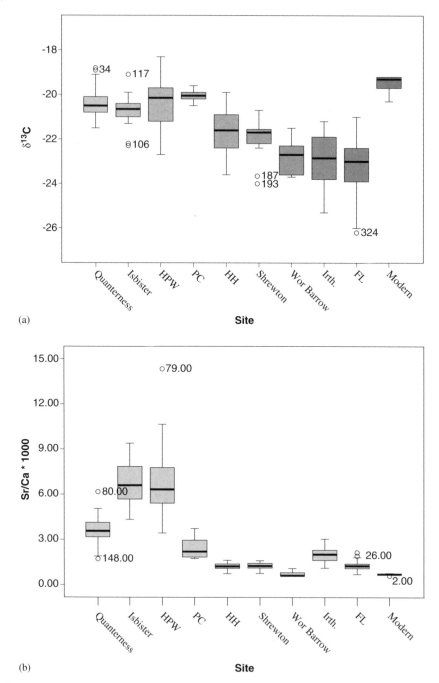

(a)

(b)

Figure 10.5   Carbon isotopes and Sr/Ca ratio from Neolithic and Bronze Age humans
in Orkney and Wessex (FL, Fussell's Lodge; HH, Hambledon Hill;
HPW, Holm of Papa Westray North; Irth, Irthlingborough; PC, Point
of Cott).

existence of large Neolithic shell middens, and concluding that '*the evidence of marine subsistence . . . is not in doubt*' for the Early Neolithic. They propose that the overemphasis of the degree of dietary shift is either due to sampling problems (*i.e.*, sampling people who, for whatever reason, did not utilize marine resources), or problems in the interpretation of the isotopic data. They re-stated the assumptions upon which the estimations of the proportion of marine food in human diet from carbon and nitrogen isotopes are based (Richards and Hedges, 1999), which may be summarized as:

(i) bone collagen composition reflects the average diet over the last 5–10 years of life, depending on the rate of bone turnover and the particular bone sampled;

(ii) isotopes in collagen reflect only protein inputs into diet, and therefore the sources of lipid and carbohydrate are invisible by this method;

(iii) the offset (fractionation) between dietary protein and bone collagen is 0–1‰ for $\delta^{13}C$ and 3‰ for $\delta^{15}N$;

(iv) variations in diet, such as variations in the ratio of protein to carbohydrate, have no effect on the above statements;

(v) the dietary 'end members' are sufficiently well-known in terms of isotopic value that intermediate proportions can be estimated {*i.e.*, pure marine [$\delta^{13}C = -12\pm1‰$, $\delta^{15}N = 4$–10 ‰] and pure C_3 terrestrial inputs [$\delta^{13}C = -20‰$, $\delta^{15}N = 10$–22‰], taking the figures given by Richards and Hedges (1999)}.

Milner *et al.* (2004) challenge some of these assumptions, but particularly the last two. They point out, for example, that $\delta^{13}C$ in marine organisms has been shown to correlate with salinity (Eriksson and Lidén, 2002), so there is the possibility that the dietary 'end members' are not necessarily fixed geographically or temporally, thus making the estimates of the proportion of marine protein in the diet suspect. They advocate, if possible, the 'total ecological reconstruction' approach – the measurement of all components of the food web, rather than just that of humans – before interpretations are made. They also commented on the assumption that dietary proteins were routed directly into the bone collagen, as shown by Ambrose and Norr (1993) in their controlled feeding experiments on rats. This was true only if the total proportion of protein in the diet was greater than 5% – on low protein diets, around 50% of the bone collagen carbon came from dietary lipids and carbohydrates. Thus, a terrestrial diet based on a large proportion of plant intake could reduce the value of $\delta^{13}C$ in the bone collagen, again making estimates of dietary proportions unreliable. On the basis of these uncertainties, they argue that the data cannot be interpreted in such a clear-cut way, and suggest that in a plant-dominated diet up to 20% of the protein could come from marine sources without shifting the $\delta^{13}C$ value from –21‰. The response from Hedges (2004) accepts that there is uncertainty in some of these assumptions (particularly related to the metabolic 'routing' of different sources of dietary carbon – see also Sillen *et al.*, 1989), but re-asserts that the overwhelming body of isotopic evidence is for a sharp change in diet towards a more terrestrial subsistence at the onset of the Neolithic. He concedes, however, that some marine protein may have been incorporated into the diet (certainly up to 20%, possibly as high as 30%), within the uncertainties of current models. This is likely to provide the impetus to carry out more detailed studies, including compound-specific isotope studies, in which individual amino acids can be targeted, and also the investigation of the dietary significance of other isotopic systems, such as δD and $\delta^{34}S$. As is often the case, further work is required!

As a consequence of the uncertainties in the routing and estimation of dietary components [and also of the suspicion that isotopic signals at the base of the food chain might change with time and space (Hedges *et al.*, 2004)], an approach has been generally adopted in archaeology which can be described as 'total ecosystem reconstruction'. In order to study the ecology of prehistoric humans, it is preferable to analyse as wide a range as possible of faunal remains in addition to those of the humans. These might be domesticates in more recent studies, or wild species in deeper time. Using uniformitarianism, it can be assumed that modern-day herbivores such as cattle (but not if intensively reared!) or deer have always been C_3 herbivores, and therefore that their isotopic signal is a marker for C_3 herbivory in the past. It is likewise with carnivores and omnivores, and also with marine species. Thus an isotopic framework can be constructed from associated fauna, which allows, by comparison, the behaviour of the humans to be predicted without recourse to uncertain theoretical models. An example from a study of later Medieval England is given in Figure 10.6 (Müldner and Richards, 2005), showing humans from three sites (the rural hospital of St Giles by Brompton Bridge,

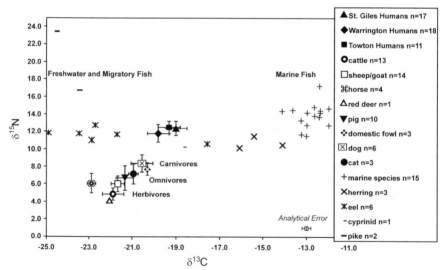

Figure 10.6   An example of 'total ecosystem reconstruction' – a trophic level diagram from three sites in Later Medieval England. (From Müldner and Richards, 2005; Figure 3, with permission. Copyright Elsevier 2005.)

the Augustinian Friary at Warrington and a mass grave from the battle of Towton, fought on a 'snowy Palm Sunday', March 29th, 1461) plotted against a range of domesticates and wild food sources. In this case, the elevation of the human signal above the level of terrestrial carnivores is interpreted as being evidence for substantial consumption of marine and freshwater protein.

One extremely promising research avenue which has been somewhat slow to take off due to experimental difficulties, but is now gathering pace, is the study of the carbon and nitrogen isotope signals in the individual amino acids which make up collagen (Hare *et al.*, 1991) – sometimes referred to as *compound-specific* amino acid isotopic analysis. Collagen consists of 17 or so different amino acids (see Chapter 8), some of which (the *essential* or *indispensable* amino acids – in humans, isoleucine, leucine, lysine, methionine, phenylalanine, threonine, tryptophan and valine) – are not capable of being manufactured from dietary components (*e.g.*, Dawber and Moore, 1980; 428). Most of the rest are referred to as *non-essential* or *dispensable*, and can be manufactured within the body from different sources of protein. A few amino acids (arginine, cysteine, glycine, glutamine and tyrosine) are referred to as *conditionally essential*, since they must be supplied from external sources if not synthesized in adequate amounts for reasons of diet, stress or illness. Using a chromatographic separation of the individual amino acids before stable isotope measurement [conventionally by gas chromatography (GC), but now high pressure liquid chromatography (HPLC) is possible (McCullagh *et al.*, 2006)] allows the $\delta^{13}C$ value of each amino acid to be measured. Clearly, the values measured from the essential amino acids are directly related to the values of these amino acids in the diet, whereas those of the non-essential amino acids (which, by definition, are synthesized in the body) may reflect an average from a range of different protein sources. In this way it is hoped that more detail can be obtained on sources of dietary protein. In particular, the compound-specific isotopic study of amino acids has been proposed as a way of identifying more sensitively the amount of marine protein in human diets (Corr *et al.*, 2005).

10.5 STATUS AND HEALTH

The specificity of dietary reconstruction from stable isotopes to a particular individual has been exploited in a series of comparative studies. It has allowed several researchers to reconstruct the relative diets of individuals within cemeteries, so enabling models of differential access to food resources according to status or gender difference to be tested (*e.g.*, Ubelaker *et al.*, 1995; Wright and Schwarcz, 1999; Katzenberg and Weber, 1999; Ambrose *et al.*, 2003). Other studies have attempted to find evidence for the weaning age of infants (Herring *et al.*, 1998; Richards *et al.*, 2002; Fuller *et al.*, 2006a), or changes in nutritional status as a result of disease (*e.g.*, Fuller *et al.*, 2005).

Ubelaker *et al.* (1995) studied human remains from six shaft tombs excavated at La Florida, Quito, Ecuador, and dated to the Chaupicruz phase (*ca.* AD 100–450). They used isotopes to assess the dietary relationship between high-status individuals and the victims of sacrifice buried in the same tombs.

Although ethnographic evidence from the 16th Century AD suggested that there might be a higher consumption of animal protein by the elite, this was not observed analytically. The only isotopic difference suggested was a higher consumption of maize by the elite class (average $\delta^{13}C$ –10.3‰ in the elite group, compared to –11.6‰ in those of lower rank), attributed to a higher consumption of *chicha* (maize beer) by the elite.

Ambrose *et al.* (2003) studied status and gender differences from the Mississippian period (*ca.* AD 1050–1150) burials in Mound 72, Cahokia (near St Louis, Illinois). The archaeological evidence showed substantial difference in status between burials, and included mass graves of young adult females showing skeletal evidence of nutritional stress. Nitrogen isotope analysis suggested that the high-status individuals consumed much more animal protein. Carbon isotopes showed, however, roughly equal maize consumption between high and low status, with the exception of the low-status females from mass graves, who appeared to show a 60% higher consumption of maize. This evidence, which corroborates the archaeological, palaeopathological and dental evidence, clearly shows significant status- and gender-based differentiation in diet (and therefore in nutritional status) and gives a remarkable insight into prehistoric social organization and inequalities.

It is to be expected that infants fed on breast milk will show isotopic enhancement compared to their mothers, since they are effectively consuming their mother's protein and are therefore on a higher trophic level (the 'carnivore' effect). On modern fingernail and hair samples, for example, the effect has been shown to be $+1‰$ for $\delta^{13}C$ and $+2$–3‰ for $\delta^{15}N$ (Fuller *et al.*, 2006b). Measurements of $\delta^{15}N$ values from rib collagen taken from infants buried at the deserted medieval village site of Wharram Percy, Yorkshire, UK, showed values which decreased to adult rib values after the age of two years, indicating the weaning was complete at this age (Richards *et al.*, 2002). In an earlier period, at Late/Sub-Roman Dorchester-on-Thames, similar evidence suggested complete cessation of breastfeeding by the age of 3–4 years (Fuller *et al.*, 2006a). It was also observed at this site that females showed a slightly lower average value of $\delta^{15}N$ (9.9±0.9‰) compared to males (10.6±0.5‰). It was suggested that either continuous lactation or pregnancy might be responsible for this, rather than reduced consumption of animal protein by women.

10.6 MOBILITY

We have alluded above to the fact that dietary reconstruction from bone can be no more than a relatively long-term average, since in life bone is constantly remodelled. In general, a dietary reconstruction based on bone collagen is likely to represent the average diet of that individual over the last few years of life – perhaps up to as much as ten years before death, depending on the particular bone used. An extension of the isotopic dietary method is to use the differential information available within a single skeleton to study human lifetime mobility. This technique has been developed and exploited most clearly on historic material from South Africa (Sealy *et al.*, 1995; Sealy, 2001; Cox *et al.*, 2001).

As with dietary studies in general, this technique has largely developed from isotopic studies of other mammals – in this case, the need to trace the movement of elephants in Africa, and to be able to provenance traded ivory (*e.g.*, Koch *et al.*, 1995).

The method is based on the observation that the continual resorption and remodelling of bone leads to different rates of bone turnover within the skeleton, which, in turn, means that different elements of the skeleton give information from different periods of time prior to death. Turnover rates in humans are not known precisely, but are likely to range up to ten years (Sealy *et al.*, 1995). Also useful in this context is that secondary teeth are formed before or in early adulthood, and are essentially unaltered after eruption, since, unlike bone, there is no major remodelling within dental tissue. Thus, the isotopic composition of the collagen in teeth will represent dietary inputs during early life (up to perhaps the age of 20 years), whereas the isotopic composition of the rib, for example, may reflect the diet over the last few years prior to death. Thus, an individual who died aged more than 30 years might be expected to show dietary differences between tooth and rib, if he or she spent their adult years in a different ecological environment to that of childhood.

Sealy *et al.* (1995) measured $\delta^{13}C$ and $\delta^{15}N$ in dentine, enamel, rib and femur or humerus from five individuals excavated in South Africa. Two were prehistoric Khoisan hunter-gatherers, whose isotopic signatures showed similar values in all tissues, indicating no significant movement outside their own ecozone. Another was an adult male from Cape Town Fort, probably a European who died at the Cape during the 17th Century AD. His isotopes also showed no variation, suggesting he was newly arrived when he died. The fourth, a European from the Dutch East India Company station at Oudespost (occupied between AD 1673 and 1732), did show significant differences between dental and bone measurements – less negative $\delta^{13}C$ values and higher $\delta^{15}N$ figures in the bone, suggestive of greater consumption of marine protein during adult life. The fifth was an African woman, assumed to have been a slave, dating to the 18th Century AD. She also showed a marked shift in all ratios measured. Before early adulthood she had $\delta^{13}C$ and $\delta^{15}N$ values from a C_4 biome, consistent with a tropical or sub-tropical inland area of southern Africa. At the end of her life, she had elevated $\delta^{15}N$ values, consistent with a more marine diet following a move to the coast. In both of these latter examples, strontium isotopes were also measured (see Section 10.8), corroborating the evidence from the light stable isotopes. This work shows that the whole of an individual's life history is isotopically encoded in the skeleton, and, if the dietary changes are sufficiently marked, can be 'read' using isotope chemistry.

10.7 OTHER ISOTOPIC APPROACHES TO DIET AND MOBILITY

There are a number of reasons why isotope archaeologists interested in diet and mobility have turned their attention to systems other than carbon and nitrogen in bone collagen. One is that collagen in bone does not survive indefinitely – it is susceptible to hydrolytic degradation, and is unlikely to survive in most environments for more than a few thousand years (Smith *et al.*, 2001). The exact time depends on a number of factors, such as average burial temperature, moisture availability, rapidity of burial, *etc.* Exceptions are likely to be extreme aridity or permafrost. In order to study diets in 'deep time' (*i.e.*, pre-Holocene and earlier), therefore, it has proved necessary to look for carbon in other more durable biominerals – the obvious candidate being dental enamel, which is the most robust material in the body. Since, however, enamel contains very little protein, it has proved necessary to look at the carbonate fraction incorporated into the mineral. A second reason for looking elsewhere is that carbon and nitrogen in collagen only give a partial view of diet. It is possible that isotopic measurements on other biomolecules in bones and teeth (*e.g.*, lipids, especially cholesterol) might give insights into other aspects of diet, such as carbohydrate. It is also possible that additional insights into diet will be given by studying other isotopes in collagen, such as those of hydrogen (Birchall *et al.*, 2005) or sulfur (Richards *et al.*, 2001).

10.7.1 Carbonates and 'Deep Time'

It has been known for some time that $\delta^{13}C$ values can be recovered from the carbonate fraction present in the apatite mineral phase of bone and teeth, providing the effects of *post-mortem* alteration (diagenesis) can be eliminated. Controlled feeding experiments have suggested that the $\delta^{13}C$ value in bone and dental carbonate (sometimes referred to as the 'apatite' value, or $\delta^{13}C_{apa}$) reflects the carbon source of the entire diet, in contrast to the $\delta^{13}C$ value in bone collagen ($\delta^{13}C_{coll}$), which, under adequate nutrition, reflects the source of protein (see Figure 10.2). The difference between the collagen and the carbonate value for an individual is called the 'dietary spacing', Δ, defined as:

$$\Delta = (\delta^{13}C_{apa} - \delta^{13}C_{coll})$$

The value of Δ reflects the difference between protein intake and total dietary intake, and thus reveals more about the balance of diet between protein and other foodstuffs. This information can be used to correct for biases in interpretation in cases where, for example, all the protein comes from a marine environment but terrestrial plants are also eaten, which would be almost invisible from the collagen values alone. It is now common to see plots of $\delta^{13}C_{coll}$ against dietary spacing Δ as another tool for the interpretation of diet, in addition to the more familiar plots of $\delta^{15}N$ against $\delta^{13}C$, as shown in Figure 10.3.

Although adsorbed carbonates on bone mineral and dentine can be easily removed by routine cleaning pre-treatment, the diagenetic fraction has proved more difficult and controversial. Attempts have been made to use sequential acid washing and density separation for bone, as described above, but, at present, the results are rather ambiguous. The carbonate fraction of dental enamel, however, has proved much more amenable, and significant progress has been made. Dental enamel is the most durable mammalian tissue, so this has allowed the isotopic dietary method to be extended back into very 'deep time' [*i.e.*, hundreds of thousands, up to tens of millions, of years (Ambrose and Krigbaum (2003)]. From the teeth of fossils from grazing animals, this has, indirectly, enabled the dating of the evolutionary emergence of C_4 plants in Pakistan and Kenya to between 10 and 15 Ma (Morgan *et al.*, 1994). Given that all trees and shrubs are C_3 plants, but many tropical grasses are C_4, it allows grazing herbivores to be distinguished from browsers in the fossil record. From the teeth of carnivores, it also allows the general nature of the prey species to be identified.

Of considerably more significance archaeologically is that this allows the diet of fossil hominids to be studied (Lee-Thorp and Sponheimer, 2006). One of the earliest so far studied, dating to between 1.8 and 1 Ma, is *Australopithecus robustus* from Swartkrans (Lee-Thorp *et al.*, 1994). The dental enamel carbonate from eight hominid samples gave $\delta^{13}C$ with an average value of $-8.5 \pm 1.0‰$. Comparison of this value with measurements from the teeth of C_4 grazers (average value *ca.* 0‰) and C_3 browsers (average –11 to –12‰) from Swartkrans indicated that the Australopithecines had a significant component (25–30%) of C_4 plant material in their diet, either obtained by eating C_4 grasses or, more likely (since the teeth show no signs of wear from the consumption of grasses), from consuming grazing animals. In contradiction to the previously held belief from dental morphometric studies that Australopithecines were vegetarian, this indicates an omnivorous diet. More recent work using a laser ablation method for measuring carbon isotopes in dental enamel [which allows within-lifetime variation to be studied (Sponheimer *et al.*, 2006)] has revealed that the diet of *Paranthropus robustus* (1.8 million years ago) changed seasonally and inter-annually, suggesting that *P. robustus* was not a dietary specialist. All of this work paints a more detailed picture of early hominins opportunistically exploiting savannah-based foods such as grasses and sedges (or animals eating these), and shows that isotopic studies on fossil dental enamel have an extremely significant role to play in understanding human evolution.

10.7.2   Other Biomolecules and Isotopes

Collagen, although by far the most abundant, is not the only biomolecule present in living bone. It is possible that some of these minor organic components might survive for longer in archaeological bone than does collagen, and therefore there has been some interest in detecting non-collagenous components in bone for dietary and dating purposes. Until recently, however, the very low levels of these other biomolecules has precluded the measurement of their $\delta^{13}C$ values in bone, but recent advances in compound-specific isotope ratio mass spectrometry has now allowed the isotopic value of cholesterol in bone to be measured (Jim *et al.*, 2004). Using the bone samples previously studied in controlled diet studies in rats (Ambrose and Norr, 2003), it proved possible to show that the $\delta^{13}C$ value of cholesterol, like that of bone mineral carbonate, is linked to the whole-diet $\delta^{13}C$ value. This too, therefore, offers the potential for an additional source of information in dietary reconstructions, and also for measurements further back in time than are possible with collagen.

Carbon and nitrogen are not the only elements in collagen for which stable isotope measurements can be made, although our understanding of the dietary routing of these other elements is less than complete. Hydrogen isotopes (δD) in most environmental systems are measured along with those of oxygen and come directly from the water consumed. Its isotopic value reflects the temperature of formation of the meteoric water (Pollard *et al.*, 2007; 171), and therefore it might have geographical significance in bone collagen. Preliminary measurements on modern fauna suggest that it is also influenced by trophic level (Birchall *et al.*, 2005). Sulfur (δ^{34}S) measurements in collagen appear to be complementary to δ^{13}C, but there is a hope that they may help to discriminate between marine, freshwater and terrestrial ecosystems, and also have some geographical significance (Richards *et al.*, 2001).

10.8 'PROVENANCING HUMANS' – STRONTIUM AND OXYGEN ISOTOPES

The use of differential dietary signals between tooth (formed early in life) and bone tissue (remodelled continuously up to death) provided a significant breakthrough in the idea of 'provenancing humans' – giving evidence for human mobility from chemical analysis, similar to the way that ceramics can be chemically 'fingerprinted' to clay source. Relying purely on the isotopes of carbon and nitrogen, however, limits the technique to individuals who have moved from one biome to another, where the second is radically different from the first – preferably from a C_4 to a C_3 biome (or *vice versa*, of course). Other more subtle moves are unlikely to be detected in this way. The level of the element strontium in bone has been of major dietary interest for many years (see Section 10.2), and it was first suggested by Ericson (1985) that measurement of the isotopic ratio of strontium (^{87}Sr/^{86}Sr) in bone might have some geographical significance. This has now been turned into a powerful tool for provenancing human and other bone samples (Bentley, 2006) – providing, of course, adequate measures can be taken to eliminate *post-mortem* contamination by strontium from the groundwater.

Strontium has four stable naturally occurring isotopes: ^{84}Sr (0.56%), ^{86}Sr (9.86%), ^{87}Sr (7.0%) and ^{88}Sr (82.58%). Of these, ^{87}Sr is radiogenic, being produced by the decay of the radioactive alkali metal ^{87}Rb, with a half life of 4.75×10^{10} years. Thus, as with lead, the isotope ratio ^{87}Sr/^{86}Sr in a rock is related to the age and the original isotopic composition of the rock. The natural variation of the ^{87}Sr/^{86}Sr ratio in minerals and rocks ranges from about 0.699 in meteorites and moon rock to greater than 0.74 in the carbonate fractions of deep-sea sediments (it is not customary to use the δ notation for Sr isotope ratios). The Sr ratio measured in biominerals should, therefore, be characteristic of the local underlying geology, since it can be assumed that the isotopic ratio is unmodified as it passes through the food chain. Movement which results in the last years of life being spent in a region of different geology to that of the early years should, therefore, show up as a different strontium isotope ratio between dental enamel and bone tissue. Moreover, because modern geological data are available for Sr isotope values, it is possible, in principle, to predict where these regions might be, since the geological values will not have changed significantly in recent times. The measurement of Sr isotopes, previously by thermal ionization mass spectrometry (TIMS), but more recently by high-resolution inductively coupled plasma mass spectrometry (ICP-MS), is discussed in Pollard *et al.* (2007; 174).

From a European archaeological perspective, the most significant application of Sr isotopes is that of Price and co-workers, who have studied the 'Neolithicization' of Europe (Bentley *et al.*, 2002), and the later spread of the 'Beaker folk' (*e.g.*, Price *et al.*, 1994; Grupe *et al.*, 1997). In both cases, the key question is whether the cultural changes seen in the archaeological record were brought about by the movement of ideas or of people. Thus, the availability of a technique which can address the question of the mobility of an individual assumes a central importance. The Neolithic, as discussed above, is characterized primarily by the adoption of farming and a more sedentary way of life, as opposed to a reliance on the harvesting of wild resources. The 'Beaker folk' are traditionally seen as a group of people identified by their burial practices (including the use of characteristic ceramic 'beakers') who spread through central and western Europe at the end of the Neolithic (*ca.* 2500 BC), and who appear to have brought with them the knowledge of metalworking. What is not clear, archaeologically, is whether the 'Beaker folk' themselves moved, or whether 'Beaker ideas' moved and were simply adopted by indigenous populations. Price *et al.* (1994) measured strontium isotope ratios in teeth and bone from eight Bell Beaker burials in Bavaria. Of these, two showed clear evidence of having moved significantly between adulthood and childhood. Although a very small sample, this suggests that mobility was common. Subsequently, Bentley *et al.* (2002) applied the same technique to the LBK (*Linearbandkeramik*) culture, traditionally believed to be the first European farmers (*ca.* 7500 BP). From three different LBK cemeteries they identified 27 non-locals amongst the 64 individuals analysed. They observed that non-local females were common in these cemeteries, and that burial practices varied between locals and non-locals. Overall, they concluded that there was a great deal of mobility during the LBK period (especially of females), but that the most likely explanation for the spread of the Neolithic was a colonization of south-eastern Europe by farmers, followed by the adoption of farming by the indigenous population in central and northern Europe.

One of the advantages of using Sr isotopes is that the data can be compared with modern geochemical maps of Sr isotope values, which gives the method a predictive capacity (the same is true of Pb isotope measurements, as discussed in Chapter 9, and which can also be used in bone studies). One further isotope which can be used in this way is $\delta^{18}O$ (the ratio $^{18}O/^{16}O$), measured in the phosphate mineral of bone or, preferably, dental enamel. The oxygen isotopic composition of biomineral is related to that of the source of ingested water (Longinelli, 1995). The oxygen and hydrogen isotope values of meteoric and terrestrial waters have been studied extensively, so maps are available of the modern variation in surface water isotopic compositions for many parts of the world. In the British Isles, for example, the oxygen isotope values in rainfall vary from –4.5‰ (against the international standard VSMOW) in the west to −8.5‰ in the northeast (Darling *et al.*, 2003). As with the other light isotopes, however (and unlike the heavier Sr and Pb isotopes), fractionation will occur as the oxygen is incorporated into the body, so that some form of calibration is required before the measured biomineral values can be compared with geo-chemical data. The calibration produced by Levinson *et al.* (1987; 369) is widely used for human tissue – for the phosphate in human teeth, this relationship is:

$$\delta_p = 0.46\,\delta_w + 19.4$$

where δ_p is the oxygen isotope ratio ($\delta^{18}O$) in tooth phosphate and δ_w is the ratio in ingested water. This equation can therefore be used to 'locate' the geographical origin of an archaeological tooth sample by calculating the value for δ_w and comparing this with modern δ_w distribution maps. This assumes, of course, that the oxygen isotope ratio of drinking water is time invariant – an assumption which must become increasingly questionable further back in time.

Oxygen isotopes, sometimes combined with carbon, nitrogen, strontium and lead isotopic measurements, have been used to study the mobility of diverse human groups such as those in Teotihuacán and Oaxaca, Mexico, dating from about 300 BC to AD 750 (Stuart-Williams *et al.*, 1996), the Kellis 2 cemetery (*ca.* AD 250) in the Dakhleh Oasis, Egypt (Dupras and Schwarcz, 2001), and also early medieval immigration into Britain (Budd *et al.*, 2004). In a recent study, Evans *et al.* (2006) used oxygen and strontium isotopes in dental enamel to chart the movement of seven individuals (three adults, a sub-adult, two juveniles and an infant) from a single grave at Boscombe Down, near Stone-henge. The three adults (the 'Boscombe Bowmen') were found to have spent their early lives in an area with higher $^{87}Sr/^{86}Sr$ than that found in Wiltshire – the nearest possible locality according to the oxygen and strontium data being Wales. The two juveniles had a signal consistent with the local environment. This is in contrast with a further two adults recovered from a nearby burial at Normanton Down and also a single individual from the ditch of Stonehenge itself, who all showed no evidence of migration. This remarkable result has thrown new and very vivid light onto the social structure at Stonehenge in the Early Bronze Age (*ca.* 2500–2000 BC).

10.9 SUMMARY

The study of human remains has long been the most direct method of under-standing the detailed lives of our ancestors as individuals. The methods of forensic science and palaeopathology have been applied for many years to determine age, sex, stature, nutritional status and cause of death, and to detect evidence of some diseases. Increasingly, DNA is being used to determine kinship. It is, however, no exaggeration to say that entire new vistas have been opened up by the application of chemical analysis to human remains. Although the measurement of $\delta^{13}C$ in bone collagen began as a means of correcting radiocarbon dates for fractionation, it has rapidly developed (along with that of $\delta^{15}N$) into a widely applicable tool to study the diets of past populations. Interpretational difficulties have been encountered along the way, particularly with the light isotopes in collagen, where questions about dietary routing have had to be studied by controlled feeding experiments in order to interpret the data. Trace element work on bone mineral has somewhat slipped out of favour, because of the difficulties associated with distinguishing between *in vivo* trace element concentrations and those resulting from *post-mortem* diagenesis. The mineral fraction, at least in dental enamel, has, however, been found to be reliable in terms of measuring a carbonate $\delta^{13}C$ signal, and this has allowed dietary reconstruction to be pushed back many millennia.

The ability to study a particular individual has allowed questions of deep archaeological significance to be addressed from skeletal analysis. Issues of wealth and status, long inferred purely from archaeological evidence (grave goods, tomb location and architecture, *etc.*), can now be considered in terms of differential access to resources. The diets of slaves can be compared with those of the elite. Following on from diet, it is possible to compare the health and nutritional status of different sectors of society – men against women, rich against poor, warriors against farmers. The interpretation of the data for women, however, needs to be approached with some caution. Does a difference in stable isotope signal really mean access to a poorer range of diets, or is it the physiological consequence of pregnancy and breastfeeding?

It has also proved possible to use the different rates of development and turnover of the human skeleton itself to address questions of mobility – to 'provenance humans'. Differences of diet – or, more recently, of location as determined by oxygen, strontium and lead isotopes – between early and late life can be used to infer migration, and therefore to begin to answer some of the really big questions in the history of our species, such as how did we come to adopt farming, or how did a knowledge of metallurgy spread. Addressing these issues directly constitutes one of the greatest challenges for archaeological chemistry, and requires a deep knowledge of geochemistry, biochemistry and physiology, as well as of archaeology. A fitting challenge for archaeological chemists!

REFERENCES

Ambrose, S.H. (1990). Preparation and characterization of bone and tooth collagen for isotopic analysis. *Journal of Archaeological Science* **17** 431–451.

Ambrose, S.H. (1993). Isotopic analysis of paleodiets: methodological and interpretative considerations. In *Investigations of Ancient Human Tissue: Chemical Analysis in Anthropology*, ed. Sandford, M.K., Gordon and Breach, Langhorne, pp. 59–130.

Ambrose, S.H. and Krigbaum, J. (2003). Bone chemistry and bioarchaeology. *Journal of Anthropological Archaeology* **22** 193–199.

Ambrose, S.H. and Norr, L. (1993). Experimental evidence for the relationship of the carbon isotope ratios of whole diet and dietary protein to those of bone collagen and carbonate. In *Prehistoric Human Bone: Archaeology at the Molecular Level*, ed. Lambert, J.B. and Grupe, G., Springer Verlag, Berlin, pp. 1–37.

Ambrose, S.H., Buikstra, J and Kreuger, H.W. (2003). Status and gender differences in diet at Mound 72, Cahokia, revealed by isotopic analysis of bone. *Journal of Anthropological Archaeology* **22** 217–226.

Antoine, S.E., Dresser, P.Q., Pollard, A.M. and Whittle, A.W.R. (1988a). Bone chemistry and dietary reconstruction in Prehistoric Britain: examples from Wessex. In *Science and Archaeology, Glasgow 1987*, ed. Slater, E.A. and Tate, J.O., BAR British Series 196, British Archaeological Reports, Oxford, pp. 369–380.

Antoine, S.E., Pollard, A.M., Dresser, P.Q. and Whittle, A.W.R. (1988b). Bone chemistry and dietary reconstruction in Prehistoric Britain: examples from Orkney, Scotland. In *Proceedings of 26th International Archaeometry Symposium, Toronto, 1988*, ed. Farquhar, R.M., Hancock, R.G.V. and Pavlish, L.A., University of Toronto, Toronto, pp. 101–106.

Arias, P. (1999). The origins of the Neolithic along the Atlantic coast of continental Europe: a survey. *Journal of World Prehistory* **13** 403–464.

Bell, L.S., Cox, G. and Sealy, J. (2001). Determining isotopic life history trajectories using bone density fractionation and stable isotope measurements: a new approach. *American Journal of Physical Anthropology* **116** 66–79.

Bentley, R.A. (2006). Strontium isotopes from the earth to the archaeological skeleton: a review. *Journal of Archaeological Method and Theory* **13** 135–187.

Bentley, R.A., Price, T.D., Lüning, J., Gronenborn, D., Wahl, J. and Fullager, P.D. (2002). Prehistoric migration in Europe: strontium isotope analysis of early Neolithic skeletons. *Current Anthropology* **43** 799–804.

Berna, F., Matthews, A. and Weiner, S. (2004). Solubilities of bone mineral from archaeological sites: the recrystallization window. *Journal of Archaeological Science* **31** 867–882.

Birchall, J., O'Connell, T.C., Heaton, T.H.E. and Hedges, R.E.M. (2005). Hydrogen isotope ratios in animal body protein reflect trophic level. *Journal of Animal Ecology* **74** 877–881.

Budd, P., Millard, A., Chenery, C., Lucy, S. and Roberts, C. (2004). Investigating population movement by stable isotope analysis: a report from Britain. *Antiquity* **78** 127–141.

Burton, J.H. and Wright, L.E. (1995). Nonlinearity in the relationship between bone Sr/Ca and diet: paleodietary implications. *American Journal of Physical Anthropology* **96** 273–282.

Corr, L.T., Sealy, J.C., Horton, M.C. and Evershed, R.P. (2005). A novel marine dietary indicator utilising compound-specific bone collagen amino acid delta C-13 values of ancient humans. *Journal of Archaeological Science* **32** 321–330.

Cox, G., Sealy, J., Schrire, C. and Morris, A. (2001). Stable carbon and nitrogen isotopic analyses of the underclass at the colonial Cape of Good Hope in the eighteenth and nineteenth centuries. *World Archaeology* **33** 73–97.

Darling, W.G., Bath, A.H. and Talbot, J.C. (2003). The O and H stable isotopic composition of fresh waters in the British Isles: 2, surface and groundwaters. *Hydrology and Earth System Sciences* **7** 183–195.

Dawber, J.G. and Moore, A.T. (1980). *Chemistry for the Life Sciences.* Macmillan, London, 2nd edn.

DeNiro, M.J. (1987). Stable isotopy and archaeology. *American Scientist* **75** 182–191.

Dupras, T.L. and Schwarcz, H.P. (2001). Strangers in a strange land: stable isotope evidence for human migration in the Dakhleh Oasis, Egypt. *Journal of Archaeological Science* **28** 1199–1208.

Ehleringer, J.R. and Monson, R.K. (1993). Evolutionary and ecological aspects of photosynthetic pathway variation. *Annual Review of Ecology and Systematics* **24** 411–439.

Ehleringer, J.R. and Rundel, P.W. (1988). Stable isotopes: history, units and instrumentation. In *Stable Isotopes in Ecological Research*, ed. Rundel, P.W., Ehleringer, J.R. and Nagy, K.A., Springer-Verlag, New York, pp. 1–15.

Ericson, J.E. (1985). Strontium isotope characterization in the study of prehistoric human ecology. *Journal of Human Evolution* **14** 503–514.

Eriksson, G. and Lidén, K. (2002). Mammalian stable isotope ecology in a Mesolithic lagoon at Skateholm. *Journal of Nordic Archaeological Science* **13** 5–10.

Evans, J.A., Chenery, C.A. and Fitzpatrick, A.P. (2006). Bronze Age childhood migration of individuals near Stonhenge, revealed by strontium and oxygen isotope tooth enamel analysis. *Archaeometry* **48** 309–321.

Ezzo, J.A. (1994). Putting the chemistry back into archaeological bone chemistry analysis – modeling potential paleodietary indicators. *Journal of Anthropological Archaeology* **13** 1–34.

Farquhar, G.D., O'Leary, M.H. and Berry, J.A. (1982). On the relationship between carbon isotope discrimination and the intercellular carbon dioxide concentration in leaves. *Australian Journal of Plant Physiology* **9** 121–137.

Fuller, B.T., Fuller, J.L., Sage, N.E., Harris, D.A., O'Connell, T.C. and Hedges, R.E.M. (2005). Nitrogen balance and delta N-15: why you're not what you eat during nutritional stress. *Rapid Communications in Mass Spectrometry* **19** 2497–2506.

Fuller, B.T., Molleson, T.I., Harris, D.A., Gilmour, L.T. and Hedges, R.E.M. (2006a). Isotopic evidence for breastfeeding and possible adult dietary differences from Late/Sub-Roman Britain. *American Journal of Physical Anthropology* **129** 45–54.

Fuller, B.T., Fuller, J.L., Harris, D.A. and Hedges, R.E.M. (2006b). Detection of breastfeeding and weaning in modern human infants with carbon and nitrogen stable isotope ratios. *American Journal of Physical Anthropology* **129** 279–293.

Gartner, L.P. and Hiatt, J.L. (1994). *Color Atlas of Histology*. Williams and Wilkins, Baltimore, 2nd edn.

Griffiths, H. (ed.) (1998). *Stable Isotopes. Integration of Biological, Ecological and Geochemical Processes*. BIOS Scientific, Oxford.

Grupe, G., Price, T.D., Schröter, P., Söllner, F., Johnson, C.M. and Beard, B.L. (1997). Mobility of Bell Beaker people revealed by strontium isotope ratios of tooth and bone: a study of southern Bavarian skeletal remains. *Applied Geochemistry* **12** 517–525.

Gulliford, A. (1996). Bones of contention: the repatriation of Native American human remains. *Public Historian* **18** 119–143.

Hancock, R.G.V., Grynpas, M.D. and Pritzker, K.P.H. (1989). The abuse of bone analyses for archaeological dietary studies. *Archaeometry* **31** 169–179.

Hare, P.E., Fogel, M.L., Stafford, T.W., Mitchell, A.D. and Hoering, T.C. (1991). The isotopic composition of carbon and nitrogen in individual amino acids isolated from modern and fossil proteins. *Journal of Archaeological Science* **18** 277–292.

Hedges, R.E.M. (2002). Bone diagenesis: an overview of processes. *Archaeometry* **44** 319–328.

Hedges, R.E.M. (2004). Isotopes and red herrings: comments on Milner *et al.* and Lidén *et al. Antiquity* **78** 34–37.

Hedges, R.E.M. and Millard, A.R. (1995). Measurements and relationships of diagenetic alteration of bone from three archaeological sites. *Journal of Archaeological Science* **22** 201–211.

Hedges, R.E.M., Stevens, R.E. and Richards, M.P. (2004). Bone as a stable isotope archive for local climatic information. *Quaternary Science Reviews* **23** 959–965.

Hedges, R.E.M., Stevens, R.E. and Koch, P.L. (2006). Isotopes in bones and teeth. In *Isotopes in Paleoenvironmental Research*, ed. Leng, M.J., Reviews in Paleonvironmental Research Vol. 10, Springer, Dordrecht, pp. 117–145.

Herring, D.A., Saunders, S.R. and Katzenberg, M.A. (1998). Investigating the weaning process in past populations. *American Journal of Physical Anthropology* **105** 425–439.

IAEA (1995). *Reference and Intercomparison Materials for Stable Isotopes of Light Elements*. IAEA TECDOC Series No. 825, International Atomic Energy Agency, Vienna. [http://www-pub.iaea.org/MTCD/publications/PDF/te_825_prn.pdf].

Jim, S., Ambrose, S.H. and Evershed, R.P. (2004). Stable carbon isotopic evidence for differences in the dietary origin of bone cholesterol, collagen and apatite: implications for their use in palaeodietary reconstruction. *Geochimica et Cosmochimica Acta* **68** 61–72.

Katzenberg, M.A. (1992). Advances in stable isotope analysis of prehistoric bones. In *The Skeletal Biology of Past Peoples: Research Methods*, ed. Saunders, S. and Katzenberg, M.A., Wiley, New York, pp. 105–120.

Katzenberg, M.A. and Harrison, R.G. (1997). What's in a bone? Recent advances in archaeological bone chemistry. *Journal of Archaeological Research* **5** 265–293.

Katzenberg, M.A. and Weber, A. (1999). Stable isotope ecology and palaeodiet in the Lake Baikal region of Siberia. *Journal of Archaeological Science* **26** 651–659.

Katzenberg, M.A., Schwarcz, H.P., Knyf, M. and Melbe, F.J. (1995). Stable isotope evidence for maize horticulture and paleodiet in southern Ontario, Canada. *American Antiquity* **60** 335–350.

Keegan, W.F. (1989). Stable isotope analysis of prehistoric diet. In *Reconstruction of Life from the Skeleton*, ed. Iscan, M.Y. and Kennedy, K.A.R., Liss, New York, pp. 223–236.

Klepinger, L.L. (1990). Magnesium ingestion and bone magnesium concentration in paleodietary reconstruction – cautionary evidence from an animal model. *Journal of Archaeological Science* **17** 513–517.

Koch, P.L., Heisinger, J., Moss, C., Carlson, R.W., Fogel, M.L. and Behrensmeyer, A.K. (1995). Isotopic tracking of change in diet and habitat use in African elephants. *Science* **267** 1340–1343.

Lajtha, K. and Michener R. (ed.) (1994). *Stable Isotopes in Ecology and Environmental Science*. Blackwell Scientific Publications, Oxford.

Lambert, J.B. and Weydert-Homer, J.M. (1993a). The fundamental relationship between ancient diet and the inorganic constituents of bone as derived from feeding experiments. *Archaeometry* **35** 279–294.

Lambert, J.B. and Weydert-Homer, J.M. (1993b). Dietary inferences from element analysis of bone. In *Prehistoric Human Bone: Archaeology at the Molecular Level*, ed. Lambert, J.B. and Grupe, G., Springer-Verlag, Berlin, pp. 217–228.

Lambert, J.B., Szpunar, C.B. and Buikstra, J.E. (1979). Chemical analysis of excavated human bone from middle and late woodland sites. *Archaeometry* **21** 115–129.

Lambert, J.B., Vlasak, S.M., Thometz, A.C. and Buikstra, J.E. (1982). A comparative study of the chemical analysis of ribs and femurs in woodland populations. *American Journal of Physical Anthropology* **59** 289–294.

Lambert, J.B., Simpson, S.V., Buikstra, J.E. and Hanson, D. (1983). Electron microprobe analysis of elemental distribution in excavated human femurs. *American Journal of Physical Anthropology* **62** 409–423.

Larsen, C.S. (1997). *Bioarchaeology: Interpreting Behaviour from the Human Skeleton*. Cambridge University Press, Cambridge.

Larsen, C.S., Schoeninger, M.J., van der Merwe, N.J., Moore, K.M. and Lee-Thorp, J.A. (1992). Carbon and nitrogen stable isotopic signatures of human dietary change in the Georgia Bight. *American Journal of Physical Anthropology* **89** 197–214.

Lee-Thorp, J. and Sponheimer, M. (2006). Contributions of biogeochemistry to understanding hominin dietary ecology. *Yearbook of Physical Anthropology* **49** 131–148.

Lee-Thorp, J.A., van der Merwe, N.J. and Brain, C.K. (1994). Diet of *Australopithecus robustus* at Swartkrans from stable carbon isotopic analysis. *Journal of Human Evolution* **27** 361–372.

Leng, M.J. (ed.) (2006). *Isotopes in Paleoenvironmental Research.* Reviews in Paleonvironmental Research Vol. 10, Springer, Dordrecht.

Levinson, A.A., Luz, B. and Kolodny, Y. (1987). Variations in oxygen isotope compositions of human teeth and urinary stones. *Applied Geochemistry* **2** 367–371.

Lidén, K., Eriksson, G., Nordqvist, B., Götherström, A. and Bendixen, E. (2004). 'The wet and the wild followed by the dry and the tame' – or did they occur at the same time? Diet in Mesolithic–Neolithic southern Sweden. *Antiquity* **78** 23–33.

Longinelli, A. (1995). Stable isotope ratios in phosphate from mammal bone and tooth as climatic indicators. In *Problems of Stable Isotopes in Tree-rings, Lake Sediments and Peat-bogs as Climatic Evidence for the Holocene*, ed. Frenzel, B., European Science Foundation, Strasbourg, pp. 57–70.

Lowenstam, H.A. and Weiner, S. (1989). *On Biomineralization.* Oxford University Press, New York.

Lubell, D., Jackes, M., Schwarcz, H., Knyf, M. and Meiklejohn, C. (1994). The Mesolithic–Neolithic transition in Portugal – isotopic and dental evidence of diet. *Journal of Archaeological Science* **21** 201–216.

McConnell, D. (1973). *Apatite. Its Crystal Chemistry, Mineralogy, Utilization, and Geologic and Biologic Occurences.* Springer-Verlag, Vienna.

McCullagh, J.S.O., Juchelka, D. and Hedges, R.E.M. (2006). Analysis of amino acid C-13 abundance from human and faunal bone collagen using liquid chromatography/isotope ratio mass spectrometry. *Rapid Communications in Mass Spectrometry* **20** 2761–2768.

Milner, N., Craig, O.E., Bailey, G.N., Pedersen, K. and Andersen, S.H. (2004). Something fishy in the Neolithic? A re-evaluation of stable isotope analysis of Mesolithic and Neolithic coastal populations. *Antiquity* **78** 9–22.

Morgan, M.E., Kingston, J.D. and Marino, B.D. (1994). Carbon isotopic evidence for the emergence of C_4 plants in the Neogene from Pakistan and Kenya. *Nature* **367** 162–165.

Müldner, G. and Richards, M.P. (2005). Fast or feast: reconstructing diet in later medieval England by stable isotope analysis. *Journal of Archaeological Science* **32** 39–48.

O'Connell, T.C. (1996). *The Isotopic Relationship between Diet and Body Proteins: Implications for the Study of Diet in Archaeology.* Unpublished D.Phil. Thesis, University of Oxford, Oxford.

Papathanasiou, A. (2003). Stable isotope analysis in Neolithic Greece and possible implications on human health. *International Journal of Osteoarchaeology* **13** 314–324.

Pate, F.D. (1994). Bone chemistry and paleodiet. *Journal of Archaeological Method and Theory* **1** 161–209.

Pate, F.D. and Hutton, J.T. (1988). The use of soil chemistry data to address post-mortem diagenesis in bone mineral. *Journal of Archaeological Science* **15** 729–739.

Pollard, A.M. (1998). Archaeological reconstruction using stable isotopes. In *Stable Isotopes. Integration of Biological, Ecological and Geochemical Processes*, ed. Griffiths, H., BIOS Scientific Publishers, Oxford, pp. 285–301.

Pollard, A.M. and Wilson, L. (2001). Global biogeochemical cycles and isotope systematics – how the world works. In *Handbook of Archaeological Sciences*, ed. Brothwell, D.R. and Pollard, A.M., Wiley, Chichester, pp. 191–201.

Pollard, A.M., Antoine, S.E., Dresser, P.Q. and Whittle, A.W.R. (1991). Methodological study of the analysis of archaeological bone. In *Archaeological Sciences 1989*, ed. Budd, P., Chapman, B., Jackson, C.M., Janaway, R.C. and Ottaway, B.S., Monograph 9, Oxbow Books, Oxford, pp. 363–372.

Pollard, A.M., Batt, C.M., Stern, B. and Young, S.M.M. (2007). *Analytical Chemistry in Archaeology*. Cambridge University Press, Cambridge.

Price, T.D. (ed.) (1989). *The Chemistry of Prehistoric Human Bone*. Cambridge University Press, Cambridge.

Price, T.D. (1991). The Mesolithic of northern Europe. *Annual Reviews of Anthropology* **20** 211–233.

Price, T.D., Grupe, G. and Schröter, P. (1994). Reconstruction of migration patterns in the Bell Beaker period by stable strontium isotope analysis. *Applied Geochemistry* **9** 413–417.

Radosevich, S.C. (1993). The six deadly sins of trace element analysis: A case of wishful thinking in science. In *Investigations of Ancient Human Tissue: Chemical Analysis in Anthropology*, ed. Sandford, M.K., Gordon and Breach, Langhorne, pp. 269–332.

Richards, M.P. and Hedges, R.E.M. (1999). Stable isotope evidence for similarities in the types of marine food used by late Mesolithic humans on the Atlantic coast of Europe. *Journal of Archaeological Science* **26** 717–722.

Richards, M.P. and Mellars, P.A. (1998). Stable isotopes and the seasonality of the Oronsay middens. *Antiquity* **72** 178–184.

Richards, M.P., Fuller, B.T. and Hedges, R.E.M. (2001). Sulphur isotopic variation in ancient bone collagen from Europe: implications for human palaeodiet, residence mobility, and modern pollutant studies. *Earth and Planetary Science Letters* **191** 185–190.

Richards, M.P., Mays, S. and Fuller, B.T. (2002). Stable carbon and nitrogen isotope values of bone and teeth reflect weaning age at the Medieval Wharram Percy site, Yorkshire UK. *American Journal of Physical Anthropology* **119** 205–210.

Richards, M.P., Schulting, R.J. and Hedges, R.E.M. (2003). Sharp shift in diet at onset of Neolithic. *Nature* **425** 366.

Sandford, M.K. (ed.) (1993). *Investigations of Ancient Human Tissue: Chemical Analysis in Anthropology*. Gordon and Breach, Langhorne.

Schlesinger, W.H. (1997). *Biogeochemistry: An Analysis of Global Change*. Academic Press, San Diego, 2nd edn.

Schoeninger, M.J. and Moore, K. (1992). Bone stable isotope studies in archaeology. *Journal of World Prehistory* **6** 247–296.

Schoeninger, M.J., DeNiro, M.J. and Tauber, H. (1983). Stable nitrogen isotope ratios of bone collagen reflect marine and terrestrial components of prehistoric human diet. *Science* **220** 1381–1383.

Schulting, R.J. and Richards, M.P. (2001). Dating women and becoming farmers; new palaeodietary and AMS data from the Breton Mesolithic cemeteries of Téviec and Hoëdic. *Journal of Anthropological Archaeology* **20** 314–344.

Schulting, R.J. and Richards, M.P. (2002). The wet, the wild and the domesticated: the Mesolithic–Neolithic transition on the west coast of Scotland. *European Journal of Archaeology* **5** 147–189.

Schwarcz, H.P. and Schoeninger, M.J. (1991). Stable isotope analyses in human nutritional ecology. *Yearbook of Physical Anthropology* **34** 283–322.

Sealy, J.C. (2001). Body tissue chemistry and paleodiet. In *Handbook of Archaeological Sciences*, ed. Brothwell, D.R. and Pollard, A.M., Wiley, Chichester, pp. 269–279.

Sealy, J., Armstrong, R. and Schrire, C. (1995). Beyond lifetime averages: tracing life histories through isotopic analysis of different calcified tissues from archaeological human skeletons. *Antiquity* **69** 290–300.

Sillen, A. (1989). Diagenesis of the inorganic phase of cortical bone. In *The Chemistry of Prehistoric Human Bone*, ed. Price, T.D., Cambridge University Press, Cambridge, pp. 211–229.

Sillen, A. and Kavenagh, M. (1982). Strontium and paleodioetary research: a review. *Yearbook of Physical Anthropology* **25** 67–90.

Sillen, A. and Sealy, J.C. (1995). Diagenesis of strontium in fossil bone: a reconsideration of Nelson *et al.* (1986). *Journal of Archaeological Science* **22** 313–320.

Sillen, A., Sealy, J.C. and van der Merwe, N.J. (1989). Chemistry and paleodietary research – no more easy answers. *American Antiquity* **54** 504–512.

Smith, C., Chamberlain, A.T., Riley, M.S., Cooper, A., Stringer, C.B. and Collins, M.J. (2001). Not just old but old and cold? *Nature* **410** 771–772.

Sponheimer, M., Passey, B.H., de Ruiter, D.J., Guatelli-Steinberg, D., Cerling, T.E. and Lee-Thorp, J.A. (2006). Isotopic evidence for dietary variability in the early hominin *Paranthropus robustus*. *Science* **314** 980–982.

Stuart-Williams, H.L.Q., Schwarcz, H.P., White, C.D. and Spence, M.W. (1996). The isotopic composition and diagenesis of human bone from Teotihuacán and Oaxaca, Mexico. *Palaeogeography Palaeoclimatology Palaeoecology* **126** 1–14.

Szarek, S.R. and Troughton, J.H. (1976). Carbon isotope ratios in crassulacean acid metabolism. *Plant Physiology* **58** 367–370.

Tauber, H. (1981). ^{13}C evidence for dietary habits of prehistoric man in Denmark. *Nature* **292** 332–333.

Taylor, H.P.Jr, O'Neil, J.R. and Kaplan, I.R. (ed.) (1991). *Stable Isotope Geochemistry: A Tribute to Samuel Epstein*. Special Publication No. 3, Geochemical Society, San Antonio.

Tieszen, L.L. and Fagre, T. (1993). Effect of diet quality and composition on the isotopic composition of respiratory CO_2, bone collagen, bioapatite and soft tissues. In *Prehistoric Human Bone: Archaeology at the Molecular Level*, ed. Lambert, J.B. and Grupe, G., Springer Verlag, Berlin, pp. 121–155.

Trickett, M., Budd, P., Montgomery, J. and Evans, J. (2003). An assessment of solubility profiling as a decontamination procedure for the $^{87}Sr/^{86}Sr$ analysis of archaeological human tissue. *Applied Geochemistry* **18** 653–658.

Trueman, C.N. (1999). Rare earth element geochemistry and taphonomy of terrestrial vertebrate assemblages. *Palaios* **14** 555–568.

Trueman, C.N., Behrensmeyer, A.K., Potts, R. and Tuross, N. (2006). High-resolution records of location and stratigraphic provenance from the rare earth element composition of fossil bones. *Geochimica et Cosmochimica Acta* **70** 4343–4355.

Tykot, R.H. (2002). Contribution of stable isotope analysis to understanding dietary variation among the Maya. In *Archaeological Chemistry: Materials, Methods and Meaning*, ed. Jakes, K.A., ACS Symposium Series 831, American Chemical Society, Washington, D.C., pp. 214–230.

Tykot, R.H. (2006). Isotope analysis and the history of maize. In *Histories of Maize: Multidisciplinary Approaches to the Prehistory, Linguistics, Biogeography, Domestication and Evolution of Maize*, ed. Staller, J.E., Tykot, R.H. and Benz, B.F., Elsevier Academic Press, Burlington, pp. 131–142.

Ubelaker, D.H., Katzenberg, M.A. and Doyon, L.G. (1995). Status and diet in precontact highland Ecuador. *American Journal of Physical Anthropology* **97** 403–411.

van der Merwe, M.J. (1992). Light stable isotopes and the reconstruction of prehistoric diets. In *New Developments in Archaeological Science*, ed. Pollard, A.M., Proceedings of the British Academy 77, Oxford University Press, Oxford, pp. 247–264.

van der Merwe, N.J. and Medina, E. (1991). The canopy effect, carbon isotope ratios and foodwebs in Amazonia. *Journal of Archaeological Science* **18** 249–259.

van der Merwe, N.J. and Vogel, J.C. (1977). ^{13}C content of human collagen as a measure of prehistoric diet in woodland North America. *Nature* **276** 815–816.

van der Merwe, N.J., Roosevelt, A.C. and Vogel, J.C. (1981). Isotopic evidence for prehistoric subsistence change at Parmana, Venezuela. *Nature* **292** 536–538.

Wada, E., Mizutani, H. and Minagawa, M. (1991). The use of stable isotopes for food web analysis. *Critical Reviews in Food Science and Nutrition* **30** 361–371.

Weiner, S. and Bar-Yosef, O. (1990). States of preservation of bones from the prehistoric sites in the Near East: a survey. *Journal of Archaeological Science* **17** 187–196.

White, C.D. and Schwarcz, H.P. (1989). Ancient Maya diet – as inferred from isotopic and elemental analysis of human bone. *Journal of Archaeological Science* **16** 451–474.

Wilson, L. (2004). *Geochemical Approaches to Understanding In Situ Diagenesis.* Unpublished PhD thesis, University of Bradford, UK.

Wright, L.E. and Schwarcz, H.P. (1999). Correspondence between stable carbon, oxygen and nitrogen isotopes in human tooth enamel and dentine: infant diets at Kaminaljuyu. *Journal of Archaeological Science* **26** 1159–1170.

CHAPTER 11

The Detection of Small Biomolecules: Dairy Products in the Archaeological Record

11.1 INTRODUCTION

Occasionally the publication of a scientific article has far-reaching consequences that impact immediately on the lives of millions. Most, however, make their appearance into print or on-line without a fanfare. The process of science goes on often regardless of other dramatic events occurring in history. On 10th March 1933, a few weeks after Hitler became Chancellor of Germany, a chemist called Johannes Grüss published the results of a chemical study of prehistoric pottery. He had applied routine wet chemical tests to encrustations of burnt food still clinging to the surface of ancient pottery sherds after hundreds of years of burial. Grüss concluded that the deposits had formed by the burning of milk fats in the pottery containers. Seventy-five years later, the chemical detection of dairy products has developed in sophistication and utilizes sensitive and selective molecular and isotopic instrumental techniques. More importantly, the recognition of dairying in the archaeological record is as current as ever and a recent publication identifying Early Neolithic dairying in Britain (Copley *et al.*, 2003) was the subject of massive media and public interest. People are always interested in food, and what our ancestors had as food on the plate, as well as the associated rituals of feasting, is a fascinating and intriguing subject (Jones, 2007).

In the early 21st Century, the field of biomolecular archaeology is undergoing a dramatic phase of expansion. A better understanding of organic preservation has stimulated a range of novel analytical investigations based on molecular and isotopic characteristics of preserved organic matter. Relatively few attempts have been made to survey the diversity of biomolecular approaches to archaeological enquiry. The book by Jones (2001) is one

Archaeological Chemistry, Second Edition
By A. Mark Pollard and Carl Heron
© The Royal Society of Chemistry 2008

exception, and we also have some excellent reviews of specific compound classes [*e.g.*, Evershed *et al.* (2001) for lipids]. In this chapter, we focus on a specific residue type – dairy products. A survey of the contemporary literature demonstrates that dairy products can be identified in a number of ways, including the chemical detection of distinctive lipid biomarkers, carbon isotope measurements undertaken on common fatty acids and immunological recognition of proteins. This chapter begins with some simple organic chemistry (Sections 11.2 and 11.3) before moving on to a discussion of recent studies (Sections 11.4 and 11.5). The following questions are relevant:

   (i) Why is the detection of dairy products important?
  (ii) Do we have to rely on sophisticated analytical techniques to enable us to identify dairying in the past?
 (iii) What can archaeologists learn from knowing whether dairying was practiced at a given point in time and space?

11.2 FATTY ACIDS: A BRIEF OVERVIEW OF CHEMISTRY AND OCCURRENCE

Fatty acids are constituents of fats and oils and belong to a class of molecule known as lipids (Coultate, 2001; 73–124). Lipids play a wide range of roles in living organisms. This includes roles in cell and tissue structure, metabolism and physiology. Lipids are often defined rather inappropriately as molecules with greater solubility in organic solvents than in water. The most common solvent used to solubilize lipids is a mixture of chloroform or dichloromethane and methanol in a 2:1 volume ratio. Some chemists (*e.g.*, Christie, 1987; 42) have advocated more specific definitions of lipids, more closely related to the properties of fatty acids and their derivatives, including sterols since they share similar functions.

   Most fatty acids are simple molecules comprising linear chains of carbon atoms with a carboxyl functionality (-COOH), although branched and cyclic molecules also occur. The number of carbon atoms varies from 1 to 50 or more. However, the so-called mid-chain fatty acids have between four and 30 carbon atoms. The simplest are the saturated fatty acids (Figure 11.1). In these molecules, all the carbon-to-carbon bonds are single. In unsaturated fatty acids, one or more carbon-to-carbon double bonds are present (Figure 11.2). Table 11.1 presents the common and systematic names for the most common saturated fatty acids, together with a simple abbreviated notation for each molecule and a brief note on their occurrence in common food sources. Table 11.2 presents similar information on unsaturated fatty acids. The majority of natural fats and oils are made up of straight- or branched-chain fatty acids with an even number of carbons in the chain (also known as *aliphatic* fatty acids). Branched-chain fatty acids occur most commonly with *iso-* or *anteiso*-methyl branching. They are synthesized by many micro-organisms and occur in adipose fats, particularly those of ruminant animals, and milk.

Figure 11.1 Structures of commonly occurring saturated fatty acids; (i) myristic acid, $C_{14:0}$ (ii) palmitic acid, $C_{16:0}$ (iii) stearic acid, $C_{18:0}$.

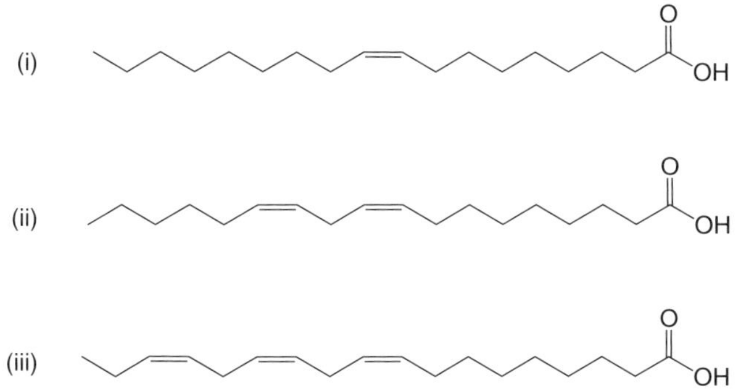

Figure 11.2 Structures of commonly occurring unsaturated fatty acids; (i) oleic acid, $C_{18:1}$ (ii) linoleic acid, $C_{18:2}$ (iii) α-linolenic acid, $C_{18:3}$.

The fatty acid composition of fats and oils varies according to specific origin and, although there are overlaps, some generalizations can be drawn. Fats of higher land animals tend to be more saturated, consisting mainly of palmitic (given as $C_{16:0}$, which identifies 16 carbons in the chain, and no double bonds), oleic ($C_{18:1}$, with one double bond) and stearic acids ($C_{18:0}$; see Tables 11.1 and 11.2 for systematic names), and may be affected by dietary intake. Table 11.3 is a simplified table of the fatty acid composition of common sources of terrestrial animal fat. The composition of milk fats is considered in Section 11.3. Marine oils contain abundant unsaturated fatty acids (C_{16} to C_{22}) with up to six double bonds (Table 11.2). Oils of plant origin are characterized by relatively low levels of saturated fatty acids. They are dominated by palmitic ($C_{16:0}$), oleic ($C_{18:1}$), linoleic ($C_{18:2}$) and linolenic ($C_{18:3}$) acids, but may also contain unusual acids such as erucic ($C_{22:1}$; see Table 11.2). That noted, some plants produce what can be classed as vegetable fats, *e.g.*, cocoa butter, which are highly saturated.

The majority of fatty acids in tissues are combined with other molecules. In food fats and oils, fatty acids are esterified to molecules of glycerol (propane-1,2,3-triol). Figure 11.3 shows how three fatty acids can combine with a single

Table 11.1 Systematic and common names of saturated fatty acids together with the shorthand designation and typical occurrence (compiled from Gunstone *et al.*, 1994; Hilditch and Williams, 1964; Perkins, 1993; Robinson, 1982). The shorthand designation denotes Cx:n where x is the number of carbon atoms and n is the number of double bonds.

Systematic name	Common name	Shorthand designation	Occurrence
Butanoic	Butyric	C4:0	Milk fat
Pentanoic	Valeric	C5:0	
Hexanoic	Caproic	C6:0	Milk fat
Heptanoic	Enanthic	C7:0	
Octanoic	Caprylic	C6:0	Milk fat; minor in other fats
Nonanoic	Pelargonic	C9:0	
Decanoic	Capric	C10:0	Milk fat, minor in other fats
Undecanoic	Undecylic	C11:0	
Dodecanoic	Lauric	C12:0	High in seed fat of Lauraceae, present in milk, vegetable oils, nuts
Tridecanoic	Tridecylic	C13:0	
Tetradecanoic	Myristic	C14:0	High in seed fat of Myristiceae, minor in most animal lipids
Pentadecanoic	Pentadecylic	C15:0	Present in bacteria, human skin, fish
Hexadecanoic	Palmitic	C16:0	Most common saturated fatty acid in plants and animals
Heptadecanoic	Margaric	C17:0	Present in bacteria, human skin, fish
Octadecanoic	Stearic	C18:0	High in tallow of ruminants
Nonadecanoic	–	C19:0	Present in bacteria, human skin
Eicosanoic	Arachidic	C20:0	Abundant in a few seed oils, widespread minor constituent
Heneicosanoic	–	C21:0	
Docosanoic	Behenic	C22:0	Abundant in a few seed oils, widespread minor constituent
Tricosanoic	–	C23:0	
Tetracosanoic	Lignoceric	C24:0	
Pentacosanoic	–	C25:0	
Hexacosanoic	Cerotic	C26:0	Present in plant and insect waxes
Heptacosanoic	Carboceric	C27:0	
Octacosanoic	Montanic	C28:0	Major component of some plant waxes
Nonacosanoic	–	C29:0	
Tricontanoic	Melissic	C30:0	Present in plant waxes
Hentricontanoic	–	C31:0	
Dotriacontanoic	Lacceric	C32:0	Wool fat, plant leaf waxes
Tritriacontanoic	Psyllic	C33:0	
Tetratricontanoic	Geddic	C34:0	
Pentatriacontanoic	Ceroplastic	C35:0	

Table 11.2 Systematic and common names of unsaturated fatty acids together with the shorthand designation and typical occurrence (compiled from Gunstone *et al.*, 1994; Hilditch and Williams, 1964; Perkins, 1993; Robinson, 1982). The shorthand designation denotes Cx:n where x is the number of carbon atoms and n is the number of double bonds (position of the last double bond in relation to terminal methyl group).

Systematic name	Common name	Shorthand designation	Occurrence
Monounsaturated fatty acids			
cis-9-Tetradecenoic	Myristoleic	C14:1 ($n = 5$)	Minor constituent of bacteria, plant lipids and whale oil
cis-9-Hexadecenoic	Palmitoleic	C16:1 ($n = 7$)	Abundant in aquatic species and in some seed oils
cis-9-Octadecenoic	Oleic	C18:1 ($n = 9$)	Abundant in nearly all fats and oils
trans-9-Octadecenoic	Elaidic	C18:1 ($n = 9$)	Ruminant fats
trans-11-Octadecenoic	*trans*-Vaccenic	C18:1 ($n = 7$)	Minor in ruminant depot and milk fats
cis-11-Octadecenoic	*cis*-Vaccenic	C18:1 ($n = 7$)	Major bacterial fatty acid
cis-6-Octadecenoic	Petroselinic	C18:1 ($n = 12$)	Major in certain plant oils (Umbelliferae)
cis-9-Eicosenoic	Gadoleic	C20:1 ($n = 11$)	High in fish oils, low in terrestrial species
cis-11-Eicosenoic	Gondoic	C20:1 ($n = 9$)	Minor in fish oils and some seed oils
cis-13-Eicosenoic	–	C20:1 ($n = 7$)	Rapeseed oil
cis-13-Docosenoic	Erucic	C22:1 ($n = 9$)	High in seed oils of Cruciferae, present in fish oils
trans-13-Docosenoic	Brassidic	C22:1 ($n = 9$)	
cis-15-Tetracosenoic	Nervonic	C24:1 ($n = 9$)	Minor in some fish oils and in specific animal tissues
Polyunsaturated fatty acids (most important are methylene-interrupted, all bonds *cis*)			
6,9-Octadecadienoic	–	C18:2 ($n = 9$)	Minor in animals
9,12-Octadecadienoic	Linoleic	C18:2 ($n = 6$)	Major in many seed oils, minor in animals
6,9,12-Octadecatrienoic	γ-Linolenic	C18:3 ($n = 6$)	Major in some seed oils, minor in animals
9,12,15-Octadecatrienoic	α-Linolenic	C18:3 ($n = 3$)	Abundant in linseed oil and in other plant oils and tissues

Table 11.2   (*Continued*).

Systematic name	Common name	Shorthand designation	Occurrence
6,9,12,15-Octadecatetraienoic	–	C18:4 ($n = 3$)	
8,11-Eicosadienoic	–	C20:2 ($n = 9$)	
5,8,11-Eicosatrienoic	–	C20:3 ($n = 9$)	
8,11,14-Eicosatrienoic	Homo-γ-linolenic	C20:3 ($n = 6$)	Present in fish oils
5,8,11,14-Eicosatetraenoic	Arachidonic	C20:4 ($n = 6$)	Abundant in animal phospholipids, marine algae and fish
8,11,14,17-Eicosatetraenoic	–	C20:4 ($n = 3$)	
5,8,11,14,17-Eicosapentaenoic	EPA	C20:5 ($n = 3$)	Abundant in fish oils and marine algae, present in animals
7,10,13,16-Docosatetraenoic	–	C22:4 ($n = 6$)	
4,7,10,13,16-Docosapentaenoic	–	C22:5 ($n = 6$)	
7,10,13,16,19-Docosapentaenoic	Elupadonic	C22:5 ($n = 3$)	Abundant in fish, minor in animal phospholipids
4,7,10,13,16,19-Docosahexaenoic	DHA	C22:6 ($n = 3$)	Abundant in fish oils, marine algae

Table 11.3   Fatty acids found in animal depot fats (adapted from deMan, 1990; 41).

Animal	C12:0	C14:0	C16:0	C18:0	C20:0	C16:1	C18:1	C18:2	C18:3	C20:1
Cow	–	6.3	27.4	14.1	–	–	49.6	2.5	–	–
Pig	–	1.8	21.8	8.9	0.8	4.2	53.4	6.6	0.8	0.8
Sheep	–	4.6	24.6	30.5	–	–	36.0	4.3	–	–
Goat	3.5	2.1	25.5	28.1	2.4	–	38.4	–	–	–

glycerol molecule resulting in a triacylglycerol (TAG) or triglyceride. The 'stereospecific numbering' (*sn*) system designates the position of the fatty acid on one of the three carbons in glycerol. In a Fischer projection of a natural L-glycerol derivative, the secondary hydroxyl group is shown to the left of C-2; the carbon atom above this then becomes C-1 and the one below becomes C-3. The prefix *sn* is placed before the stem name of the compound. Fats comprise a higher proportion of saturated fatty acid moieties in their TAGs and, as a result, are solids at room temperature. Oils, with a higher proportion of unsaturated moieties, tend to be liquids at room temperature. The distribution of fatty acids within the TAGs is not random. In most animal fats, for example,

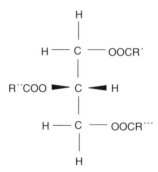

Figure 11.3 Fischer projection of a triacylglycerol. The fatty acyl moieties are represented by R′, R″ and R‴ positioned at *sn*-1, *sn*-2 and *sn*-3 respectively.

palmitic acid is found to a greater extent on the first or third carbon. An exception is pig depot fat in which palmitic acid occupies the second carbon atom to a large extent. Such positional specificity may account for variations in the physical properties of fats with similar degrees of total unsaturation.

There are various ways of determining the type and amount of fatty acids in foods. Today, most archaeological investigations are carried out using gas chromatography-mass spectrometry (GC-MS; see Chapter 2). Lipid extracts of archaeological samples often comprise free fatty acids together with intact TAGs and partially hydrolysed mono- and diacylglycerols with one and two fatty acyl moieties, respectively. In other cases, the fatty acids have been released in their entirety and are present in the free state. Chemical derivatization of fatty acids is usually undertaken to produce methylated or trimethylsilylated derivatives more amenable to chromatographic separation (Evershed, 1992a). The mass spectra of these components once separated are easily recognized (Evershed, 1992b). In simple terms, the fatty acids present in a sample can be compared with published tables of fatty acid composition in common fats and oils. However, for a number of reasons this approach is not straightforward. Although there are differences, food fats and oils contain similar abundances of common fatty acids. The fatty acid composition of food varies according to many factors including, in animals, the age and sex, the diet fed to the animal and the body part from which the sample is taken. In plants, soil type and climate play an important role. In addition to these factors, the preferential degradation or loss of particular fatty acids means that ancient samples can have fatty acid distributions markedly different to those in the original tissue. For example, double bonds joining carbon atoms are susceptible to oxidation. Shorter-chain fatty acids are more soluble than their longer chain counterparts. Short-term simulation experiments carried out in conjunction with archaeological investigations have confirmed that these trends do influence the fatty acid profile of aged samples (Evershed and Charters, 1995; Dudd *et al.*, 1998). As we shall see, the impact of degradation interferes with attempts to identify specific fats and oils based on fatty acid compositions alone.

Sterols are minor constituents of most fats. Those of animal origin contain cholesterol and traces of other sterols, whereas plants contain phytosterols, of which β-sitosterol is the most common. Sterols occur in the free form or, after esterification to fatty acids through the 3-OH group, as steryl esters. The presence of sterols in archaeological residues can be a useful indicator of a plant or animal origin or an indicator of both if cholesterol and phytosterols are detected in the same sample. That noted, cholesterol is a potential contaminant of all archaeological samples subjected to handling.

11.3 DAIRY CHEMISTRY

Milk is mostly water containing globules of milk fat and solids. The solids comprise protein, sugar, minerals (especially calcium) and vitamins and range from about 8.0% to 10% by weight in domesticated cow's milk. Milk fat makes up around 2.5–5.5%. The fatty acids in cow's milk originate either from microbial activity in the rumen, and are transported to the secretory cells via the blood and lymph, or from direct synthesis in the secretory cells. TAGs dominate milk fat (>98%) with small amounts of cholesterol and phospholipids making up most of the remainder. Of the fatty acids present, palmitic, stearic and oleic make up more than half. Shorter-chain fatty acids produced in the rumen, from butyric ($C_{4:0}$) to capric ($C_{10:0}$) acids, are also found in milk fat. Butyric acid causes the rancid taste in butter when it is released from the glycerol backbone by hydrolysis. The short-chain fatty acids are diagnostic of a dairy source. Table 11.4 shows the typical range of major fatty acids found in milk fat. Milk fat also contains a wide range of trace fatty acids, and around 3% of the total fatty acids are as *trans* isomers. In terms of the relationship between the fatty acids and their positions on the TAG molecules, 98.1% by weight of butyric acid ($C_{4:0}$) and 93.2% of caproic acid ($C_{6:0}$) in cow's milk is

Table 11.4 Averaged composition of the major fatty acids in bovine milk fat (adapted from Jensen, 2000; 297).

Fatty acid	Notation	Weight (%)
Butyric	C4:0	2–5
Caproic	C6:0	1–5
Caprylic	C8:0	1–3
Capric	C10:0	2–4
Lauric	C12:0	2–5
Myristic	C14:0	8–14
Pentadecanoic	C15:0	1–2
Palmitic	C16:0	22–35
Palmitoleic	C16:1	1–3
Margaric	C17:0	0.5–1.5
Stearic	C18:0	9–14
Oleic	C18:1	20–30
Linoleic	C18:2	1–3
Linolenic	C18:3	0.5–2

located in the *sn*-3 position (Jensen, 2000; 298). This has implications for the loss of these components by hydrolysis. Reduced steric hindrance means these moieties are more susceptible to hydrolysis during the processing of the milk fats or during burial. Secondly, the short-chain fatty acids are more water-soluble than their longer-chain counterparts. Therefore, following enzymatic or chemical hydrolysis, the short-chain fatty acids are more liable to be lost through leaching in the burial environment.

The most abundant milk protein is casein, of which there are several different kinds, usually designated α-, β-, and κ-casein. The different caseins relate to small differences in their amino acid sequences. Casein micelles in milk have diameters less than 300 nm. Disruption of the casein micelles occurs during the preparation of cheese. Lactic acid increases the acidity of the milk until the micelles crosslink and a curd develops. The liquid portion, known as whey, containing water, lactose and some protein, is removed. Addition of the enzyme rennet (chymosin) speeds up the process by hydrolysing a specific peptide bond in κ-casein. This opens up the casein and encourages further cross-linking.

Lactose is a disaccharide made up of glucose and galactose (both mono-saccharides). It appears to be present in the milk of nearly all mammals, although apparently it is not found in the milk of pinnipeds (seals and sea lions) due to the high fat content. It is abundant in human milk (*ca.* 7% w/v) and comprises around 4.8 to 5.2% w/v in cow's milk. One of its most important functions is its utilization as a fermentation substrate. Lactobacilli in the milk produce lactic acid from lactose, which is the starting point for producing butter and cheese. Therefore, processed dairy products contain little or no lactose and can be tolerated by those who do not produce the enzyme lactase. Babies and young children produce lactase which cleaves the disaccharide lactose in the mother's milk into glucose and galactose, which can then be absorbed in the intestines. However, the majority of adults lose the ability to produce lactase and are unable to absorb milk sugar. These individuals are lactose intolerant if they display certain symptoms – often described as a gassy, bloated or nauseous state.

11.4   ARCHAEOLOGICAL INVESTIGATIONS OF DAIRYING

There is considerable interest concerning the origin of dairying in Europe and other parts of the world. This is not simply aimed at documenting the first use and consumption of dairy products, but to explore how and why dairying was adopted, how it spread and what impact it had on populations in terms of their economic and social organization. For example, was dairying practiced in the Early Neolithic in Europe when domesticated plants and animals were first introduced? In an influential paper published in 1981 Andrew Sherratt considered, using indirect evidence based on iconography, faunal remains and pottery shapes, that dairying was introduced in the Late Neolithic in the 4th Millennium BC. According to Harrison (1985; 75), '*Sherratt argued that animal power harnessed for traction and transport, as well as the exploitation of secondary products such as milk, cheese and wool, represents a set of innovations which were not part of the original complex of animal and plant domestication in Eurasia. They appeared three to four thousand years after the introduction of agriculture to Europe.*' Sherratt's view was that the use of animals for milking rather than for meat (primary products) was more efficient in terms of energy intake and in drawing upon renewable resources. His work emphasized the social changes that populations were undergoing at this time and sought to document these by understanding the revolutions taking place in managing domesticated animals.

Faunal remains recovered from archaeological excavations have long been used to assess the economic and social status of prehistoric populations. Legge (1981) considered that a bias towards females in the adult faunal assemblage together with a high neonatal cull, presumably of males, since sex is difficult to determine in young animals, indicated a dairy economy. There is a voluminous literature on the interpretation of faunal assemblages and readers are referred to Halstead (1988) for a thorough review of the application and the challenge of applying 'mortality models' to recognizing dairying in the past. According to Halstead (1988; 5), there are problems of 'equifinality: *a 'milk' mortality pattern might be the result of processes unrelated to management of dairy products.*' Using pottery and faunal evidence, Bogucki (1984) has argued that dairying came to Europe as part of a fully developed Neolithic package with the first farmers (LBK or Linear Pottery culture). Bogucki associated distinctive perforated pottery vessels of Early Neolithic date, described in Greenfield (1988; 586) as 'widespread but rare', with separating curds from whey. The suggestion of dairying in the Early Neolithic in Europe does not necessarily contradict Sherratt's view of a period of intensification in the Late Neolithic. As Bogucki (1984; 27) suggests, '*the roots of these systems, however, lie several millennia earlier, during the colonization of Europe by the Linear Pottery culture.*'

The contribution of chemistry to these debates has only impacted significantly in the past decade or so. Nevertheless, there is a long history of attempts to characterize the organic residues associated with pottery vessels in particular. Following the work of Grüss (1933), lipid analysis of residues in pottery has been used by a number of other researchers to suggest the presence of dairy products in Neolithic pottery (*e.g.*, Rottländer and Schlichtherle, 1979; Needham and Evans, 1987; Bourgeois and Gouin, 1995; Jones, 1999; 63; Agozzino *et al.*, 2001). In each case, the distribution of fatty acids in the residues was assumed to be associated with dairy products, either through the presence of short-chain fatty acids or by determining the ratios of common fatty acids. Presentation and interpretation of the analytical data in some of these papers is rather rudimentary or dubious, such as the claim that the presence of a single fatty acid is sufficient to signify milk (Agozzino *et al.*, 2001; 443).

Few, if any, studies appear to demonstrate the persistence of short-chain fatty acids in archaeological residues, either esterified to glycerol or in the free state (see, however, Mirabaud *et al.*, 2007). Short-term laboratory incubation experiments provide empirical data demonstrating the speed at which hydrolysis takes place (Dudd and Evershed, 1998; Dudd *et al.*, 1998). Once released from the glycerol backbone, these fatty acids are both more soluble and thus prone to leaching and to microbial degradation. The solubility of caproic acid ($C_{6:0}$) is more than three orders of magnitude greater than that of palmitic acid ($C_{16:0}$). The preferential loss of these diagnostic molecules means that a degraded milk-fat residue is not dissimilar to a distribution given by a degraded adipose-tissue fat (Challinor *et al.*, 1998; 144–146: Dudd and Evershed, 1998; 1479). Whilst, on occasions, the relative abundance of untypical fatty acids can be very useful in identifying certain substances (*e.g.*, Copley *et al.*, 2001), most archaeological lipid residues are dominated by palmitic and stearic acids and offer little clue as to their origin without further investigation.

Figure 11.4 shows a total ion current (TIC) chromatogram of a lipid residue extracted from a potsherd of Early Neolithic date from Ecsegfalva Hungary (Craig *et al.*, 2007; 354). The sample is dominated by palmitic and stearic acids, although a wider range of saturated fatty acids is present. A trace of oleic acid remains and cholesterol is present, which points to an animal source (the samples were not handled from the point of excavation to analysis in the laboratory). The fate of unsaturated fatty acids is not entirely clear. Hydrolysis means the splitting of the fatty acids esterified to the hydroxyl groups of glycerol. In most cases, the rate of hydrolysis is independent of the nature of the fatty acids released. There are exceptions – fatty acids with chain lengths less than 11 carbon atoms, especially the very short-chain fatty acids of milk fats, are cleaved more rapidly than the fatty acids of normal chain length

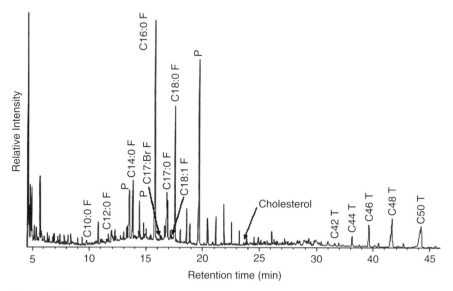

Figure 11.4 Total ion current (TIC) chromatogram of lipid residue extracted from a potsherd of Early Neolithic date (Ecsegfalva, Hungary). Cx:y F refer to fatty acids with carbon number (x) and number of unsaturations (y). Cx T refer to triacylglycerols with number of carbon atoms (x). P = plasticizer contamination. (Craig *et al.*, 2007, 354; Figure 18.1, by permission.)

(14–18 carbon atoms) due to the increased polarity of their esters. Hydrolysis of TAGs occurs a result of the action of enzymes which can be naturally present in foods or may be contributed by micro-organisms invading the residue. Evidence of hydrolysis is indicated by the abundant free fatty acids released from the glycerol backbone. Nevertheless, it is remarkable that TAGs, albeit in low abundance, survive in pottery around 7000 years old. The distribution (from C42) indicates ruminant fat. However, the fatty acid distribution is likely to be shared by many degraded fats and oils and further isotopic analysis is required.

The main pathways of the breakdown of fatty acids in biological systems involve oxidation at various points along the chain or oxidation at certain double bonds of specific unsaturated fatty acids (Coultate, 2001; 90–98). The main forms of oxidation are termed α-, β- and ω-. They are named depending on which carbon of the chain is attacked. Of these β-oxidation is the most general and prevalent. Degradation proceeds by the liberation of two-carbon (acetyl-CoA) fragments from the chain. The enzymes responsible for oxidation are widely found in plants, animals and micro-organisms.

One of the characteristic reactions of lipids exposed to atmospheric oxygen is the formation of peroxides. There are three separate processes that form the basis of a chain-reaction process – *initiation, propagation* and *termination. Initiation* involves the reaction of free radicals (metal ions, reactive oxygen and so on) with the fatty acid molecules. *Propagation* is the reaction with molecular oxygen which gives rise to a peroxide radical and these can react with other molecules. *Termination* reactions may lead to the formation of both high and low molecular weight products of the peroxidation reactions. Oxidation of fatty acids leads to the formation of dicarboxylic acids and hydroxy acids. In the case of oils with a high percentage of polyunsaturated fatty acid moieties, the TAGs will undergo oxidative polymerisation forming a complex cross-linked network.

Heated fats and oils will undergo many of the same reactions as noted above, together with other specifically thermal effects. Non-enzymatic hydrolysis will occur when fats are heated in the presence of water, although prolonged heating or high temperatures may be required for the reaction to proceed rapidly. It is now well-established that unsaturated fatty acids are severely depleted or even absent in ancient residues. Saturated fatty acids, on the other hand, tend to preserve reasonably well. To date, few archaeological studies have investigated the degradation products of fatty acids and whether these products may yield further information regarding the origin of the lipid. One such is Regert *et al.* (1998), who reported dicarboxylic acids (diacids) in pottery residues from the Neolithic waterlogged site at Chalain, France, and the arid 6th Century AD settlement at Qasr Ibrîm, Egypt. They showed that solvent extraction (*e.g.*, chloroform and methanol) of powdered potsherds and/or visible residues showed lipid molecules (tri-, di- and monoacylglycerols as well as free fatty acids), but no apparent degradation products. They tried an alternative extraction method known as alkaline hydrolysis (or saponification). This involves treating the powdered sample with 0.5 M NaOH (sodium hydroxide). This is a more powerful extractant, not least because it breaks ester bonds (thus any acylglycerols will be hydrolysed as part of the extraction step). Following alkaline hydrolysis, the results showed:

(i) a series of dicarboxylic acids with azelaic (C_9) acid as the most abundant;
(ii) a series of hydroxy fatty acids.

Most of these molecules are degradation products of the original unsaturated fatty acids in the fats and oils contained in the vessel. Given the waterlogged nature of the burial environment at Chalain, it was suggested that these degradation products in the free state would have been lost as groundwater percolated through the vessel wall. The fact that they are only seen following alkaline hydrolysis suggests that degraded lipids might be preserved as an insoluble (in typical organic solvents) polymeric matrix but linked by ester bonds which are broken by alkaline hydrolysis. This was put to the test with the samples from Qasr Ibrîm. Here, free diacids did survive in the free state and were seen in the chromatograms of solvent extracts alone. The very dry burial conditions were thought to limit their loss during burial.

The problem of resolving similarities in composition of many degraded fats and oils has been alleviated somewhat by the introduction of gas chromatography-combustion-isotope ratio mass spectrometry (GC-C-IRMS; Meier-Augenstein, 2002; Evershed *et al.*, 2001; Evershed *et al.*, 2002a). This technique is able to determine the carbon isotope ($\delta^{13}C$) ratio of single fatty acids and other lipid molecules and was first applied to archaeology in the early 1990s (Evershed *et al.*, 1994). Initially, the technique proved useful in discriminating between ruminant and non-ruminant adipose tissues (Evershed *et al.*, 1997). The characterization of dairy products followed shortly afterwards in a paper by Dudd and Evershed (1998). Using GC-C-IRMS they determined the stable carbon isotope ratios of the two major fatty acids (palmitic and stearic) in archaeological lipid residues from Iron Age/Romano-British and late Saxon/early Medieval pottery sherds from Stanwick and West Cotton (Northamptonshire, UK), respectively. These data were plotted alongside determinations undertaken on reference fats. The latter were obtained from pig adipose (four samples), chicken adipose (eight samples), ruminant adipose (three cow and six sheep) and milk fat (six cows and one sheep) with all the animals fed only on C_3 diets. In subsequent papers, the number and type of reference samples has been increased (Copley *et al.*, 2003; 1526). The four groups separated reasonably well with some of the archaeological residues plotting within the ranges obtained for the reference groups. Others plot between the ruminant and non-ruminant fields suggested mixing of these products. The more depleted (*i.e.*, more negative) $\delta^{13}C$ values for stearic acid compared with palmitic acid in milk are explained by the direct incorporation, following biohydrogenation, of dietary C_{18} fatty acids into the milk since the mammary gland cannot synthesize $C_{18:0}$ (Copley *et al.*, 2003). The $C_{18:0}$ in adipose fat derives from *de novo* biosynthesis from acetate and this is most likely to come from carbohydrate in the diet which has a higher $\delta^{13}C$ value (Copley *et al.*, 2003). In contrast, palmitic acid in milk is synthesized *de novo* by the animal, resulting in slightly less negative values for this component compared to values for stearic acid.

This approach was expanded in a subsequent series of articles pointing to evidence of dairying in prehistoric pottery from sites in southern Britain (Copley *et al.*, 2003; 2005a, 2005b, 2005c, 2005d). Around 1000 potsherds from 14 sites were sampled and dairy residues were found in each assemblage, including Early Neolithic sites. The data were plotted as $\Delta^{13}C$ values (defined as $\delta^{13}C_{18:0} - \delta^{13}C_{16:0}$) and this showed that ruminant dairy fats are distinguished from adipose fats by displaying $\Delta^{13}C$ values of less than $-3.3‰$. Figure 11.5 shows a plot of the $\delta^{13}C$ values for stearic ($\delta^{13}C_{18:0}$) and palmitic ($\delta^{13}C_{16:0}$) acids prepared from lipid extracts of pottery sherds from three Neolithic pottery assemblages in the UK (Copley *et al.*, 2005c). The ellipses indicate the $\delta^{13}C$ values of the reference animal fats. Sherds plotting in between the ellipses represent the mixing of animal products in the vessel. The implication of this research is that dairying was already well established soon after the introduction of agriculture. The faunal evidence from the Neolithic sites tends to support this conclusion [for a brief summary see Copley *et al.* (2005d; 900–903)], although the caveats in interpreting the animal bone data referred to above remain.

Out of 108 Neolithic sherds studied (Copley *et al.*, 2005d), the proportion with predominantly dairy residues is high (25%). This has led some to suggest that heated milk fats may have been used to seal the interior walls of the pot to allow the vessel to store other liquids. Messing (1957; 59) describes the water-proofing of permeable pots in Ethiopia in the following way: '*The pot is reheated for half an hour on the usual hearth used for cooking, which consists of three one-foot-high rocks. It is then placed on the ground and cold milk is quickly poured in and rapidly swished about until the pot is cold. The milk is then thrown away. This is considered the best method by the craftswomen.*' Thus an extracted dairy residue could derive from milk fat used to seal a pot or from processing, storage or consumption of dairy products. The challenge of equifinality can impact on interpreting molecular and isotopic information too. Further evidence (Copley *et al.*, 2005d), based on short-term burial experiments, indicates that milk fats decay much faster than butter, possibly due to the higher sugar content, and the more labile fats in milk encouraging rapid bacterial attack. Although further experimental work is required, for example, to explore the degradation of *heated* milk fats in contact with ceramics, this could indicate that by the Early Neolithic milk was being processed into storable products such as cheese or butter. Moreover, residues of these products are likely to survive during long-term burial. The opportunity of storing surplus dairy produce would have been important as plentiful supplies of milk would subside during the winter months. Indeed, this could be one feature underlying the adoption and development of dairying.

The data presented above tend to support the idea that dairying was introduced simultaneously with, or soon after, the adoption of cereal agriculture and domestication of animals. Can other evidence be deployed to support or refute this hypothesis? Detection of casein using immunochemistry has also

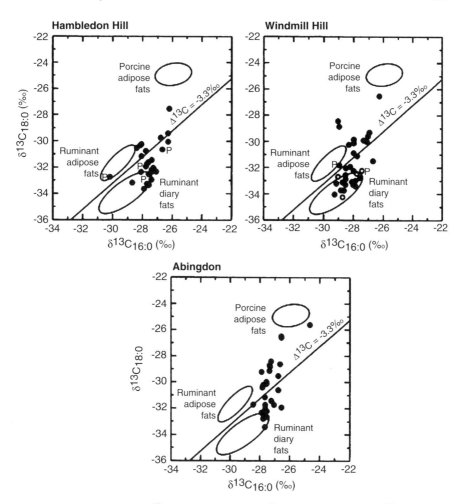

Figure 11.5 Plot of the $\delta^{13}C$ values of stearic ($\delta^{13}C_{18:0}$) and (palmitic $\delta^{13}C_{16:0}$) acids prepared from lipid extracts of pottery sherds from three Neolithic pottery assemblages in the UK. The ellipses indicate the $\delta^{13}C$ values of the reference animal fats. Sherds plotting in between the ellipses represent the mixing of animal products in the vessel. $\Delta^{13}C$ ($= \delta^{13}C_{18:0} - \delta^{13}C_{16:0}$) values of lower than $-3.3‰$ indicate dairy fats. The black filled circles represent extracts containing triacylglycerol distributions indicative of degraded adipose fats. The open circles represent those with typical degraded dairy fat TAG distributions, whilst the circles with crosses within them represent sherds that yielded no TAGs. A 'P' adjacent to the circle denotes that plant lipids were also detected, whilst a 'B' represents the presence of beeswax. (Copley *et al.*, 2005c; Figure 3a, by permission.)

been undertaken and compared with data obtained using single compound isotope analysis of fatty acids. Using a novel digestion-and-capture immuno-assay (Craig and Collins, 2000), a monoclonal antibody for bovine α_{s1}-casein was tested against sherds of Bronze Age and Iron Age date from two sites on

South Uist in the Outer Hebrides. A number of positive results were obtained and these compared well with the analysis of the fatty acids. However, some samples gave a negative result for α_{s1}-casein but a positive one for milk fat based on the $\Delta^{13}C$ value. The possibility exists that some of the residues represent sheep milk (which would give a negative result for bovine α_{s1}-casein), or that in some cases the casein molecule is degraded with consequent depletion of the immunological response.

This combined methodology has been applied to a much earlier assemblage of pottery from the Early Neolithic of south-east Europe. In this work, Craig *et al.* (2005) sought to explore the presence of dairy residues in pottery from 6th Millennium BC sites in the Danube basin in Hungary and Romania. Forty-one sherds from Ecsegfalva, Hungary, were extracted of which only seven retained detectable lipid residues, whereas five of the eight sherds sampled from Schela Cladovei in Romania produced lipid residues. The plotting of the $\Delta^{13}C$ values (defined as above) showed that seven of the residues had values less than $-3.3‰$ indicating that dairy lipids were present (Figure 11.6). However, analysis of over 80 Neolithic sherds using antibodies specific for the bovine form of

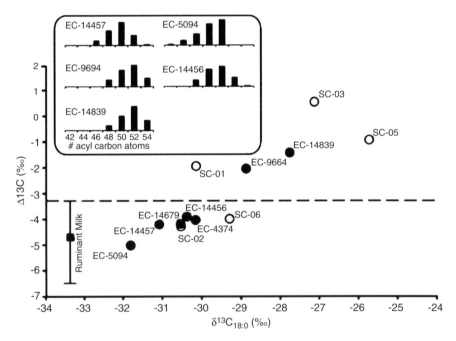

Figure 11.6   Plot of the difference (Δ value) between the $\delta^{13}C$ values of stearic ($C_{18:0}$) and palmitic ($C_{16:0}$) acid against the $\delta^{13}C$ value of stearic acid ($C_{18:0}$). The data points represent samples extracted from Early Neolithic pottery in the study published by Craig *et al.*, 2005. Δ values below $-3.3‰$ are interpreted as ruminant dairy. The inset shows the range and abundance of triacylglycerols preserved in the Early Neolithic vessels from Ecseg-falva, Hungary. (from Craig *et al.*, 2005; Figure 3b, by permission of Antiquity Publications Ltd.)

α_{s1}-casein produced negative results, including those sherds with diary lipids. This hints at the loss of the immunological response due to protein alteration, although the possibility that the milk derived from sheep and/or goats also merits further consideration.

Other investigations include the work of Spangenberg *et al.* (2006). Late Neolithic pottery from the site of Arbon Bleiche 3 on the shore of Lake Constance, Switzerland, was sampled and both animal and plant products were found to occur, some of which match the compound-specific isotope pattern for dairy products. This study included a wide range of compound-specific carbon isotope determinations on modern contemporary fats. Data obtained on modern calf and lamb adipose samples showed that the values plot close to the reference field for pig adipose and not the reference field for ruminant adipose. This apparent anomaly is explained by the observation that young (suckling) ruminants mainly utilize glucose carbon for fatty acid synthesis, due to the high sugar content in the mother's milk. After weaning, ruminants use acetate as the main source of carbon in fatty acid synthesis and the adipose fatty acids become more depleted, since the $\delta^{13}C$ value for acetate is lower than that of glucose (Vernon, 1981).

Carbon isotope ratios determined on single fatty acids have greatly enhanced the resolution of lipid-residue identification. Nevertheless, the validity of the '$\Delta^{13}C$' model does require further testing. For example, during the early stages of lactation or during nutritional stress, dairy cows make use of adipose tissue reserves to produce milk. The impact of this on the carbon isotope ratios of the fatty acids has not been addressed. The possibility that other animal sources might overlap with the reference field for milk fatty acid isotope ratios needs to be given further consideration (*e.g.*, see data on reference deer fat samples in Evershed *et al.*, 2002b and that reported in Craig *et al.*, 2005; 889).

11.5 DNA, LACTASE PERSISTENCE AND EARLY DAIRYING

Simoons (1969; 1970) advanced the idea that lactase persistence in humans is closely linked to dairying. His 'culture–evolution' hypothesis suggested that groups which kept cattle and other dairy animals would gain an advantage if adults retained the ability to digest milk. In general, the distribution of adult lactase persistence and dairying shows a positive relationship. A good example is northern Europe where an estimated 90% of the population is lactose tolerant. In areas with little or no dairying tradition, such as China, Oceania and parts of Africa, few adults can digest lactose. Others (*e.g.*, Holden and Mace, 1997) have used phylogenetic analysis to test the relationship between lactase persistence and pastoralism. Beja-Pereira *et al.* (2003) have studied geographic patterns of variation in genes encoding the six milk proteins in 70 native European cattle breeds. They found a high degree of correspondence in north-central Europe between high diversity in cattle milk genes, locations of European Neolithic cattle farming sites and present-day lactose tolerance in European populations.

Other selective forces may also have been at work. Flatz (1987) has suggested that calcium absorption was a factor in northern Europe. Lactose is known to facilitate calcium absorption in the intestine. The northerly climate frequently prevented skin exposure to sunlight, thereby reducing the body's production of vitamin D. With little vitamin D available, calcium was poorly absorbed and conditions such as rickets could result. The ability to digest lactose would not only allow adults to use an excellent source of calcium, but the lactose would also facilitate its absorption.

There is growing interest in the potential of ancient DNA to address the antiquity of lactase persistence in humans. Lactase is encoded by a single gene (*LCT*) located on chromosome 2. Burger *et al.* (2007) have explored the evidence of ancient DNA in a small sample of human remains dated to the European Mesolithic and Neolithic (6th Millennium BC). Interestingly, this study suggested that early Europeans were not able to digest lactose on the basis of the absence of the 13.910*T allele that controls the ability to digest milk. It is this allele which is associated with lactase persistence in modern Europeans. This led the researchers to conclude that the consumption and tolerance of milk would have been very rare or absent at the time. Although the sample is very small, this study challenges the theory that Early Neolithic people had the genetic adaptation to produce lactase beyond their first 2–4 years in order to digest lactose sugar present in fresh milk. Instead, the results suggest that humans evolved tolerance to milk in this area within the last 8000 years. Today around 80% of southern Europeans cannot tolerate lactose even though the first dairy farmers in Europe probably lived in those areas – indeed, the evidence of Craig *et al.* (2005) points towards dairy products in the Early Neolithic of south-eastern Europe. According to Burger *et al.* (2007), it is unlikely that fresh milk consumption was widespread in Europe before frequencies of the 13.910*T allele had risen appreciably during the millennia after the onset of farming. The authenticity of ancient human DNA can be extremely difficult to verify since both contamination from modern DNA and DNA alteration are pervasive problems (Willerslev and Cooper, 2005; 6–9). The research team did adhere to stringent laboratory procedures and one could envisage that in any contaminating DNA, the presence of the 13.910*T allele would be the more likely scenario. The apparent absence of the 13.910*T allele in early farmers could be explained by the conversion of milk into cheese, butter or yoghourt which has little or no lactose. Nevertheless, it is encouraging that ancient DNA can contribute to this debate alongside evidence from lipids and proteins. Direct tracing of lactose in pottery vessels is highly unlikely given its solubility and the likelihood of its rapid decay by micro-organisms.

11.6 SUMMARY: WHERE DO WE GO FROM HERE?

Single-compound carbon isotope analysis has proved extremely valuable in documenting a hitherto 'invisible' dietary component in European prehistory. Although some chemical analyses had been conducted by the time Andrew Sherratt wrote of the 'secondary products revolution', it is only in the past decade that systematic molecular and isotopic investigations have been carried out using satisfactory sample sizes. Further work is needed to distinguish milk from cows, sheep or goat (*cf.* the recent application of Nano ESI MS and MS/MS in Mirabaud *et al.*, 2007), to differentiate milk and processed dairy products, such as butter, cheese and yoghourt, and also to resolve questions of whether the dairy residues identified in pots represent sealants or residues of milk processing to make a specific fermented product. Lipid analysis and laboratory experimentation will continue to develop, but the application of protein sequencing (proteomics) could enable species distinction if proteins survive. The relative degradation of milk versus butter has been noted above (Copley *et al.*, 2005d). Other evidence, such as clay tablets impressed with signs and symbols provide further indications. Those from Uruk in Iraq date to the late 4th and 3rd Millennia BC. The 'proto-cuneiform' on these tablets has been studied from a number of perspectives, including animal husbandry (Green, 1980). The milking of cows, sheep and goats has been recognized, although it is more difficult to determine the type of dairy product produced. According to Green (1980; 9) '*Milk, cream, butter, and cheese can probably be excluded, since the archaic signs for these are well attested. Most likely some fermented product like buttermilk, yogurt, or curds is specified ... Although the significant difference in dairy production for ewes and she-goats might indicate different types of produce, this could be merely a difference in consistencies or a reflection of greater milk productivity of goats. An instance of confusion of signs (text no. 2) as well as mention of only one type of produce for the combined sheep/goat herd (no. 48) suggest that the produce was similar from both species.*' Overall, this evidence points to a long history of processed dairy products.

The testing of new isotope systems holds great promise for complementing the approaches outlined here. Chu *et al.* (2006) have explored the measurement of calcium isotopes to bones and teeth, indicating that dairy products contain calcium which is isotopically lighter than that in other food sources. This application is not only complementary to lipid and protein analysis, but offers the possibility of quantifying how much dairy produce was actually consumed by humans. As Craig (2003; 89) has emphasized, '*the key issue in defining a dairy economy is not only to provide evidence for the utilization of dairy products but also to assess the scale of production and the significance of this activity.*' One way of addressing significance is to evaluate patterns of human consumption.

REFERENCES

Agozzino, P., Avellone, G., Donato, I.D. and Filizzola, F. (2001). Mass spectrometry for cultural heritage knowledge: gas chromatographic/mass spectrometric analysis of organic remains in Neolithic potsherds. *Journal of Mass Spectrometry* **36** 443–444.

Beja-Pereira, A., Luikart, G., England, P.R., Bradley, D.G., Jann, O.C., Giorgio Bertorelle, G., Chamberlain, A.T., Nunes, T.P., Metodiev, S., Ferrand, N. and Erhardt, G. (2003). Gene–culture co-evolution between cattle milk protein genes and human lactase genes. *Nature Genetics* **35** 311–313.

Bogucki, P.I. (1984). Linear pottery ceramic sieves and their economic implications. *Oxford Journal of Archaeology* **3** 15–30.

Bourgeois, G. and Gouin, P. (1995). Résultats d'une analyse de traces organiques fossiles dans une 'faisselle' Harappéenne. *Paléorient* **21** 125–144.

Burger, J., Kirchner, M., Bramanti, B., Haak, W. and Thomas, M.G. (2007). Absence of the lactase-persistence-associated allele in early Neolithic Europeans. *Proceedings of the National Academy of Sciences, USA* **104** 3736–3741.

Challinor, C., Brown, L. and Heron, C. (1998). Molecular information from ceramics. In *Old Scatness Broch, Shetland: Retrospect and Prospect*, ed. Nicholson, R.N. and Dockrill, S.J., Bradford Archaeological Sciences Research 5/North Atlantic Biocultural Organisation Monograph No. 2, University of Bradford, Bradford, pp. 139–149.

Christie, W.W. (1987). *HPLC and Lipids*. Pergamon Press, Oxford.

Chu, N.C., Henderson, G.M., Belshaw, N.S. and Hedges, R.E.M. (2006). Establishing the potential of Ca isotopes as proxy for consumption of dairy products. *Applied Geochemistry* **21** 1656–1667.

Copley, M.S., Rose, P.J., Clapham, A., Edwards, D.N., Horton, M.C. and Evershed, R.P. (2001). Processing palm fruits in the Nile Valley – biomolecular evidence from Qasr Ibrîm. *Antiquity* **75** 538–542.

Copley, M.S., Berstan, R., Dudd, S.N., Docherty, G., Mukherjee, A.J., Straker, V., Payne, S. and Evershed, R.P. (2003). Direct chemical evidence for widespread dairying in prehistoric Britain. *Proceedings of the National Academy of Sciences, USA* **100** 1524–1529.

Copley, M.S., Berstan, R., Dudd, S.N., Straker, V., Payne, S. and Evershed, R.P. (2005a). Dairying in antiquity. I. Evidence from absorbed lipid residues dating to the British Iron Age. *Journal of Archaeological Science* **32** 485–503.

Copley, M.S., Berstan, R., Straker, V., Payne, S. and Evershed, R.P. (2005b). Dairying in antiquity. II. Evidence from absorbed lipid residues dating to the British Bronze Age. *Journal of Archaeological Science* **32** 505–521.

Copley, M.S., Berstan, R., Mukherjee, A.J., Dudd, S.N., Straker, V., Payne, S. and Evershed, R.P. (2005c). Dairying in antiquity. III. Evidence from absorbed lipid residues dating to the British Neolithic. *Journal of Archaeological Science* **32** 523–546.

Copley, M.S., Berstan, R., Dudd, S.N., Aillaud, S., Mukherjee, A.J., Straker, V., Payne, S. and Evershed, R.P. (2005d). Processing of milk products in pottery vessels through British prehistory. *Antiquity* **79** 895–908.

Coultate, T.P. (2001). *Food: The Chemistry of its Components*. Royal Society of Chemistry, Cambridge.

Craig, O.E. (2003). Dairying, dairy products and milk residues: potential studies in European prehistory. In *Food, Culture and Identity in the Neolithic and Early Bronze Age*, ed., Parker Pearson, M., British Archaeological Reports IS 1117, Archaeopress, Oxford, pp. 89–96.

Craig, O.E. and Collins, M.J. (2000). An improved method for the immuno-logical detection of mineral bound protein using hydrofluoric acid and direct capture. *Journal of Immunological Methods* **236** 89–97.

Craig, O.E., Chapman, J., Heron, C., Willis, L.H., Bartosiewicz, L., Taylor, G., Whittle, A. and Collins, M.J. (2005). Did the first farmers of central and eastern Europe produce dairy foods? *Antiquity* **79** 882–894.

Craig, O.E., Heron, C., Willis, L., Yusof, N. and Taylor, G. (2007). Organic residue analysis of pottery vessels. In *The Early Neolithic on the Great Hungarian Plain: Investigations of the Körös Culture Site of Ecsegfalva 23,* ed. Whittle, A., County Békés, Volume 1, Archaeological Institute of the Hungarian Academy of Sciences, Budapest, pp. 349–359.

deMan, J. (1990). *Principles of Food Chemistry*. Van Nostrand Reinhold, London.

Dudd, S.N. and Evershed, R.P. (1998). Direct demonstration of milk as an element of archaeological economies. *Science* **282** 1478–1480.

Dudd, S.N., Regert, M. and Evershed, R.P. (1998). Assessing microbial lipid contributions during laboratory degradations of fats and oils and pure triacylglycerols absorbed in ceramic potsherds. *Organic Geochemistry* **29** 1345–1354.

Evershed, R.P. (1992a). Gas chromatography of lipids. In *Lipid Analysis: A Practical Approach*, ed. Hamilton, R.J. and Hamilton, S., IRL Press, Oxford, pp. 113–151.

Evershed, R.P. (1992b). Mass spectrometry of lipids. In *Lipid Analysis: A Practical Approach*, ed. Hamilton, R.J. and Hamilton, S., IRL Press, Oxford, pp. 263–308.

Evershed, R.P. and Charters, S. (1995). Interpreting lipid residues in archae-ological ceramics: preliminary results from laboratory simulations of vessel use and burial. *Materials Research Society Symposia Proceedings* **352** 85–95.

Evershed, R.P., Arnot, K.I., Collister, J., Eglinton, G. and Charters, S. (1994). Application of isotope ratio monitoring gas chromatography-mass spectro-metry to the analysis of organic residues of archaeological origin. *Analyst* **119** 909–914.

Evershed, R.P., Mottram, H.R., Dudd, S.N., Charters, S., Stott, A.W., Gibson, A.M., Conner, A., Blinkhorn, P.W. and Reeves, V. (1997). New criteria for the identification of animal fats preserved in archaeological pottery. *Naturwissenschaften* **84** 402–406.

Evershed, R.P., Dudd, S.N., Lockheart, M.J. and Jim, S. (2001). Lipids in archaeology. In *Handbook of Archaeological Sciences*, ed. Brothwell, D.R. and Pollard, A.M., Wiley, Chichester, pp. 331–349.

Evershed, R.P., Dudd, S.N., Copley, M.S., Berstan, R., Stott, A.W., Mottram, H., Buckley, S.A. and Crossman, Z.A. (2002a). Chemistry of archaeological animal fats. *Accounts of Chemical Research* **35** 660–668.

Evershed, R.P., Dudd, S.N., Copley, M.S. and Mukerjee, A. (2002b). Identi-fication of animal fats via compound specific $\delta^{13}C$ values of individual fatty acids: assessments of results for reference fats and lipid extracts of archaeological pottery vessels. *Documenta Praehistorica* **21** 73–96.

Flatz, G. (1987). Genetics of lactose digestion in humans. *Advances in Human Genetics* **16** 1–77.

Green, M.W. (1980). Animal husbandry at Uruk in the Archaic Period. *Journal of Near Eastern Studies* **39** 1–35.

Greenfield, H.J. (1988). The origins of milk and wool production in the Old World: a zooarchaeological perspective from the Central Balkans. *Current Anthropology* **29** 573–593.

Grüss, J. (1933). Über Milchreste aus der Hallsattzeit und andere Funde. *Forschungen und Fortschritte* **9** 105–106.

Gunstone, F.D., Harwood, J.L. and Padley, F.B. (1994). *The Lipid Handbook*. Chapman and Hall, London, 2nd edn.

Halstead, P. (1988). Mortality models and milking: problems of optimality, uniformitarianism and equifinality reconsidered. *Anthropozoologica* **27** 3–20.

Harrison, R.J. (1985). The 'Policultivo Ganadero' or the Secondary Products Revolution in Spanish agriculture, 5000–1000 BC. *Proceedings of the Prehistoric Society* **51** 75–102.

Hilditch, T.P. and Williams, P.N. (1964). *The Chemical Constitution of Natural Fats*. Chapman and Hall, London, 4th edn.

Holden, C. and Mace, R. (1997). Phylogenetic analysis of the evolution of lactose digestion in adults. *Human Biology* **69** 605–628.

Jensen, R.G. (2000). The composition of bovine milk lipids: January 1995 to December 2000. *Journal of Dairy Science* **85** 295–350.

Jones, A. (1999). The world on a plate: ceramics, food technology and cosmology in Neolithic Orkney. *World Archaeology* **31** 55–77.

Jones, M.K. (2001). *The Molecule Hunt*. Penguin, London.

Jones, M. (2007). *Feast: Why Humans Share Food*. Oxford University Press, Oxford.

Legge, A.J. (1981). Aspects of cattle husbandry. In *Farming Practice in British Prehistory*, ed. Mercer, R., Edinburgh University Press, Edinburgh, pp. 169–181.

Meier-Augenstein, W. (2002). Stable isotope analysis of fatty acids by gas chromatography-isotope ratio mass spectrometry. *Analytica Chimica Acta* **465** 63–79.

Messing, S. (1957). Further comments on resin-coated pottery: Ethiopia. *American Anthropologist* **59** 134.

Mirabaud, S., Rolando, C. and Regert, M. (2007). Molecular criteria for discriminating adipose fat and milk from different species by nano ESI MS and MS/MS of their triacylglycerols: application to archaeological remains. *Analytical Chemistry* **79** 6182–6192.

Needham, S. and Evans, J. (1987). Honey and dripping: Neolithic food residues from Runnymede Bridge. *Oxford Journal of Archaeology* **6** 21–28.

Perkins, E.G. (ed.) (1993). *Analyses of Fats, Oils and Derivatives*. AOCS Press, Champaign, Illinois.

Regert, M., Bland, H.A., Dudd, S.N., van Bergen, P.F. and Evershed, R.P. (1998). Free and bound fatty acid oxidation products in archaeological ceramic vessels. *Proceedings of the Royal Society of London B: Biological Sciences* **265** 2027–2032.

Robinson, P.G. (1982). Common names and abbreviated formulae for fatty acids (letter to the Editor). *Journal of Lipid Research* **23** 1251–1253.

Rottländer, R.C.A. and Schlichtherle, H. (1979). Food identification of samples from archaeological sites. In *Proceedings of the 22nd Symposium on Archaeometry*, ed. Aspinall, A. and Warren, S.E., University of Bradford. Bradford, pp. 218–223.

Sherratt, A.G. (1981). Plough and pastoralism: aspects of the secondary products revolution. In *Pattern of the Past: Studies in Honour of David Clarke*, ed. Hodder, I., Isaac, G. and Hammond, N., Cambridge University Press, Cambridge, pp. 261–305.

Simoons, F.J. (1969). Primary adult lactose intolerance and the milking habit: a problem in biological and cultural interrelations. I. Review of the medical research. *American Journal of Digestive Diseases* **14** 819–836.

Simoons, F.J. (1970). Primary adult lactose intolerance and the milking habit: a problem in biological and cultural interrelations. II. A culture historical hypothesis. *American Journal of Digestive Diseases* **15** 695–710.

Spangenberg, J.E., Jacomet, S. and Schibler, J. (2006). Chemical analyses of organic residues in archaeological pottery from Arbon Bleiche 3, Switzerland – evidence for dairying in the late Neolithic. *Journal of Archaeological Science* **33** 1–13.

Vernon, R.G. (1981). Lipid metabolism in the rumen. In *Lipid Metabolism in the Adipose Tissue of Ruminant Animals*, ed. Christie, W.W., Pergamon, Oxford, pp. 279–362.

Willerslev, E. and Cooper, A. (2005) Ancient DNA. *Proceedings of the Royal Society B* **272** 3–16.

CHAPTER 12

Summary – Whither Archaeological Chemistry?

12.1 HISTORICAL SUMMARY

The application of analytical chemistry to archaeological artefacts and sites has a long and distinguished history, going back well into the 18th Century. The pace of this activity has increased rapidly since about 1950, reflecting a growth in the availability of instrumental methods of analysis, starting with optical emission spectroscopy and neutron activation analysis, and culminating in the application of the latest techniques for sub-nanogram analysis of biomolecular remains. The range of materials studied has varied enormously, from the more obviously durable such as obsidian, flint, glass and pottery, to skin, hair and faint traces of putative blood on stone surfaces. Only a few have been considered in any detail here.

The analysis of archaeological material has, in general, been regarded as a specialist pursuit. Several universities and museums have gradually established well-equipped analytical laboratories to deal with some, but inevitably not all, of the materials which are encountered archaeologically. These laboratories have been able to dedicate some of the smaller pieces of analytical equipment [such as atomic absorption spectroscopy (AAS), X-ray fluorescence (XRF), up to scanning electron microscopy (SEM) and, occasionally, mass spectrometers] solely to the study of archaeological material. Larger equipment, such as high-resolution inductively coupled plasma mass spectroscopy (ICP-MS), has tended to be shared, or used on a commercial basis, and there is every indication that this is becoming increasingly common. The justification for this level of instrumental dedication has traditionally been the special nature of archaeological material – often small, occasionally valuable, but mostly precious in the sense that it has to be returned to a museum or reference collection essentially undamaged, since it may be required for other types of study. In the early days of instrumental analysis, this was certainly a severe

Archaeological Chemistry, Second Edition
By A. Mark Pollard and Carl Heron

restriction – sample requirements were usually measured in terms of grams, and instrumental modifications were often necessary to accommodate smaller samples. Great value was placed on techniques such as XRF, which could be used in a virtually non-destructive mode, but at the cost of restricting the analysis to surface layers.

The past 10–20 years has seen a great change in analytical capabilities. Most modern instruments, whatever the technique involved, routinely deal with samples in the milligram range or smaller, and are designed to run automatically under computer control, theoretically giving astonishing capacity for the potential throughput of samples. Additionally, other scientific disciplines, such as forensic science or biochemical palaeontology, have developed an interest in handling samples with characteristics very similar to those of archaeological material. In fact, many disciplines have always had similar requirements to archaeology, but only belatedly has this come to be recognized by either side. It is probably now true to say that archaeological chemistry is no longer a unique area of endeavour, requiring specialist equipment, although the problems to be solved often remain specific to archaeology.

One overriding concern in archaeological chemistry is the degraded or altered nature of the sample. Post-depositional alteration cannot be discounted in the discussion of any material, even the most apparently inert, such as stone or gold. The increasing focus on material of a biological or organic nature drastically increases this concern. It simply cannot be assumed that any such material will have survived unaltered over archaeological time. The general rule has to be that archaeological samples *are* altered in some way from their original state. This gives rise to the conviction which has governed the compilation of this book, that a full understanding of the archaeological material can only be obtained from a thorough knowledge of the structure and behaviour of the material concerned. In this respect, for many materials, we still have some way to go.

12.2 THE ARCHAEOLOGICAL RELEVANCE OF CHEMICAL APPLICATIONS

An obvious and reasonable question to ask after so many years is 'what has been the value of all this effort?' In terms of the study of archaeological materials, a great deal has been learnt about subjects such as the history of technology, the exploitation of raw materials in antiquity and the long-term stability of such materials. Without doubt, our knowledge of ancient materials has been vastly enhanced by this cumulative effort. But what have been the archaeological benefits? If one asks the question 'what has been the greatest impact of the natural sciences on archaeology over the past 100 years?', again without doubt the predominant answer would be 'the development of scientific techniques of dating', of which by far the most important has been radiocarbon dating. Chemical studies, probably the second most important contribution, have traditionally concentrated largely on the field of 'provenance studies', which has been justified in terms of reconstructing ancient trade and exchange patterns. The extent to which these have contributed to the wider archaeological narrative is sometimes difficult to assess, but in some geographical areas (such as the Mediterranean) the impact appears to have been considerable. The past 20 years or so has seen an explosion in the chemical analysis of human bones being applied to the study of diet, nutrition, status and mobility. Although popularity does not necessarily equate directly with relevance, it seems certain that these applications are likely to have a significant influence on the direction of archaeological thinking over the next few years. The recognition of archaeological science in general, and of archaeological chemistry in particular, as a valuable component of archaeological research is a tribute to the success of the many pioneers in these fields.

Nevertheless, there has been a constant debate within the discipline of archaeology about the value of the scientific approach to understanding material culture. Undoubtedly, this debate is clouded by deep issues of modern cultural identity (in the sense of C.P. Snow's 'two cultures'), including various factors such as English class structure and financial considerations, but it is clearly a matter of great concern when a leading archaeologist can use the phrase '*Why archaeologists don't care about archaeometry*' in the title of a book review (Dunnell, 1993). It has to be said that in some cases traditional scientific applications to archaeology have failed to deliver answers to the questions which are of immediate interest to mainstream archaeologists. The reasons for these failings are many – perhaps the original question was framed in terms having little archaeological relevance, or, quite commonly, the scientific interpretations have failed to take into account other strands of archaeological evidence, or, more generally, they have not been integrated within an appropriate theoretical framework. In this respect archaeological chemistry differs significantly from almost all other applications of chemistry – the results can only be fully utilized if they harmonize with other studies of the social and cultural context of the archaeological problem. In short, human behaviour, in all its glorious complexity and irrationality, is almost always involved somewhere along the line.

On the other hand, some archaeologists can be criticized for focusing (understandably, perhaps) solely on the archaeological value of scientific studies of archaeological material. Many of the greatest benefits of archaeological science have been felt outside the strictly archaeological sphere. A classic example might be radiocarbon dating itself, which has revolutionized late Quaternary geology and archaeology, but has also provided a wealth of data on the changes in solar flux over the past 10 000 years, which has been of great value to solar scientists. There is a strong view in some quarters that archaeology (or at least certain aspects of it) should recognize this commonality of interest and take its place amongst the other 'historical sciences', such as Quaternary geology, palaeobiology and historical environmental sciences (Pollard, 1995).

More specifically, there has been a feeling that some aspects of archaeological chemistry have unconsciously limited its usefulness to archaeology by focusing excessively on questions of provenance (Budd *et al.*, 1996), interpreted to mean the geological source of raw materials for the manufacture of inorganic artefacts. Mainstream archaeology re-evaluated and largely discarded the simplistic notions of diffusionism as an explanation for cultural development during the 1960s and 1970s, and chemical provenance studies played a significant role in this process. Despite contributing to the fall of diffusionism, archaeological chemistry has sometimes failed to move into the more complex theoretical frameworks which have grown to replace it. In many ways, the concept of provenance in its simplest sense is largely archaeologically irrelevant – a knowledge of where some raw material or traded object came from is of limited value, unless questions about the social and economic structure of the supply and exchange networks of such materials can be addressed using these data. At the very least, it becomes merely a component of a wider discussion. We have, however, seen signs of reconciliation between scientific data and past social and economic activity (*e.g.*, Beck and Shennan, 1991; Bradley and Edmonds, 1993; Budd *et al.*, 1995; Jones, 2002). Such developments, and they continue to increase in number, have gone a long way towards countering the negative view of archaeological science in some quarters. Indeed, some recent dialogues tend towards a rapprochement, particularly around the concept of materiality and the opportunities this presents in bringing together theoretical and scientific perspectives on materials and material culture [see Jones (2004) and the commentaries on this paper by Boivin, Gosden, Bray and Pollard, Killick and many others published in *Archaeometry* in 2005)].

In fact, it now seems likely that the pendulum may have swung the other way, and archaeological chemistry is capable of setting parts of the archaeological agenda. The ability to provenance Roman glass sources using Sr and Nd isotopes, for example, has not only supported the assertions of the classical writers – that glass production was restricted to a few sites in the Eastern Mediterranean – but is beginning to frame new questions about supply and consumption for glass specialists (Freestone *et al.*, 2003). Perhaps even more pertinently, the use of isotopic systems to 'provenance' humans from their bone chemistry has allowed questions of fundamental importance to archaeology to be addressed directly for the first time. It has long been debated, for example, whether the adoption of agriculture in the European Neolithic was the result of diffusion of the 'idea' of farming or of the population replacement of Mesolithic hunter-gatherers by migrations of pastoral Neolithic peoples. Bone chemistry has the potential to tell us whether individuals were buried in the same place as they were brought up, thus allowing the identification of 'immigrants' in the archaeological record (Bentley, 2006). This, when combined with many other sources of evidence [DNA (ancient and modern), linguistics and the archaeological record of material culture], offers the opportunity to answer these fundamental questions for the first time.

12.3 WHITHER ARCHAEOLOGICAL CHEMISTRY?

Archaeological chemistry is now undoubtedly capable of utilizing many of the most sophisticated analytical techniques currently available. Sampling restrictions no longer need cause undue worries on the part of museum curators (although small-scale sampling poses questions of analytical homogeneity, *etc.*). Although sometimes perceived as expensive in archaeological terms, good-quality analytical facilities are now widely available. The real restrictions, therefore, to good archaeological chemistry are now more in terms of the quality of thinking rather than the practicalities. Careful construction of relevant archaeological questions, and intelligent interpretation of results within a sound theoretical framework, should lead to further a better integration of chemical studies within archaeology.

A good example of the way in which archaeology, chemistry and geology must interact to answer 'real' questions is provided by the long-standing debate about the origin and transport of the bluestones at Stonehenge, on Salisbury Plain in southern England. It has been established beyond reasonable doubt by petrology and geochemistry that the geological origin of these stones lies in the Preseli Hills of west Wales. What is at issue is how the stones were transported to Stonehenge. The traditional archaeological view is that human action some 4000 years ago was responsible for the movement of the stones, although archaeological evidence for long-distance transport of the megaliths used in other stone circles is insubstantial (Thorpe and Williams-Thorpe, 1991). The visualization of groups of 'Stone Age Men' rolling these huge rocks on tree trunks for hundreds of kilometres, or building barges to float them up the Severn estuary, has been one of the most pervasive in British prehistory. The work of Williams-Thorpe and Thorpe (1992), involving petrology, geochemistry and evaluation of the glacial geomorphology, has suggested that glacial action around 400 000 years ago may actually have been the agent by which large quantities of this rock was transported to Salisbury Plain. The archaeological implications of these two opposing models of how the stones reached Stonehenge are immense, in terms of the way we interpret the significance of this major prehistoric monument. What is abundantly clear is that no single approach in isolation can possibly hope to provide a definitive answer.

There remain, of course, many great challenges for archaeological chemistry. In most countries, archaeological research over the past few decades has moved away from the policy of total excavation for sites threatened by development, erosion or sheer population pressure. Many countries have adopted a legal framework of 'preservation *in situ*' (Corfield *et al.*, 1997; Bowsher, 2004). Planning regulations require mitigation strategies such that development can occur in such a way that known archaeological deposits are not damaged (or minimally damaged) by building work. Implicit in this is the assumption that the safest way of preserving archaeological remains is to leave them in the ground, since they have survived, in some cases, for several millennia in that state. It is recognized, however, that development, even if it does not directly destroy buried remains, may change the local burial conditions in many ways – perhaps significantly altering the water table, or the flow of water through the deposits or increasing the compaction of the soil and thus altering the redox (Eh) balance. The question which is now being asked within the legal framework of the planning regulations is 'what effect will these changes have on the state of preservation of the buried material?' In some cases, we can make educated guesses. Considerable effort has, for instance, been put into the study of the preservation environment of waterlogged wood, so we have some idea of the effect of drying on such objects. More generally, however, we know very little about the detailed effects of variations in burial conditions on a wide range of materials, such as bone, metalwork, *etc.* This requires a detailed knowledge of the deterioration mechanisms of the materials themselves, but also an ability to predict the changes resulting from variations in soil conditions. This requires an understanding of the interaction between soil (strictly, burial medium), groundwater and archaeological object, which in turn requires a very wide range of chemical and physical knowledge.

In terms of the conservation and preservation of archaeological materials (and the cultural heritage in general), chemistry therefore has a tremendous role to play in understanding the mechanisms of corrosion for a vast range of materials. This leads to the development of strategies for the control of deterioration so that the objects (or buildings, or entire cities and landscapes) can be stabilized for long-term preservation, display, study or storage, thus contributing substantially to the sustainability of the huge economic resource that is our cultural heritage. Such studies are likely to move rapidly to the top of the political and (therefore!) scientific agenda in the next few years, and form, we believe, a very great challenge and opportunity for archaeological chemistry. In many ways, the archaeological demand for qualified archaeological chemists or archaeologists with considerable chemical knowledge has never been greater.

REFERENCES

Beck, C.W. and Shennan, S. (1991). *Amber in Prehistoric Britain*. Oxbow, Oxford.

Bentley, R.A. (2006). Strontium isotopes from the Earth to the archaeological skeleton: a review. *Journal of Archaeological Method and Theory* **13** 135–187.

Bowsher, D. (ed.) (2004). Preserving Archaeological Remains in Situ? Proceedings of the 2nd Conference 12–14 September 2001, Museum of London Archaeology Service, London.

Bradley, R. and Edmonds, M. (1993). *Interpreting the Axe Trade: Production and Exchange in Neolithic Britain*. Cambridge University Press, Cambridge.

Budd, P., Pollard, A.M., Scaife, B. and Thomas, R.G. (1995). Oxhide ingots, recycling and the Mediterranean metals trade. *Journal of Mediterranean Archaeology* **8** 1–32.

Budd, P., Haggerty, R., Pollard, A.M., Scaife, B. and Thomas, R.G. (1996). Rethinking the quest for provenance. *Antiquity* **70** 168–174.

Corfield, M., Hinton, P., Nixon, T. and Pollard, M. (eds) (1997). *Preserving Archaeological Remains in-Situ*. Museum of London Archaeology Service, London.

Dunnell, R.C. (1993). Why archaeologists don't care about archaeometry. *Archeomaterials* **7** 161–165.

Freestone, I.C., Leslie, K.A., Thirlwall, M. and Gorin-Rosen, Y. (2003). Strontium isotopes in the investigation of early glass production: Byzantine and early Islamic glass from the Near East. *Archaeometry* **45** 19–32.

Jones, A. (2002). *Archaeological Theory and Scientific Practice.* Cambridge University Press, Cambridge.

Jones, A. (2004). Archaeometry and materiality: materials-based analysis in theory and practice. *Archaeometry* **46** 327–338.

Pollard, A.M. (1995). Why teach Heisenberg to archaeologists? *Antiquity* **69** 242–247.

Thorpe, R.S. and Williams-Thorpe. O. (1991). The myth of long-distance megalith transport. *Antiquity* **65** 64–73.

Williams-Thorpe, O. and Thorpe, R.S. (1992). Geochemistry, sources and transport of the Stonehenge bluestones. In *New Developments in Archaeological Science,* ed. Pollard, A.M., Proceedings of the British Academy 77, Oxford University Press, Oxford, pp. 133–161.

The Structure of the Atom, and the Electromagnetic Spectrum

In the early part of the 20th Century, a simple model of atomic structure became accepted, now known as the *Bohr model* of the atom. This stated that most of the mass of the atom is concentrated in the *nucleus*, which consists of *protons* (positively charged particles) and *neutrons* (electrically neutral particles, of approximately the same mass). The number of protons in the nucleus is called the *atomic number*, the value of which defines the element. The number of protons plus neutrons in the nucleus is called the *atomic mass* (see Appendix 2 for more discussion of nuclear structure and isotopes). Electrical neutrality in the atom is maintained by a number of negatively charged electrons (equal numerically to the number of protons, but with only a fraction of the mass) circling the nucleus like a miniature solar system. This gives a picture of the atom as having a very small dense nucleus at the centre containing most of the mass, surrounded by a nebulous 'cloud' of electrons some distance away (on the atomic scale of things). Rutherford demonstrated that this was a plausible model by firing relatively heavy α particles (the nuclei of the helium atom, with two protons and two neutrons) at thin metal foils. Most passed straight through, indicating that the majority of the atom was empty space. A few were deflected on transit through the foil, but some were reflected backwards, hinting that somewhere within the atom was a small but very heavy core, since it would require a relatively massive object to reflect a fast-moving α particle.

This experiment established the nuclear model of the atom. A key point derived from this is that the electrons circling the nucleus are in fixed stable orbits, just like the planets around the sun. Furthermore, each 'orbital' or 'shell' contains a fixed number of electrons – additional electrons are added to the next stable orbital above that which is full. This stable orbital model is a departure from classical electromagnetic theory (which predicts unstable orbitals, in which the electrons spiral into the nucleus and are destroyed), and can only be explained by quantum theory. The fixed numbers for each orbital were determined to be two in the first level, eight in the second level, eight in the third level (but extendible to 18) and so on. Using this simple model, chemists derived the systematic structure of the Periodic Table (see Appendix 5), and began to understand factors such as atomic sizes, the shapes of molecules and the underlying reasons behind the observed periodicity of chemical properties. Table A1.1 shows how the electronic configuration (the number and distribution of the orbital electrons) of most of the elements can be understood by simply following the above rules for filling up the successive electron orbitals to keep pace with the increasing number of protons in the nucleus. The orbitals are labelled 1, 2, 3, *etc.*, but as can be seen from Table A1.1, sub-divisions occur increasingly with increasing number. For example, the second orbital is split into two sub-shells, labelled *s* and *p* – the 2*s* shell is full when it contains only two electrons, but the 2*p* shell can accommodate six (making eight in total for the second orbital). This splitting is a result of quantum mechanical considerations, and each sub-shell has a characteristic shape in space (*e.g.*, the *s*-orbitals are spherical, whereas the *p*-orbitals are elliptical figure-of-eights – see Pollard *et al.*, 2007; 240), which is important in descriptions of chemical bonding and the shapes of molecules. A full discussion of this fascinating area can be found in any basic inorganic chemistry textbook (*e.g.*, Cotton *et al.*, 1995).

It is important to realize that, in any atom, although the orbital electrons may fill only, say, the first two or three orbitals (the exact number of electrons depending on the number of protons in the nucleus), the other unfilled energy levels still exist, and under certain circumstances an electron from a lower state can be promoted up to one of the unfilled energy levels. For example, the element sodium (Na), with 11 protons in the nucleus, and therefore 11 orbital electrons, has the electronic configuration $1s^2 2s^2 2p^6 3s^1$, which means that it has two 1*s*-electrons, two 2*s*-electrons, six 2*p*-electrons and one 3*s*-electron. This is the lowest energy configuration the neutral sodium atom can have, since all of the possible first and second orbitals are full, and the final spare electron has gone into the 3*s* level, which is the next available orbital. This configuration is termed the *ground state*, meaning it has the lowest possible energy, and is the configuration listed for each element in Table A1.1. In the case of sodium, it is possible to promote the outer (3*s*) electron up to one of the unfilled higher orbitals, such as the 4*p*, by supplying some energy to this outer electron, although not all 'promotions' are 'allowed' – certain *selection rules* apply, as discussed below. With the outer electron temporarily up in this higher state, the atom is said to be *excited*, and it is to be expected that the atom will return to its ground state as soon as possible, since this is energetically the most stable condition. It is possible, of course, to remove the electron from the atom completely if enough energy is supplied to do so, when the atom is then said to be *ionized*. Because it is now one electron deficient, it has one unbalanced nuclear positive charge, and hence the resulting ion carries a single positive charge – in this case, it has become the sodium *ion*, symbolized by Na^+.

Although the exact spacing between the *energy levels* (as these electronic orbitals are called) for each atom is different, it is possible to represent schematically the relative sequence of electronic orbitals, since this is the same for all atoms, as shown in Figure A1.1. It can be seen here that complications set in with the fourth set of orbitals, because the 3*d* level has an energy slightly

Table A1.1 Electronic configuration of the elements. Elements in square brackets (e.g., [He]) imply that the electronic configurations of the inner orbitals are identical to those of the element in brackets. Thus silver (Ag, atomic number 47) has a configuration of $[Kr]4d^{10}5s^1$, which if written out in full would be $1s^22s^22p^63s^23p^63d^{10}4s^24p^64d^{10}5s^1$, giving 47 electrons in all. For the heavier elements (atomic number above 55), the alternative notation K, L, M is used to denote the inner shells corresponding to orbitals 1, 2 and 3 respectively. This notation is common in X-ray spectroscopy (see p. 33). (Adapted from Lide, 1990.)

Element	1s	2s	2p	3s	3p	3d	4s	4p	4d	4f	5s	5p	5d	5f	5g
1 H	1														
2 He	2														
3 Li	[He]	1													
4 Be		2													
5 B		2	1												
6 C		2	2												
7 N		2	3												
8 O		2	4												
9 F		2	5												
10 Ne		2	6												
11 Na			[Ne]	1											
12 Mg				2											
13 Al				2	1										
14 Si				2	2										
15 P				2	3										
16 S				2	4										
17 Cl				2	5										
18 Ar				2	6										
19 K					[Ar]		1								
20 Ca							2								
21 Sc						1	2								
22 Ti						2	2								
23 V						3	2								
24 Cr						5	1								
25 Mn						5	2								
26 Fe						6	2								
27 Co						7	2								
28 Ni						8	2								
29 Cu						10	1								
30 Zn						10	2								
31 Ga						10	2	1							
32 Ge						10	2	2							
33 As						10	2	3							
34 Se						10	2	4							
35 Br						10	2	5							
36 Kr						10	2	6							
37 Rb								[Kr]			1				
38 Sr											2				
39 Y									1		2				
40 Zr									2		2				
41 Nb									4		1				
42 Mo									5		1				
43 Tc									5		2				
44 Ru									7		1				
45 Rh									8		1				
46 Pd									10						
47 Ag									10		1				
48 Cd									10		2				
49 In									10		2	1			
50 Sn									10		2	2			
51 Sb									10		2	3			
52 Te									10		2	4			
53 I									10		2	5			
54 Xe									10		2	6			

Table A1.1   (*Continued*).

Element	K	L	M	4s	4p	4d	4f	5s	5p	5d	5f	5g	6s	6p	6d	6f	6g	6h	7s
54 Xe	2	8	18	2	6	10		2	6										
55 Cs									[Xe]				1						
56 Ba													2						
57 La										1			2						
58 Ce							1			1			2						
59 Pr							3						2						
60 Nd							4						2						
61 Pm							5						2						
62 Sm							6						2						
63 Eu							7						2						
64 Gd							7			1			2						
65 Tb							9						2						
66 Dy							10						2						
67 Ho							11						2						
68 Er							12						2						
69 Tm							13						2						
70 Yb							14						2						
71 Lu							14			1			2						
72 Hf							14			2			2						
73 Ta							14			3			2						
74 W							14			4			2						
75 Re							14			5			2						
76 Os							14			6			2						
77 Ir							14			7			2						
78 Pt							14			9			1						
79 Au							14			10			1						
80 Hg							14			10			2						
81 Tl							14			10			2	1					
82 Pb							14			10			2	2					
83 Bi							14			10			2	3					
84 Po							14			10			2	4					
85 At							14			10			2	5					
86 Rn							14			10			2	6					
87 Fr														[Rn]					1
88 Ra																			2
89 Ac															1				2
90 Th															2				2
91 Pa												2			1				2
92 U												3			1				2
93 Np												4			1				2
94 Pu												6							2
95 Am												7							2
96 Cm												7			1				2
97 Bk												9							2
98 Cf												10							2
99 Es												11							2
100 Fm												12							2
101 Md												13							2
102 No												14							2
103 Lr												14			1				2

$[Xe] = 1s^2 2s^2 2p^6 3s^2 3p^6 3d^{10} 4s^2 4p^6 4d^{10} 5s^2 5p^6$
$[Rn] = 1s^2 2s^2 2p^6 3s^2 3p^6 3d^{10} 4s^2 4p^6 4d^{10} 4f^{14} 5s^2 5p^6 5d^{10} 6s^2 6p^6$

lower than the 4*p* level. This results in a whole series of elements with full 1*s*-, 2*s*-, 2*p*-, 3*s*-, 3*p*- and 4*s*-orbitals, but instead of filling the 4*p* level, they 'pause' to fill up the empty 3*d* levels, which can accommodate up to 10 electrons. These are the so-called *d-block* or *transition* metals (Sc, Ti, V, Cr, Mn, Fe, Co, Ni, Cu and Zn), whose chemical properties are intimately related to the behaviour of these $3d$ electrons. Some of these properties, such as being highly coloured in the ionic state, are important in archaeological contexts, since these are the natural pigments and colorants (of glass and glazes) in antiquity. Some of these properties are further explained in Chapter 5.

As noted above, not all possible transitions between energy levels are theoretically allowed. Each energy level is uniquely characterized by a set of *quantum numbers*. The integer used to define the energy level in the above discussion (1, 2, 3, *etc.*) is called the *principal quantum number*, n. The sub-levels described by the letters (*s, p, d, f, etc.*) are associated with the *second quantum number*, given the symbol l, with $l = 1$ synonymous with s, $2 = p$, *etc.* The multiplicity of levels associated with each sub-level (*i.e.*, the number of horizontal lines for each orbital in Figure A1.1) is defined by a third quantum number m_l, which has values $0, \pm 1 \ldots \pm l$. Thus, s-orbitals only have one sub-level, p-orbitals have three (with m_l values 0 and ± 1), d-orbitals have five, *etc.* The selection rules can

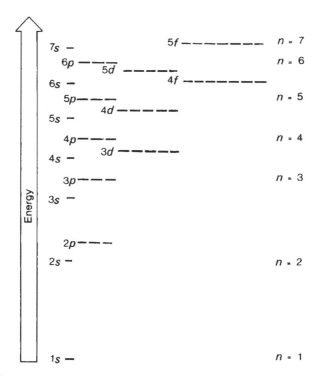

Figure A1.1   Approximate energy level diagram for electronic orbitals in a multi-electron atom. Each horizontal line can accommodate two electrons (paired as so-called spin-up and spin-down electrons), giving the rules for filling the orbitals – two in the s-levels, 6 in the p-levels, 10 in the d-levels. Note that the 3d-orbital energy is lower than the 4p, giving rise to the d-block or transition elements. (From Brady, 1990; Figure 7.10. Copyright 1990 John Wiley & Sons, Inc. Reprinted by permission of the publisher.)

be most simply stated as follows:

$$\Delta n \geq 1$$

$$\Delta l = \pm 1$$

Thus, transitions such as $2p \rightarrow 1s$, $3p \rightarrow 1s$, $3d \rightarrow 2p$ are allowed, whereas $2s \rightarrow 1s$, $3d \rightarrow 1s$, $3d \rightarrow 3d$ are strictly forbidden. As discussed in Chapter 5, however, transitions within the d-orbitals (so-called d–d band transitions) do occur, and are important in the consideration of the colour developed by transition metals in solution. Other 'forbidden' transitions also occur.

A second concept of fundamental importance to instrumental chemical analysis is that of the *electromagnetic spectrum*, and the relationship between particle energy and wavelength. One of the great unifications of 19th Century science was the realization that everything from radio waves down through visible light to X-rays was essentially manifestations of the same thing – *electromagnetic radiation*, simply differentiated by having a different wavelength. Thus it was realized that visible light, for example, was simply composed of oscillating electric and magnetic fields, capable of travelling through a vacuum at a particular speed (the speed of light), with a wavelength of between 400 and 700 nm. Figure A1.2 shows the correspondence between wavelength and type of electromagnetic wave. Radio waves have a very long wavelength, with a correspondingly low frequency, which is given by the reciprocal relationship:

$$\lambda = c/\upsilon$$

where λ is the wavelength, c is the speed of light (normally taken as close to $3 \times 10^8 \, \mathrm{m \, s^{-1}}$) and υ (Greek nu) is the frequency. X-rays and γ rays have very

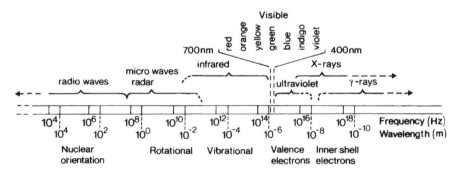

Figure A1.2   The electromagnetic spectrum, showing the reciprocal relationship between wavelength (in metres) and frequency (in Hertz, or cycles per second). The visible band is a very narrow region, with wavelengths between 400 and 700 nm. (Adapted from Physical Science Study Committee, 1960; Figure 31–27, reprinted by permission of D.C. Heath and Company.)

short wavelengths and very high frequencies. As can be seen, visible light only occupies a very small region of the total electromagnetic spectrum, but is obviously important to us because it is the wavelength range to which our eyes are sensitive. A second discovery of the early 20th Century was less intuitively obvious – that of *particle–wave duality*. Simply put, this states that electromagnetic radiation can be thought of as *either* radiation characterized by its wavelength (as described above), *or* as a stream of particles (called *quanta*) with a fixed energy. The energy of the quantum (E) is in units of electron volts (eV), and is proportional to the frequency (or wavelength) of the wave; the relationship involves Planck's constant, h ($=6.626 \times 10^{-34}$ J s):

$$E = hc/\lambda = h\upsilon$$

This is an elegant explanation for one of the early conundrums posed by quantum theory, which noted that in some experiments visible light behaves as if it were a wave [*e.g.*, Young's slits, which demonstrates diffraction of a wave front– see any classical text on optics, such as Jenkins and White (1976) or Pollard *et al.* (2007; 280)], but sometimes it exhibits behaviour which can only be explained by assuming that it is manifested as a stream of particles [*e.g.*, the photoelectric effect – see, for example, Caro *et al.* (1978) or Pollard *et al.* (2007; 280)]. It would appear, therefore, that light (or any electromagnetic radiation) can behave as either a particle of energy E or a wave of wavelength λ, depending on the experiment that we carry out to observe it! Particle–wave duality is one of the central tenets of quantum theory.

REFERENCES

Brady, J.E. (1990). *General Chemistry*. John Wiley, New York, 5th edn.

Caro, D.E., McDonnell, J.A. and Spicer, B.M. (1978). *Modern Physics*. Edward Arnold, London, 3rd edn.

Cotton, F.A., Wilkinson, G. and Gaus, P.I. (1995). *Basic Inorganic Chemistry*. John Wiley, New York, 3rd edn.

Jenkins, F.A. and White, H.E. (1976). *Fundamentals of Optics*. McGraw- Hill, New York, 4th edn.

Lide, D.R. (ed.) (1990). *CRC Handbook of Chemistry and Physics*. CRC Press, Boca Raton, Florida, 71st edn.

Physical Science Study Committee (1960). *Physics*. D.C. Heath, Boston.

Pollard, A.M, Batt, C.M., Stern, B. and Young, S.M.M. (2007). *Analytical Chemistry in Archaeology*. Cambridge University Press, Cambridge.

APPENDIX 2
Isotopes

Atoms are made up of a positively charged nucleus surrounded by a 'cloud' of orbital electrons carrying an equal negative charge – classically this is envisaged as being similar to the solar system, with the sun occupying the position of the atomic nucleus, and the planets the orbiting electrons. The nucleus contains both positively charged particles (protons) and electrically neutral particles (neutrons), which are roughly the same weight as the protons. This weight is given the value one on the *atomic mass unit* scale, also termed a *Dalton* after the man who put forward the modern atomic theory in the 19th Century. On this classical model, it is impossible to explain the structure of the nucleus itself (this requires quantum mechanics and subatomic physics), but it is conventionally stated that the neutrons act as some form of 'glue' which prevents the protons in the nucleus flying apart as a result of electrical repulsion between the positively charged particles.

The number of protons in the nucleus is called the *atomic number*, and is given the symbol Z. It is this number which gives all the elements their different chemical characteristics, and distinguishes one element from another. Z varies continuously from 1, which is the lightest of all elements, hydrogen, up to 102, which is the heaviest naturally occurring element (nobelium). Heavier synthetic elements of atomic number 103–111 have been named [rutherfordium (Rf) to roentgenium (Rg) – see http://www.webelements.com/ for details]. More exotically still, elements have been reported above 112 (Uub, ununbium), up to perhaps 115 (Uup, ununpentium, reported in 2004) and even higher. However, many of these unstable heavier elements have only been made in single figure atom numbers, and for a few milliseconds! Since Z gives the number of units of positive charge in the nucleus, it also dictates the number of electrons orbiting the nucleus. In a neutral atom, the number of electrons is identical to the number of protons in the nucleus, since the charge on the proton and electron is identical but opposite. The familiar chemical symbols (*e.g.*, H, C, O, Pb, *etc.*) are effectively a shorthand code for the proton number – thus the symbol Pb stands for 'that element with 82 protons in its nucleus'.

All elements except the simplest form of hydrogen have some neutrons in their nucleus. For the lightest elements, the number of neutrons is the same (or roughly the same) as the number of protons (*e.g.*, helium, He, has two protons and two neutrons), but as the number of protons increases there appears to be a need for an excess of neutrons in order to hold the nucleus together. Thus gold (Au, atomic number 79) needs 118 neutrons in its nucleus, giving it a total weight on the atomic scale of 197 Daltons. The number of neutrons is given the symbol N, and the combined number of protons plus neutrons is given the symbol A, and is referred to as the *atomic mass number*. Protons and neutrons are collectively termed *nucleons*, and the following simple relationship holds:

$$A = N + Z$$

As stated above, all elements have a unique proton number, but a few also have a unique number of neutrons (at least in naturally occurring forms), and therefore a unique atomic weight – examples are gold ($Z = 79$, $N = 118$, giving $A = 197$), bismuth (Bi; $Z = 83$, $N = 126$, $A = 209$) and at the lighter end of the scale, fluorine (F; $Z = 9$, $N = 10$, $A = 19$) and sodium (Na; $Z = 11$, $N = 12$, $A = 23$). Such behaviour is, however, rare in the Periodic Table, where the vast majority of natural stable elements can exist with two or more different neutron numbers in their nucleus. These are termed *isotopes*. Isotopes of the same element have the same number of protons in their nucleus (and hence orbital electrons, and hence chemical properties), but different numbers of neutrons, and hence different atomic weights. That is one reason why the quoted atomic weights of many elements are not whole numbers – they are averages of two or more different isotopes, with the exact value of the average depending on the relative abundance of the various isotopes. For example, copper has a quoted atomic mass of 63.54 Daltons, as a result of having two isotopes – one of mass 63 with a relative abundance of 69%, and one of mass 65, with an abundance of 31%. It is conventional to denote the various isotopes by using the atomic weight as a preceding superscript – thus these two isotopes of copper are ^{63}Cu and ^{65}Cu. Strictly speaking, the atomic number should also be given as a preceding subscript, but for most purposes that is unnecessary since the symbol 'Cu' is synonymous with the element whose atomic number is 29. (Note that even elements which have only one isotope do not have an atomic weight which is exactly a whole number, because the actual mass of a nucleus is never exactly equal to the sum of the masses of its constituents, due to the relativistic conversion of a small proportion of the mass into energy, known as the *binding energy*.)

For the lightest elements there are usually only two isotopes which are stable, and the lighter of the two is usually the most common. Thus carbon has two stable isotopes, ^{12}C (abundance 99%) and ^{13}C (abundance 1%). A third isotope, ^{14}C, is extremely rare (fractional abundance 10^{-12}), radioactively unstable, but, of course, vitally important in archaeology! Similarly, nitrogen has two stable isotopes (^{14}N and ^{15}N, with abundances 99.6% and 0.4%, respectively). Oxygen has three stable isotopes (^{16}O, ^{17}O and ^{18}O, with abundances 99.76%, 0.04% and 0.20%, respectively), and is typical of the slightly heavier (but still 'light') elements. As stated above, isotopes of the same element chemically all behave identically, but, because they have different weights, processes involving diffusion, or transport across membranes, can all give rise to *fractionation* – a systematic change in the abundance ratio of two isotopes. For these light isotopes, fractionation is particularly marked in biological processes, since the relative weight difference between the individual isotopes is large. Fractionation in light isotopes is now widely used as a natural isotopic marker system in biogeochemical cycles (*e.g.*, Lajtha and Michener, 1994; Griffiths, 1998; Leng, 2006), and has found extensive application in archaeology as a marker for dietary reconstruction (*e.g.*, van der Merwe, 1992; Pollard *et al.*, 2007; 180: see Chapter 10). By the time the transition metals, and the heavier elements in general, are reached, the situation becomes more complex, in that some metals have a large number of stable isotopes. Lead has four (^{204}Pb, ^{206}Pb, ^{207}Pb and ^{208}Pb, with average natural abundances of 1.3%, 26.3%, 20.8% and 51.5%, respectively). The most prolific in this respect is tin (Sn), with ten stable isotopes (112, 114, 115, 116, 117, 118, 119, 120, 122 and 124), with abundances varying from 0.4% up to 33%.

In addition to these stable isotopes, many elements have one or more radioactively unstable isotopes which are produced either as a result of specific nuclear processes (such as ^{14}C, produced by the interaction of cosmic radiation with ^{14}N) or as daughter nuclides during the radioactive decay of heavier unstable elements. Lead, for example, in addition to the four stable isotopes, has at least a further 30 unstable isotopes, ranging from ^{178}Pb up to ^{214}Pb, and with half lives (see Chapter 9) which vary from a few milliseconds up to 3×10^5 years (Emsley, 1998; see also 'Table of Nuclides', http://atom.kaeri.re.kr/). In fact, of the total number of nuclides known (now said to number around 2500) only 260 are stable, suggesting that radioactive instability is the rule rather than the exception. The term 'stable' perhaps needs some clarification – it is conventional to refer to a number of isotopes as 'stable', when in fact they are known to be radioactively unstable, but with an extremely long half life. An example is ^{204}Pb, estimated to have a half life of at least 1.4×10^{17} years, and thus justifiably termed 'stable' on the geological timescale. Observations suggest that in order to maintain stability as the atomic number increases, the ratio of neutrons to protons has to increase from 1:1 for the very light elements up to about 3:1 for the heaviest. Too few or too many neutrons lead to nuclear instability. The reasons for this radioactive instability of some nuclei and not others are rather unclear, but it is has been observed that certain combinations of nucleon numbers are more stable than others, giving rise to the concept of '*magic numbers*' in nuclear stability (Faure, 1986; 15). Over half of the stable nuclides known have even numbers of both Z and N. Odd/even and even/odd combinations contribute just over another 100, but only four nuclei with an odd/odd configuration are stable. Furthermore, it has been noted that nuclei which can be imagined as being multiples of the helium nucleus (^4He, $A = 2$, $N = 2$), such as ^{12}C, ^{16}O, ^{32}S and ^{40}Ca, are all particularly stable, which gives some indication of the robustness of the helium nucleus in nuclear physics (see Chapter 9), and gives rise to the concept of magic numbers. Standard tables of nuclides exist, for example Littlefield and Thorley (1979; Appendix C), which lists the nuclear configuration, natural abundances, decay process and half life of all known nuclides, although new information is constantly being added and can best be accessed via websites such as http://atom.kaeri.re.kr/.

REFERENCES

Emsley, J. (1998). *The Elements*. Clarendon Press, Oxford, 3rd edn.

Faure, G. (1986). *Principles of Isotope Geology*. John Wiley, New York, 2nd edn.

Griffiths, H. (ed.) (1998). *Stable Isotopes. Integration of Biological, Ecological and Geochemical Processes*. BIOS Scientific, Oxford.

Lajtha, K. and Michener, R. (ed.) (1994). *Stable Isotopes in Ecology and Environmental Science*. Blackwell Scientific Publications, Oxford.

Leng, M.J. (ed.) (2006). *Isotopes in Paleoenvironmental Research*. Reviews in Paleonvironmental Research Vol. 10, Springer, Dordrecht.

Littlefield, T.A. and Thorley, N. (1979). *Atomic and Nuclear Physics*. Van Nostrand Rheinhold, New York, 3rd edn.

Pollard, A.M., Batt, C.M., Stern, B. and Young, S.M.M. (2007). *Analytical Chemistry in Archaeology*. Cambridge University Press, Cambridge.

van der Merwe, N.J. (1992). Light stable isotopes and the reconstruction of prehistoric diets. In *New Developments in Archaeological Science*, ed. Pollard, A.M., Proceedings of the British Academy 77, Oxford University Press, Oxford, pp. 247–264.

Fundamental Constants

Description	Symbol	Value
Avogadro constant	$\mathcal{N}_A$	$6.022169 \times 10^{23}\,\text{mol}^{-1}$
Faraday constant	F	$96486.70\,\text{C}\,\text{mol}^{-1}$
Charge of electron	e	$1.6021917 \times 10^{-19}\,\text{C}$
Mass of electron	m_e	$9.109558 \times 10^{-31}\,\text{kg}$
Mass of proton	m_p	$1.672614 \times 10^{-27}\,\text{kg}$
Mass of neutron	m_n	$1.674920 \times 10^{-27}\,\text{kg}$
Planck constant	h	$6.626196 \times 10^{-34}\,\text{J}\,\text{s}$
Speed of light in vacuum	c	$2.9979250 \times 10^8\,\text{m}\,\text{s}^{-1}$
Rydberg constant	R_∞	$1.09737312 \times 10^7\,\text{m}^{-1}$
	R_H	$1.09677578 \times 10^7\,\text{m}^{-1}$
Bohr magneton	μ_B	$9.274096 \times 10^{-24}\,\text{A}\,\text{m}^2$
Gas constant	R	$8.31434\,\text{J}\,\text{K}^{-1}\,\text{mol}^{-1}$
Boltzmann constant	k	$1.380622 \times 10^{-23}\,\text{J}\,\text{K}^{-1}$

Atomic Number and Approximate Atomic Weights (based on $^{12}C = 12.000$) of the Elements

Name	Symbol	Atomic Number	Atomic Weight
Actinium	Ac	89	(227)
Aluminium	Al	13	26.98
Americium	Am	95	(243)
Antimony	Sb	51	121.7
Argon	Ar	18	39.94
Arsenic	As	33	74.92
Astatine	At	85	(210)
Barium	Ba	56	137.3
Berkelium	Bk	97	(247)
Beryllium	Be	4	9.012
Bismuth	Bi	83	209.0
Boron	B	5	10.81
Bromine	Br	35	79.90
Cadmium	Cd	48	112.4
Calcium	Ca	20	40.08
Californium	Cf	98	(251)
Carbon	C	6	12.01
Cerium	Ce	58	140.1
Cesium	Cs	55	132.9
Chlorine	Cl	17	35.45
Chromium	Cr	24	52.00
Cobalt	Co	27	58.93
Copper	Cu	29	63.54
Curium	Cm	96	(247)
Dysprosium	Dy	66	162.5

(*Continued*).

Name	Symbol	Atomic Number	Atomic Weight
Einsteinium	Es	99	(252)
Erbium	Er	68	167.2
Europium	Eu	63	152.0
Fermium	Fm	100	(257)
Fluorine	F	9	19.00
Francium	Fr	87	(223)
Gadolinium	Gd	64	157.2
Gallium	Ga	31	69.72
Germanium	Ge	32	72.59
Gold	Au	79	197.0
Hafnium	Hf	72	178.4
Helium	He	2	4.003
Holmium	Ho	67	164.9
Hydrogen	H	1	1.008
Indium	In	49	114.8
Iodine	I	53	126.9
Iridium	Ir	77	192.2
Iron	Fe	26	55.84
Krypton	Kr	36	83.80
Lanthanum	La	57	138.9
Lawrencium	Lr	103	(260)
Lead	Pb	82	207.2
Lithium	Li	3	6.941
Lutetium	Lu	71	175.0
Magnesium	Mg	12	24.31
Manganese	Mn	25	54.94
Mendelevium	Md	101	(258)
Mercury	Hg	80	200.5
Molybdenum	Mo	42	95.94
Neodymium	Nd	60	144.2
Neon	Ne	10	20.17
Neptunium	Np	93	(237)
Nickel	Ni	28	58.70
Niobium	Nb	41	92.91
Nitrogen	N	7	14.01
Nobelium	No	102	(259)
Osmium	Os	76	190.2
Oxygen	O	8	16.00
Palladium	Pd	46	106.4
Phosphorus	P	15	30.97
Platinum	Pt	78	195.0
Plutonium	Pu	94	(244)

(*Continued*).

Name	Symbol	Atomic Number	Atomic Weight
Polonium	Po	84	(209)
Potassium	K	19	39.09
Praeseodymium	Pr	59	140.9
Promethium	Pm	61	(147)
Protactinium	Pa	91	231.0
Radium	Ra	88	226.0
Radon	Rn	86	(222)
Rhenium	Re	75	186.2
Rhodium	Rh	45	102.9
Rubidium	Rb	37	85.47
Ruthenium	Ru	44	101.0
Samarium	Sm	62	150.4
Scandium	Sc	21	44.96
Selenium	Se	34	78.96
Silicon	Si	14	28.08
Silver	Ag	47	107.9
Sodium	Na	11	22.99
Strontium	Sr	38	87.62
Sulfur	S	16	32.06
Tantalum	Ta	73	180.9
Technetium	Tc	43	(98)
Tellurium	Te	52	127.6
Terbium	Tb	65	158.9
Thallium	Tl	81	204.3
Thorium	Th	90	232.0
Thulium	Tm	69	168.9
Tin	Sn	50	118.6
Titanium	Ti	22	47.90
Tungsten	W	74	183.8
Uranium	U	92	238.0
Vanadium	V	23	50.94
Xenon	Xe	54	131.3
Ytterbium	Yb	70	173.0
Yttrium	Y	39	88.91
Zinc	Zn	30	65.38
Zirconium	Zr	40	91.22

Atomic weights in brackets are approximate.

Periodic Table of the Elements

REPRODUCED WITH THE KIND PERMISSION OF GLAXO WELLCOME PLC.

1	2		3 4d, 4d, 5d	4	5	6	7	8	9	10	11	12	13	14	15	16	17	18
1s	**H** 1 1.0079																	**He** 2 4.0026
2s	**Li** 3 6.941	**Be** 4 9.01218											**B** 5 10.81	**C** 6 12.011	**N** 7 14.0067	**O** 8 15.9994	**F** 9 18.9984	**Ne** 10 20.179
3s	**Na** 11 22.98977	**Mg** 12 24.305											**Al** 13 26.9815	**Si** 14 28.0855	**P** 15 30.9738	**S** 16 32.06	**Cl** 17 35.453	**Ar** 18 39.948
4s	**K** 19 39.0983	**Ca** 20 40.08	**Sc** 21 44.9559	**Ti** 22 47.88	**V** 23 50.9415	**Cr** 24 51.996	**Mn** 25 54.938	**Fe** 26 55.847	**Co** 27 58.9332	**Ni** 28 58.69	**Cu** 29 63.546	**Zn** 30 65.38	**Ga** 31 69.72	**Ge** 32 72.59	**As** 33 74.9216	**Se** 34 78.96	**Br** 35 79.904	**Kr** 36 83.80
5s	**Rb** 37 85.4678	**Sr** 38 87.62	**Y** 39 88.9059	**Zr** 40 91.22	**Nb** 41 92.9064	**Mo** 42 95.94	**Tc** 43 (98)	**Ru** 44 101.07	**Rh** 45 102.9055	**Pd** 46 106.42	**Ag** 47 107.868	**Cd** 48 112.41	**In** 49 114.82	**Sn** 50 118.69	**Sb** 51 121.75	**Te** 52 127.60	**I** 53 126.9045	**Xe** 54 131.29
6s	**Cs** 55 132.9054	**Ba** 56 137.33	#**La** 57 138.9055	**Hf** 72 178.49	**Ta** 73 180.9479	**W** 74 183.85	**Re** 75 186.207	**Os** 76 190.2	**Ir** 77 192.22	**Pt** 78 195.08	**Au** 79 196.9665	**Hg** 80 200.59	**Tl** 81 204.383	**Pb** 82 207.2	**Bi** 83 208.9804	**Po** 84 (209)	**At** 85 (210)	**Rn** 86 (222)
7s	**Fr** 87 (223)	**Ra** 88 226.0254	***Ac** 89 227.0278															

( ) mass numbers of most stable isotope

LANTHANUM SERIES

4f	**Ce** 58 140.12	**Pr** 59 140.9077	**Nd** 60 144.24	**Pm** 61 (145)	**Sm** 62 150.36	**Eu** 63 151.96	**Gd** 64 157.25	**Tb** 65 158.9254	**Dy** 66 162.50	**Ho** 67 164.9304	**Er** 68 167.26	**Tm** 69 168.9342	**Yb** 70 173.04	**Lu** 71 174.967

* ACTINIUM SERIES

5f	**Th** 90 232.0381	**Pa** 91 231.0359	**U** 92 238.0389	**Np** 93 237.0482	**Pu** 94 (244)	**Am** 95 (243)	**Cm** 96 (247)	**Bk** 97 (247)	**Cf** 98 (251)	**Es** 99 (252)	**Fm** 100 (257)	**Md** 101 (258)	**No** 102 (259)	**Lr** 103 (260)

Subject Index

Page references to *figures*, *tables* and *text boxes* are shown in *italics*